Frontiers in Clinical Drug Research - CNS and Neurological Disorders

(Volume 10)

Edited by

Atta-ur-Rahman, *FRS*

Kings College
University of Cambridge, Cambridge
UK

&

Zareen Amtul

The University of Windsor
Department of Chemistry and Biochemistry
Windsor, ON
Canada

Frontiers in Clinical Drug Research - CNS and Neurological Disorders

Volume # 10

Editor: Prof. Atta-ur-Rahman, *FRS* & Dr. Zareen Amtul

ISSN (Online): 2214-7527

ISSN (Print): 2451-8883

ISBN (Online): 978-981-5040-67-8

ISBN (Print): 978-981-5040-68-5

ISBN (Paperback): 978-981-5040-69-2

need for a court order if at any point you breach any terms of this License Agreement. In no event will any delay or failure by Bentham Science Publishers in enforcing your compliance with this License Agreement constitute a waiver of any of its rights.

3. You acknowledge that you have read this License Agreement, and agree to be bound by its terms and conditions. To the extent that any other terms and conditions presented on any website of Bentham Science Publishers conflict with, or are inconsistent with, the terms and conditions set out in this License Agreement, you acknowledge that the terms and conditions set out in this License Agreement shall prevail.

Bentham Science Publishers Pte. Ltd.
80 Robinson Road #02-00
Singapore 068898
Singapore
Email: subscriptions@benthamscience.net

CONTENTS

PREFACE

Neurodegenerative and neurological disorders cost patients and societies trillions of dollars in terms of health care and out-of-pocket costs every year. This significant economical and neurodegenerative burden of brain disorders is expected to grow exponentially, especially in low- and middle-income countries in the next decade, if left unattended. There is a desperate need to focus our research strategies on epidemiology, disease frequency, and the understanding of the associated risk factors and possible outcomes to develop the therapeutics faster.

Volume 10 of our book series *Frontiers in Clinical Drug Research - CNS and Neurological Disorders* presents a set of creative and ground-breaking research endeavors of established as well as emerging researchers in the field. Our eminent authors have reviewed, evaluated, commented, and provided their valued feedback regarding the current central nervous system disorders to keep our readers updated.

For instance, **Chapter 1** compares the potency of therapeutic preparations derived from medicinal plants with the currently available pharmaceutical treatments for Parkinson's, Alzheimer's, and Huntington's Diseases. The chapter specifically focuses on the active ingredients, molecular targets, and challenges to developing plant-based pharmaceuticals. **Chapter 2** reviews the biochemical and physiological mechanism(s) of Parkinson's disease pathogenesis. The chapter also highlights the emerging therapies, such as stem cell progenitor cells transplantation, and the targeting of neurogenesis, apoptosis, neuroinflammation, mitochondrial dysfunction, and oxidative damage to treat Parkinson's disease. **Chapter 3** summarizes the translational limitations of using neurotrophic factors in various preclinical and clinical trials to combat neurodegeneration and neuropathic pain in Parkinson's disease. The author also discusses the current alternatives with improved translational perspectives, such as mutant proteins, small molecules, and peptides that target receptors for neurotrophic factors. **Chapter 4** highlights the advantages of using functional near-infrared spectroscopy to explore the neural bases of executive functions in children suffering from attention deficit hyperactivity disorder. The authors also discuss the utility of the technique in managing the accessibility and tolerance of body motion while measuring the sensitive cortical oxyhemoglobin concentration changes associated with neuronal activation. **Chapter 5** analyses the translational potential of a non-CpG, and non-coding oligodeoxynucleotide (IMT504) as an anti-nociceptive to treat chronic pain. The chapter describes the immunomodulating, and anti-inflammatory activities of the oligodeoxynucleotide, and explores its future potential as a multi-target pain medication. **Chapter 6** reviews the application, efficacy, and development of current anticonvulsants in chronic pain management. **Chapter 7** evaluates the behavioral, morphological, and nootropic effects of testosterone treatment on brain functioning. The chapter emphasizes the limitations of testosterone clinical application in mitigating aging-related cognitive decline.

In a nutshell, the current volume brings together another collection of exciting and cutting-edge research works of prominent scientists from all corners of the world. It is anticipated that the scholarly compendium of timely review articles will serve as the harbinger of new brain therapeutics in the near future.

We are grateful for the timely efforts made by the editorial personnel, especially Mr. Mahmood Alam (Editorial Director Publications), and Ms. Asma Ahmed (Senior Manager Publications) at Bentham Science Publishers.

Atta-ur-Rahman, *FRS*
Honorary Life Fellow
Kings College
University of Cambridge
Cambridge
UK

&

Zareen Amtul
Fulbright, AvH, OMHF, CIHR fellow
The University of Windsor
Department of Chemistry and Biochemistry
Windsor, ON
Canada

List of Contributors

Adejoke Y Onaolapo Behavioural Neuroscience/Neurobiology Unit, Department of Anatomy, Faculty of Basic Medical Sciences, Ladoke Akintola University of Technology, Ogbomoso, Oyo State, Nigeria

Akanksha Mishra Neuroscience and Ageing Biology Division, CSIR- Central Drug Research Institute, Lucknow 226031, U.P., India
Department of Cell Biology and Anatomy, New York Medical College, Valhalla, NY 01595, USA

Akari Inoue Applied Cognitive Neuroscience Laboratory, Chuo University, Tokyo, Japan

Alejandro Montaner Instituto de Ciencia y Tecnología "Dr. César Milstein", CONICET- Fundación Pablo Cassará, Buenos Aires, Argentina

Candelaria Leiguarda Instituto de Investigaciones en Medicina Traslacional (IIMT), CONICET- Universidad Austral, Buenos Aires, Argentina

Carlos H. A. Jesus Department of Pharmacology, Biological Sciences Sector, Federal University of Parana, Curitiba, Brazil

Eder Gambeta Department of Pharmacology, Biological Sciences Sector, Federal University of Parana, Curitiba, Brazil

Erika I. Araya Department of Pharmacology, Biological Sciences Sector, Federal University of Parana, Curitiba, Brazil

Geetha V Department of Biochemistry, Central Food Technological Research Institute, Mysore, India

Ippeita Dan Applied Cognitive Neuroscience Laboratory, Chuo University, Tokyo, Japan
Center for Development of Advanced Medical Technology, Jichi Medical University, Tochigi, Japan

Joelle M. Turnes Department of Pharmacology, Biological Sciences Sector, Federal University of Parana, Curitiba, Brazil

Juliana G. Chichorro Department of Pharmacology, Biological Sciences Sector, Federal University of Parana, Curitiba, Brazil

Julia Rubione Instituto de Investigaciones en Medicina Traslacional (IIMT), CONICET- Universidad Austral, Buenos Aires, Argentina

Masako Nagashima-Kawada Department of Pediatrics, Jichi Medical University, Tochigi, Japan

Mailín Casadei Instituto de Investigaciones en Medicina Traslacional (IIMT), CONICET- Universidad Austral, Buenos Aires, Argentina

Marcelo J. Villar Instituto de Investigaciones en Medicina Traslacional (IIMT), CONICET- Universidad Austral, Buenos Aires, Argentina

María Florencia Coronel Instituto de Investigaciones en Medicina Traslacional (IIMT), CONICET- Universidad Austral, Buenos Aires, Argentina

Moumita Das   Department of Biochemistry, Central Food Technological Research Institute, Mysore, India
                       Academy of Scientific and Innovative Research(AcSIR), Ghaziabad, India

Mayookha V.P   Department of Biochemistry, Central Food Technological Research Institute, Mysore, India
                       Academy of Scientific and Innovative Research(AcSIR), Ghaziabad, India

Olakunle J Onaolapo Behavioural Neuroscience/Neuropharmacology Unit, Department of Pharmacology, Faculty of Basic Medical Sciences, Ladoke Akintola University of Technology, Ogbomoso, Oyo State, Nigeria

Pablo R. Brumovsky Instituto de Investigaciones en Medicina Traslacional (IIMT), CONICET-Universidad Austral, Buenos Aires, Argentina

Parul   Neuroscience and Ageing Biology Division, CSIR- Central Drug Research Institute, Lucknow 226031, U.P., India

Pratibha Tripathi Neuroscience and Ageing Biology Division, CSIR- Central Drug Research Institute, Lucknow 226031, U.P., India

Shubha Shukla   Neuroscience and Ageing Biology Division, CSIR- Central Drug Research Institute, Lucknow 226031, U.P., India
                       Academy of Scientific and Innovative Research (AcSIR), Ghaziabad 201002, India

Sonu Singh   Neuroscience and Ageing Biology Division, CSIR- Central Drug Research Institute, Lucknow 226031, U.P., India
                       Department of Neuroscience, School of Medicine, University of Connecticut (Uconn) Health Center 263 Farmington Avenue L-4078, Farmington CT 06030, USA

Suresh Kumar G.   Department of Biochemistry, Central Food Technological Research Institute, Mysore, India
                       Academy of Scientific and Innovative Research(AcSIR), Ghaziabad, India

Takahiro Ikeda   Department of Pediatrics, Jichi Medical University, Tochigi, Japan

Takanori Yamagata Department of Pediatrics, Jichi Medical University, Tochigi, Japan

Tatsuya Tokuda   Applied Cognitive Neuroscience Laboratory, Chuo University, Tokyo, Japan

Vanessa B. P. Lejeune Department of Pharmacology, Biological Sciences Sector, Federal University of Parana, Curitiba, Brazil

Virendra Tiwari   Neuroscience and Ageing Biology Division, CSIR- Central Drug Research Institute, Lucknow 226031, U.P., India
                       Academy of Scientific and Innovative Research (AcSIR), Ghaziabad 201002, India

Yulia A. Sidorova Institute of Biotechnology, HiLIFE, Viikinkaari 5D, FI-00014 University of Helsinki, Helsinki, Finland

Yukifumi Monden   Department of Pediatrics, Jichi Medical University, Tochigi, Japan

CHAPTER 1

Neurodegenerative Disease: Prevention and Treatment Through Plant Extracts Therapy

Mayookha V.P[1,2], Geetha V[1], Moumita Das[1,2] and Suresh Kumar G.[1,2,*]

[1] *Department of Biochemistry, Central Food Technological Research Institute, Mysore, India*

[2] *Academy of Scientific and Innovative Research(AcSIR), Ghaziabad, India*

Abstract: Neurodegenerative diseases such as Parkinson's, Alzheimer's, Huntington's *etc.* have their root in damaged nerve cells followed by the loss of their functions. Though the exact reason for different neurodegenerative diseases is still unknown, degradation and accumulation of proteins in neurons, oxidative stress, inflammation, defects in mitochondria, genetic mutation *etc.*, are said to be the general factors that leads to this disease. The old ages are the worst affected group due to the rise in human life expectancies. Although there is no complete cure for this disease, some drug treatments have been found to be useful for reducing few of the physical or mental symptoms associated with neurodegenerative diseases. In fact, most neurodegenerative disorders show multiple symptoms;a promising result can be achieved only through the combination of different compounds like natural plant extracts which have many disease targets. Antioxidants are proven to have the capacity to act against the oxidative stress developed in the cells. So, antioxidant rich plant extracts can be utilized for the treatment of neurological related disorders. Several studies are being carried out on the effect of various plant extracts on the neurodegenerative disease prevention, management as well as treatment. This chapter will discuss the *in vitro, in vivo* and clinical studies conducted on the effect of various plant extracts for the treatment and prevention of different neurodegenerative disease conditions.

Keywords: Alzheimer's Disease, Huntington's Disease, Neurodegenerative Disease, Parkinson's Disease.

INTRODUCTION

Neurodegenerative diseases include various types of disorders which result from the damage of neurons followed by their functional loss. Since the last 10 years, not many drugs were introduced for the treatment of neurodegenerative disease. Alzheimer's disease, Parkinson's disease, Huntington's disease, amyotrophic lateral sclerosis, frontotemporal dementia and the spinocerebellar ataxias are some

---

[*] **Corresponding author Suresh Kumar G.:** Department of Biochemistry, Central Food Technological Research Institute, Mysore, India; Email-sureshg@cftri.res.in.

Atta-ur-Rahman & Zareen Amtul (Eds.)

major neurodegenerative diseases. They show different pathophysiological conditions, where some leads to memory and cognitive damages, and others may disturb a person's locomotor as well as speech and breath functions. Among these, Alzheimer's disease (AD) is a chronic, age-dependent neurodegenerative disease which is the major cause of dementia in older individuals and in the Western world, it is on the fourth place for the cause of death. AD starts with brain atrophy, leading to gradual damage of CNS. Specific neurofibrillary tangles and neural plaques are observed in post mortem. Another neurodegenerative disorder affecting millions of the elderly population is Parkinson's disease (PD), with most cases arising after the age of fifty. The noticeable symptoms are movement disorders, including tremor, inflexibility, slow motion, and trouble while walking and in posture. Due to slow dysfunction of dopamine-generating neurons in the basal ganglia substantia nigra, a midbrain area, the motor symptoms of PD result in gradual loss of muscle co-ordination and balance. Later, there may be cognitive and behavioral difficulties, with dementia usually arising in the early stages of the disorder, while the most commonly observed symptom is depression. Sensory, sleep and mental disorders are other symptoms. A neurodegenerative genetic disorder that impacts muscle function and contributes to cognitive impairment and psychiatric disorders is Huntington's disease (HD). It is distinguished by irregular repetitive writhing movements called chorea. Although the physical symptoms of HD can occur from children to elder individuals at any age, the typical symptoms start between the ages of 35 and 44. Efficient therapies should be in immediate action, but only a profound understanding of the reasons and mechanisms of each disease can come with them. The chapter discusses various aspects of medicinal plants, their phytochemicals, along with traditional uses, important bioactivities, psychological and clinical proof on effectiveness and safety.This chapter also emphasises on the promising evidences through animal studies as well as clinical trials.

Alzheimer's Disease (AD)

Dementia is an ailment in which there is an immense reduction of mental and cognitive capacities of an individual. The affected individual is unable to function independently due to memory loss as a consequence of disease progression. Several reversible or irreversible causes are responsible for dementia. The most common reversible causes include substance abuse, subdural hematoma, removable tumors, and central nervous system (CNS) disorders (Table 1). Irreversible causes of dementia are neurodegenerative diseases such as Alzheimer's disease (AD), Parkinson's disease (PD) and Huntington's disease (HD).

Table 1. Most common forms of dementia [9].

DEMENTIA FORM	NEUROPATHOLOGY	SYMPTOMS	DEMENTIA CASES (%)
AD-RELATED DEMENTIA	Formation of Aβ plaques and neurofibrillary tangles.	Memory deficits, depression, poor judgment or evidence of mental confusion.	50-80
VASCULAR DEMENTIA	Decreased blood flow to the brain, hypoperfusion, oxidative stress.	Similar to AD, but less affected memory	20-30
DEMENTIA WITH LEWY BODIES	Aggregation of α-Synuclein in neurons and glial cells (cortical Lewy bodies)	Similar to AD and less to PD, hallucinations, tremors,impaired attention.	<5
FRONTOTEMPORAL DEMENTIA	Accumulation of MAP tau, atrophy of frontal andtemporal lobes.	Changes in social behavior, difficulties with language.	5-10

Alzheimer's disease (AD) is the most common form of dementia, mostly found in old age. This clinical disorder affects almost 5% population among the people over 65 years of age and 25% of the population over 80 years. AD is mainly characterised by impairment of cumulative memory (especially anterograde amnesia), aphasia, agnosia and also impairment of other higher cognitive functions and insight, and by various personality and behavioural/ neuropsychiatric characteristics. Alzheimer's disease was first reported in the year 1907 by Alois Alzheimer. He has used one newly emerged histological method for staining to recognize the elements of neurofibrillary tangles in the brain tissue. This identification remarkably became [together with β-amyloid (Aβ) plaques] the pathological pioneer for diagnosing the disease in the post-mortem condition. The formation of plaques and tangles in the brain of an affected individual is recognized as the crucial characteristic feature of the Alzheimer's disease condition. Progressive loss in the connectivity between the nerve cells (neurons) in the brain is another important characteristic of this disease. The main function of the neurons is the transmission of messages in different parts of the brain and from the brain to different parts of the body like muscles and different organs. Various complicated changes occuring in the brain contribute to the Alzheimer's disease. Initial damage is seen in the hippocampus area of the brain that is mainly responsible for forming memories. As the neurons degenerate, auxiliary parts of the brain get affected. At the end stage of this disease condition, the damage becomes severe, leading to shrinkage of the brain tissues remarkably. Alzheimer's disease and other forms of dementia mainly cause immense dysfunction of the brain leading to several neurodegenerative conditions like neuronal injury, synaptic failure and neuronal death. Neuro fibrillary tangles (NFT) and neurophil threads are the main pathological hallmarks of AD [1].

Various hypotheses have been proposed to establish the foundation of multi factorial syndrome on the basis of several causative aspects [1], such as cholinergic hypothesis, Aβ hypothesis, tau hypothesis and inflammation hypothesis [2]. In the past two decades Aβ hypotheses has been persuaded in the scientific community [3]. In recent times, several studies have emphasized the functional importance of Aβ oligomers in causing impairment of synapses and also suggested this as a fundamental signal that may damage the integrity of brain functions [3 - 6].

The amyloid cascade theory has hypothesized that amyloid precursor protein (APP) is cleaved subsequently by the enzymes α-secretase followed by β and γ-secretases, which lead to an imbalance in the synthesis and clearance of Aβ peptide [7]. This imbalance causes spontaneous aggregation of Aβ peptides into soluble oligomers and also amalgamate to form insoluble fibrils of beta-sheet conformation which are eventually deposited in diffuse senile plaques [3, 8].

The synergistic activities of both neurons and astrocytes raise the production of Aβ42 oligomers [5, 10]. It has been noticed that an increase in the production of Aβ42 oligomers gives rise to oxidative stress, enhanced hyper-phosphorylation which induces toxicity on the synapses and mitochondria [1, 2]. The increase in the formation of Aβ42 senile plaques enhances the activation of microglia [8], which consequently stimulates the production and release of pro-inflammatory cytokines, including IL-1β, TNF-α, and IFN-γ. Further, these cytokines give rise to the surrounding astrocyte-neuron producing an ample amount of Aβ42 oligomers, thus stimulating higher production of Aβ42 and also its dispersal. Oligodendroglia (OLGs) associated with neurons–astrocyte complex; Aβ oligomers also result in its destruction. In the brains of AD patients, aggregation of Aβ oligomers enhances neuronal and vascular degeneration [5]. It increases the oxidative stress to which OLGs are particularly susceptible. They have a low level of reduced glutathione (GSH) content and also a high concentration of iron as a result of decreased oxygen radicals scavenging ability [6]. It has also been observed that Aβ42 oligomers may lead to damage of cholesterol rich membranes, specifically found in OLGs and myelin [5].

CURRENT PHARMACEUTICAL TREATMENT

Various drugs which are (semi-) synthetic have been recognized worldwide for the treatment of different forms of dementia especially Alzheimer's disease. Few most extensively used curatives for AD related dementia are- selective reversible acetylcholinesterase (AChE) inhibitor donepezil, the non-selective butyrylcholinetserase (BuChE) and AChE inhibitor rivastigmine, as well as the N-methyl-D-aspartate (NMDA) receptor antagonist memantine [10, 11]. Most prominent drugs approved for clinical use in AD are Memantine, Rivastigmine, Donepezil, *etc.* (Table **2**). FDA approved a combined drug donepezil-memantine in the year 2014 from the brand NamzaricR for treating moderate to severe AD affected people, who are recommended with clinically effective dose of 10mg/day donepezil hydrochloride [12]. There have been reports on various side effects when this combination of medicines are taken, such as muscle problems, slow heartbeat, fainting, increased stomach acid levels, nausea, vomiting and also seizures. Few of the studies also have suggested that the above mentioned drugs did not ameliorate the agitation condition among the patients who are severely affected [13, 14]. Therefore, an immense investigation on the natural ingredients with potential bioactives for treating different dementia conditions demands the attention of the scientific community.

Table 2. Different therapeutic targets for treating AD.

Therapeutic Approaches	Drug Targets
Targeting Ab protein (anti-amyloid approach)	• Targeting amyloid transport. • Modulation of secretase enzymes. • Targeting amyloid aggregation. • Targeting amyloid clearance. • Amyloid based vaccination therapy.
Targeting tau protein	• Inhibition of tau phosphorylation. • Targeting microtubule stabilization. • Blocking tau oligomerization. • Enhancing tau degradation.
Targeting intracellular signaling cascades	• Inhibition of phosphodiestrase (PDE).
Modulating levels of neurotransmitter	• Acetylcholinesterase inhibitors (AChEIs). • Modulation of GABAergic neurons. • NMDA receptor antagonism. • Modulation of serotonin receptor. • Histaminergic modulators. • Modulation of adenosine receptor.
Targeting mitochondrial dysfunction	• Supressing the production of ROS and minimizing ROS injury.
Targeting oxidative stress	• Administration of antioxidants including vitamins (E,C and carotenoids), phytochemicals and synthetic compounds.
Anti-inflammatory therapy	• Targeting γ–secretase, Rho-GTPases and PPAR. • Targeting cyclooxygenase (COX) inhibition.

(Table 2) cont.....

Therapeutic Approaches	Drug Targets
Other pharmaco-therapeutic strategies	• Cholesterol lowering drugs. • Neuroprotective gonadotropin hormones. • Neurogenesis. • Epigenesis. • Caspase inhibitors. • Modulators of NOS. • Nucleic acid drugs.

CURRENT DRUG FROM PLANT ORIGIN AGAINST DEMENTIA

Galantamine is the only drug originated from plants, that is advised widely for treating mild-to-moderate AD and AD related dementia. It has been proven to be effective through many clinical trials. Galantamine is also known as galanthamine. It is an alkaloid (isoquinoline) synthesized from the plants that belong to Amaryllidaceae family. galantamine was first discovered by a Bulgarian Chemist D. Paskov and his team in the year of 1956. It was first separated from bulbs of *Galanthus nivalis* (common snowdrop). The efficacy of this drug is very prominent and also has been enlisted in the treatment guidelines for Alzheimer's disease/dementia in USA and Europe. There are many countries worldwide who have approved this drug, including Canada, in the European Union (except for The Netherlands, under the name NivalinR in 2000), Japan, Korea, Mexico, Singapore, South Africa, Thailand, *etc.* The plant-derived galantamine is a well-established medicine for the treatment of dementia. The main mechanism of galantamine is to modulate the acetylcholine signaling and inhibition of oxidative damage. Galantamine affects the brain's cholinergic system through its distinctive duplex mode of action. It inhibits the AChE enzyme reversibly by competitive inhibition. It also allosterically enhances the activity of the nicotinic acetylcholine receptor (nAChR) [15].

Galantamine has a protective role in mitochondrial dysfunction. The changes that may occur in morphology and the membrane potential of mitochondria (MMP) which is instigated by Aβ25/35 or hydrogen peroxide treatment, can be reversed by galantamine [16]. Oxidative stress is mainly developed by the toxic effect of reactive oxygen species (ROS). These are mainly produced in mitochondrial membrane during electron transport. The neuroprotection efficacy of galantamine is mainly through protecting the mitochondria and also inhibiting AChE activity, thus it helps in decreasing oxidative damage to cells [17]. P-glycoprotein is present in brain's vascular endothelium. It is a well-known transporter which resists multiple drugs targeted for brain. It effectively prevents many drugs crossing blood brain barrier and efflux back into the blood stream [18]. Galantamine may reduce the potency of this protein and allow many other drugs co-administered with it to reach the brain more easily [18]. Neurotoxicity induced cognitive impairment has been reported in the rat model, which includes mitochondrial dysfunction, oxidative damage *etc.* Galantamine helps in reversing this condition by enhancing the neuro protective potency of rofecoxib (an anti-inflammatory COX-2 inhibitor) and caffeic acid (a plant derived phenol) [19]. In a similar way, galantamine increases the antioxidant efficacy of melatonin, a brain sleep hormone. It has been reported that galantamine treatment delayed the onset of various behavioral symptoms of dementia consistently like anxiety, euphoria, depression, irritability, delusions and unusual motor behavior [20].

Overview of Treatment Strategies Based on Medicinal Plants

Dementias are basically known to be complex diseases, as shown in modern research having several molecular mechanisms involved in the pathogenesis of the disease conditions. This multiple mechanisms approach evolved new strategies to treat these pathologies: treatment of dementia should have a holistic perspective, including several underlying molecular targets instead of focusing in any particular target (Fig. **1**). Plant and plant extracts comprises of several bioactives that are hypothesized to target additively or synergistically on multiple molecular mechanisms [21] (Table **3**). Several herbal medicines have been utilized for a long time for treating dementia related neurodegenerative disorders. But unfortunately, bioactive components of these herbs are poorly narrated. Similarly, we still know very little about how these compounds interact with each other and with prescribed medicines [22]. The extensive research on these emerging issues will be crucial to find out therapeutic approaches that are devoid of harmful side effects.

Table 3. Neuroprotective activities exerted by various plants and active ingredients [23].

S.No	Activity/Mechanism	Active Ingredients/ Plants
A)	Anti-inflammatory	Flavonoids from plant sources
B)	Antioxidant	*Bacopamonniera* *Curcuma longa* *Thymus vulgaris*
C)	Nicotinic receptor stimulation	*Lobelia inflata,* *Nicotianatabacum*
D)	Inhibitors of acetylcholinesterase: i) alkaloids	*Physostigmavenenosum* *Galanthusnivalis* *Huperziaserrata*
	ii) terpenoids and other phytochemicals	*Melissa officinalis* *Rosmarinusofficinalis*
E)	Phyto-oestrogens	Soy isoflavones

(Table 3) cont.....

S.No	Activity/Mechanism	Active Ingredients/ Plants
F)	**Antiamyloid aggregation**	*Ginkgo biloba*

Important Medicinal Plants Used for the Treatment of Neurodegenerative Disorders

The following medicinal plants have been extensively studied and experimented for the treatment of neurodegenerative disorders (Table **4**).

Table 4. Summary of Plants and their active ingredients with traditional use tested recently *in vitro*, *in vivo* and in clinical trials for Alzheimer's disease (AD).

Plant Name	Bioactive Compound	Traditional Utility	Molecular Target	References
Panaxnoto ginseng	Ginsenoside Rg1	Improvement in learning & memory function Used in Traditional Chinese medicine (TCM)	Secretase activity	[57]
P. notoginseng	Ginsenoside	Improvement in learning & function of memory; used in TCM	Neprilysin	[38]
Ginkgo biloba	Fresh plant extract	Used in TCM; improvement in memory loss and also ameliorating respiratory disorders	DemTec cognition score	[58]
Dipsacus asper	Wall Akebiasaponin D	Upgrade the kidney function; used in TCM	Aβ toxicity	[59]
Paeonia suffruticosa Andrews	1,2,3,4,6-penta-*O*-galloyleta-d-glucopyranose	TCM; treatment of inflammatory and pyretic disorders	Development and stabilization of Aβ-fibril; improve long-term memory impairment *in vivo*	[60]
Polygala tenuifolia Willd	Tenuifolin (extract)	TCM; improvement of memory loss	Secretase activity; morphological plasticit	[61, 62]
Radix salviaemiltiorrhizae **(Dashen)**	Triterpenoids; Tanshinone	TCM; treat heart conditions and stroke	AChE activity; Aβ toxicity *in vivo* and *in vitro*; NOS	[63 - 65]

(Table 4) cont.....

Plant Name	Bioactive Compound	Traditional Utility	Molecular Target	References
Bacopa monnieri	Bogenines, Steroids, Triterpene	Ayurvedic medicine, improve intelligence and memory	Ameliorates ACh deficits	[66]
Salvia officinalis	Essential oils, containing cineole, thujone and others	Mediterranean, antiinflammatory agent	Anti-inflammation	[67]
Melissa officinalis	Terpenes, tannins, Eugenol, Rosmarinic acid	Mediterranean, used as anxiolytic or mild sedative agent	AChE inhibition *in vitro*	[68, 69]
Murrayakoenigii	Cabazolalkaloids,saponins	Indian flavour	Antiamnestic, reduction of choli-nesterase activity	[70]
Cassia obtisufolia	Obtusifolin	Eastern medicine, used as a topical analgesic and anti-inflammatory natural medicine.	AChE inhibition; Mitochondrial protection; Calcium stabilization	[71, 72]
Centella asiatica	Triterpen glycosides, Saponies	Ayurveda, anxiolytic agent and cerebral tonic	Reducing Aβ *in vivo*	[73]
Fungus *Ganoderma Lucidum*	Ganoderic acid (Triterpen Glycoside	TCM, as anti-tumor, immunomodulatory and immunotherapeutic agent	Preserving synaptic density; preserving Aβ-induced apoptosis	[74]
Desmodium gangeticum	Aminoglucosyl-glycerolipids, Cerebrosides	Ayurveda, treatment of neurological disorders	Reserved amnesia, AChE inhibition	[75]
Lycium barbarum	Polysaccharides	TCM; used as anti-tumor, immunomodulatory, anti-hypertension agent	Reverses Aβ and homocysteine induced apoptosis	[76, 77]

Ginkgo Biloba

Ginkgo biloba (Coniferae) is a well-known traditional Chinese medicine which has been mostly used in treating respiratory diseases. Traditionally it has been used in Iran for the improvement of memory loss related to irregularities in blood circulation. This herb has been studied extensively for its promising role in treating cognitive disorders [23]. It is utilized mainly for improving memory impairment along with the alleviation of dementia and Alzheimer's disease condition [24]. The two major phytochemical components thought to be responsible for neuroprotective function of Gingko are: terpene lactones (ginkgolides and bilobalide) and flavonoids (flavonols and flavone glycosides) [25, 26]. Ginkgo leaf extract contains about 24% flavonoids and 6% terpene lactones [27]. The most widely investigated herbal medicines for the treatment of cognitive impairment include the Gb extract known as EGb761. Several clinical trials are done to explore the potential of this standardized Gb extract for treating Alzheimer's disease condition and also AD related dementia [25].

Gb extract targets multiple mechanisms for maintaining brain functions including regulation of oxidative stress through its potential antioxidant activity [24, 28 - 30]. The role of the extract in neuroprotection *via* modulating oxidative stress is likely to be modulation of circulating glucocorticoid levels, Aβ aggregation, ion homeostasis and growth factors synthesis. Moreover, the role played by bioactives from Gingko is well accepted in the scientific community in maintaining mitochondrial function. Several *in vitro* studies have reported the protective role of Gingko constituents to maintain mitochondrial membrane potential (MMP) from various toxicants and oxidative stress [23]. Gingko extract can be beneficial in various aspects of mitochondrial morphology such as fission [31], swelling [32], and coupling [33]. It is also reported to interact with mitochondrial electron transport chains. It was fascinating to find that there was a significant improvement in the oxidative phosphorylation efficiency in cells overexpressing amyloid precursor protein (APP) than in control cells [34]. This indicates the effectiveness of Ginkgo extract, especially in AD therapy. EGb761 may prevent dysfunction of neurovascular unit known to be one of the pathologies associated with AD [35]. In summary, Ginkgo extract exhibits neuroprotective effect through its antioxidative and/or antiplatelet activities. Clinical studies confirm the effectiveness of Ginkgo extract for dementia treatment.

Panax Ginseng C.A. Meyer (Ginseng)

Panax ginseng (root powder) has been used in Oriental medicine since thousands of years for treating a variety of disorders including pragmatic utilization for the treatment of cerebrovascular disease. The dried root of this plant has been used traditionally as a medicine mainly in China and Korea. There are various species of Panax, including *P. ginseng* (Oriental ginseng), *P. japonicus* (Japanese ginseng), *P. quinquefolius* (American ginseng), *P. trifolius*, *P. notoginseng* (Burkill) and *P. major* [23]. *Panax ginseng CA Meyer* is the most often used and extensively researched species among the other varieties of ginseng. Few active components of this plant like Rg1, Rg3 and ginseng polysaccharides have been explored for their therapeutic potential [36, 37].Ginsenoside Rg3 has a backbone of steroid with aliphatic side chains, including carbohydrate part. Rg3 is mainly responsible for the neuroprotective potential of ginseng. Heat treatment of the roots of ginseng at high temperature produces Rg3 [38, 39]. The bioactive components of ginseng are reported to target few of the mechanisms related to dementia-like amyloid-β metabolism, oxidative stress, neuro-inflammation and acetylcholine signaling. However,the exact effectiveness of ginseng on dementia patients remain undefined.

Curcuminoids from Genus Curcuma

Around 80 species belong to the genus Curcuma (commonly termed as Turmeric) and is one of the largest genera from the Zingiberaceae family [23]. The major bioactive phytochemical of Curcuma genus is curcumin. Curcumin is well known as a traditional Indian medicine for its effectiveness in treating several disorders like anorexia, hepatic diseases, cold, cough and others. Curcumin has a great potential in neuroprotective actions through its various properties like antioxidant, anti-neuro-inflammatory, anti-proliferative, anti-amyloidogenic and neuro-regulative effects [39]. The major functions of Curcumin involve inhibiting the activity of tumor necrosis factor (TNF), preventing the formation of Aβ plaques and also protecting the brain cells from noxious agents [40]. Diets enriched with Curcumin strengthens memory and hippocampal neurogenesis in aged rats. Curcumin modulated expression of several genes are specifically involved in cell growth and synaptic plasticity [41]. The neuroprotective efficacy of curcumin is mainly because of its anti-inflammatory, antioxidant and lipophilic functions. *In-vitro* studies have shown that curcumin provides protection to mitochondria from several noxious factors like oxidative stress and rotenone (inhibitor of electron transport chain) [42, 43]. Various studies have shown that aging generally leads to mitochondrial loss and also diminished oxidative activity in rodent brains [44, 45]. Treatment with curcumin ameliorates this condition and also it has shown to improve the hippocampal-dependent memory of Aβ-infused rats [46].

Glycyrrhiza Species

Licorice (liquorice) belongs to the genus Glycyrrhiza, is a well-known member of Fabaceae family and comprises around 30 species. Many plants of this genus are enduring herbs originating from Mediterranean region, Asia, Southern Russia and Iran [47]. The licorice roots and also rhizomes are naturally sweet, so utilized all over the world as a sweetener. It has been traditionally used in herbal therapeutics because of its protective role, mainly in autoimmune hepatitis C, jaundice, peptic ulcer and skin diseases such as atopic dermatitis and inflammation induced hyperpigmentation [47, 48]; several studies have suggested the pharmacological importance of licorice roots in many disease conditions. Important properties include its anticancer, antioxidative, anti-inflammatory, antiviral, antimicrobial, hepatic and cardioprotective effects [47, 49].

The main bioactive phytochemical of Glycyrrhiza glabra (liquorice) root are - triterpene saponin glycyrrhizin (glycyrrhizic acid) and the phenolic type compound isoliquiritigenin. Other important ingredients also include several isoflavonoid derivatives such as shinpterocarpin, glabrone, glabridin, galbrene, lico-isoflavones A and B [47]. The active ingredients of licorice have potent antioxidant activities. Various species of Glycyrrhiza were studied and have been shown to possess neuroprotective ability, thus may help in treating neurodegenerative disorders such as PD, AD and dementia. The effectiveness of the extract of G. inflate was investigated in the cell model of spinocerebellar ataxia type 3 (SCA3) also known as Machado-Joseph disease (MJD) and results have shown to reduce oxidative stress through upregulation of PPARGC1A and the NFE2L2-ARE pathway [50]. MPP+ (1-methyl-4-phenylpyridinium) is a neurotoxic compound which is mainly involved in interfering mitochondrial oxidative phosphorylation [51]. This component arises several conditions like cytotoxicity, ROS generation and downregulation of glutathione (GSH). Glycyrrhizin is known to reverse these conditions induced by MPP+ [51]. Brain cells are prone to oxidative damage. Licorice extract mainly reduces the oxidative stress in the brain cells by acting on the mitochondrial function. The active compound is oliquiritigenin from licorice is known to be responsible for this activity [52]. Licorice also reverses the damage in the brain cells by improving the neuronal function, preventing the memory impairment connected to dementia condition. Apart from the antioxidative effect of licorice, it is also associated with anti-inflammatory properties known to enhance memory in dementia condition [23]. In summary, the extract of *Glycyrrhiza* is shown to possess anti-inflammatory and antioxidative properties and also has a modulating role in glutamate signaling and apoptosis.

Camellia Sinensis Kuntze

One of the most considerably consumed beverages worldwide includes *Camellia sinensis Kuntze* (green tea) brew [53]. It has been reported in animal as well as human studies that consuming green tea regularly helps in improving cognitive functions and also inhibits impairment of memory. The compound in green tea which is mainly responsible for these functions is-epigallocatechin-3-gallate. Consumption of green tea on a daily basis has been speculated to minimize the risk of developing age-related dementia and AD [54]. Some of the clinical studies have reported that L-theanine exhibited improvement in the cognitive functions and mood in amalgamation with caffeine in healthy human individuals, although, the results of few studies on the effects of L-theanine alone on mood remained elusive. Green tea catechins have the potential to modulate the activity of P-glycoprotein, thus influencing the easier availability of co-administered components to the brain. In summary, green tea extract showed antiapoptotic and antioxidative functions. It has the potential to inhibit Aβ plaque formation directly [23]. Various human studies incorporated the integrity to the hypothesis of the effectiveness of green tea in modulating human cognitive function and also its importance in treating dementia condition.

Moringa Oleifera

Moringa oleifera (MO) is included in the family *Moringaceae*. It is commonly found in almost everywhere in Asian and African countries. The fruit and leaves of *Moringa oleifera* have shown to possess anti-inflammatory and hypotensive effect and also consumed as a food ingredient by many people. It has been discovered recently that *Moringa oleifera* leaf extract is not toxic when taken even in higher concentration levels. It enhances memory through nootropics function and also delivers important antioxidants which includes vitamin C and E that helps in combating oxidative stress in AD. Several investigations have shown that the memory loss caused by monoamines is modified by leaf extracts of *Moringa oleifera* [55, 56]. Various lines of evidence also have shown that colchicines-induced AD can be altered by ethanolic extract of *Moringa oleifera* by modifying the brain monoamines (norepinephrine, dopamine and serotonin) and electrical activity in a rat model [55].

Fig. (1). Neuroprotective role of some extensively studied plants and their bioactive components in Alzeimer's disease.

Parkinson's Disease

Parkinson's disease (PD) is a multifactorial condition which is age-related and involves the neurodegeneration of dopaminergic neurons in substantia nigra, which is neuropathologically recognized. The nerve cells (neurons) in the brain gradually break down or die in Parkinson's disease. Many of the effects are due to the loss of neurons that forms - dopamine; a chemical messenger in the brain. It induces abnormal brain activity as dopamine levels decline, leading to impaired movement and other symptoms of Parkinson's disease.

In PD study, the development of symptomatic therapies has been partially successful, but a number of inadequacies remain in the therapeutic strategies for the disease. Symptoms of Parkinson's generally start out progressively and get worse over time. People can have trouble walking and talking as the disease progresses. They may also experience mental and behavioral changes, issues with sleep, depression, trouble with memory, and exhaustion. Parkinson's disease appears in both men and women. The disorder, however, affects about 50 percent more men than women.

The cause of Parkinson's disease is unclear, but it appears that many factors play a role, including:

a. ***Genes***: Unique genetic mutations that may induce Parkinson's disease have been identified by researchers.
b. ***Environmental Causes:*** The risk of Parkinson's disease can be increased by exposure to certain chemicals or environmental factors.
c. ***Lewy Bodies':*** Microscopic markers of Parkinson's disease are clumps of particular substances inside brain cells called lewy bodies, researchers assume that these Lewy bodies are significant key to the cause of Parkinson's disease.
d. ***Alpha-synuclein Found Inside Lewy Bodies:*** While several substances are present in Lewy bodies, researchers agree that the normal and widespread protein called alpha-synuclein (α- synuclein) is a significant one.
e. ***Low Dopamine Levels***: Parkinson's disease is also caused by the low or dropping levels of dopamine, a neurotransmitter. This occurs when dopamine producing cells die in the brain, dopamine is involved in transmitting signals to the part of the brain that regulates motion and co-ordination, low levels of dopamine can make it more difficult for individuals to regulate their movements. Some of the causes of PD are as shown in Fig. (**2**).

Fig. (2). Factors responsible for Parkinson's disease.

Pathophysiology of PD

PD pathogenesis is a multifactorial mechanism in which complex reactions such as inflammation, neurotoxicity of glutamate, increased iron and nitric oxide levels, depletion of endogenous antioxidants, decreased development of neurotrophic factors, increased expression of apoptotic proteins and blood circulation dysfunction lead to neuronal degeneration. Pathologically, PD is the gradual depletion of dopaminergic neurons in the brain area of substantia nigra pars compacta. Intracytoplasmic protein inclusions called Lewy bodies (LBs) and dystrophic neurites (Lewy neurites) are present in PD patients' surviving neurons. Genetic factors such as alpha-synuclein or parkin gene mutations and environmental factors such as neurotoxic contaminants have also been suggested for the initiation of PD.

Oxidative stress in substantia nigra is suspected to play a major role in the loss of neurons that create dopamine (DA). The other characteristic histopathological finding is the presence of deposits called lewy bodies in the surviving neurons which is the characteristic deficiency of DA in the substantia nigra was observed in the brain of PD patients after post mortem. Abnormal protein folding in PD is characterized by about 80 percent loss of the neurotransmitter dopamine in the corpus striatum.and DA in the corpus striatum region of the brain. Also, PD results because of more than 50% dopaminergic (DA-ergic) loss of neurons in the substantia nigra pars compacta (SNpc) region of the brain. The causative mechanisms of PD also includes oxidative stress and inflammation. Postural dysfunction in PD patients is due to the massive and gradual death of dopaminergic neurons in the substantia nigra. The existence of intra-neuronal protein aggregates which are known as lewy bodies, indicates the cellular inability to clear abnormal proteins, is also one of the main pathological characteristics of the disease.

Dopamine is a dietary amino acid (tyrosine) neurotransmitter and has essential roles in a number of motor, cognitive, motivational, and neuroendocrine functions. The key class of medications used to treat PD symptoms are DA receptor agonists (*e.g.*, bromocriptin, cabergoline, pergolide, rotigotine, apomorphine, ropinirole, and pamipexole). Due to the gradual loss of neuronal cells in the brain that synthesize it, PD symptoms arise in response to decreased levels of the chemical messenger DA. Increased levels of free radicals, oxidative stress, inflammation, mitochondrial dysfunction, and alpha-synuclein aggregation are the neurochemical events related to the pathology of PD. In addition, PD has also recorded increased concentrations of redox active metals such as iron and copper, decreased levels of glutathione and increased lipid peroxidation.

Pharmacological treatments currently available have only modest symptomatic relief for PD patients and have no success in reversing the underlying neuropathological changes associated with this disorder. There is also a clinical need to have therapeutic agents to recognize the agents that may enhance the deleterious processes associated with PD or slow them down. Relevant molecular or pharmacological effects, which are likely to contribute to the production of neuroprotective agents against PD, have been observed increasingly in natural products [78]. However, there is no complete understanding of the pathogenesis and etiology of PD. Important cellular causes of dopaminergic cell death, including neuroinflammation, oxidative stress, mitochondrial dysfunction and excitotoxicity have been outlined by comprehensive analysis of different models imitating key features of PD. Neurotoxic models have proven to be a valuable method for the development of new therapeutic strategies and also for the assessment of the effectiveness and adverse effects of PD symptomatic therapy.

Several natural plant-derived products have the ability to be used as PD treatment drugs. They have been shown to play roles that would have the desired effects, and some of these or their derivatives have been introduced into clinical use. In order to delay or reverse the underlying neuronal degeneration observed in PD, plant-derived natural products and their constituents have been shown to have an impact on the regulation of the levels of dopaminergic neurotransmission and motor function. The anti-PD effects of these natural products are due to their regulation ability to reduce reactive oxygen species and neuroinflammation, production of dopamine, excitotoxicity, metal homeostasis, mitochondrial function, and cellular signaling pathways, all of which are disrupted in the brain of PD patients are regulated by the plant derived extracts. (Fig. **3**).

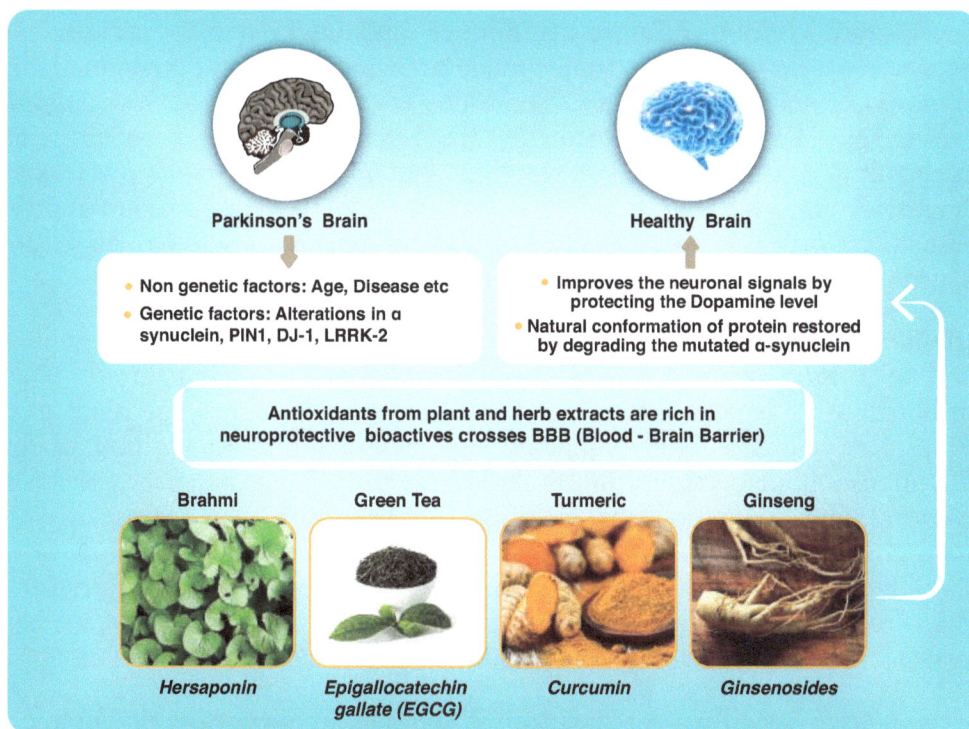

Fig. (3). Effect of various plant extracts on PD.

The aggregation or fibril formation of α-syn oligomers is inhibited by phytochemicals and plant extracts. They also direct the development of α-syn oligomer into its unstructured form and also facilitate non-toxic pathways and can therefore be useful drugs for PD and synucleinopathy. The key advantage of phytochemicals is their structural diversity, which provides knowledge for drug discovery and production. There are several groups of phytochemicals, including saponins, lignins, glycosides, carotenoids, *etc.*

Structurally-diverse and secondary plant nitrogen-containing metabolites are alkaloids that are defensive agents against neurodegenerative diseases. Plant bioactive derivatives such as flavonoids, stilbenoids and alkaloids have strong anti-oxidant and anti-inflammatory properties which are of great importance for PD treatment. These phytochemicals, which occur naturally, can also promote mitochondrial function and act as major cognitive enhancers. In addition, these compounds act as inhibitors of alpha- synuclein aggregation, activation of c-Jun N-terminal kinase (JNK) and development of monoamine oxidase (MAO) and are dopaminergic neuron agonists.

Ginsenosides: Ginsenosides are triterpinoid saponins unique to the species of *Panax ginseng,* eliciting a pleotrophic mode of action from these compounds. The inhibitory DA uptake activity of ginsenosides has the ability to act as antagonists for N-methyl-D-aspartate receptor (NMDA) and to defend neurons against mitochondrial dysfunction and elevation of glutamate and excitotoxicity caused by methyl-4-phenyl-1,2,3,6-tetrahydropyridine (MPTP). Ginsenoside's ability to minimize the influx of calcium and free radical generation and oxidative stress can also play a role in modifying the effects of ginseng in PD.

Caffeine: Caffeine is an antagonist for Adenosine 2A receptor present in the beans of *Coffea arabica* and *Coffea canephora* plants, which are widely distributed in Asia and Africa. In MPTP-induced PD mice, caffeine exerts neuroprotective effects against dopaminergic neuronal loss. In addition, caffeine in PD mouse models induces motor deficit reversal. It has been shown that the behavioral and neurobiochemical effects of caffeine cause a decrease in apomorphine-induced rotation and improved motor control. The amount of DA and its metabolites have also been shown to recover after administration of caffeine in the experimental depletion of dopaminergic neurotransmissions using neurotoxin 6-hydroxydopamine (6-OHDA).

Ginkgo biloba: Specific phytochemicals found only in the *Ginkgo biloba* tree are ginkgolides and bilobalides. In C57BL/6J mice, EGb761, a well-defined blend of active compounds extracted from the leaves, exerts a protective effect against MPTP-induced oxidative stress. In the striatum and substantia nigra pars compacta, mice receiving EGb761 recovered striatal DA levels and tyrosine hydroxylase. The neuroprotective effect of EGb761 against the neurotoxicity of MPTP is correlated with its free radical scavenging activity, lipid peroxidation blockage, and reduction of radical output of superoxides [79]. Additionally, *G. biloba* extract has an inhibitory effect on the activity of MAOs in rat mitochondria, indicating that its neuroprotective effects on dopaminergic neurons may be due to its inhibitory effects on monoamine oxidase.

Polygala: Polygala root extract (PRE) consists of xanthones, saponins and oligosaccharide esters [80] and has been documented to have a neuroprotective impact on dopaminergic neurons in both *in vivo* and *in vitro* PD models with 6-hydroxydopamine (6-OHDA) mediated neurotoxicity. The potential mechanism of action is due to decreased production of ROS and nitric oxide (NO) and altered activity of caspase-3 [81]. Furthermore, through binding to norepinephrine transporter proteins, oligosaccharide derivatives of PRE act against clinical depression. In addition, the 3,4,5-trimethoxycinnamic acid (TMCA) present in PRE exerts anti-stress effects by norepinephrine suppression.

Uncaria rhynchophylla: As a traditional medicine, *Uncaria rhynchophylla* is used to treat convulsive seizures, tremors, and hypertension. Rhynchophylline, corynoxeine, corynantheine, and hirsutine are the major alkaloids, with catechin and epicatechin being the main flavonoids. All animals with depleted DA activity using 6-OHDA have been shown to have a cytoprotective effect [82]. Dopaminergic neuronal loss and apomorphine mediated rotation were enhanced by Uncaria rhynchophylla extract (URE). In the meantime, a substantial decrease was observed in ROS and caspase 3 activity and a remarkable maintenance of cell viability and GSH levels was observed in PC12 neurotoxic cells.

Bacopa monniera: *Bacopa monniera* is a popular medicinal plant commonly used for the treatment of anxiety, memory disorders, and epilepsy. *Bacopa moniera* extract (BME) has been shown to exert a dose-dependent protective effect in 6-OHDA-lesioned PD rat models, as determined by major behavioral activity improvements and restoration of activity levels of GSH, SOD and catalase and decreased lipid peroxidation. Its antioxidant, free radical scavenging properties, and DA-enhancing effects are due to the potential mechanism of action of BME [83].

Cassia obtusifolia L: *Cassia obtusifolia L,* is an annual plant that is commonly consumed and widely distributed in Korea and China as roasted tea. In the substantia nigra and striatum of MPTP-induced PD mice and dopaminergic neurons *in vitro*, *Cassiae semen* (sicklepod) seed extract (CSE) has been shown to defend against dopaminergic neuronal degeneration. CSE supplementation has been shown to reduce cell damage and attenuate ROS generation and mitochondrial membrane depolarization in 6-OHDA mediated PC12 cells. MPP+, the neurotoxic metabolite of MPTP, causes neuronal dopaminergic loss by inhibiting respiratory complex 1 activity in dopaminergic neuron mitochondria [84].

The most abundant group of polyphenols with a well-established anti-parkinsonian effect are flavonoids. Citrus flavanone naringenin administration to rodents for four days prior to 6-OHDA injury resulted in a substantial decrease in the loss of tyrosine hydroxylase (TH)-positive cells as well as loss of DA and its metabolites 3,4-dihydroxyphenylacetic acid (DOPAC) and homovanillic acid (HVA) in the brain [85]. Since tyrosine hydroxylase catalyzes L-3,--dihydroxyphenylalanine (L-DOPA) formation, which is the rate-limiting stage in dopamine biosynthesis, TH expression is one of the most widely used markers for the identification of DA neurons. The cellular mechanisms underlying the neuroprotective effects of naringenin have been studied and shown to activate the Keap1/Nrf2/ARE axis readily, thus inhibiting oxidative stress in mice with 6-OHDA lesions [86].

Pre-treatment with naringenin glycoside, naringin, has also been shown to provide neuroprotection in the PD experimental model [87]. By reducing neuro-inflammation and microglial activation as well as increasing the mammalian target of rapamycin complex 1 (mTORC1) activity, DA neurons in the mouse brain were protected.

The excitatory amino acid neurotransmitters and related receptors that are present in the CNS, such as the N-methyl-D-aspartate (NMDA) receptor, are gaining increasing attention. Natural products have been made that either bind to the receptors or influence the levels of the transmitters. The key therapeutic response to PD was to increase DA levels either by inhibiting monoamine oxidase (MAO), which metabolizes DA to compounds that are less active, or by increasing DA precursor concentrations by administering L-hydroxyphenylalanine (LDOPA). Strategic drug discovery that have advanced progress in the clinical treatment of PD patients have centered on the alleviation of motor symptoms by the use of dopaminergic mimetics, the development of novel nondopaminergic drugs for symptomatic improvement, and finally, the discovery of neuroprotective compounds with disease-modifying effects in PD.

There are cell lines of neuronal lineage that, when differentiated into dopaminergic neurons, have the potential to serve as a human cellular model for PD. Cell culture models have advantages over animal models since they can be based on human genomes, enabling pathophysiological characteristics to be directly investigated in far less time, less labor intensive, and these techniques can be developed for high-throughput screening of therapeutic compounds. The ability of transplantation of fetal dopaminergic cells to defend and restore the damaged nigrostriatal dopaminergic pathway has been studied by several researchers in clinical trials and in animal models [88]. The presynaptic protein alpha-Synuclein (α-syn) controls the release of neurotransmitters from the brain's synaptic vesicles. Like Lewy bodies, α-syn aggregates are features of both intermittent and familial PD forms.

Several major stages of fibrillation, aggregation and oligomerization have undergone by the aggregates. With disease development, therapeutic drug effects decrease and alleviate only symptomatic actions. Therefore, novel therapeutic techniques are required that can either inhibit or postpone the progression of PD. Literature illustrates the close link between α-syn and etiopathogenesis and PD progression. Studies show that α-syn is a significant therapeutic target and an important method of disease modification is the inhibition of its aggregation, oligomerization, and fibrillation. Several hypotheses for the death of dopaminergic cells in the SN compact and mitochondrial complex defect associated with the electron transport chain defect have been proposed [89].

Accumulating evidences indicate that the function of the mitochondrial complex I decreases partially in PD. Approximately 100 percent of molecular oxygen is consumed during cellular respiration by the mitochondria, and strong oxidants are formed as a by-product, including hydrogen peroxide and superoxide radicals. By inhibiting the mitochondrial complex I, which can generate toxic hydroxyl radicals or react with nitric oxide and produce peroxynitrites, and reactive oxygen species (ROS) production also increases. These molecules can also damage nucleic acids, proteins and lipids. A lot of research has also shown that ROS plays a part in the degeneration of dopaminergic neurons in PD patients' brain tissues. In the brain tissue of patients with PD, elevated levels of lipid peroxidation, glutathione depletion and increased protein oxidation are reported. Dopamine oxidation contributes to the formation of dopamine quinone, which is able to modify proteins directly.

The main culprit is suspected to be mitochondrial dysfunction and oxidative insult in PD. The current therapy available for PD relies mainly on Levodopa, which provides the ability to delay disease progression to some degree, but has several side effects. It is obvious that redox stabilization and replenishment of mitochondrial function seem to be an important therapeutic approach against PD as both are necessary for optimal neuronal functioning. By enhancing mitochondrial function and alleviating oxidative stress, certain natural and synthetic products exhibit neuroprotective and anti-apoptotic potential.

In drug development, there are many factors that are taken into account: toxicity, the ability to cross BBB, source, dose, efficacy, structure, combinatorial effect, clinical studies, molecular mechanism of action, *etc.* In view of recent scientific advances around the world, the medicinal properties of plants have been studied due to their potent pharmacological activities, low toxicity and economic viability. Ayurveda has a therapeutic specialty called rasayana, which, by optimization of homeostasis, prevents diseases and counteracts the aging process.

Antioxidant Property of the Plant Extracts

Antioxidants have been proposed as a possible treatment choice for neurological conditions to counter cellular oxidative stress within the nervous system. However, awareness about other beneficial molecular activities that natural extracts exhibit in addition to antioxidant activity has led to more research into the possible use of natural extracts for neurodegenerative disease prevention or treatment/management. Plant extracts show antioxidant properties, prevention of apoptosis, inhibition of DA-transporter activity, prevention of microglial activation, anti-inflammation, reduction of nitric oxide synthesis, inhibition of monoamine oxidase (MAO), which are the main modes of neuroprotection.

Plants have been an excellent source of medicine since ancient times. For thousands of years, plants have played a notable role in sustaining human health and enhancing the quality of life, as valuable components of medicines, seasonings, drinks, cosmetics, and dyes. The emphasis on plant science has increased worldwide in modern times, and a vast body of evidence has been gathered to demonstrate the tremendous potential of medicinal plants used in different conventional systems. Since time immemorial, plant-derived natural products have developed their own niche in the treatment of neurological diseases. Plant derived molecules that emphasize the behavioral, cellular, or biochemical aspects of neuroprotection found in cellular or animal models of the disease have been thoroughly investigated. In a dynamic interplay of various pathways in both normal aging and neurodegenerative diseases, free radicals play a crucial role [90]. Free radicals and other reactive oxygen species, collectively known as ROS, are continuously produced under pathological conditions by normal physiological processes. They are simultaneously degraded by enzymatic and non-enzymatic antioxidant defense mechanisms to non-reactive forms.

Active microglia increase in the striatum and SN in patients with PD. Microglial cells are macrophages in the brain which responds under various unfavorable conditions *via* rapid hypertrophic proliferation and expression of a number of cytokines. A variety of pro-inflammatory cytokines are released by active microglial upregulating cell surface markers, such as macrophage antigen complex 1. A number of these cytokines contribute to the inflammation phase, including IL-1, IL-6 and TNF-a [91]. Attenuating cytotoxicity caused by MPTP and improving cell viability are some of the most widely recorded mechanisms of action for plant extracts. The production of reactive oxygen species (ROS) results in the possible collapse of the mitochondrial transmembrane and dysfunction of the mitochondrial respiratory chain complex-1, which eventually leads to increased cytosolic Ca^{++} and mitochondrial cytochrome C concentrations, activating signalling pathways for apoptosis. The decrease in cellular free radicals therefore prevents apoptosis and preserves the role of mitochondria [92]. Mitochondrial quality control and dynamics are important for optimum neuronal functioning, with neuronal cell death triggered by any alteration. Brain autopsy samples from PD patients and animal models showed a decline in mitochondrial function of specifically complex I activity. Furthermore, mitochondria are active in mitophagy, thus keeping defective organelles under check.

To treat PD, which has been studied in animal studies or clinical trials, quite a few herbal medicines and herbal formulations have been used. While no model has been able to recapitulate all the pathological characteristics of PD to date, preclinical PD research models (biochemical, cellular, and animal) have made a major contribution to our understanding of human PD. In order to mimic PD *in vitro* and *in vivo*, three neurotoxins, 6-hydroxydopamine (6-OHDA), 1-methyl-4-phenyl-1,2,3,6-tetrahydropyridine (MPTP) and rotenone, are the most effective agents to date.

The dopamine transporter takes up 6-OHDA and produces free radicals. MPTP is converted into MPP+ by monoamine oxidase B and then taken up by the transporter of dopamine and can be stored by mitochondria, leading to inhibition of complex I and free radical generation. Rotenone is a direct inhibitor of complex I, which results in the production of free radicals as well. C57/BL6 mice, Wistar rat and PC12 and SH-SY5Y cells could serve as useful models for assessing the efficacy and side effects of *in vivo* or *in vitro* symptomatic PD treatments, respectively. Significant cellular actions of cell death have been established through comprehensive analysis of these models, including oxidative stress, mitochondrial dysfunction, excitotoxicity, neuro-inflammation and nitric oxide, *etc.*

As plants generate large quantities of antioxidants to escape the oxidative stress induced by photons and oxygen, they are a possible source of new antioxidant activity compounds. In order to find active and therapeutically beneficial compounds from plants- Ayurveda, the oldest medicinal method in the world, offers plenty of leads about the plant derived bioactive compounds. In Indian traditional medicine, polyherbal preparations may have antioxidant activity stemming from individual plants and may function synergistically to avoid aging and associated degenerative diseases. Such widely used plants have been tested and their capacities for antioxidants have been compared. In the 21st century, additional studies to isolate active principles from these plants would be of great medicinal significance.

By using the neurotoxin 1-methyl-4-phenyl-1,2,3,6-tetrahydropyridine (MPTP), the best-characterized animal model of PD has been developed. Remarkable clinical symptoms similar to intermittent PD in human results after injection of MPTP. MPTP crosses the blood-brain barrier after systemic administration and is metabolized in astrocytes by MAO-B to its active metabolite,1-methyl-4-pheyl-pyridinium ion (MPP+) [93]. Due to its affinity with DA transporters, MPP+ is selectively absorbed by DA-ergic neurons, resulting in selective toxicity to DA-ergic neurons. Infusion of MPP+ into the median forebrain bundle (MFB) or striatum or SNpc in rats resulted in a substantial loss of SNpc DA-ergic neurons, ensuring improvements in the lesion's behavioral, neurochemical, and biochemical aspects [94]. The first agent used to model PD was the neurotoxin 6 hydroxydopamine (6-OHDA).

The PD 6-OHDA model is based on the production of free radicals and the effective reversible inhibition of the Complex-I and -IV mitochondrial electron transport chain (ETC). Rotenone is a naturally occurring plant-derived compound that has been used in vegetable gardens as an insecticide and also for the reduction or sampling of fish stocks in lakes or reservoirs. Rotenone is considered to be a high-affinity selective inhibitor of the inner mitochondrial membrane ETC Complex-I enzyme involved in oxidative phosphorylation. More than 20 years ago, it was shown that rotenone MFB infusion caused harm to the nigrostriatal pathway dopaminergic neurons. Several years later, it was recorded that by sparing SNpc but damaging both nucleus caudate putamen (NCP) and globus pallidus, systemic administration of rotenone created a unique pattern of CNS damage [95].

Mucuna pruriens: In clinical trials, powdered seed formulations of *Mucuna pruriens* showed beneficial effects on patients with Parkinson's disease, with a quicker onset of action and also without the concomitant rise in dyskinesia. A commercial preparation of *Mucuna pruriens*, Zandopa (HP-200), is available for the treatment of PD [96]. Scientific research has shown that levodopa is present in *Mucuna pruriens*, which could provide a long-term improvement in PD [97].

Resveratrol: Resveratrol is a natural phytoestrogen contained predominantly in grapes and red wine and is noted for its neuroprotective effects by reducing oxidative stress. In the MPTP-mediated PD mouse model at the molecular level, resveratrol downregulates α-syn expression mediating miR-214, and neurotoxicity induced by MPP+ in neuroblastoma cells [98]. Resveratrol prevents the formation of fibrils and oligomers and also stabilizes or disaggregates the α-syn oligomers and may thus be potentially useful compounds for the production of pharmaceuticals and medicines.

Ficus religiosa: *Ficus religiosa. L* (Moraceae) commonly referred to as "Pimpala" or "Pipal" tree is a tall, narrowly deciduous, heart-shaped tree with spreading branches and gray bark, without aerial roots from the branches [99]. Hindus and Buddhists consider the tree sacred. It is known by many vernacular names in India, the most frequently used being Asvatthah (Sanskrit), Sacred Fig (Bengali), Peepali (Hindi). In animals with Parkinson's disease, the petroleum ether extract of the *Ficus religiosa* plant proved to be an antioxidant and showed a promising impact.

Saponins: The abundant group of secondary plant metabolites are saponins, which can be categorized as steroids, triterpenoids, and glycosides [100]. Oleuropein (present in olive oil), resveratrol (present in grapes), curcumin (present in turmeric), astaxanthin (a carotenoid present in fruit and vegetables), catechins (present in black and green tea), and lycopene (present in tomatoes) are common polyphenols in foods [101]. These polyphenols prevent the formation and fibrillation of α-syn aggregation of amyloid protofilaments and fibrils and have also shown protective effects in neurodegenerative diseases. Polyphenols are an essential class of compounds that can effectively inhibit α-syn assembly. Several of them also inhibited α-synfilament assembling with IC50 value even with very low [102, 103] concentration.

Neuroprotective Activity of Bioactive Compounds from Herbs

In traditional medicine, 22,500 medicinal herbs are used throughout China, of which only a few have been successfully investigated for possible production into herbal formulations for the treatment of PD in animal experiments or clinical trials [104]. Regardless of our success in understanding the pathogenesis of PD, traditional medicines have not achieved sufficient results for pharmacological therapies. It is therefore possible that the use of bioactive compounds from natural sources can give rise to more suitable potential candidates for PD preventive therapy. Comprehensive research into the development of novel compounds for neuroprotective drugs has shown that natural products, such as plant extracts and their bioactive compounds, may have considerable potential as lead candidates for neuroprotection in the treatment of PD. Herbs and extracts function as antioxidants to relieve oxidative stress, dopamine transporter (DAT) or MAO-B inhibitors to reduce neurotoxicity, scavenge free radicals, chelate harmful metals, modulate cell survival genes and signals of apoptosis, and even improve blood circulation in the brain [105]. Within the sense of ancient herbal medicine systems, Asian countries, such as India, China, Japan, and Korea, have used various combinations of herbal formulations to treat PD. By deriving patient-specific cell lines from individuals with sporadic variants of PD and even those with recognized disease-causing mutations, cell culture techniques make it possible to reproduce few of the features.

Although there are several other therapies available, such as monoamine oxidase-B (MAO-B) inhibitors that delay the catabolism of dopamine, antioxidants that minimize free radical damage, growth factors that rescue dopamine neurons, and tissue transplants, there is still a significant unmet medical need for PD. Pathogenesis and development of PD was associated with inflammation, oxidative stress, free radical toxicity, decreased growth factors, and apoptosis in several areas [106]. There is growing evidence that herbs and extracts attenuate the substantia nigra degeneration of dopamine neurons and improve Parkinsonism caused by 1-methyl-4-phenyl-1,2,3,6-tetrahydropyridine (MPTP) and 6-hydroxydopamine neurotoxins (6-OHDA).

Ginseng and Ginsenoside: Ginseng is *Panax ginseng* and *Panax notoginseng* (Araliaceae), whose dried root and rhizome is centered on the belief that it is a general tonic for the promotion of strength, wellness, and longevity, ginseng is a valuable herb in herbal medicine that has been used for over several centuries. Many forms of diseases, including ischemia, anemia, diabetes mellitus, gastritis, and insomnia, have been treated with ginseng aqueous extract [107]. There are more than 30 ginsenosides, of which ginsenosides Rb1, Rd, Re, and Rg1 are the key active ingredients responsible for its vibrant pharmaceutical actions [108].

In SH-SY5Y human neuroblastoma cells, the aqueous extract of *Panax ginseng* has been investigated for its protective effects against the cellular model of parkinsonism such as 1-methyl-4-phenylpyridine (MPP+) mediated cytotoxicity. The study showed that *Panax ginseng* aqueous extract reduced overproduction of reactive oxygen species (ROS), cytochrome C release and caspase-3 activation, increased Bax/Bcl-2 ratio, and thus increased cell survival in SH-SY5Y cells treated with MPP+ [109]. In addition to *Panax ginseng*, saponins, obtained by induction of thioredoxin-1 from *Panax notoginseng*, have a very potent neuroprotective effect on MPP+ mediated toxicity to PC12 cells and Kunming mice [110].

In the MPTP model, ginsenoside Rg1 had a protective impact on apoptosis in substantia nigra neurons, and pretreatment with ginsenoside Rg1 prevented the loss of Nissl staining neurons and tyrosine hydroxylase-positive neurons and decreased the percentage of TUNEL-positive cells and caspase-3 activity [111]. Pretreatment with different doses of Rg1 prevented the decrease of glutathione and superoxide dismutase activation, attenuated JNK and c-Jun phosphorylation in substantia nigra neurons and defended against MPTP-induced substantia nigra neuronal death. The antioxidant properties of Rg1 can contribute to the neuroprotective effect against MPTP toxicity, along with the blocking of JNK signals. Neurotrophin-like effects on survival and neurite development of cultured embryonic mesencephalic dopamine neurons affected by MPP+ toxicity was also found to be Ginsenosides Rb1 and Rg1. Compared to regulation, Ginsenoside Rb1 increased mesencephalic dopamine neuron survival by 19 percent, where MPP+ significantly decreased the amount of dopamine neurons and seriously impacted neuronal processes. Ginsenosides Rb1 and Rg1 counteracted neuronal degeneration and reduced the duration and number of dopamine neurite neurons when compared to the dopamine alone treated group [112].

Green tea: Green tea is rich in polyphenols, a group of neuroprotective chemical compounds, and tea polyphenols can protect against PD and potentially minimize the occurrence of PD. In both *in vitro* and *in vivo* , the neuroprotective role of green tea polyphenols has been established. The key biological effects identified by green tea polyphenols include antioxidants, Dopamine Transporter (DAT) inhibition that decreases the absorption of dopamine and methyl-4-phenylpyridine, free radical scavenging, and iron chelating activity. These biological effects have indicated a possible therapeutic use of green tea polyphenols in cell culture and animal model studies in the prevention and treatment of PD. To investigate the neuroprotective properties of molecules, the MPTP model of PD is widely used. Neuroprotective effects of polyphenols of green tea on neurons of nigral dopamine have been established. Polyphenol (---epigallocatechin-3-gallate (EGCG) present in green tea has attenuated the degeneration of dopamine neurons in the substantia nigra *via* their antioxidant activity [113].

Interestingly, 3H-dopamine and 1-methyl-4-phenylpyridinium (MPP+) prevented the uptake of green tea polyphenols and protected dopamine neurons from toxicity *via* inhibition of DAT activity. *in vitro* and *in vivo* studies of DAT-pCDNA3-transfected Chinese Hamster Ovary (DAT-CHO) cells and mouse striatal synaptosomes have observed the inhibitory effects of polyphenols on 3H-dopamine uptake. By downregulating the inducible NO synthase and TNF-alpha expression, EGCG potently inhibited LPS-activated microglial secretion of nitric oxide (NO) and tumor necrosis factor-alpha (TNF-alpha). In both the human dopaminergic cell line SH-SY5Y and in primary rat mesencephalic cultures, EGCG exerted strong defense against microglial activation-induced neuronal injury. Green tea extract EGCG prevented oxidative stress in animal models of PD, especially nitric oxide-mediated MPTP-induced PD in mice by inhibiting nNOS in substantia nigra. Oral EGCG administration retained striatal dopamine levels, prevented dopamine neuron loss and reduced nNOS expression in substantia nigra [114]. Neurodegeneration caused by MPTP and 6-OHDA in rodents and non-human primates is associated with increased iron and alpha-synuclein presence in substantia nigra.

Green tea polyphenol EGCG had an effect on substantia nigra on iron and alpha-synuclein, and EGCG, as well as brain-permeable VK-28 series derivative iron chelators, prevented iron and alpha-synuclein accumulation in substantia nigra of MPTP-treated mice [115]. In addition, green tea polyphenols have been shown to attenuate activation and cell death of 6-OHDA-induced nuclear factor-kappaB (NFkappaB) in neuronal cultures. The application of green tea extract prior to 6-OHDA inhibited both nuclear NF-kappaB translocation and cultured neuron binding activity. The neuroprotective function may also be due to the prevention of NF-kappaB-promoting nuclear translocation and activation of cell death [116]. Green tea polyphenols are also known as therapeutic agents aimed at altering brain aging and the progression of PD with natural antioxidants. Amyloid precursor protein (APP) secretion was modulated by EGCG and defended against

Abeta toxicity in human SHSY5Y neuroblastoma and rat PC12 cells. EGCG rescued PC12 cells in a dose-dependent manner against Abeta toxicity. EGCG has also showed protective effects against neurotoxicity caused by Abeta and it also controlled the secretory processing of non-amyloidogenic APP through the PKC pathway [117].

Catechins, (-)-Epigallocatechin-3-Gallate (EGCG) and theaflavins contained in black and green tea, is a defensive agent against neurodegenerative diseases [118 - 120]. Theaflavins that inhibit the development of amyloid-b and fibrillogenesis of α-syn are present in fermented black tea [121]. These bioactive compounds smoothen the progress of the α-syn and amyloid-b assembly into harmless, spherical aggregates that are unable to undergo changes in amyloid plaque formation. EGCG binds directly to and inhibits the conversion of native unfolded polypeptides into toxic intermediates [122]. For targeted drug delivery, the combination of EGCG with its unique α-syn proteolytic peptide sequences was well developed and α-syn fibrillation was found to be prevented. Such data, in combination, indicates that EGCG may be a successful therapy for neurodegenerative diseases and a good candidate for inclusion in diets and the development of pharmaceutical drugs.14 polyphenolic compounds and theaflavin-containing black tea extract were investigated and found to be scutellarein, baicalein, myricetin, nordihydroguaiaretic acid, and (-)-epigallocatechin-3-gallate (EGCG) derived from black tea extract. Green tea extract has been the perfect active ingredient that can be used to examine its direct effect on the inhibition of α-syn oligomer formation in experimental PD models.

Polygonum cuspidatum: A perennial herb, *Polygonum cuspidatum* is mainly used in traditional Chinese medicine and other Asian cultures. The neuroprotective potential of *P. cuspidatum*-derived resveratrol (RES) in 6-OHDA induced mice has been shown in recent studies. Its antioxidant-lowering and antiapoptotic abilities exert a defensive effect [123]. In another study, dopaminergic neuronal loss and neurobehavioral defects escaped following 6-OHD injection in male wistar rats pretreated with RES. This effect is likely to be caused by upregulation of the status of antioxidant enzymes and mitigation of deprivation of DA.

Pueraria thomsonii Benth: is an herbal remedy enriched with puerarin, daidzin, daidzein, and genistein isoflavonoids. The neuroprotective effect of genistein in dopaminergic neurons has been documented in PC12 cells differentiation with the 6- OHDA mediated nerve growth factor (NF). There was an inhibitory effect of genistein and daidzin on caspase 3 and caspase 8 activation, thereby preventing apoptosis [124].

Baicalein and baicalin: Baicalein and baicalin are flavonoids present in the herbs *Scutellaria baicalensis* and *Bupleurum scorzonerifolfium* (S/B) in large concentrations. These compounds have been shown to attenuate substantia nigra with iron-induced lipid peroxidation and DA depletion. They can also increase GSH levels, impede alpha-synuclein aggregation, and decrease mitochondrial stress and apoptosis caused by iron. Alpha-synuclein disaggregation by baicalein is mainly because of its functionality with 3-OH group attached to the benzene ring, which may act as a molecular target. A substantial decrease in nitric oxide synthase and inflammatory markers was also observed in the substantia nigra of S/B remedy-treated neurotoxic rat brain. In addition, baicalein-ameliorated 6-OHDA mediated toxicity in SHSY-5Y cells by reducing phosphor-JNK and caspase activity [125]. In MPTP induced PD mice, Baicalein pretreatment improved behavior, increased DA and melatonin in the striatum, this can be shown as a potential mechanism of action [126].

Curcumin: Acombination of curcumin and b-cyclodextrin showed that the synergistic inhibition of α-syn aggregation in study groups often degraded the aggregates to monomers at very low concentrations [127, 128].The curcumin's ability to act as a therapeutic agent in PD was further improved by a well-balanced arrangement of the phenolic groups, benzene rings, and versatility. The phenolic structure classes assist to enhance the interactions of curcumin with α-synmonomers and oligomers. Transfected PC12 cells with recombinant plasmids, a-syn-pEGFP-A53T downregulated α-syn expression or oligomer formation *via* the regulation of mitochondrial membrane potential mediated by apoptosis [129]. The *in vitro* α-syneffect of curcumin was also reconfirmed through *in vivo* studies using genetic mouse models of synucleinopathy. The increase in phosphorylated forms of α-syn at the cortical presynaptic terminals was also seen by curcumin, but it was noted that it had no direct effect on the aggregation of α-syn . However, motor and behavioral efficiency improved [130]. Curcumin is less soluble and also less stable because it has restricted oral bioavailability, so many nanoformulations and structural analogs have been increasingly developed to boost its solubility, stability and also increase its oral bioavailability [131 - 133]. For targeted delivery, nanoformulation containing curcumin and piperine with glyceryl mono-oleate nanoparticles coated with multiple surfactants was produced to boost its bioavailability in the brain [134]. It has been shown that nanoformulation decreases oxidative stress and apoptosis, and also prevents oligomerization and fibrillation of α-syn and thus induces autophagy. There was another nanoformulation prepared by sol-oil chemistry with lactoferrin, which improves rotenone-induced neurotoxicity in dopaminergic SK-N-SH cells [135], which showed better bioavailability, improved cell viability, even attenuated oxidative stress, and thus reduced tyrosine hydroxylase and α-syn expression. Furthermore, nanoformulation enhanced motor deficits and improved the expression of dopamine and tyrosine hydroxylase with the promotion of α-syn clearance in the PD mouse model [136]. In the rat model of lipopolysaccharide-induced PD, curcumin also inhibited α-syn aggregates in dopaminergic neurons and thereby attenuated oxidative stress, apoptosis, inflammation, and motor deficits [137]. The anti-Parkinson's effects of various plants, herb bioactives' in *in-vivo* and *in-vitro* are shown as in the Table **5**.

Table 5. Mechanism of action of various plant extracts using *in-vitro* and *in-vivo* models for PD Treatment.

Source	Model Used for the Study	Mechanism of Action	Reference
Triple herbal extract DA-9805	Male C57BL/6 mice and SH-SY5Y human neuroblastoma cells	Neuroprotective effect *via* amelioration of mitochondrial damage.	[138]
Sida cordifolia Evolvulus alsinoides and *Cynodon dactylon* extract	Rats	Antioxidant, anti-inflammatory and neuro-regenerative effects	[139]
Aqueous extract of *S. cordifiloia* and its subfractions	Rotenone-induced rat model of PD	Protection from behavioral and oxidative stress related damages, dopamine loss caused by prolonged rotenone exposure.	[140]
EGb761; from *Ginkgo biloba* L.	MPP + induced *in-vivo* mice models	Antioxidant potential	[141]
	1-methyl-4-phenyl-1,2,3,6 tetrahydropyridine (MPTP) and 6- hydroxydopamine (6-OHDA) induced PC12 cells	Reduction of cell apoptosis, Increased Bcl-2 expression by inhibiting caspase-3 activity.	[142]
Tussilagone from *Tussilago farfara*	BV-2 microglial cells.	Anti-inflammation and inhibition of nitric oxide and PGE 2 production in LPS activated microglia Suppressing the expression of protein and mRNA of inducible nitric oxide synthase and COX-2.	[143]
Green tea polyphenols	SH-SY5Y cells	Protected against 6-OHDA-induced toxicity. Suppressing the ROS-nitric oxide pathway and preventing Ca^{2+} accumulation in the cells.	[144]
Black tea polyphenol-theaflavin	Mice	Neuroprotection in the MPTP and probenecid induced model	[145]

(Table 5) cont.....

Source	Model Used for the Study	Mechanism of Action	Reference
Delphinium denudatum	1-methyl-4-phenyl-1,2,3,6-tetrahydropyridine (MPTP) and 6- hydroxydopamine (6-OHDA) rat.	Prevents lipid peroxidation and depletion of GSH content in the SNpc area. Increased SOD and catalase activities	[146]
Olive *(OleaeuropaeaL.)* leaf extract	6-OHDA-induced PC12 cells.	Antioxidant and antiapop-totic effects	[147]
SF-6 compound from *Indigo feratictoriaL.* Plant	6-OHDA model of the disease *in-vitro* and *in-vivo*	Scavenging free radicals; Protect SH-SY5Y cells from toxicity caused by α-synuclein, 6-OHDA or Hydrogen peroxide	[148]
Cinnamomum sp. Cinnamon extract	*Drosophila* expressing mutant A53T α-synuclein in the nervous system	The extract attenuated behavioral abnormalities and α-synuclein aggregat-ion in the diseased flies	[149]
Aqueous extract of *Selaginella delicatula (Alston plant)*	Rotenone induced PD in *Drosophila*	The extract protected against rotenone induced lethality, motor dysfunc-tion; oxidative stress, mitochondrial dysfunction and neurotoxicity The extract normalized the activity of NADH-cytochrome *c* reductase and succinate dehydrogenase enzymes	[150]
Sesame Seed Oil *(Sesamum indicum L.)*	1-methyl-4-phenyl-1,2,3,6-tetrahydropyridine (MPTP) and 6- hydroxydopamine (6-OHDA). 6-OHDA-induced PD in mice	Dopamine level and antioxidant enzymes were significantly increased. Inhibited NADPH oxidase dependent inflammatory pathway and protect against 6-OHDA toxicity	[151]
Petroleum ether extract of *Ficus religiosa* leaves	6- hydroxydopamine (6-OHDA) induced experimental rat animal models.	Improved motor perfor-mance and also signi-ficantly attenuated oxida-tive damage. Significantly attenuated the motor defects and also protected the brain from oxidative stress.	[152]

(Table 5) cont.....

Source	Model Used for the Study	Mechanism of Action	Reference
Green tea polyphenols and the major component EGCG	PC12 cells	EGCG inhibited lipopolysaccharide (LPS)-induced microglial activation, which may play a pivotal role in inflammatory damage of dopamine neurons in PD. Showed protective effect on 6-OHDA-induced apoptosis *via* antioxidant action.	[153]
Ginsenosides Rb1 and Rg1 of ginseng	MPTP induced toxicity in mice.	Ginsenoside Rg1 reduced MPTP-induced substantia nigra neuronal loss by suppressing oxidative stress and MPTP toxicity	[154]
Flavonoid Baicalein	Primary midbrain neuron cultures from E-14 rat embryos.	Neuroprotective effect of baicalein on dopamine neurons it possessed anti-inflammatory and antioxidant properties. Production of TNF-alpha and free radicals such as NO and superoxide by LPS stimulation were also attenuated by baicalein at a concentration dependent pattern.	[155]
Ginsenoside Rg1	PC12 cells induced by dopamine.	Rg1 protected PC12 cells against apoptosis by inhibiting the activation of caspase-3 and regulating the ratio of Bcl-2 to Bax expression. Rg1, the percentage of apoptotic cells and caspase-3 activity decreased, and the percentage of cells with positive Bcl-2 protein increased.	[156]

(Table 5) cont.....

Source	Model Used for the Study	Mechanism of Action	Reference
Oil from *ganoderma lucidum* spores	MPTP animal model of PD.	Treatment with ganoderma spores oil increased surviving dopamine neurons in the substantia nigra and levels of dopamine in the striatum, attenuated involuntary motor sympotoms of mice.	[157]
Flavonoid Wogonin (5,7- dihydroxy---methoxy flavone) from the root of *Scutellaria baicalensis* Georgi	PC12 cultured cells.	Wogonin inhibited inflammatory activation of cultured brain microglia by diminishing lipopolysaccharide-induced TNFalpha, interleukin-1beta, and NO production. Inhibition of inflammatory activation of microglia by wogonin led to the reduction in microglial cytotoxicity	[158]
Baicalein- flavone from the roots of *Scutellaria baicalensis* Georgi	SHSY5Y and HeLa cells	Baicalein prevents α-syn fibrillation and protects against neurotoxicity by preventing α-syn oligomer formation Baicalein interacts with α-syn through a tyrosine residue.	[159 - 162]

(Table 5) cont.....

Source	Model Used for the Study	Mechanism of Action	Reference
Curcumin from turmeric	Rats	Enhanced LRRK2 mRNA and protein expression, inhibited oligomerization of mutant α-syn into higher molecular weight aggregates. Possesses antioxidant, anti-inflammatory, and antiapoptotic properties.	[163 - 165]
	PC12 cells	Strongly bound to the hydrophobic nonamyloid-b component of α-syn ameliorates A53T mutant α-syn-induced PD. Involved in macroautophagy, a process in the degradation pathway that clears proteins in cells by activating the mTOR/p70S6K signaling pathway.	[166, 167]
Bacopa monnieri **(Brahmi)**	Transgenic PD model in C. elegans	Supplementation attenuates α-synuclein aggregation and slows down degeneration of DA-ergic neuron	[168]

The etiology of PD is not well known and the development of neurodegenerative disorders have been shown by a lot of research in this area. The current therapeutic strategy against PD relies mainly on treatment with Levodopa, which can delay the progression of the disease to a limited degree with associated side effects. However, there is no possible treatment available that is capable of preventing development or providing PD with a full cure. The development of a potential drug for the treatment of PD is, therefore, of utmost importance. In drug production, there are many factors that are taken into account: toxicity, the ability to cross BBB, source, dose, efficacy, structure, combinatorial effect, clinical studies, molecular mechanism of action, *etc.* Therefore, the focus of the present review was on discussing the function of natural plant extracts that were studied for PD. There are, however, many plants that have been shown to be helpful for neurodegenerative diseases. However, due to their minimal toxicity and good anti-inflammatory and anti-oxidative ability, only a few plants were chosen for the current study. Several plants have been found to share a very similar mechanism of action that includes redox stabilization and mitochondrial function replenishment. This is significant, in particular, because oxidative stress and mitochondrial dysfunction are known to be mainly involved in PD pathology. They have also been shown to provide similar advantages in both familial (transgenic animal model or human patient) and intermittent forms of PD (toxin-induced *e.g.* MPTP, PQ, Maneb, Rotenone). In addition, it has been found therapy with few herbal extracts as well. Natural herbs offer better protection compared to levodopa treatment. These results indicate the need for a comparative review of these herbs with medicinal potential. These herbal extracts have also been found to improve mitochondrial function and thus provide protection for DA-ergic neurons in the SN region of the brain. In order to explore the underlying molecular mechanism and associated signaling pathways, however, further studies are needed.

Huntington's Disease

Huntington's disease (HD) is a progressive neurodegenerative autosomal-dominant inherited condition, associated with short involuntary movements and gradual loss of higher neuronal functions, like cognition accompanied with certain psychological symptoms. While the disorder has the ability to occur from infancy to old age at any period, it typically starts between the ages of 35 and 50 and progresses steadily. Symptoms comprise weakened mental capacity weakening that leads to personality change, like depression and suicidal thoughts. Relatively uncommon signs of HD are loss of psychomotor functions due to poor muscle control and irregular jerky involuntary movements, aggressive activity and dementia development [169]. HD is triggered by a normal CAG trinucleotide polyglutamine expansion at the IT 15(HTT) gene N-terminus, because it codes for the protein huntingtin (Htt) and is situated close to the chromosome 4 (4p16.3)tip, near to the telomere [170]. Htt is usually extensively upregulated in the brain and is reported in other organs such as the liver, heart, and lungs. Even though specific role of Htt has yet to be uncovered, embryogenesis and growth have been found to be involved. Htt plays a major role in the postembryonic development of few brain areas as well as in the existence of neurons [171]. The major changes seen in the body during HD condition are shown in Fig. (**4**).

During HD condition, transcription of the mutant huntingtin occurs due to the amplified CAG repeats present in the HTT gene. The number of CAG repeats in a healthy individual is around 10-30, while there are 38-120 repeats in patients with HD. The progress and earlier initiation of the disease is reflected in number of repeats of these CAGs. The repetition of the CAG will result in protein aggregation inside the striatal neurons which is further degraded into long and short parts. They impair cell function in a variety of ways as short reactive pieces misfold and clump together and aggregate in the nerve cells, which tends to be a major reason for neurodegeneration [170].

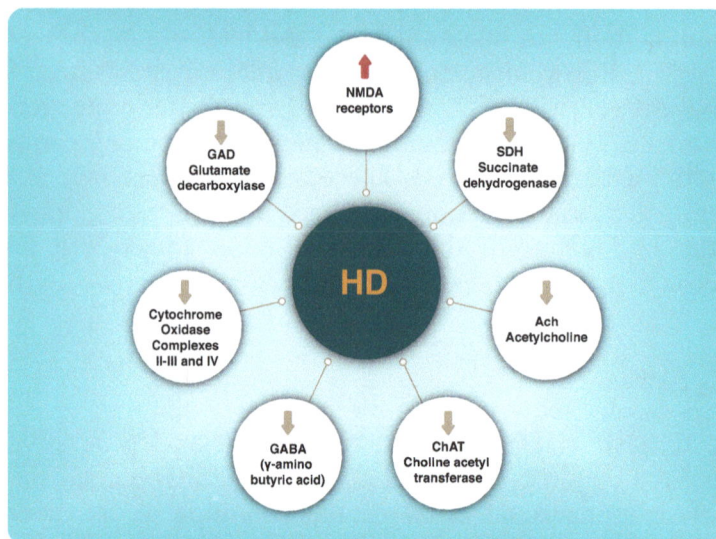

Fig. (4). Changes during HD condition in the human body.

A broad array of different neurotoxicity mechanisms has been proposed, which include proteasome activity inhibition, caspase activation, transcriptional pathway dysregulation, and enhanced ROS generation. The affected areas in HD are primarily in the striatum corpus (caudate nucleus and putamen) and cortex (frontal and temporal cortices). Many other nuclei, such as the globuspallidus, thalamus, hypothalamus, subthalamic nucleus, substantia nigra, and cerebellum are also influenced [172, 173]. The neuropathological characteristic of HD is the caudate nuclei disintegration of the basal ganglia located in the lateral ventricle brain, whereas in the putamen, it is leastconspicuous. As the disease advances, there is a rise in caudate neuronal loss, and both astrocytes and microglia in the caudate grey matter are partially substituted. No major gliosis is found in the early phases. The relatively weak signals from the subthalamic nuclei of the striatum thus decreases the initiation and movement modulation, leading to the distinctive moves of the disease [174].

Plant Extracts Proven to be Effective for Treatment in *in-vitro*, *in-vivo* and Clinical Studies

Nature is the greatest computational physician and undoubtedly has solutions to many human diseases. Several types of plants that growing worldwide have a significant therapeutic effect on the body. There are preventive or therapeutic effects on different neurodegenerative disorders from the natural compounds with the influence of anti-oxidant, anti-inflammation, anti-apoptosis, calcium antagonization, and neurofunctional control [175, 176]. Many of the plants and phytochemicals, which have demonstrated potency against neuronal dysfunction induced by3-NP, a commonly used HD animal model, are addressed below:

a). Bacopa Monnieri

Bacopa monnieri or *Herpestis monniera*, commonly known as Brahmi grows in subcontinent India and is given much importance in Ayurveda. It has been used for treating anxiety, epilepsy, insomnia, as well as memory booster from ancient age [177, 178]. The key bioactive components detected in the plant are tri-terpenoid saponins, which are from dammarane group as well as Bacosides A and B [177, 179]. Other than these main components, the presence of different Saponins comprising bacopasaponin A-G were also found [180] with jujubogenin, pseudojujubogenin, bacopaside I-V, X, N1 and N2 [181]. Brahmine, herpestine, and monnierin are some of the other components which are present in the plant [182]. Many number of studies have proved the memory boosting effects of the plant [183, 184]. *Bacoside A*. is shown to possess memory enhancing property among different components present [185]. There are different mechanisms like metal ions chelation, free radical scavenging, improved antioxidative defense enzymes which are supporting in the neuroprotective and memory enhancing effects of BM extracts [186]. In addition to this, it also exhibits anti-stress [187], antiulcerogenic antidepressant [188], anxiolytic [189], hepatoprotective [190] effects. Succinate dehydrogenase (SDH), which is a mitochondrial enzyme and the electron transport chain complexes II-III inactivates 3-NP [191]. It also raises the ROS, MDA and FFAs levels indicating that oxidative stress plays a crucial role in neurotoxicity development [192]. BM leaf powder intake considerably lowered basal levels of many oxidative markers, increased antioxidant molecules linked to thiol and antioxidant enzyme activity, indicating its antioxidant capacity. One research has shown that dietary BM supplements contribute to substantial defence against brain oxidative damage caused by neurotoxicants [11, 180]. The research indicates that BM may be beneficial in HD treatment because of the powerful antioxidant effect and the inhibitory activity against stress-induced neuronal disorders.

b). Ginkgo Biloba (Maidenhair Tree, Family: Ginkgoaceae)

Ginkgo biloba L., a Chinese traditional medicinal plant [193] renowned as the "living fossil" [194] is the oldest living species in the world. The leaf of the plant contains trilactonicditerpenes: Ginkgolide A-C, Ginkgolide J-M; a trilactonicsesquiterpene: Bilobalide; flavonoids such asisorhamnetins, quercetin, kaempferol, and biflavonoids (amentoflavone, bilobetin, 5-methoxybilobetol, sciadopitysin, ginkgetin, andisoginkgetin); and proanthocyanidins [195, 196]. Ginkgo leaf extract has displayed therapeutic potential towards disorders such as Alzheimer's, cardiovascular diseases, cancer, stress, tinnitus, geriatric problems such as vertigo, macular degeneration associated with age, and psychiatric disorders such as schizophrenia [197]. These robust effects of the Ginkgo leaf are because of its antioxidant effect, anti-platelet activating factor (Anti-PAF) activity, beta amyloid peptide (Aβ) aggregation inhibition which helps in inhibiting the development of AD [198], reduced peripheral benzodiazepine receptor expression for stress relief and activation of relaxing factor originated from endothelium which enhances blood circulation [199]. There was a 3-NP induced neurobehavioral deficits improvement due to *G. biloba* extract given in a concentration of 100 mg/kg, i.p. for 15 days [200] along with the striatal MDA level reduction. There was a down as well as up-regulation of striatal glyceraldehyde-3-phosphate dehydrogenase and Bcl-xl expression levels by the action of standardized *G. biloba* extract (EGb 761), respectively. These results, along with histopathological analysis gives an evidence for neuroprotective effect of EGb 761 in HD [42].

c). *Withania Somnifera*

Withania somnifera, belonging to solanaceae family generally called Ashwagandha, is an Ayurvedic medicine which is used for the endurance and strength enhancement [201]. Many studies have indicated that WS can increase circulating cortisol and physical performance, on the other hand it is helpful for reducing fatigue as well as refractory depression [202]. It is also having a role in the modulation of different neurotransmitter receptor systems belonging to CNS [203]. GABAergic's role in the pathogensis of HD has been well recorded and the GABAergic system has been identified in WS. In 3-NP treated animals, WS root extract pre-treatment markedly increased cognitive function, reclaimed activity of the acetyl cholinesterase enzyme, and glutathione enzyme level [204, 205]. Due to its GABAergic and antioxidant action, the root extract of WS showed a potential neuroprotective effect toward 3-NP-induced neurotoxicity in rats and made it an effective result in the therapy of HD [55, 206].

d). *Curcuma Longa*

Curcuma longa (turmeric), is a perennial herb from the family Zingiberaceae. Turmeric contains curcuminiods, sesquiterpenes, starch as well as some essential oil. Majority of the curcuminiods are diarylheptanoid, and the curcumin is found to be a strong bioactive component derivative of it. The remaining two curcuminoids are called as desmethoxycurcumin, and bis-desmethoxycurcumin. The scavenging activity of superoxide, hydroxyl radicals, metal chelating effects [207], anti-mutagenic activity and antioxidant enzymes'induction ability are considered to be the major reason for the antioxidant capacity of turmeric [208]. Its ability to inhibit cyclooxygenase (COX)-2 upregulation related to its anti-inflammatory potential. Prolonged curcumin treatment had an impact on body weight improvement, reversed motor deficits, and elevated SDH activity in 3-NP rats. The enhanced 3-NP-induced motor and cognitive damage together with a strong antioxidant potential points out that curcumin can act as a key component in the HD treatment [209].

e) Panax Ginseng C. A. Meyer and Panax Quinquefolium L (Ginsenosides)

Ginseng root is a renowned herbal medicine and among them Asian ginseng (*Panax ginseng C. A. Meyer*) and American ginseng (*Panaxquinquefolium L.*) are the widely used ginseng species [210].The major active component present in the plant is 'ginsenosides' which is a tetracyclic dammarane triterpenoid saponin glycoside. Based on the structural variations, Ginsenosidesare categorized into three groups: the panaxadiols (Rb1-Rb3, Rc, Rd, Rg3, Rh2, and Rs1), panaxatriols (Re, Rf, Rg1-2, and Rh1) and oleanolic acid derivatives [211]. There are studies which states that ginseng along with its bioactive components are having effect against aging and some of neurodegenerative diseases [212]. It helps in the reduction of lipid peroxidation, excitotoxicity inhibition, as well as over-influx of Ca^{2+} into neurons, retains cellular ATP levels, protect the structural integrity of neurons, and also enhance cognitive performance [57]. Through blocking Ca^{2+} influx *via* glutamate receptors, Rb1 and Rg3 showed defensive effects on cortical neurons against glutamate-induced cell death. In rat hippocampal neurons, ginseng saponins are found to have inhibitory effect on both NMDA as well as hike in glutamate-induced Ca^{2+} levels [213]. Ginsenosides(Rb1,Rb3,Rd) have shown neuroprotective effect against 3-N--induced striatal neuronal impairment [214, 215]. On the other hand, ginsenoside (Rb1, Rc, Rg5) have found to be effective in protecting the medium spiny neurons from glutamate-induced apoptosis in animal model. It has been assumed that neuroprotective effect of ginsenosides could be because of their ability to hinder glutamate-induced Ca^{2+} responses in spinal neuronal cells [216]. These findings strongly support that ginseng and ginsenosides can be utilized for the development of new HD treatment therapeutics.

f). Centella asiatica (syn. Hydrocotyleasiatica)

Centella asiatica, from the Umbelliferae family, has been used in the Indian traditional medicine, Ayurveda because of its ability to boost memory and effectiveness towards age related neurodegenerative disorders. Researchers have found its neuro-pharmacological effects in various aspects, including increased neurite elongation and hastening of neuro regeneration [217]. It also exhibits anti-oxidant property [218, 219]. The major chemical components from *Centella asiatica*, are triterpenoid saponins comprising asiaticoside, asiatic acid, madecassoside, and madecassic acid [220, 221]. The presence of brahmoside and brahminoside were detected in minor quantities [220, 222]. Other than this, betullic acid, triterpene acids, brahmic, and isobrahmic acid were also found in the plant [220, 222]. *Centella asiatica*, reduced the 3-NP-induced GSH level depletion, endogenous antioxidants, and total thiols in the striatum and other brain parts [222]. It has also shown protective effects against 3-NP-induced mitochondrial dysfunctions, such as, mitochondrial viability and SDH activity reduction [222]. The output of above mentioned researches undoubtedly pointing out the protective effect of *Centella asiatica* against oxidative stress induced neuronal damages along with its memory boosting power which will be helpful for treating HD-related disorders.

Challenges on the Usage of Plant Extracts for Drug Preparation

Despite the advancement of plant drug development initiatives over many years, ongoing attempts are facing several obstacles. In order to stay ahead with other drug discovery activities, the researchers of natural products and   pharmaceutical companies would need to constantly enhance the functionality of compounds that reach the process of drug development [223].

Various methods for plant drug development can be summarised as random screening subjected to bio-chemical analysis followed by safety as well as toxicological studies as shown in Fig. (**5**). The most commonly faced challenge is the lesser availability of the interested compound. This can be rectified by synthetic or semi-synthetic compound or by modifying the biosynthetic path-way of the compound through genetic engineering. The herbal drug development process is connected with quite a lot of hurdles. Crude plant extracts are generally prepared as tablets, capsules, as well as oral liquids. Due to issues faced in absorption, therapeutic effectiveness and poor compliance, these formulations are not effective. The dosage type of the tablet or capsule needs crude herbs to be powdered, and particle size influences the mixing, compression and filling operation. Moreover, due to the processing of huge amounts, high moisture content and the intrinsic nature of natural resources, uniformity is hard to achieve. Owing to their hygroscopic character, low solubility and stickiness, crude extracts are hard to develop in conventional tablet formulations. As drug development from plants is time-consuming, it is important to employ quicker and effective plant selection strategies, bioassay analysis, compound isolation and component development [224].

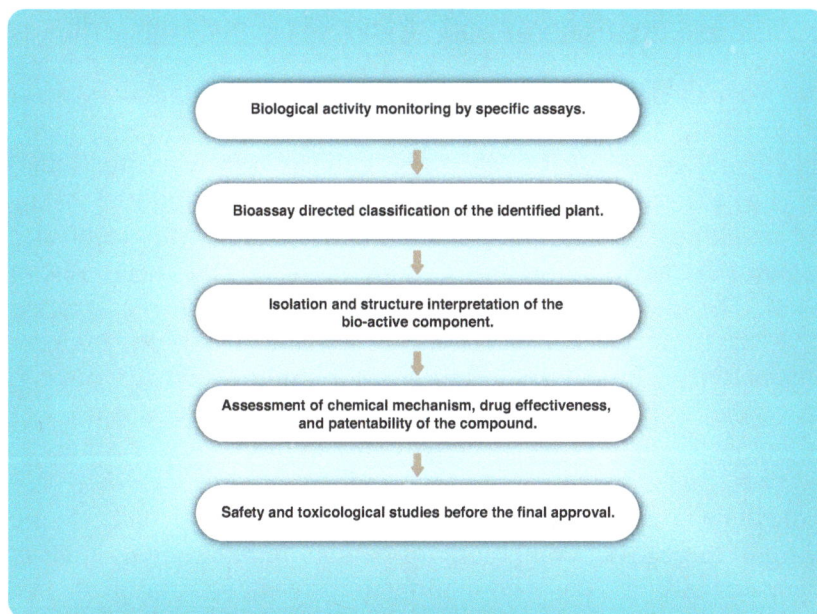

Fig. (5). Steps involved in drug development from plant extracts.

Innovative approaches are required to improve the plant collection process, particularly with regard to the legal and financial problems involving profit agreements [225, 226]. For all drug discovery projects, the model, evaluation and implementation of effective, clinically applicable bioassays are challenging processes [227, 228]. While it can be difficult to design anticipated assays, compound and extract libraries can be tested for biological activity once a screening assay is in place. Solubility is the prevalent issue encountered during extract screening and the screening of extract libraries is also troublesome, but modern techniques such as extract pre-fractionation can mitigate many of these problems [229 - 231]. Obstacles in the screening process through bioassays remain an important concern for the potential discovery of drugs from medicinal plants. Using sophisticated techniques such as LC-NMR and LC-MS, the active compound isolation speed can be improved. The production of drugs from plant-isolated active ingredients is meeting specific challenges. In general, natural products are usually isolated in small amounts that are inadequate for conducting effective optimization, development and clinical trials. Therefore, alliances with synthetic and medicinal chemists need to be formed to broaden the scope of their semi-synthesis or complete synthesis [232 - 234]. The compound synthesis from natural extracts can also be enhanced by developing libraries of natural ingredients that integrate the characteristics of bioactive components with synthetic chemistry.

Scope and Future of the Usage of Plant Extracts for the Preparation of Drugs

Pure natural products or plant extracts exhibiting antioxidant activity have shown fairly positive results through *in-vitro* and *in-vivo* animal models, their success rates in human subjects are still unclear and indicate minimal effectiveness. This may partly be attributed to the reason that almost single substances are investigated in clinical trials. In comparison, in studies prior to clinical trials, analysis of plant extracts comprising a range of secondary metabolites is much more frequent. The integration of various active ingredients in extracts can have additive or synergistic effects, resulting in enhanced antioxidant or disorder modification activity.Clinical trial findings for phytochemicals have typically been highly variable, possibly based on the way these tests are done. A broad number of subjects of varying genetic and environmental backgrounds and sometimes even different disease signs and stages are explored in clinical studies.It may be essential looking more closely, not just at the broad understanding of the entire demograohy of patient, but at an individual person or specific groups of individuals, who demonstrate substantial progress and decide why they may respond to the medication, if others do not. This could lead to a deeper understanding of the possible use of antioxidants in certain patient groups, either of a certain genetic or environmental history, but this would be correlated with higher study costs, which will also contribute to the betterment of the neurological condition.

In particular, clinical trials for the majority of natural plant extract have only focused at behavioural or cognitive changes in patients, with least studies having explicitly directly tested molecular markers of disease or oxidative stress. There is also a need for more exploration of multi-target drugs and drug cocktails. The one drug-one disease method that is commonly employed for several various treatments of diseases undermines this idea. However, since neurological disorders can be linked to various cellular dysfunctions induced by multiple external and internal causes, a multi-target drug method may be a smarter method ahead, particularly because of the complex nature of these disorders.

CONCLUSION

In conclusion, plant extracts have shown many beneficial effects in neurodegenerative disease models, both *in-vitro* and *in-vivo*. Though, there is a lack of good findings in adaptation to clinical trials, yet more studies into the pathology and neurodegenerative disease mechanisms could help to clarify the inefficiency of plant extract in clinical studies. In addition to this, it is necessary to better understand the processes that cause the disease. The precise cause of the condition is still only partly known, if not fully unknown, for most neurodegenerative disorders. For both a deeper understanding of the neurological condition and the quest for new treatments, a continuous investigation into modified animal models for these diseases may assist. Antioxidants from natural products or plant extracts may be a perfect choice for possible pre-symptom care, since they are most often unrelated to side effects and are widely tolerated by people.

CONSENT FOR PUBLICATION

Not Applicable.

CONFLICT OF INTEREST

The authors declare no conflict of interest, financial or otherwise.

ACKNOWLEDGEMENTS

Declared none.

REFERENCE

[1]   Kumar A, Dogra S. Neuropathology and therapeutic management of Alzheimer's disease – an update. Drugs Future 2008; 33(5): 433-46.
[http://dx.doi.org/10.1358/dof.2008.033.05.1192677]

[2]   Kurz A, Perneczky R. Novel insights for the treatment of Alzheimer's disease. Prog Neuropsychopharmacol Biol Psychiatry 2011; 35(2): 373-9.
[http://dx.doi.org/10.1016/j.pnpbp.2010.07.018] [PMID: 20655969]

[3]   Hardy J. The amyloid hypothesis for Alzheimer's disease: a critical reappraisal. J Neurochem 2009; 110(4): 1129-34.
[http://dx.doi.org/10.1111/j.1471-4159.2009.06181.x] [PMID: 19457065]

[4]   Anand R, Gill KD, Mahdi AA. Therapeutics of Alzheimer's disease: Past, present and future. Neuropharmacology 2014; 76(Pt A): 27-50.
[http://dx.doi.org/10.1016/j.neuropharm.2013.07.004] [PMID: 23891641]

[5]   Dal Prà I, Chiarini A, Gui L, *et al.* Do astrocytes collaborate with neurons in spreading the "infectious" aβ and Tau drivers of Alzheimer's disease? Neuroscientist 2015; 21(1): 9-29.
[http://dx.doi.org/10.1177/1073858414529828] [PMID: 24740577]

[6]   Galimberti D, Ghezzi L, Scarpini E. Immunotherapy against amyloid pathology in Alzheimer's disease. J Neurol Sci 2013; 333(1-2): 50-4.
[http://dx.doi.org/10.1016/j.jns.2012.12.013] [PMID: 23299047]

[7]   Salomone S, Caraci F, Leggio GM, Fedotova J, Drago F. New pharmacological strategies for treatment of Alzheimer's disease: focus on disease modifying drugs. Br J Clin Pharmacol 2012; 73(4): 504-17.
[http://dx.doi.org/10.1111/j.1365-2125.2011.04134.x] [PMID: 22035455]

[8]   Rosenmann H. Immunotherapy for targeting tau pathology in Alzheimer's disease and tauopathies. Curr Alzheimer Res 2013; 10(3): 217-28.

[http://dx.doi.org/10.2174/1567205011310030001] [PMID: 23534533]

[9]   Abbott A. Dementia: a problem for our age. Nature 2011; 475(7355): S2-4.
      [http://dx.doi.org/10.1038/475S2a] [PMID: 21760579]

[10] Blennow K, de Leon MJ, Zetterberg H. Alzheimer's disease. Lancet 2006; 368(9533): 387-403.
      [http://dx.doi.org/10.1016/S0140-6736(06)69113-7] [PMID: 16876668]

[11] Winblad B, Amouyel P, Andrieu S, *et al.* Defeating Alzheimer's disease and other dementias: a
      priority for European science and society. Lancet Neurol 2016; 15(5): 455-532.
      [http://dx.doi.org/10.1016/S1474-4422(16)00062-4] [PMID: 26987701]

[12] http://www.alz.org

[13] Howard RJ, Juszczak E, Ballard CG, *et al.* Donepezil for the treatment of agitation in Alzheimer's
      disease. N Engl J Med 2007; 357(14): 1382-92.
      [http://dx.doi.org/10.1056/NEJMoa066583] [PMID: 17914039]

[14] Fox C, Crugel M, Maidment I, *et al.* Efficacy of memantine for agitation in Alzheimer's dementia: a
      randomised double-blind placebo controlled trial. PLoS One 2012; 7(5): e35185.
      [http://dx.doi.org/10.1371/journal.pone.0035185] [PMID: 22567095]

[15] Tewari D, Stankiewicz AM, Mocan A, *et al.* Ethnopharmacological approaches for dementia therapy
      and significance of natural products and herbal drugs. Front Aging Neurosci 2018; 10: 3.
      [http://dx.doi.org/10.3389/fnagi.2018.00003] [PMID: 29483867]

[16] Liu X, Xu K, Yan M, Wang Y, Zheng X. Protective effects of galantamine against Abeta-induced
      PC12 cell apoptosis by preventing mitochondrial dysfunction and endoplasmic reticulum stress.
      Neurochem Int 2010; 57(5): 588-99.
      [http://dx.doi.org/10.1016/j.neuint.2010.07.007] [PMID: 20655346]

[17] Tsvetkova D, Obreshkova D, Zheleva-Dimitrova D, Saso L. Antioxidant activity of galantamine and
      some of its derivatives. Curr Med Chem 2013; 20(36): 4595-608.
      [http://dx.doi.org/10.2174/09298673113209990148] [PMID: 23834167]

[18] Namanja HA, Emmert D, Pires MM, Hrycyna CA, Chmielewski J. Inhibition of human P-glycoprotein
      transport and substrate binding using a galantamine dimer. Biochem Biophys Res Commun 2009;
      388(4): 672-6.
      [http://dx.doi.org/10.1016/j.bbrc.2009.08.056] [PMID: 19683513]

[19] Kumar A, Prakash A, Pahwa D. Galantamine potentiates the protective effect of rofecoxib and caffeic
      acid against intrahippocampal Kainic acid-induced cognitive dysfunction in rat. Brain Res Bull 2011;
      85(3-4): 158-68.
      [http://dx.doi.org/10.1016/j.brainresbull.2011.03.010] [PMID: 21439356]

[20] Romero A, Egea J, García AG, López MG. Synergistic neuroprotective effect of combined low
      concentrations of galantamine and melatonin against oxidative stress in SH-SY5Y neuroblastoma
      cells. J Pineal Res 2010; 49(2): 141-8.
      [http://dx.doi.org/10.1111/j.1600-079X.2010.00778.x] [PMID: 20536682]

[21] Long F, Yang H, Xu Y, Hao H, Li P. A strategy for the identification of combinatorial bioactive
      compounds contributing to the holistic effect of herbal medicines. Sci Rep 2015; 5: 12361.
      [http://dx.doi.org/10.1038/srep12361] [PMID: 26198093]

[22] Zhou X, Seto SW, Chang D, *et al.* Synergistic effects of chinese herbal medicine: a comprehensive
      review of methodology and urrent research. Front Pharmacol 2016; 7: 201.
      [http://dx.doi.org/10.3389/fphar.2016.00201] [PMID: 27462269]

[23] Eckert GP. Traditional used plants against cognitive decline and Alzheimer disease. Front Pharmacol
      2010; 1: 138.
      [http://dx.doi.org/10.3389/fphar.2010.00138] [PMID: 21833177]

[24] Zhou X, Cui G, Tseng HHL, *et al.* Vascular contributions to cognitive impairment and treatments with

traditional chinese medicine. Evid Based Complement Alternat Med 2016; 2016: 9627258.
[http://dx.doi.org/10.1155/2016/9627258] [PMID: 28042305]

[25]   Solfrizzi V, Panza F. Plant-based nutraceutical interventions against cognitive impairment and dementia: meta-analytic evidence of efficacy of a standardized *Gingko biloba* extract. J Alzheimers Dis 2015; 43(2): 605-11.
[http://dx.doi.org/10.3233/JAD-141887] [PMID: 25352453]

[26]   IARC Working Group. Some Drugs and Herbal Products Lyon: *Ginkgo Biloba.* International Agency for Research and Cancer (IARC) Monographs, WHO . 2016.

[27]   Isah T. Rethinking Ginkgo biloba L.: Medicinal uses and conservation. Pharmacogn Rev 2015; 9(18): 140-8.
[http://dx.doi.org/10.4103/0973-7847.162137] [PMID: 26392712]

[28]   Butterfield DA, Swomley AM, Sultana R. Amyloid β-peptide (1-42)-induced oxidative stress in Alzheimer disease: importance in disease pathogenesis and progression. Antioxid Redox Signal 2013; 19(8): 823-35.
[http://dx.doi.org/10.1089/ars.2012.5027] [PMID: 23249141]

[29]   Dávila D, Fernández S, Torres-Alemán I. Astrocyte resilience to oxidative stress induced by insulin-like growth factor I (IGF-I) involves preserved AKT (protein kinase B) activity. J Biol Chem 2016; 291(23): 12039.
[http://dx.doi.org/10.1074/jbc.A115.695478] [PMID: 27261528]

[30]   Alam MM, Okazaki K, Nguyen LTT, *et al.* Glucocorticoid receptor signaling represses the antioxidant response by inhibiting histone acetylation mediated by the transcriptional activator NRF2. J Biol Chem 2017; 292(18): 7519-30.
[http://dx.doi.org/10.1074/jbc.M116.773960] [PMID: 28314773]

[31]   Wang C, Wang B. Ginkgo biloba extract attenuates oxidative stress and apoptosis in mouse cochlear neural stem cells. Phytother Res 2016; 30(5): 774-80.
[http://dx.doi.org/10.1002/ptr.5572] [PMID: 26799058]

[32]   Zhou X, Wang HY, Wu B, *et al.* Ginkgolide K attenuates neuronal injury after ischemic stroke by inhibiting mitochondrial fission and GSK-3β-dependent increases in mitochondrial membrane permeability. Oncotarget 2017; 8(27): 44682-93.
[http://dx.doi.org/10.18632/oncotarget.17967] [PMID: 28591721]

[33]   Schwarzkopf TM, Koch KA, Klein J. Neurodegeneration after transient brain ischemia in aged mice: beneficial effects of bilobalide. Brain Res 2013; 1529: 178-87.
[http://dx.doi.org/10.1016/j.brainres.2013.07.003] [PMID: 23850645]

[34]   Rhein V, Giese M, Baysang G, *et al.* Ginkgo biloba extract ameliorates oxidative phosphorylation performance and rescues abeta-induced failure. PLoS One 2010; 5(8): e12359.
[http://dx.doi.org/10.1371/journal.pone.0012359] [PMID: 20808761]

[35]   Zlokovic BV. Neurovascular pathways to neurodegeneration in Alzheimer's disease and other disorders. Nat Rev Neurosci 2011; 12(12): 723-38.
[http://dx.doi.org/10.1038/nrn3114] [PMID: 22048062]

[36]   Song L, Xu MB, Zhou XL, Zhang DP, Zhang SL, Zheng GQ. A Preclinical systematic review of ginsenoside-Rg1 in experimental Parkinson's disease. Oxid Med Cell Longev 2017; 2017: 2163053.
[http://dx.doi.org/10.1155/2017/2163053] [PMID: 28386306]

[37]   Sun M, Ye Y, Xiao L, Duan X, Zhang Y, Zhang H. Anticancer effects of ginsenoside Rg3 (Review). Int J Mol Med 2017; 39(3): 507-18. [Review].
[http://dx.doi.org/10.3892/ijmm.2017.2857] [PMID: 28098857]

[38]   Yang L, Hao J, Zhang J, *et al.* Ginsenoside Rg3 promotes beta-amyloid peptide degradation by enhancing gene expression of neprilysin. J Pharm Pharmacol 2009; 61(3): 375-80.
[http://dx.doi.org/10.1211/jpp.61.03.0013] [PMID: 19222911]

[39]   Chin D, Huebbe P, Pallauf K, Rimbach G. Neuroprotective properties of curcumin in Alzheimer's disease--merits and limitations. Curr Med Chem 2013; 20(32): 3955-85.
[http://dx.doi.org/10.2174/09298673113209990210] [PMID: 23931272]

[40]   Belkacemi A, Doggui S, Dao L, Ramassamy C. Challenges associated with curcumin therapy in Alzheimer disease. Expert Rev Mol Med 2011; 13: e34.
[http://dx.doi.org/10.1017/S1462399411002055] [PMID: 22051121]

[41]   Dong S, Zeng Q, Mitchell ES, *et al.* Curcumin enhances neurogenesis and cognition in aged rats: implications for transcriptional interactions related to growth and synaptic plasticity. PLoS One 2012; 7(2): e31211.
[http://dx.doi.org/10.1371/journal.pone.0031211] [PMID: 22359574]

[42]   Daverey A, Agrawal SK. Curcumin alleviates oxidative stress and mitochondrial dysfunction in astrocytes. Neuroscience 2016; 333: 92-103.
[http://dx.doi.org/10.1016/j.neuroscience.2016.07.012] [PMID: 27423629]

[43]   Ramkumar M, Rajasankar S, Gobi VV, *et al.* Neuroprotective effect of Demethoxycurcumin, a natural derivative of Curcumin on rotenone induced neurotoxicity in SH-SY 5Y Neuroblastoma cells. BMC Complement Altern Med 2017; 17(1): 217.
[http://dx.doi.org/10.1186/s12906-017-1720-5] [PMID: 28420370]

[44]   Dkhar P, Sharma R. Effect of dimethylsulphoxide and curcumin on protein carbonyls and reactive oxygen species of cerebral hemispheres of mice as a function of age. Int J Dev Neurosci 2010; 28(5): 351-7.
[http://dx.doi.org/10.1016/j.ijdevneu.2010.04.005] [PMID: 20403421]

[45]   Eckert GP, Schiborr C, Hagl S, *et al.* Curcumin prevents mitochondrial dysfunction in the brain of the senescence-accelerated mouse-prone 8. Neurochem Int 2013; 62(5): 595-602.
[http://dx.doi.org/10.1016/j.neuint.2013.02.014] [PMID: 23422877]

[46]   Hoppe JB, Haag M, Whalley BJ, Salbego CG, Cimarosti H. Curcumin protects organotypic hippocampal slice cultures from $A\beta$1-42-induced synaptic toxicity. Toxicol. *in vitro* 2013; 27(8): 2325-30.
[http://dx.doi.org/10.1016/j.tiv.2013.10.002] [PMID: 24134851]

[47]   Asl MN, Hosseinzadeh H. Review of pharmacological effects of Glycyrrhiza sp. and its bioactive compounds. Phytother Res 2008; 22(6): 709-24.
[http://dx.doi.org/10.1002/ptr.2362] [PMID: 18446848]

[48]   Callender VD, St Surin-Lord S, Davis EC, Maclin M. Postinflammatory hyperpigmentation: etiologic and therapeutic considerations. Am J Clin Dermatol 2011; 12(2): 87-99.
[http://dx.doi.org/10.2165/11536930-000000000-00000] [PMID: 21348540]

[49]   Waltenberger B, Mocan A, Šmejkal K, Heiss EH, Atanasov AG. Natural Products to counteract the epidemic of cardiovascular and metabolic disorders. Molecules 2016; 21(6): 1-33.
[http://dx.doi.org/10.3390/molecules21060807] [PMID: 27338339]

[50]   Chen CM, Weng YT, Chen WL, *et al.* Aqueous extract of Glycyrrhiza inflata inhibits aggregation by upregulating PPARGC1A and NFE2L2-ARE pathways in cell models of spinocerebellar ataxia 3. Free Radic Biol Med 2014; 71: 339-50.
[http://dx.doi.org/10.1016/j.freeradbiomed.2014.03.023] [PMID: 24675225]

[51]   Yim SB, Park SE, Lee CS. Protective effect of glycyrrhizin on 1-methyl-4-phenylpyridinium-induced mitochondrial damage and cell death in differentiated PC12 cells. J Pharmacol Exp Ther 2007; 321(2): 816-22.
[http://dx.doi.org/10.1124/jpet.107.119602] [PMID: 17314199]

[52]   Yang EJ, Min JS, Ku HY, *et al.* Isoliquiritigenin isolated from Glycyrrhiza uralensis protects neuronal cells against glutamate-induced mitochondrial dysfunction. Biochem Biophys Res Commun 2012; 421(4): 658-64.

[http://dx.doi.org/10.1016/j.bbrc.2012.04.053] [PMID: 22538371]

[53]   Goenka P, Sarawgi A, Karun V, Nigam AG, Dutta S, Marwah N. Camellia sinensis (Tea): Implications and role in preventing dental decay. Pharmacogn Rev 2013; 7(14): 152-6.
       [http://dx.doi.org/10.4103/0973-7847.120515] [PMID: 24347923]

[54]   Mandel SA, Amit T, Weinreb O, Youdim MBH. Understanding the broad-spectrum neuroprotective action profile of green tea polyphenols in aging and neurodegenerative diseases. J Alzheimers Dis 2011; 25(2): 187-208.
       [http://dx.doi.org/10.3233/JAD-2011-101803] [PMID: 21368374]

[55]   Ganguly R, Guha D. Alteration of brain monoamines & EEG wave pattern in rat model of Alzheimer's disease & protection by Moringa oleifera. Indian J Med Res 2008; 128(6): 744-51.
       [PMID: 19246799]

[56]   Ganguly R, Guha D. Protective role of an Indian herb, Moringa oleifera in memory impairment by high altitude hypoxic exposure: Possible role of monoamines. Biog Amines 2006; 20: 121-33.

[57]   Wang YH, Du GH. Ginsenoside Rg1 inhibits beta-secretase activity *in vitro* and protects against Abeta-induced cytotoxicity in PC12 cells. J Asian Nat Prod Res 2009; 11(7): 604-12.
       [http://dx.doi.org/10.1080/10286020902843152] [PMID: 20183297]

[58]   Bäurle P, Suter A, Wormstall H. Safety and effectiveness of a traditional ginkgo fresh plant extract - results from a clinical trial. Forsch Komplement Med 2009; 16(3): 156-61.
       [http://dx.doi.org/10.1159/000213167] [PMID: 19657199]

[59]   Zhou YQ, Yang ZL, Xu L, Li P, Hu YZ. Akebia saponin D, a saponin component from Dipsacus asper Wall, protects PC 12 cells against amyloid-beta induced cytotoxicity. Cell Biol Int 2009; 33(10): 1102-10.
       [http://dx.doi.org/10.1016/j.cellbi.2009.06.028] [PMID: 19615455]

[60]   Fujiwara H, Iwasaki K, Furukawa K, *et al.* Uncaria rhynchophylla, a Chinese medicinal herb, has potent antiaggregation effects on Alzheimer's beta-amyloid proteins. J Neurosci Res 2006; 84(2): 427-33.
       [http://dx.doi.org/10.1002/jnr.20891] [PMID: 16676329]

[61]   Lv J, Jia H, Jiang Y, *et al.* Tenuifolin, an extract derived from tenuigenin, inhibits amyloid-beta secretion *in vitro*. Acta Physiol (Oxf) 2009; 196(4): 419-25.
       [http://dx.doi.org/10.1111/j.1748-1716.2009.01961.x] [PMID: 19208093]

[62]   Naito R, Tohda C. Characterization of anti-neurodegenerative effects of *Polygala tenuifolia* in Abeta(25-35)-treated cortical neurons. Biol Pharm Bull 2006; 29(9): 1892-6.
       [http://dx.doi.org/10.1248/bpb.29.1892] [PMID: 16946504]

[63]   Lin HQ, Ho MT, Lau LS, Wong KK, Shaw PC, Wan DC. Anti-acetylcholinesterase activities of traditional Chinese medicine for treating Alzheimer's disease. Chem Biol Interact 2008; 175(1-3): 352-4.
       [http://dx.doi.org/10.1016/j.cbi.2008.05.030] [PMID: 18573242]

[64]   Yin Y, Huang L, Liu Y, *et al.* Effect of tanshinone on the levels of nitric oxide synthase and acetylcholinesterase in the brain of Alzheimer's disease rat model. Clin Invest Med 2008; 31(5): E248-57.
       [http://dx.doi.org/10.25011/cim.v31i5.4871] [PMID: 18980714]

[65]   Liu T, Jin H, Sun QR, Xu JH, Hu HT. The neuroprotective effects of tanshinone IIA on β-amyloi--induced toxicity in rat cortical neurons. Neuropharmacology 2010; 59(7-8): 595-604.
       [http://dx.doi.org/10.1016/j.neuropharm.2010.08.013] [PMID: 20800073]

[66]   Uabundit N, Wattanathorn J, Mucimapura S, Ingkaninan K. Cognitive enhancement and neuroprotective effects of *Bacopa monnieri* in Alzheimer's disease model. J Ethnopharmacol 2010; 127(1): 26-31.
       [http://dx.doi.org/10.1016/j.jep.2009.09.056] [PMID: 19808086]

[67]   Akhondzadeh S, Noroozian M, Mohammadi M, Ohadinia S, Jamshidi AH, Khani M. *Salvia officinalis* extract in the treatment of patients with mild to moderate Alzheimer's disease: a double blind, randomized and placebo-controlled trial. J Clin Pharm Ther 2003; 28(1): 53-9.
[http://dx.doi.org/10.1046/j.1365-2710.2003.00463.x] [PMID: 12605619]

[68]   Akhondzadeh S, Noroozian M, Mohammadi M, Ohadinia S, Jamshidi AH, Khani M. *Melissa officinalis* extract in the treatment of patients with mild to moderate Alzheimer's disease: a double blind, randomised, placebo controlled trial. J Neurol Neurosurg Psychiatry 2003; 74(7): 863-6.
[http://dx.doi.org/10.1136/jnnp.74.7.863] [PMID: 12810768]

[69]   Dastmalchi K, Ollilainen V, Lackman P, *et al.* Acetylcholinesterase inhibitory guided fractionation of *Melissa officinalis* L. Bioorg Med Chem 2009; 17(2): 867-71.
[http://dx.doi.org/10.1016/j.bmc.2008.11.034] [PMID: 19070498]

[70]   Vasudevan M, Parle M. Antiamnesic potential of *Murraya koenigii* leaves. Phytother Res 2009; 23(3): 308-16.
[http://dx.doi.org/10.1002/ptr.2620] [PMID: 18844259]

[71]   Drever BD, Anderson WG, Riedel G, *et al.* The seed extract of Cassia obtusifolia offers neuroprotection to mouse hippocampal cultures. J Pharmacol Sci 2008; 107(4): 380-92.
[http://dx.doi.org/10.1254/jphs.08034FP] [PMID: 18719316]

[72]   Kim DH, Hyun SK, Yoon BH, *et al.* Gluco-obtusifolin and its aglycon, obtusifolin, attenuate scopolamine-induced memory impairment. J Pharmacol Sci 2009; 111(2): 110-6.
[http://dx.doi.org/10.1254/jphs.08286FP] [PMID: 19834282]

[73]   Dhanasekaran M, Holcomb LA, Hitt AR, *et al. Centella asiatica* extract selectively decreases amyloid beta levels in hippocampus of Alzheimer's disease animal model. Phytother Res 2009; 23(1): 14-9.
[http://dx.doi.org/10.1002/ptr.2405] [PMID: 19048607]

[74]   Lai CS, Yu MS, Yuen WH, So KF, Zee SY, Chang RC. Antagonizing beta-amyloid peptide neurotoxicity of the anti-aging fungus *Ganoderma lucidum*. Brain Res 2008; 1190: 215-24.
[http://dx.doi.org/10.1016/j.brainres.2007.10.103] [PMID: 18083148]

[75]   Joshi H, Parle M. Antiamnesic effects of *Desmodium gangeticum* in mice. Yakugaku Zasshi 2006; 126(9): 795-804.
[http://dx.doi.org/10.1248/yakushi.126.795] [PMID: 16946593]

[76]   Yu MS, Leung SK, Lai SW, *et al.* Neuroprotective effects of anti-aging oriental medicine *Lycium barbarum* against beta-amyloid peptide neurotoxicity. Exp Gerontol 2005; 40(8-9): 716-27.
[http://dx.doi.org/10.1016/j.exger.2005.06.010] [PMID: 16139464]

[77]   Ho YS, Yu MS, Yang XF, So KF, Yuen WH, Chang RC. Neuroprotective effects of polysaccharides from wolfberry, the fruits of *Lycium barbarum*, against homocysteine-induced toxicity in rat cortical neurons. J Alzheimers Dis 2010; 19(3): 813-27.
[http://dx.doi.org/10.3233/JAD-2010-1280] [PMID: 20157238]

[78]   Essa MM, Vijayan RK, Castellano-Gonzalez G, Memon MA, Braidy N, Guillemin GJ. Neuroprotective effect of natural products against Alzheimer's disease. Neurochem Res 2012; 37(9): 1829-42.
[http://dx.doi.org/10.1007/s11064-012-0799-9] [PMID: 22614926]

[79]   White HL, Scates PW, Cooper BR. Extracts of Ginkgo biloba leaves inhibit monoamine oxidase. Life Sci 1996; 58(16): 1315-21.
[http://dx.doi.org/10.1016/0024-3205(96)00097-5] [PMID: 8614288]

[80]   Choi JG, Kim HG, Kim MC, *et al.* Polygalae radix inhibits toxin-induced neuronal death in the Parkinson's disease models. J Ethnopharmacol 2011; 134(2): 414-21.
[http://dx.doi.org/10.1016/j.jep.2010.12.030] [PMID: 21195155]

[81]   Cheng MC, Li CY, Ko HC, Ko FN, Lin YL, Wu TS. Antidepressant principles of the roots of Polygala tenuifolia. J Nat Prod 2006; 69(9): 1305-9.

[http://dx.doi.org/10.1021/np060207r] [PMID: 16989524]

[82] Wang Y, Xu H, Fu Q, Ma R, Xiang J. Protective effect of resveratrol derived from Polygonum cuspidatum and its liposomal form on nigral cells in parkinsonian rats. J Neurol Sci 2011; 304(1-2): 29-34.
[http://dx.doi.org/10.1016/j.jns.2011.02.025] [PMID: 21376343]

[83] Shobana C, Kumar RR, Sumathi T. Alcoholic extract of Bacopa monniera Linn. protects against 6-hydroxydopamine-induced changes in behavioral and biochemical aspects: a pilot study. Cell Mol Neurobiol 2012; 32(7): 1099-112.
[http://dx.doi.org/10.1007/s10571-012-9833-3] [PMID: 22527857]

[84] Shahpiri Z, Bahramsoltani R, Hosein Farzaei M, Farzaei F, Rahimi R. Phytochemicals as future drugs for Parkinson's disease: a comprehensive review. Rev Neurosci 2016; 27(6): 651-68.
[http://dx.doi.org/10.1515/revneuro-2016-0004] [PMID: 27124673]

[85] Xia N, Fang F, Zhang P, *et al.* A knockin reporter allows purification and characterization of mDA neurons from heterogeneous populations. Cell Rep 2017; 18(10): 2533-46.
[http://dx.doi.org/10.1016/j.celrep.2017.02.023] [PMID: 28273465]

[86] Lou H, Jing X, Wei X, Shi H, Ren D, Zhang X. Naringenin protects against 6-OHDA-induced neurotoxicity *via* activation of the Nrf2/ARE signaling pathway. Neuropharmacology 2014; 79: 380-8.
[http://dx.doi.org/10.1016/j.neuropharm.2013.11.026] [PMID: 24333330]

[87] Kim HD, Jeong KH, Jung UJ, Kim SR. Naringin treatment induces neuroprotective effects in a mouse model of Parkinson's disease *in vivo*, but not enough to restore the lesioned dopaminergic system. J Nutr Biochem 2016; 28: 140-6.
[http://dx.doi.org/10.1016/j.jnutbio.2015.10.013] [PMID: 26878791]

[88] Wang Q, Liu Y, Zhou J. Neuroinflammation in Parkinson's disease and its potential as therapeutic target. Transl Neurodegener 2015; 4(1): 19.
[http://dx.doi.org/10.1186/s40035-015-0042-0] [PMID: 26464797]

[89] Gaki GS, Papavassiliou AG. Oxidative stress-induced signaling pathways implicated in the pathogenesis of Parkinson's disease. Neuromolecular Med 2014; 16(2): 217-30.
[http://dx.doi.org/10.1007/s12017-014-8294-x] [PMID: 24522549]

[90] Niranjan R. The role of inflammatory and oxidative stress mechanisms in the pathogenesis of Parkinson's disease: focus on astrocytes. Mol Neurobiol 2014; 49(1): 28-38.
[http://dx.doi.org/10.1007/s12035-013-8483-x] [PMID: 23783559]

[91] Du T, Li L, Song N, Xie J, Jiang H. Rosmarinic acid antagonized 1-methyl-4-phenylpyridinium (MPP+)-induced neurotoxicity in MES23.5 dopaminergic cells. Int J Toxicol 2010; 29(6): 625-33.
[http://dx.doi.org/10.1177/1091581810383705] [PMID: 20966113]

[92] Michel PP, Hirsch EC, Hunot S. Understanding dopaminergic cell death pathways in Parkinson disease. Neuron 2016; 90(4): 675-91.
[http://dx.doi.org/10.1016/j.neuron.2016.03.038] [PMID: 27196972]

[93] Javitch JA, D'Amato RJ, Strittmatter SM, Snyder SH. Parkinsonism-inducing neurotoxin, N-methyl-4-phenyl-1,2,3,6 -tetrahydropyridine: uptake of the metabolite N-methyl-4-phenylpyridine by dopamine neurons explains selective toxicity. Proc Natl Acad Sci USA 1985; 82(7): 2173-7.
[http://dx.doi.org/10.1073/pnas.82.7.2173] [PMID: 3872460]

[94] Sindhu KM, Banerjee R, Senthilkumar KS, *et al.* Rats with unilateral median forebrain bundle, but not striatal or nigral, lesions by the neurotoxins MPP+ or rotenone display differential sensitivity to amphetamine and apomorphine. Pharmacol Biochem Behav 2006; 84(2): 321-9.
[http://dx.doi.org/10.1016/j.pbb.2006.05.017] [PMID: 16820197]

[95] Ferrante RJ, Schulz JB, Kowall NW, Beal MF. Systemic administration of rotenone produces selective damage in the striatum and globus pallidus, but not in the substantia nigra. Brain Res 1997; 753(1): 157-62.

[http://dx.doi.org/10.1016/S0006-8993(97)00008-5] [PMID: 9125443]

[96]   Li XZ, Zhang SN, Liu SM, Lu F. Recent advances in herbal medicines treating Parkinson's disease. Fitoterapia 2013; 84: 273-85.
[http://dx.doi.org/10.1016/j.fitote.2012.12.009] [PMID: 23266574]

[97]   Lieu CA, Kunselman AR, Manyam BV, Venkiteswaran K, Subramanian T. A water extract of Mucuna pruriens provides long-term amelioration of parkinsonism with reduced risk for dyskinesias. Parkinsonism Relat Disord 2010; 16(7): 458-65.
[http://dx.doi.org/10.1016/j.parkreldis.2010.04.015] [PMID: 20570206]

[98]   Wang ZH, Zhang JL, Duan YL, Zhang QS, Li GF, Zheng DL. MicroRNA-214 participates in the neuroprotective effect of Resveratrol *via* inhibiting α-synuclein expression in MPTP-induced Parkinson's disease mouse. Biomed Pharmacother 2015; 74: 252-6.
[http://dx.doi.org/10.1016/j.biopha.2015.08.025] [PMID: 26349993]

[99]   Björklund A, Dunnett SB, Brundin P, *et al.* Neural transplantation for the treatment of Parkinson's disease. Lancet Neurol 2003; 2(7): 437-45.
[http://dx.doi.org/10.1016/S1474-4422(03)00442-3] [PMID: 12849125]

[100]   Dinda B, Debnath S, Mohanta BC, Harigaya Y. Naturally occurring triterpenoid saponins. Chem Biodivers 2010; 7(10): 2327-580.
[http://dx.doi.org/10.1002/cbdv.200800070] [PMID: 20963775]

[101]   da Costa IM, de Paiva JRLC, de Queiroz DB, *et al.* Supplementation with herbal extracts to promote behavioral and neuroprotective effects in experimental models of Parkinson's disease: a systematic review. Phytotherapy Research 2017. May 22; 31: 959-70.

[102]   Kumar S, Okello EJ, Harris JR. Experimental inhibition of fibrillogenesis and neurotoxicity by amyloid-beta (Aβ) and other disease-related peptides/proteins by plant extracts and herbal compounds InProtein Aggregation and Fibrillogenesis in Cerebral and Systemic Amyloid Disease. Dordrecht: Springer 2012; pp. 295-326.

[103]   Velander P, Wu L, Henderson F, Zhang S, Bevan DR, Xu B. Natural product-based amyloid inhibitors. Biochem Pharmacol 2017; 139: 40-55.
[http://dx.doi.org/10.1016/j.bcp.2017.04.004] [PMID: 28390938]

[104]   Ip PS, Tsim KW, Chan K, Bauer R. Application of complementary and alternative medicine on neurodegenerative disorders: current status and future prospects. Evidence-Based Complementary and Alternative Medicine 2012.

[105]   Pan T, Jankovic J, Le W. Potential therapeutic properties of green tea polyphenols in Parkinson's disease. Drugs Aging 2003; 20(10): 711-21.
[http://dx.doi.org/10.2165/00002512-200320100-00001] [PMID: 12875608]

[106]   Dauer W, Przedborski S. Parkinson's disease: mechanisms and models. Neuron 2003; 39(6): 889-909.
[http://dx.doi.org/10.1016/S0896-6273(03)00568-3] [PMID: 12971891]

[107]   Radad K, Gille G, Moldzio R, Saito H, Rausch WD. Ginsenosides Rb1 and Rg1 effects on mesencephalic dopaminergic cells stressed with glutamate. Brain Res 2004; 1021(1): 41-53.
[http://dx.doi.org/10.1016/j.brainres.2004.06.030] [PMID: 15328030]

[108]   Hu S, Han R, Mak S, Han Y. Protection against 1-methyl-4-phenylpyridinium ion (MPP+)-induced apoptosis by water extract of ginseng (Panax ginseng C.A. Meyer) in SH-SY5Y cells. J Ethnopharmacol 2011; 135(1): 34-42.
[http://dx.doi.org/10.1016/j.jep.2011.02.017] [PMID: 21349320]

[109]   Luo FC, Wang SD, Li K, Nakamura H, Yodoi J, Bai J. Panaxatriol saponins extracted from Panax notoginseng induces thioredoxin-1 and prevents 1-methyl-4-phenylpyridinium ion-induced neurotoxicity. J Ethnopharmacol 2010; 127(2): 419-23.
[http://dx.doi.org/10.1016/j.jep.2009.10.023] [PMID: 19857566]

[110]   Luo FC, Wang SD, Qi L, Song JY, Lv T, Bai J. Protective effect of panaxatriol saponins extracted

from Panax notoginseng against MPTP-induced neurotoxicity *in vivo*. J Ethnopharmacol 2011; 133(2): 448-53.
[http://dx.doi.org/10.1016/j.jep.2010.10.017] [PMID: 20951784]

[111] Chen XC, Zhou YC, Chen Y, Zhu YG, Fang F, Chen LM. Ginsenoside Rg1 reduces MPTP-induced substantia nigra neuron loss by suppressing oxidative stress. Acta Pharmacol Sin 2005; 26(1): 56-62.
[http://dx.doi.org/10.1111/j.1745-7254.2005.00019.x] [PMID: 15659115]

[112] Hussain G, Rasul A, Anwar H, *et al.* Role of plant derived alkaloids and their mechanism in neurodegenerative disorders. Int J Biol Sci 2018; 14(3): 341-57.
[http://dx.doi.org/10.7150/ijbs.23247] [PMID: 29559851]

[113] Levites Y, Weinreb O, Maor G, Youdim MB, Mandel S. Green tea polyphenol (-)-epigallocatechi--3-gallate prevents N-methyl-4-phenyl-1,2,3,6-tetrahydropyridine-induced dopaminergic neurodegeneration. J Neurochem 2001; 78(5): 1073-82.
[http://dx.doi.org/10.1046/j.1471-4159.2001.00490.x] [PMID: 11553681]

[114] Li R, Huang YG, Fang D, Le WD. (-)-Epigallocatechin gallate inhibits lipopolysaccharide-induced microglial activation and protects against inflammation-mediated dopaminergic neuronal injury. J Neurosci Res 2004; 78(5): 723-31.
[http://dx.doi.org/10.1002/jnr.20315] [PMID: 15478178]

[115] Mandel S, Maor G, Youdim MB. Iron and α-synuclein in the substantia nigra of MPTP-treated mice: effect of neuroprotective drugs R-apomorphine and green tea polyphenol (-)-epigallocatechi--3-gallate. J Mol Neurosci 2004; 24(3): 401-16.
[http://dx.doi.org/10.1385/JMN:24:3:401] [PMID: 15655262]

[116] Aktas O, Prozorovski T, Smorodchenko A, *et al.* Green tea epigallocatechin-3-gallate mediates T cellular NF-κ B inhibition and exerts neuroprotection in autoimmune encephalomyelitis. J Immunol 2004; 173(9): 5794-800.
[http://dx.doi.org/10.4049/jimmunol.173.9.5794] [PMID: 15494532]

[117] Weinreb O, Mandel S, Amit T, Youdim MB. Neurological mechanisms of green tea polyphenols in Alzheimer's and Parkinson's diseases. J Nutr Biochem 2004; 15(9): 506-16.
[http://dx.doi.org/10.1016/j.jnutbio.2004.05.002] [PMID: 15350981]

[118] Jha NN, Kumar R, Panigrahi R, *et al.* Comparison of α-synuclein fibril inhibition by four different amyloid inhibitors. ACS Chem Neurosci 2017; 8(12): 2722-33.
[http://dx.doi.org/10.1021/acschemneuro.7b00261] [PMID: 28872299]

[119] Xu Q, Langley M, Kanthasamy AG, Reddy MB. Epigallocatechin gallate has a neurorescue effect in a mouse model of Parkinson disease. J Nutr 2017; 147(10): 1926-31.
[http://dx.doi.org/10.3945/jn.117.255034] [PMID: 28835392]

[120] Pervin M, Unno K, Ohishi T, Tanabe H, Miyoshi N, Nakamura Y. Beneficial effects of green tea catechins on neurodegenerative diseases. Molecules 2018; 23(6): 1297.
[http://dx.doi.org/10.3390/molecules23061297] [PMID: 29843466]

[121] Caruana M, Neuner J, Högen T, *et al.* Polyphenolic compounds are novel protective agents against lipid membrane damage by α-synuclein aggregates *in vitro*. Biochim Biophys Acta 2012; 1818(11): 2502-10.
[http://dx.doi.org/10.1016/j.bbamem.2012.05.019] [PMID: 22634381]

[122] Sivanesam K, Andersen NH. Modulating the Amyloidogenesis of α-Synuclein. Curr Neuropharmacol 2016; 14(3): 226-37.
[http://dx.doi.org/10.2174/1570159X13666151030103153] [PMID. 26517049]

[123] Khan MM, Ahmad A, Ishrat T, *et al.* Resveratrol attenuates 6-hydroxydopamine-induced oxidative damage and dopamine depletion in rat model of Parkinson's disease. Brain Res 2010; 1328: 139-51.
[http://dx.doi.org/10.1016/j.brainres.2010.02.031] [PMID: 20167206]

[124] Lin CM, Lin RD, Chen ST, *et al.* Neurocytoprotective effects of the bioactive constituents of Pueraria

thomsonii in 6-hydroxydopamine (6-OHDA)-treated nerve growth factor (NGF)-differentiated PC12 cells. Phytochemistry 2010; 71(17-18): 2147-56.
[http://dx.doi.org/10.1016/j.phytochem.2010.08.015] [PMID: 20832831]

[125]   Cheng Y, He G, Mu X, *et al.* Neuroprotective effect of baicalein against MPTP neurotoxicity: behavioral, biochemical and immunohistochemical profile. Neurosci Lett 2008; 441(1): 16-20.
[http://dx.doi.org/10.1016/j.neulet.2008.05.116] [PMID: 18586394]

[126]   Lin YJ, Hou YC, Lin CH, *et al.* Puerariae radix isoflavones and their metabolites inhibit growth and induce apoptosis in breast cancer cells. Biochem Biophys Res Commun 2009; 378(4): 683-8.
[http://dx.doi.org/10.1016/j.bbrc.2008.10.178] [PMID: 19013426]

[127]   Gautam S, Karmakar S, Batra R, *et al.* Polyphenols in combination with β-cyclodextrin can inhibit and disaggregate α-synuclein amyloids under cell mimicking conditions: A promising therapeutic alternative. Biochim Biophys Acta Proteins Proteomics 2017; 1865(5): 589-603.
[http://dx.doi.org/10.1016/j.bbapap.2017.02.014] [PMID: 28238838]

[128]   Gautam S, Karmakar S, Bose A, Chowdhury PK. β-cyclodextrin and curcumin, a potent cocktail for disaggregating and/or inhibiting amyloids: a case study with α-synuclein. Biochemistry 2014; 53(25): 4081-3.
[http://dx.doi.org/10.1021/bi500642f] [PMID: 24933427]

[129]   Chen YM, Liu SP, Lin HL, *et al.* Irisflorentin improves α-synuclein accumulation and attenuates 6-OHDA-induced dopaminergic neuron degeneration, implication for Parkinson's disease therapy. Biomedicine (Taipei) 2015; 5(1): 4.
[http://dx.doi.org/10.7603/s40681-015-0004-y] [PMID: 25705584]

[130]   Spinelli KJ, Osterberg VR, Meshul CK, Soumyanath A, Unni VK. Curcumin treatment improves motor behavior in α-synuclein transgenic mice. PLoS One 2015; 10(6): e0128510.
[http://dx.doi.org/10.1371/journal.pone.0128510] [PMID: 26035833]

[131]   Gadad BS, Subramanya PK, Pullabhatla S, Shantharam IS, Rao KS. Curcumin-glucoside, a novel synthetic derivative of curcumin, inhibits α-synuclein oligomer formation: relevance to Parkinson's disease. Curr Pharm Des 2012; 18(1): 76-84.
[http://dx.doi.org/10.2174/138161212798919093] [PMID: 22211690]

[132]   Kundu P, Das M, Tripathy K, Sahoo SK. Delivery of dual drug loaded lipid based nanoparticles across the blood–brain barrier impart enhanced neuroprotection in a rotenone induced mouse model of Parkinson's disease. ACS Chem Neurosci 2016; 7(12): 1658-70.
[http://dx.doi.org/10.1021/acschemneuro.6b00207] [PMID: 27642670]

[133]   Taebnia N, Morshedi D, Yaghmaei S, Aliakbari F, Rahimi F, Arpanaei A. Curcumin-loaded amine-functionalized mesoporous silica nanoparticles inhibit α-synuclein fibrillation and reduce its cytotoxicity-associated effects. Langmuir 2016; 32(50): 13394-402.
[http://dx.doi.org/10.1021/acs.langmuir.6b02935] [PMID: 27993021]

[134]   Bollimpelli VS, Kumar P, Kumari S, Kondapi AK. Neuroprotective effect of curcumin-loaded lactoferrin nano particles against rotenone induced neurotoxicity. Neurochem Int 2016; 95: 37-45.
[http://dx.doi.org/10.1016/j.neuint.2016.01.006] [PMID: 26826319]

[135]   Zhang N, Yan F, Liang X, *et al.* Localized delivery of curcumin into brain with polysorbate 80-modified cerasomes by ultrasound-targeted microbubble destruction for improved Parkinson's disease therapy. Theranostics 2018; 8(8): 2264-77.
[http://dx.doi.org/10.7150/thno.23734] [PMID: 29721078]

[136]   Sharma N, Nehru B. Curcumin affords neuroprotection and inhibits α-synuclein aggregation in lipopolysaccharide-induced Parkinson's disease model. Inflammopharmacology 2018; 26(2): 349-60.
[http://dx.doi.org/10.1007/s10787-017-0402-8] [PMID: 29027056]

[137]   Caruana M, Vassallo N. Tea polyphenols in Parkinson's disease InNatural Compounds as Therapeutic Agents for Amyloidogenic Diseases. Cham: Springer 2015; pp. 117-37.
[http://dx.doi.org/10.1007/978-3-319-18365-7_6]

[138] Jeong JS, Piao Y, Kang S, *et al.* Triple herbal extract DA-9805 exerts a neuroprotective effect *via* amelioration of mitochondrial damage in experimental models of Parkinson's disease. Sci Rep 2018; 8(1): 15953.
[http://dx.doi.org/10.1038/s41598-018-34240-x] [PMID: 30374025]

[139] Auddy B, Ferreira M, Blasina F, *et al.* Screening of antioxidant activity of three Indian medicinal plants, traditionally used for the management of neurodegenerative diseases. J Ethnopharmacol 2003; 84(2-3): 131-8.
[http://dx.doi.org/10.1016/S0378-8741(02)00322-7] [PMID: 12648805]

[140] Khurana N, Gajbhiye A. Ameliorative effect of Sida cordifolia in rotenone induced oxidative stress model of Parkinson's disease. Neurotoxicology 2013; 39: 57-64.
[http://dx.doi.org/10.1016/j.neuro.2013.08.005] [PMID: 23994302]

[141] Rojas P, Rojas C, Ebadi M, Montes S, Monroy-Noyola A, Serrano-García N. EGb761 pretreatment reduces monoamine oxidase activity in mouse corpus striatum during 1-methyl-4-phenylpyridinium neurotoxicity. Neurochem Res 2004; 29(7): 1417-23.
[http://dx.doi.org/10.1023/B:NERE.0000026406.64547.93] [PMID: 15202774]

[142] Kang X, Chen J, Xu Z, Li H, Wang B. Protective effects of Ginkgo biloba extract on paraquat-induced apoptosis of PC12 cells. Toxicol *in vitro* 2007; 21(6): 1003-9.
[http://dx.doi.org/10.1016/j.tiv.2007.02.004] [PMID: 17509817]

[143] Lim HJ, Lee HS, Ryu JH. Suppression of inducible nitric oxide synthase and cyclooxygenase-2 expression by tussilagone from Farfarae flos in BV-2 microglial cells. Arch Pharm Res 2008; 31(5): 645-52.
[http://dx.doi.org/10.1007/s12272-001-1207-4] [PMID: 18481023]

[144] Guo S, Bezard E, Zhao B. Protective effect of green tea polyphenols on the SH-SY5Y cells against 6-OHDA induced apoptosis through ROS-NO pathway. Free Radic Biol Med 2005; 39(5): 682-95.
[http://dx.doi.org/10.1016/j.freeradbiomed.2005.04.022] [PMID: 16085186]

[145] Anandhan A, Tamilselvam K, Radhiga T, Rao S, Essa MM, Manivasagam T. Theaflavin, a black tea polyphenol, protects nigral dopaminergic neurons against chronic MPTP/probenecid induced Parkinson's disease. Brain Res 2012; 1433: 104-13.
[http://dx.doi.org/10.1016/j.brainres.2011.11.021] [PMID: 22138428]

[146] Ahmad M, Yousuf S, Khan MB, *et al.* Protective effects of ethanolic extract of Delphinium denudatum in a rat model of Parkinson's disease. Hum Exp Toxicol 2006; 25(7): 361-8.
[http://dx.doi.org/10.1191/0960327106ht635oa] [PMID: 16898164]

[147] Pasban-Aliabadi H, Esmaeili-Mahani S, Sheibani V, Abbasnejad M, Mehdizadeh A, Yaghoobi MM. Inhibition of 6-hydroxydopamine-induced PC12 cell apoptosis by olive (*Oleaeuropaea* L.) leaf extract is performed by its main component oleuropein. Rejuvenation Res 2013; 16(2): 134-42.
[http://dx.doi.org/10.1089/rej.2012.1384] [PMID: 23394606]

[148] Rajendra Kopalli S, Koppula S, Shin KY, *et al.* SF-6 attenuates 6-hydroxydopamine-induced neurotoxicity: an *in vitro* and *in vivo* investigation in experimental models of Parkinson's disease. J Ethnopharmacol 2012; 143(2): 686-94.
[http://dx.doi.org/10.1016/j.jep.2012.07.032] [PMID: 22902248]

[149] Shaltiel-Karyo R, Davidi D, Frenkel-Pinter M, Ovadia M, Segal D, Gazit E. Differential inhibition of α-synuclein oligomeric and fibrillar assembly in parkinson's disease model by cinnamon extract. Biochim Biophys Acta 2012; 1820(10): 1628-35.
[http://dx.doi.org/10.1016/j.bbagen.2012.04.021] [PMID: 22575665]

[150] Girish C, Muralidhara . Propensity of Selaginella delicatula aqueous extract to offset rotenone-induced oxidative dysfunctions and neurotoxicity in Drosophila melanogaster: Implications for Parkinson's disease. Neurotoxicology 2012; 33(3): 444-56.
[http://dx.doi.org/10.1016/j.neuro.2012.04.002] [PMID: 22521218]

[151] Ahmad S, Khan MB, Hoda MN, *et al.* Neuroprotective effect of sesame seed oil in 6-hydroxydopamine induced neurotoxicity in mice model: cellular, biochemical and neurochemical evidence. Neurochem Res 2012; 37(3): 516-26.
[http://dx.doi.org/10.1007/s11064-011-0638-4] [PMID: 22089932]

[152] Bhangale JO, Acharya SR. Anti-Parkinson activity of petroleum ether extract of *Ficus religiosa* (L.) leaves. Adv Pharmacol Sci 2016; 2016: 9436106.
[http://dx.doi.org/10.1155/2016/9436106] [PMID: 26884755]

[153] Nie G, Jin C, Cao Y, Shen S, Zhao B. Distinct effects of tea catechins on 6-hydroxydopamine-induced apoptosis in PC12 cells. Arch Biochem Biophys 2002; 397(1): 84-90.
[http://dx.doi.org/10.1006/abbi.2001.2636] [PMID: 11747313]

[154] Li FQ, Wang T, Pei Z, Liu B, Hong JS. Inhibition of microglial activation by the herbal flavonoid baicalein attenuates inflammation-mediated degeneration of dopaminergic neurons. J Neural Transm (Vienna) 2005; 112(3): 331-47.
[http://dx.doi.org/10.1007/s00702-004-0213-0] [PMID: 15503194]

[155] Radad K, Gille G, Moldzio R, Saito H, Ishige K, Rausch WD. Ginsenosides Rb1 and Rg1 effects on survival and neurite growth of MPP+-affected mesencephalic dopaminergic cells. J Neural Transm (Vienna) 2004; 111(1): 37-45.
[http://dx.doi.org/10.1007/s00702-003-0063-1] [PMID: 14714214]

[156] Chen XC, Zhu YG, Zhu LA, *et al.* Ginsenoside Rg1 attenuates dopamine-induced apoptosis in PC12 cells by suppressing oxidative stress. Eur J Pharmacol 2003; 473(1): 1-7.
[http://dx.doi.org/10.1016/S0014-2999(03)01945-9] [PMID: 12877931]

[157] Lee H, Kim YO, Kim H, *et al.* Flavonoid wogonin from medicinal herb is neuroprotective by inhibiting inflammatory activation of microglia. FASEB J 2003; 17(13): 1943-4.
[http://dx.doi.org/10.1096/fj.03-0057fje] [PMID: 12897065]

[158] Gasiorowski K, Lamer-Zarawska E, Leszek J, Parvathaneni K, Bhushan Yendluri B, Blach-Olszewska Z, *et al.* Flavones from root of Scutellaria baicalensis Georgi: drugs of the future in neurode generation? CNS & Neurological Disorders-Drug Targets (Formerly Current Drug Targets- CNS & Neurological Disorders) 2011 Mar 1; 10(2): 184-91.

[159] Sashourpour M, Zahri S, Radjabian T, Ruf V, Pan-Montojo F, Morshedi D. A study on the modulation of alpha-synuclein fibrillation by Scutellaria pinnatifida extracts and its neuroprotective properties. PLoS One 2017; 12(9): e0184483.
[http://dx.doi.org/10.1371/journal.pone.0184483] [PMID: 28957336]

[160] Caruana M, Högen T, Levin J, Hillmer A, Giese A, Vassallo N. Inhibition and disaggregation of α-synuclein oligomers by natural polyphenolic compounds. FEBS Lett 2011; 585(8): 1113-20.
[http://dx.doi.org/10.1016/j.febslet.2011.03.046] [PMID: 21443877]

[161] Lu JH, Ardah MT, Durairajan SS, *et al.* Baicalein inhibits formation of α-synuclein oligomers within living cells and prevents Aβ peptide fibrillation and oligomerisation. ChemBioChem 2011; 12(4): 615-24.
[http://dx.doi.org/10.1002/cbic.201000604] [PMID: 21271629]

[162] Y Kim J, Han Y. Curcuminoids in neurodegenerative diseases. Recent Patents CNS Drug Discov 2012; 7(3): 184-204.
[http://dx.doi.org/10.2174/157488912803252032]

[163] Ji HF, Shen L. The multiple pharmaceutical potential of curcumin in Parkinson's disease. CNS & Neurological Disorders-Drug Targets (Formerly Current Drug Targets-CNS & Neurological Disorders) 2014 Mar 1; 13(2): 369-73.

[164] Ortiz-Ortiz MA, Morán JM, Ruiz-Mesa LM, *et al.* Curcumin exposure induces expression of the Parkinson's disease-associated leucine-rich repeat kinase 2 (LRRK2) in rat mesencephalic cells. Neurosci Lett 2010; 468(2): 120-4.

[http://dx.doi.org/10.1016/j.neulet.2009.10.081] [PMID: 19879924]

[165] Liu Z, Yu Y, Li X, Ross CA, Smith WW. Curcumin protects against A53T alpha-synuclein-induced toxicity in a PC12 inducible cell model for Parkinsonism. Pharmacol Res 2011; 63(5): 439-44.
[http://dx.doi.org/10.1016/j.phrs.2011.01.004] [PMID: 21237271]

[166] Jiang TF, Zhang YJ, Zhou HY, *et al.* Curcumin ameliorates the neurodegenerative pathology in A53T α-synuclein cell model of Parkinson's disease through the downregulation of mTOR/p70S6K signaling and the recovery of macroautophagy. J Neuroimmune Pharmacol 2013; 8(1): 356-69.
[http://dx.doi.org/10.1007/s11481-012-9431-7] [PMID: 23325107]

[167] Singh PK, Kotia V, Ghosh D, Mohite GM, Kumar A, Maji SK. Curcumin modulates α-synuclein aggregation and toxicity. ACS Chem Neurosci 2013; 4(3): 393-407.
[http://dx.doi.org/10.1021/cn3001203] [PMID: 23509976]

[168] Jadiya P, Khan A, Sammi SR, Kaur S, Mir SS, Nazir A. Anti-Parkinsonian effects of Bacopa monnieri: insights from transgenic and pharmacological Caenorhabditis elegans models of Parkinson's disease. Biochem Biophys Res Commun 2011; 413(4): 605-10.
[http://dx.doi.org/10.1016/j.bbrc.2011.09.010] [PMID: 21925152]

[169] Kulkarni SK, Akula KK, Dhir A. Effect of Withania somnifera Dunal root extract against pentylenetetrazol seizure threshold in mice: possible involvement of GABAergic system 2008.

[170] Polidori MC, Mecocci P, Browne SE, Senin U, Beal MF. Oxidative damage to mitochondrial DNA in Huntington's disease parietal cortex. Neurosci Lett 1999; 272(1): 53-6.
[http://dx.doi.org/10.1016/S0304-3940(99)00578-9] [PMID: 10507541]

[171] Bhatnagar M, Sisodia SS, Bhatnagar R. Antiulcer and antioxidant activity of *Asparagus racemosus* Willd and Withania somnifera Dunal in rats. Ann N Y Acad Sci 2005; 1056(1): 261-78.
[http://dx.doi.org/10.1196/annals.1352.027] [PMID: 16387694]

[172] Singh G, Sharma PK, Dudhe R, Singh S. Biological activities of *Withania somnifera.* Ann Biol Res 2010; 1(3): 56-63.

[173] Mirjalili MH, Moyano E, Bonfill M, Cusido RM, Palazón J. Steroidal lactones from Withania somnifera, an ancient plant for novel medicine. Molecules 2009; 14(7): 2373-93.
[http://dx.doi.org/10.3390/molecules14072373] [PMID: 19633611]

[174] Kapoor LD. . Handbook of Ayurvedic medicinal plants: Herbal reference library. CRC press; 2000. Nov 10.

[175] Wu PF, Zhang Z, Wang F, Chen JG. Natural compounds from traditional medicinal herbs in the treatment of cerebral ischemia/reperfusion injury. Acta Pharmacol Sin 2010; 31(12): 1523-31.
[http://dx.doi.org/10.1038/aps.2010.186] [PMID: 21127495]

[176] Sandhya S, Vinod KR, Kumar S. Herbs used for brain disorders. Hygeia J Drugs Med 2010; 2: 38-45.

[177] Gohil KJ, Patel JA. A review on Bacopa monniera: current research and future prospects. International Journal of Green Pharmacy 2010; 4(1) [IJGP].
[http://dx.doi.org/10.4103/0973-8258.62156]

[178] Shinomol GK, Muralidhara . Bacopa monnieri modulates endogenous cytoplasmic and mitochondrial oxidative markers in prepubertal mice brain. Phytomedicine 2011; 18(4): 317-26.
[http://dx.doi.org/10.1016/j.phymed.2010.08.005] [PMID: 20850955]

[179] Bammidi SR, Volluri SS, Chippada SC, Avanigadda S, Vangalapati M. A review on pharmacological studies of Bacopa monniera. J Chem Biol Phys Sci 2011; 1(2): 250. [JCBPS].

[180] Mahato SB, Garai S, Chakravarty AK. Bacopasaponins E and F: two jujubogenin bisdesmosides from Bacopa monniera. Phytochemistry 2000; 53(6): 711-4.
[http://dx.doi.org/10.1016/S0031-9422(99)00384-2] [PMID: 10746885]

[181] Murthy PB, Raju VR, Ramakrisana T, *et al.* Estimation of twelve bacopa saponins in Bacopa monnieri extracts and formulations by high-performance liquid chromatography. Chem Pharm Bull (Tokyo)

2006; 54(6): 907-11.
[http://dx.doi.org/10.1248/cpb.54.907] [PMID: 16755069]

[182] Kapoor R, Srivastava S, Kakkar P. Bacopa monnieri modulates antioxidant responses in brain and kidney of diabetic rats. Environ Toxicol Pharmacol 2009; 27(1): 62-9.
[http://dx.doi.org/10.1016/j.etap.2008.08.007] [PMID: 21783922]

[183] Vollala VR, Upadhya S, Nayak S. Learning and memory-enhancing effect of Bacopa monniera in neonatal rats. Bratisl Lek Listy 2011; 112(12): 663-9.
[PMID: 22372329]

[184] Roodenrys S, Booth D, Bulzomi S, Phipps A, Micallef C, Smoker J. Chronic effects of Brahmi (Bacopa monnieri) on human memory. Neuropsychopharmacology 2002; 27(2): 279-81.
[http://dx.doi.org/10.1016/S0893-133X(01)00419-5] [PMID: 12093601]

[185] Vohora D, Pal SN, Pillai KK. Protection from phenytoin-induced cognitive deficit by Bacopa monniera, a reputed Indian nootropic plant. J Ethnopharmacol 2000; 71(3): 383-90.
[http://dx.doi.org/10.1016/S0378-8741(99)00213-5] [PMID: 10940574]

[186] Russo A, Izzo AA, Borrelli F, Renis M, Vanella A. Free radical scavenging capacity and protective effect of *Bacopa monniera* L. on DNA damage. Phytother Res 2003; 17(8): 870-5.
[http://dx.doi.org/10.1002/ptr.1061] [PMID: 13680815]

[187] Chowdhuri DK, Parmar D, Kakkar P, Shukla R, Seth PK, Srimal RC. Antistress effects of bacosides of *Bacopa monnieri*: modulation of Hsp70 expression, superoxide dismutase and cytochrome P450 activity in rat brain. Phytother Res 2002; 16(7): 639-45.
[http://dx.doi.org/10.1002/ptr.1023] [PMID: 12410544]

[188] Sairam K, Dorababu M, Goel RK, Bhattacharya SK. Antidepressant activity of standardized extract of Bacopa monniera in experimental models of depression in rats. Phytomedicine 2002; 9(3): 207-11.
[http://dx.doi.org/10.1078/0944-7113-00116] [PMID: 12046860]

[189] Shanker G, Singh HK. Anxiolytic profile of standardized Brahmi extract. Indian J Pharmacol 2000; 32(152): 5.

[190] Sumathy T, Subramanian S, Govindasamy S, Balakrishna K, Veluchamy G. Protective role of Bacopa monniera on morphine induced hepatotoxicity in rats. Phytother Res 2001; 15(7): 643-5.
[http://dx.doi.org/10.1002/ptr.1007] [PMID: 11746853]

[191] Andreassen OA, Ferrante RJ, Hughes DB, *et al.* Malonate and 3-nitropropionic acid neurotoxicity are reduced in transgenic mice expressing a caspase-1 dominant-negative mutant. J Neurochem 2000; 75(2): 847-52.
[http://dx.doi.org/10.1046/j.1471-4159.2000.0750847.x] [PMID: 10899963]

[192] Kim GW, Copin JC, Kawase M, *et al.* Excitotoxicity is required for induction of oxidative stress and apoptosis in mouse striatum by the mitochondrial toxin, 3-nitropropionic acid. J Cereb Blood Flow Metab 2000; 20(1): 119-29.
[http://dx.doi.org/10.1097/00004647-200001000-00016] [PMID: 10616800]

[193] Nakanishi K. Terpene trilactones from Gingko biloba: from ancient times to the 21st century. Bioorg Med Chem 2005; 13(17): 4987-5000.
[http://dx.doi.org/10.1016/j.bmc.2005.06.014] [PMID: 15990319]

[194] McKenna DJ, Jones K, Hughes K. Efficacy, safety, and use of ginkgo biloba in clinical and preclinical applications. Altern Ther Health Med 2001; 7(5): 70-86, 88-90.
[PMID: 11565403]

[195] Smith JV, Luo Y. Studies on molecular mechanisms of Ginkgo biloba extract. Appl Microbiol Biotechnol 2004; 64(4): 465-72.
[http://dx.doi.org/10.1007/s00253-003-1527-9] [PMID: 14740187]

[196] DeFeudis FV, Drieu K. Ginkgo biloba extract (EGb 761) and CNS functions: basic studies and clinical applications. Curr Drug Targets 2000; 1(1): 25-58.

[http://dx.doi.org/10.2174/1389450003349380] [PMID: 11475535]

[197]   Ramassamy C, Longpré F, Christen Y. Ginkgo biloba extract (EGb 761) in Alzheimer's disease: is there any evidence? Curr Alzheimer Res 2007; 4(3): 253-62.
[http://dx.doi.org/10.2174/156720507781077304] [PMID: 17627482]

[198]   Mahadevan S, Park Y. Multifaceted therapeutic benefits of Ginkgo biloba L.: chemistry, efficacy, safety, and uses. J Food Sci 2008; 73(1): R14-9.
[http://dx.doi.org/10.1111/j.1750-3841.2007.00597.x] [PMID: 18211362]

[199]   Popovich DG, Kitts DD. Generation of ginsenosides Rg3 and Rh2 from North American ginseng. Phytochemistry 2004; 65(3): 337-44.
[http://dx.doi.org/10.1016/j.phytochem.2003.11.020] [PMID: 14751305]

[200]   Mahdy HM, Tadros MG, Mohamed MR, Karim AM, Khalifa AE. The effect of Ginkgo biloba extract on 3-nitropropionic acid-induced neurotoxicity in rats. Neurochem Int 2011; 59(6): 770-8.
[http://dx.doi.org/10.1016/j.neuint.2011.07.012] [PMID: 21827809]

[201]   Winters M. Ancient medicine, modern use: Withania somnifera and its potential role in integrative oncology. Altern Med Rev 2006; 11(4): 269-77.
[PMID: 17176166]

[202]   Singh B, Chandan BK, Gupta DK. Adaptogenic activity of a novel withanolide-free aqueous fraction from the roots of Withania somnifera Dun. (Part II). Phytother Res 2003; 17(5): 531-6.
[http://dx.doi.org/10.1002/ptr.1189] [PMID: 12748992]

[203]   Singh B, Saxena AK, Chandan BK, Gupta DK, Bhutani KK, Anand KK. Adaptogenic activity of a novel, withanolide-free aqueous fraction from the roots of Withania somnifera Dun. Phytother Res 2001; 15(4): 311-8.
[http://dx.doi.org/10.1002/ptr.858] [PMID: 11406854]

[204]   Kumar P, Kumar A. Possible neuroprotective effect of Withania somnifera root extract against 3-nitropropionic acid-induced behavioral, biochemical, and mitochondrial dysfunction in an animal model of Huntington's disease. J Med Food 2009; 12(3): 591-600.
[http://dx.doi.org/10.1089/jmf.2008.0028] [PMID: 19627208]

[205]   Kumar P, Kumar A. Effects of root extract of Withania somnifera in 3-Nitropropionic acid-induced cognitive dysfunction and oxidative damage in rats. International Journal of Health Research 2008; 1(3)
[http://dx.doi.org/10.4314/ijhr.v1i3.55359]

[206]   Zhou S, Lim LY, Chowbay B. Herbal modulation of P-glycoprotein. Drug Metab Rev 2004; 36(1): 57-104.
[http://dx.doi.org/10.1081/DMR-120028427] [PMID: 15072439]

[207]   Daniel S, Limson JL, Dairam A, Watkins GM, Daya S. Through metal binding, curcumin protects against lead- and cadmium-induced lipid peroxidation in rat brain homogenates and against lead-induced tissue damage in rat brain. J Inorg Biochem 2004; 98(2): 266-75.
[http://dx.doi.org/10.1016/j.jinorgbio.2003.10.014] [PMID: 14729307]

[208]   Ghoneim AI, Abdel-Naim AB, Khalifa AE, El-Denshary ES. Protective effects of curcumin against ischaemia/reperfusion insult in rat forebrain. Pharmacol Res 2002; 46(3): 273-9.
[http://dx.doi.org/10.1016/S1043-6618(02)00123-8] [PMID: 12220971]

[209]   Vajragupta O, Boonchoong P, Watanabe H, Tohda M, Kummasud N, Sumanont Y. Manganese complexes of curcumin and its derivatives: evaluation for the radical scavenging ability and neuroprotective activity. Free Radic Biol Med 2003; 35(12): 1632-44.
[http://dx.doi.org/10.1016/j.freeradbiomed.2003.09.011] [PMID: 14680686]

[210]   Lü JM, Yao Q, Chen C. Ginseng compounds: an update on their molecular mechanisms and medical applications. Curr Vasc Pharmacol 2009; 7(3): 293-302.
[http://dx.doi.org/10.2174/157016109788340767] [PMID: 19601854]

[211] Rausch WD, Liu S, Gille G, Radad K. Neuroprotective effects of ginsenosides. Acta Neurobiol Exp (Warsz) 2006; 66(4): 369-75.
[PMID: 17265697]

[212] Radad K, Gille G, Liu L, Rausch WD. Use of ginseng in medicine with emphasis on neurodegenerative disorders. J Pharmacol Sci 2006; 100(3): 175-86.
[http://dx.doi.org/10.1254/jphs.CRJ05010X] [PMID: 16518078]

[213] Kim S, Ahn K, Oh TH, Nah SY, Rhim H. Inhibitory effect of ginsenosides on NMDA receptor-mediated signals in rat hippocampal neurons. Biochem Biophys Res Commun 2002; 296(2): 247-54.
[http://dx.doi.org/10.1016/S0006-291X(02)00870-7] [PMID: 12163009]

[214] Lian XY, Zhang Z, Stringer JL. Protective effects of ginseng components in a rodent model of neurodegeneration. Ann Neurol 2005; 57(5): 642-8.
[http://dx.doi.org/10.1002/ana.20450] [PMID: 15852378]

[215] Kim JH, Kim S, Yoon IS, *et al.* Protective effects of ginseng saponins on 3-nitropropionic acid-induced striatal degeneration in rats. Neuropharmacology 2005; 48(5): 743-56.
[http://dx.doi.org/10.1016/j.neuropharm.2004.12.013] [PMID: 15814108]

[216] Wu J, Jeong HK, Bulin SE, Kwon SW, Park JH, Bezprozvanny I. Ginsenosides protect striatal neurons in a cellular model of Huntington's disease. J Neurosci Res 2009; 87(8): 1904-12.
[http://dx.doi.org/10.1002/jnr.22017] [PMID: 19185022]

[217] Soumyanath A, Zhong YP, Gold SA, *et al.* Centella asiatica accelerates nerve regeneration upon oral administration and contains multiple active fractions increasing neurite elongation in-vitro. J Pharm Pharmacol 2005; 57(9): 1221-9.
[http://dx.doi.org/10.1211/jpp.57.9.0018] [PMID: 16105244]

[218] Zainol MK, Abd-Hamid A, Yusof S, Muse R. Antioxidative activity and total phenolic compounds of leaf, root and petiole of four accessions of Centella asiatica (L.). Urban Food Chemistry 2003; 81(4): 575-81.
[http://dx.doi.org/10.1016/S0308-8146(02)00498-3]

[219] Hussin M, Abdul-Hamid A, Mohamad S, Saari N, Ismail M, Bejo MH. Protective effect of Centella asiatica extract and powder on oxidative stress in rats. Food Chem 2007; 100(2): 535-41.
[http://dx.doi.org/10.1016/j.foodchem.2005.10.022]

[220] Singh B, Rastogi RP. A reinvestigation of the triterpenes of Centella asiatica. Phytochemistry 1969; 8(5): 917-21.
[http://dx.doi.org/10.1016/S0031-9422(00)85884-7]

[221] Randriamampionona D, Diallo B, Rakotoniriana F, *et al.* Comparative analysis of active constituents in Centella asiatica samples from Madagascar: application for *ex situ* conservation and clonal propagation. Fitoterapia 2007; 78(7-8): 482-9.
[http://dx.doi.org/10.1016/j.fitote.2007.03.016] [PMID: 17560738]

[222] Shinomol GK, Muralidhara . Prophylactic neuroprotective property of Centella asiatica against 3-nitropropionic acid induced oxidative stress and mitochondrial dysfunctions in brain regions of prepubertal mice. Neurotoxicology 2008; 29(6): 948-57.
[http://dx.doi.org/10.1016/j.neuro.2008.09.009] [PMID: 18930762]

[223] Dickson M, Gagnon JP. Key factors in the rising cost of new drug discovery and development. Nat Rev Drug Discov 2004; 3(5): 417-29.
[http://dx.doi.org/10.1038/nrd1382] [PMID: 15136789]

[224] Koehn FE, Carter GT. The evolving role of natural products in drug discovery. Nat Rev Drug Discov 2005; 4(3): 206-20.
[http://dx.doi.org/10.1038/nrd1657] [PMID: 15729362]

[225] osenthal, J., Curtain has fallen on hopes of legal bioprospecting. Nature 2002; 416: 15.
[http://dx.doi.org/10.1038/416015a]

[226] Soejarto DD, Gyllenhaal C, Fong HH, *et al.* The UIC ICBG (University of Illinois at Chicago International Cooperative Biodiversity Group) Memorandum of Agreement: a model of benefit-sharing arrangement in natural products drug discovery and development. J Nat Prod 2004; 67(2): 294-9.
[http://dx.doi.org/10.1021/np0304363] [PMID: 14987071]

[227] Knowles J, Gromo G. A guide to drug discovery: Target selection in drug discovery. Nat Rev Drug Discov 2003; 2(1): 63-9.
[http://dx.doi.org/10.1038/nrd986] [PMID: 12509760]

[228] Kramer R, Cohen D. Functional genomics to new drug targets. Nat Rev Drug Discov 2004; 3(11): 965-72.
[http://dx.doi.org/10.1038/nrd1552] [PMID: 15520818]

[229] Walters WP, Namchuk M. Designing screens: how to make your hits a hit. Nat Rev Drug Discov 2003; 2(4): 259-66.
[http://dx.doi.org/10.1038/nrd1063] [PMID: 12669025]

[230] Butler MS. The role of natural product chemistry in drug discovery. J Nat Prod 2004; 67(12): 2141-53.
[http://dx.doi.org/10.1021/np040106y] [PMID: 15620274]

[231] Koehn FE, Carter GT. The evolving role of natural products in drug discovery. Nat Rev Drug Discov 2005; 4(3): 206-20.
[http://dx.doi.org/10.1038/nrd1657] [PMID: 15729362]

[232] Ley SV, Baxendale IR. New tools and concepts for modern organic synthesis. Nat Rev Drug Discov 2002; 1(8): 573-86.
[http://dx.doi.org/10.1038/nrd871] [PMID: 12402498]

[233] Federsel HJ. Logistics of process R&D: transforming laboratory methods to manufacturing scale. Nat Rev Drug Discov 2003; 2(8): 654-64.
[http://dx.doi.org/10.1038/nrd1154] [PMID: 12904815]

[234] Lombardino JG, Lowe JA III. The role of the medicinal chemist in drug discovery--then and now. Nat Rev Drug Discov 2004; 3(10): 853-62.
[http://dx.doi.org/10.1038/nrd1523] [PMID: 15459676]

CHAPTER 2

Emerging Novel Approaches and Recent Advances in Parkinson's Disease Treatment and Diagnosis

Virendra Tiwari[1,2,*], **Akanksha Mishra**[1,4,*], **Sonu Singh**[1,3], **Parul** [1], **Pratibha Tripathi**[1] and **Shubha Shukla**[1,2,#]

[1] *Neuroscience and Ageing Biology Division, CSIR- Central Drug Research Institute, Lucknow 226031, U.P., India*

[2] *Academy of Scientific and Innovative Research (AcSIR),Ghaziabad 201002, India*

[3] *Department of Neuroscience, School of Medicine, University of Connecticut (Uconn) Health Center 263 Farmington Avenue L-4078 Farmington CT 06030, USA*

[4] *Department of Cell Biology and Anatomy, New York Medical College, Valhalla, NY 01595, USA*

Abstract: Parkinson's disease (PD) is the second most common neurodegenerative disorder after Alzheimer's disease (AD). Recent data suggest that more than 6 million people worldwide have PD. The death of midbrain dopaminergic (DAergic) neurons in substantia nigra pars compacta (SNpc) and ventral tegmental area (VTA), and loss of their terminal in the striatum region lead to motor and non-motor impairments in PD. Currently, several pharmacological drugs acting as dopamine agonists, dopamine reuptake inhibitors, and dopamine breakdown inhibitors are available for the treatment of PD patients. However, all these treatments only provide symptomatic relief without halting the process of neurodegeneration. They are widely associated with unwanted side effects such as orthostatic hypotension, hallucinations, somnolence, levodopa (L-DOPA)-induced dyskinesia, and impulse control disorder. This article discusses the biochemical and physiological mechanism(s) of PD pathogenesis and how the emergence of the new target and drug identification are advancing our knowledge for the development of novel PD therapies. The current data obtained from animal models and clinical research suggest that early treatment may be one of the effective strategies in PD management. Recent stem cell progenitor cell transplantation studies have added an extra layer in future treatment approaches by showing very promising results and outcomes in PD patients. Apart from classical dopamine receptor modulators as drug treatments, recent data suggest that PD pathogenesis can also be managed by targeting neurogenesis, apoptotic pathway, neuroinflammation, mitochondrial biology, and oxidative damage. Therefore, identification of reliable disease markers and putative drug targets to halt the process of neurodegeneration and genetic profiling of individuals will help in finding new avenues in PD treatment.

[#] **Corresponding Author Dr. Shubha Shukla:** Principal Scientist, Neuroscience and Ageing Biology Division, CSIR-Central Drug Research Institute, Sector 10, Jankipuram Extension, Sitapur Road, Lucknow-226031, Uttar Pradesh, India; Tel: +91 522 2772450; Fax: +91 522 2771941; E-mail: shubha_shukla@cdri.res.in
[*] These authors have equal contribution.

Keywords: Dopamine, Dopamine Agonists, Neurodegeneration, Parkinson's Disease, Pharmacological Treatment.

INTRODUCTION

Parkinson's disease (PD) is an age-related CNS disorder characterized by depletion of dopaminergic neurons in the substantia nigra (SN) region of the brain, which leads to a marked decline of dopamine (DA) neurotransmitters in the striatum. Due to deficiency of DA supply to the striatum, an imbalance occurs between neurotransmitters acetylcholine (ACh) and dopamine (DA), subsequently resulting in classical PD symptoms manifested as tremor, akinesia, postural hypotension [1 - 3]. The principal approaches applied in the management of PD symptoms by drugs that either prompt the release of dopamine or increase dopamine concentrations at receptors [4, 5]. Widespread downregulation of dopaminergic neurons increases with precarious onset in middle age to late adulthood. The mechanisms involved in the downregulation of dopaminergic neurons in PD patients are oxidative stress, mitochondrial dysfunction, neuroinflammation, excitotoxicity, imbalance of intracellular calcium, and iron homeostasis [6, 7]. Neurotoxin-based animal models have demonstrated themselves to be a reliable tool for investing in novel therapies and increasing the effectiveness of symptomatic treatment of PD [8]. The present therapeutic strategy against PD is mainly to restore the optimal dopamine(DA) level and its correlated signaling pathways in which levodopa, a precursor of dopamine(DA) is given to the PD patients [9]. Levodopa is given with carbidopa, a peripheral decarboxylase inhibitor that decreases the peripheral breakdown of L-Dopa and reduces the side effects of L-Dopa, mainly cardiovascular and gastrointestinal problems [10]. Another drug therapy for PD is monoamine oxidase B inhibitors(MAO-B), which inhibit DA metabolism which reinforces mitochondrial dysfunction and oxidative stress [11]. Further, a novel therapeutic option for PD is catechol-o-methyltransferase (COMT) inhibitors, which are administered with L-Dopa. COMT inhibitors prolong the action of L-Dopa, thus enhancing the duration of time a patient could take to benefit from L-Dopa [12, 13]. Another therapeutic approach for PD is dopamine receptor agonists, which directly stimulate the postsynaptic receptors in the striatum. They decrease the metabolism of dopamine and reduce free radical formation in the striatum [14]. Glutamate receptor antagonist is also given in PD, like amantadine, an antiviral drug that decreases motor and non-motor symptoms in PD patients, and akinesia is specifically treated by amantadine [15, 16].

Nowadays phytochemically active plant extracts and their secondary metabolites are being extensively used in PD therapy as these have substantial neuroprotective potential in PD [17]. Some important medicinal plants used in, which help in the mitigation of disease progression, are *Ginkgo biloba, Bacopa monnieri, Curcuma longa, Camellia sinensis, green tea catechins, etc* [9, 18 - 21]. A study has shown that *Mucuna pruriens* contains Levodopa, which provides a long-term improvement in PD, with decreased risk of bradykinesia in the 6-OHDA lesioned model of PD [22], and has higher efficacy than levodopa [23].

Currently, the scientific community is focused on stem cell therapy for the treatment of PD. Stem cell transplantation is an easily accessible target for *in vitro* as well as *in vivo* systems. Neural stem cells have the innate ability to proliferate, migrate and differentiate into multiple cell lineages such as astrocytes, oligodendrocytes, and neuronal cells. A recent study demonstrated reprogramming of human adipose-derived stem cells into neuronal-like cells. These reprogrammed neuronal-like cells have therapeutic potential towards PD by enhancing mitochondrial functioning [24]. The present review explains the major recent advances that mainstay the development of novel dopaminergic and non-dopaminergic drugs along with phytoactive compounds for PD.

BIOCHEMICAL AND PHYSIOLOGICAL MECHANISM OF PD

Types of Dopaminergic Neurons, Systems, and Functions

In the mammalian central nervous system, dopamine (DA) is supplied through the dopaminergic neurons present in the midbrain [25]. Substantia nigra pars compacta region of the brain is more vulnerable to dopaminergic neurodegeneration due to the presence of dark pigment neuromelanin that binds with redox-active iron, thus making the microenvironment more 'harsh' [26]. Multiple functions of the brain are controlled by dopaminergic neurons, such as voluntary movement, mood, addiction, stress, and reward behavior [27]. Several transcription factors such as Nurr1, Lmx1b, and Pitx3 have been identified to be involved in the generation of the dopaminergic neurons by the development of the mesencephalic dopaminergic system [28 - 30]. The exact cause of selective dopaminergic neurodegeneration in the substantia nigra pars compacta is still elusive in the case of PD [31, 32]. Several neurotransmitters are known to be present in the brain, but one of the most thoroughly studied is DA [33]. Midbrain dopaminergic system plays an important role in many mental and neurological disorders by producing a significant amount of DA [34, 35]. Dopaminergic neurons are a group of cells that are anatomically and functionally heterogeneous [34, 36]. Diencephalon, mesencephalon, and olfactory bulbs are the main centers for dopaminergic neurons and approximately 90% of the total dopaminergic cell population resides in the ventral mesencephalic region [34, 36 - 39].

Four Major Pathways in the Brain, Related to the Dopaminergic System

Nigrostriatal Pathway

There are several dopaminergic systems, but the best-known system is nigrostriatal, which originates from substantia nigra (zona compacta) and extends into the dorsal striatum (caudate-putamen) [34]. The voluntary movement of the body is controlled through the nigrostriatal pathway [40]. Nigrostriatal system degeneration is found in PD, and DA plays an important role in motor and learning skills [41, 42]. The antipsychotic drugs show their extrapyramidal effects by blocking the striatal dopamine receptors (Fig. **1**) [43, 44].

Mesolimbic Pathway

Other than this, the dopaminergic mesolimbic system also exists and originates from the ventral tegmental area (VTA) and emotion-based behavior is controlled by these systems [45, 46]. The mesolimbic dopaminergic system is composed of the cells present in VTA that project into the nucleus accumbens, olfactory tubercle, and also innervate the septum, amygdala, and hippocampus [47, 48]. DA controls the emotion and reward system in this pathway and is released in pleasurable conditions [49, 50]. Thus, in the brain, particularly in the nucleus accumbens and prefrontal cortex, the release of DA is stimulated by food, sex, and several drugs of abuse (Fig. **1**) [47, 51].

Mesocortical Pathway

The mesocortical dopaminergic system is composed of cells present in the medial VTA (A10 region) and projected to the prefrontal, cingulate and perirhinal cortex [52, 53]. DA release in the mesocortical pathway controls the cognitive and emotional behavior [49, 54]. Because the two dopaminergic systems, mesolimbic and mesocortical systems overlap thus they are collectively known as mesocorticolimbic systems [34]. Working memory and attention are improved in the presence of DA, especially in the prefrontal area of the brain. This balance is so delicate that the memory suffers under conditions of abnormal levels (Fig. **1**) [55, 56].

Tuberoinfundibular Pathway

The origin of this pathway takes place from the hypothalamus (arcuate and paraventricular nuclei) and goes into the pituitary gland (the median eminence). In this pathway, the release of prolactin is inhibited by DA. As a neuroendocrine inhibitor, DA limits the prolactin secretion from the anterior pituitary gland [57,

58]. Occasionally, DA is called a prolactin-inhibiting factor (PIF), prolactin-inhibiting hormone (PIH), or prolactostatin (Fig. **1**) [49].

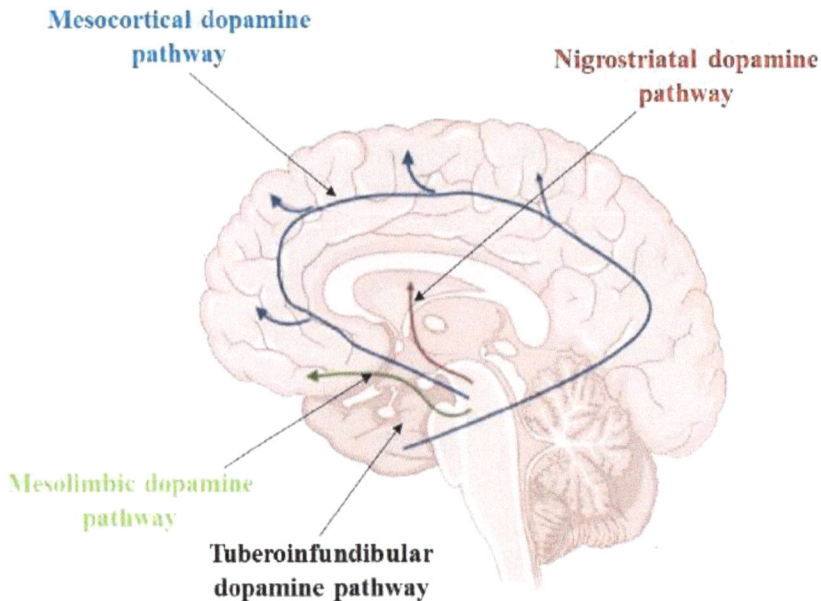

Fig. (1). Brain dopaminergic pathways.

Pathophysiology of PD

The movement and posture of the body are mainly controlled by the midbrain region and basal ganglia. The motor coordination is mainly regulated by caudate and putamen, external and internal segment of the globus pallidus, subthalamic nucleus, and substantia nigra, the major parts of the basal ganglia circuit [59, 60]. A balanced ratio of ACh and DA is found in the normal condition, but in the case of PD, the level of DA gets reduced. DA depletion causes the hyper-activation of the indirect pathway and hypo-activation of the direct pathway thus, for motor function DA plays a key role in the direct and indirect pathways (Fig. **2**) [61, 62]. The striatum receives projections from all the areas of the cerebral cortex and acts as an input division of the basal ganglia. Thalamus receives non-reciprocating projections from approximately half of the striatal medium spiny neurons (MSNs) present in basal ganglia output nuclei, GPi (internal the segment of the globus pallidus), and SNpr (substantia nigra pars reticulata). The remaining half MSNs project to the intervening nucleus, GPe (external segment of the globus pallidus), instead of output nuclei. Although being GABAergic, based on connectivity and neuromodulator production, MSNs form two different populations [61, 62].

The direct pathway that contains substance P and dynorphin and projects to the outer nuclei (GPi/SNpr) directly whereas the other one is projected to GPe and forms an indirect pathway that contains enkephalin. Basal ganglia output nuclei exert thalamic inhibition and depend on opposing actions of direct and indirect pathways (Fig. **2**) [63, 64]. In the case of MPTP-induced Parkinsonism, it has been observed that tonic discharge rate was increased in outer nuclei (GPi/SNpr) and subthalamic nuclei (STN) while in GPe the discharge rate was decreased [65, 66]. This suggests that hypokinesia or rigidity in parkinsonism occurs due to excessive inhibition of thalamic targets receiving projections from outer nuclei (GPi/SNpr) [65, 67].

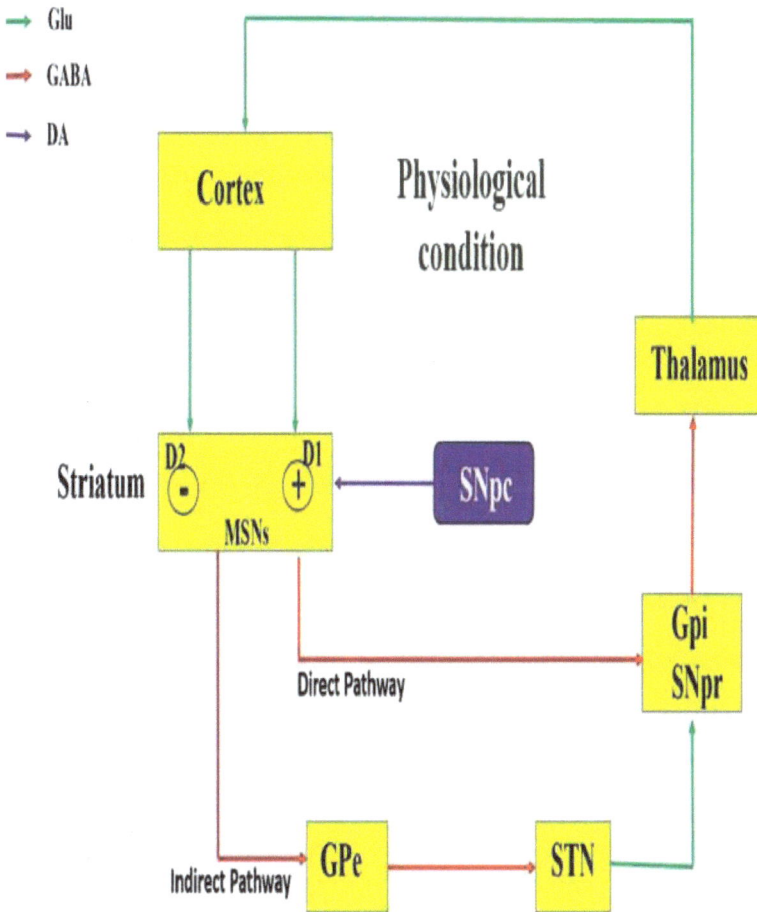

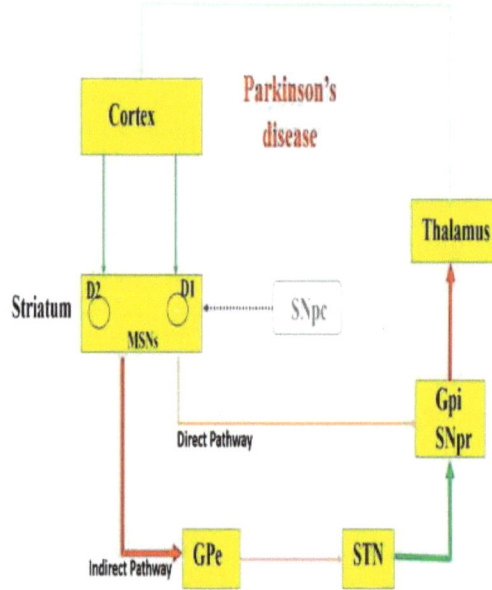

Fig. (2). Elucidation of direct and indirect pathways of basal ganglia circuit. The dopamine from substantia nigra pars compacta (SNpc) activates (+) D1 expressing neurons in the striatum and inhibits (-) D2 expressing neurons in the striatum. Thalamus receives projections from output nuclei (the internal segment of the globus pallidus (GPi) and substantia nigra pars reticulata and completes the cortico-basal ganglia-thalamo-cortical loop in physiological conditions. During PD condition, DA deficiency over activates the indirect pathway by driving excessive glutamate to the GPi and SNpc and reduced activity of inhibitory GABAergic direct pathway and activity of the same output nuclei get enhanced. Excessive inhibitory neurotransmitter GABA reduces the motor thalamus that ultimately hampers the activity of the supplementary motor cortex. Abbreviations: DA, dopamine; GABA, gamma-aminobutyric acid; Glu, glutamate; GPe, an external segment of the globus pallidus; GPi, internal segment of the globus pallidus; SNpc, substantia nigra pars compacta; SNpr, substantia nigra pars reticulata; STN, subthalamic nucleus.

Involvement of Extrastriatal DA in PD

In comparison to the striatum, the loss of DA level is less in other subcortical regions, if it happens then undoubtedly it also contributes to motor disabilities [68, 69].

Basal Ganglia Output Nuclei

Other than the caudate putamen nucleus of basal ganglia that are DA rich, a noticeable amount of DA and its receptor is also carried by the lateral, medial globus pallidus, reticular zone of substantia nigra, and subthalamic nuclei as well [70, 71]. Therefore, in addition to the striatum, other parts of basal ganglia also take care of the continuous flow of DA and this fact has been supported by many studies [72].

Nucleus Accumbens Region

Neuronal perikarya loss is also seen in the ventral tegmental area, the dopaminergic innervations mostly originated from this area of nucleus accumbens, thus a marked reduction in DA and HVA level is also seen in nucleus accumbens in the case of PD [73]. However, DA depletion in nucleus accumbens is comparatively less severe than in the striatum but akin to the striatum, the ratio of HVA to DA is shifted towards HVA in nucleus accumbens. In the case of both limbic and motor functions, the DA from accumbens coordinates thus it also accompanies biochemical pathologies related to Parkinsonism [74].

Role of DA

DA is an important neurotransmitter that is involved in balance and motor coordination. DA is produced predominantly in the substantia nigra pars compacta (SNpc) region of the basal ganglia in the brain. In a healthy brain, the striatal neurons' excitability is regulated by DA [75]. In the case of PD, the level of DA decreases significantly which is responsible for the excessive firing of the striatal neurons in the absence of their inhibition and ultimately results in the appearance of hallmark features of PD such as tremor, rigidity, and bradykinesia [76].

PD Related non-DAergic Changes

In the beginning, PD pathologies originate due to disturbing levels of DA in the nigrostriatal pathway but in later stages of the disease, the disturbance spreads to the level of other neurotransmitters also [77]. The post-mortem report of the brain of dead PD patients showed biochemical changes in the level of non-dopaminergic neurotransmitters such as GABA, glutamate, neurotensin, and others [78].

Although the marked reduction in the level of striatal DA in PD that is always common is comparable to other non-DAergic neurotransmitters, the range of their level varies from no change to significant loss among PD patients [79, 80]. Thus changes in the level of non-DAergic neurotransmitters are not common among PD patients, instead, they may vary among individuals depending upon the situation and severity of the disease [74].

Role of Serotonin

Several motor and non-motor symptoms of PD are affected by the loss of serotonin (5-HT) and play an important role in disease development [81, 82]. Previous studies based on toxin-induced models have shown the effects of a concomitant decrease in the level of serotonin and its receptors in the cortex and anterior cingulate [83]. In PD patients the binding capacity of the prefrontal cortex (PFC) and 5-HT transporter (SERT) also gets reduced. The severity of resting tremors in PD patients is correlated with the loss of serotonergic receptor (HT1A) (~25%) in the median raphe nucleus [84]. This suggests the relevance of 5-HT projections in the midbrain more than the DA-projections in substantia nigra for PD initiation [84, 85]. Similarly, other studies have also shown a close association between the level of 5-HT and depression [86].

Role of Acetylcholine

Acetylcholine is a significant neurotransmitter that plays an important role in cognition and in PD its level is decreased. A broad-band of cell clusters which are cholinergic in nature, are present in the basal forebrain subventricular region and commonly known as nucleus basalis of Meynert (nbM) [87, 88]. Neuronal loss occurs in nbM in a different pattern which strongly supports the role of the cholinergic system in PD. Role of the cholinergic system was supported after the presence of Lewy bodies (LB) and neurodegeneration in nbM of post-mortem PD brain tissue [88].

Role of GABA and Ca^{2+} System

The Ca^{2+} influx is controlled by an inhibitory neurotransmitter, gamma aminobutyric acid (GABA) directly through GABAergic receptors and astrocyte network working in an indirect manner. At the cellular and systemic levels, the neuronal activity is stabilized by the Ca^{2+}/GABA mechanism [7]. Ca^{2+}-excitotoxicity is reported in the case of PD due to the impaired Ca^{2+}-buffering system, which ultimately leads to the death of dopaminergic neurons in SNpc [89]. Recent studies have suggested that ~80% of PD patients have olfactory problems due to the death of DAergic neurons in the olfactory bulb. The Ca^{2+}/GABA system regulates glial cell-derived neurotrophic factor (GDNF) which controls the function of DAergic neurons in the midbrain and the olfactory bulb regions of the brain [7].

Research on the Origin and Biosynthesis of DA

Monoamine neurotransmitter, DA, is produced in dopaminergic neurons in different brain regions such as the ventral tegmental area of substantia nigra, and hypothalamus (arcuate nucleus) (Fig. **3**) [85]. Other than the brain regions, DA is also known to be present in the periphery to perform different functions such as in the kidney it supports renal vasodilation, diuresis, and natriuresis [90]. The five different DA receptors such as D1, D2, D3, D4, and D5, are the G-protein coupled receptor superfamily members [91, 92]. DA was discovered long before its importance as a neurotransmitter was reported and originally synthesized in 1910 [93]. For the first time, DA was found in an organism, in a plant *Sarothamnus scoparius* as a pigment building metabolite. Later on, it was isolated from sympathetic ganglia, other tissues of animals, and also in invertebrates [94].

Initially, DA was believed to be a precursor for epinephrine (E) and norepinephrine (NE) like catecholamine neurotransmitters that also acted as an intermediate for the degradation of tyrosine [95]. Recognition of DA as an independent neurotransmitter was only established after the discovery of the first DA receptor. Arvid Carlsson together with Eric Kandel and Paul Greengard shared the Nobel Prize in 2000 in medicine and physiology for DA measurement with new techniques and for making an understanding of DA as an independent neurotransmitter, for their research in the 1950s in the Cholinergic (CAergic) neurotransmission field [94].

Although in the brain, DA is an important neurotransmitter, mesenteric organs are involved in the production of a considerable amount of DA. In 1939, Blaschko postulated about DA biosynthesis by the classical pathway [96]. DA biosynthesis takes place in the cytosol of CAergic neurons by a two-step process and it starts with L-tyrosine hydroxylation at the phenol ring to yield DOPA in presence of tyrosine hydroxylase (TH) [94]. The regulation of this oxidation depends on a cofactor, tetrahydrobiopterin which is synthesized from guanosine triphosphate (GTP) by GTP cyclohydrolase (GTPCH). Ultimately DA synthesis takes place after decarboxylation of DOPA in the presence of DA decarboxylase or aromatic amino acid decarboxylase (AADC) [97, 98].

In addition to the classical pathway involved in DA biosynthesis, another pathway exists in rats *in vivo* which is mediated through cytochrome P450. In this pathway, decarboxylation of tyrosine takes place into tyramine and ultimately Cyp2D proteins hydroxylate them. Although it contributes an almost negligible amount of total DA synthesis, only under specific conditions it becomes important [99]. Alternate DA synthesis may also take place through tyrosine hydroxylation in the presence of tyrosinase and CAergic neuronal uptake of DOPA [94]. Peripheral eumelanins and pheomelanins biosynthesis normally take place in the presence of tyrosinase. DA sequestration takes place in synaptic vesicles in the CAergic neurons *via* vesicular monoamine transporter 2 (VMAT2) [94]. The slightly acidic environment of vesicles protects DA from oxidation and also minimizes the chances of oxidative stress in the cytosol. For maintaining the DA homeostasis three enzymes TH, GTPCH and AADC are essentially required, which prevents the system from excessive burden of oxidative stress [100, 101].

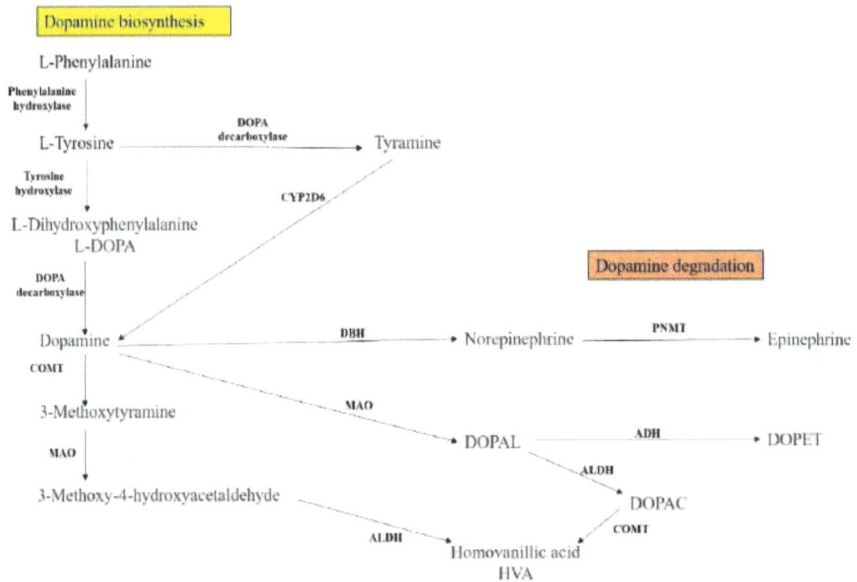

Fig. (3). Dopamine biosynthesis and degradation pathway. Abbreviations: ADH, Alcohol dehydrogenases; ALDH, aldehyde dehydrogenase; COMT, Catechol-O-methyltransferase; MAO, Monoamine oxidase; DBH, Dopamine beta-hydroxylase; PNMT, Phenylethanolamine N-methyltransferase; DOPAL, 3,4-Dihydroxyphenylacetaldehyde; DOPAC, 3,4-Dihydroxyphenylacetic acid.

Role of the Tyrosine Hydroxylase Enzyme in DA Metabolism

DA biosynthesis is strongly regulated by the tyrosine hydroxylase (TH) enzyme at the first stage. The aromatic amino acid monooxygenases, tryptophan hydroxylase, and phenylalanine hydroxylase together constitute TH [38]. The four identical subunits of TH get activated in the presence of BH4, ferrous ion, and O_2 for the synthesis of DA from tyrosine [102 - 104]. Four isoforms of TH are present in humans due to the alternative splicing of exon 2. All four isoforms of TH consist of an N-terminal regulatory domain (~150AA), a central catalytic domain (~300AA), and the C-terminal part, coding for a leucine zipper domain and tetramer formation depends on it [74]. In addition to few covalent modifications, some proteins such as DJ-1, α-synuclein (α-syn), VMAT-2, AADC, BH4, and GTPCH play important roles in the stability of TH by interacting with the N-terminal domain. The activity, stability, and intracellular localization of TH which affects the DA production, is maintained by these interactions [94, 105].

Intracellular O_2 concentration also plays an important role in the production and stability of DA. The normal concentration of O_2 in the brain is ~ 1-5%, in comparison to this the level of oxygen in the environment is 20%. Excessive oxygen concentration may elevate the level of ROS in the system and also increase the availability and activity of TH [74].

Role of BH4 and GTPCH

For pterin-dependent aromatic amino acid monooxygenases and NO synthase, BH4 (6R-L-erythro-5,6,7,8-tetrahydrobiopterine) acts as a cofactor and can facilitate the substrate hydroxylation by directly reacting with molecular oxygen [94]. The high level of BH4 could also be toxic for cells by lowering the level of TH and inhibition of the first and the fifth complexes of electron transport chain [74]).

GTP cyclohydrolase I (GTPCH) catalyzes the first rate-limiting reaction for the production of BH4 [38, 101]. In serotonergic neurons, a high concentration of GTPCH is reported and a few experimental studies have shown that substantia nigra has higher GTPCH immunoreactivity [98]. Phenylalanine induces the activity of GTPCH while as a feedback inhibition the activity of GTPCH is inhibited by BH4. Phosphorylation helps in interaction of TH and GTPCH, while BH4 mediated the inhibition of GTPCH is prevented by TH [106]. As long as the TH remains phosphorylated it maintains the active state of GTPCH and leads to DA production. Though the feedback inhibition of GTPCH by BH4 can be prevented by interaction between TH and GTPCH, in presence of GTPCH, DA can inhibit the activity of TH. These two enzymes form complexes and also regulate the level of DA inside the cell [98, 106].

Role of Aromatic Amino Acid Decarboxylase (AADC)

Blaschko was probably the first person who described the AADC, later on, Schales and Schales also described this enzyme. A question related to the specificity of AADC related to DOPA synthesis was already asked by Blaschko [94]. Studies have reported that in presence of a cofactor, pyridoxal phosphate (vitamin B6), decarboxylation of many aromatic L-amino acids such as L-DOPA, L-tyrosine, L-tryptophan, and L-histidine are catalyzed by AADC [107]. Thus, AADC is not only specific for DOPA synthesis, but also mediates the synthesis of a few other neurotransmitters. The activity of AADC depends on the level of DA and in the presence of DA receptor antagonists the activity of AADC increases while in presence of receptor agonists, it remains in a suppressed form [108]. In PD patients if DOPA is used as a drug, AADC regulates the synthesis of DA [109]. In such cases DOPA enters into the neurons through vesicles (VMAT2) and the excessive DA present in the cytosol is degraded by MAO and COMT, minimizing the incidence of oxidative stress. More studies on AADC are presently going on and detailed knowledge about the human AADC would be beneficial for future clinical studies in PD management [110].

Process of DA Degradation

DA degradation in cytosol starts when enough DA has been supplied to the excited dopaminergic nerve terminals (Fig. **3**). At dopaminergic terminals, the pre-and postsynaptic terminals having DA receptors, come in contact with DA, and the remaining DA can either be reabsorbed by DAergic neurons and recycled or degraded by glial cells after their uptake [74, 111]. DAT helps in the neuronal reuptake and VMAT2 sequesters them into synaptic vesicles [111]. MAO causes the oxidative deamination of excessive DA present in cytosol and produces hydrogen peroxide and 3,4-dihydroxyphenylacetaldehyde (DOPAL) [112]. Alcohol dehydrogenase (ADH) or aldehyde dehydrogenase (ALDH) are the enzymes that can cause reduction of aldehyde into the corresponding alcohol 3,4-dihydroxyphenylethanol (DOPET) or can produce carboxylic acid 3,4-dihydroxyphenylacetic acid (DOPAC) by oxidation [113].

Under normal conditions corresponding carboxylic acid production takes place from DOPAL (98). The predominating product after reduction of NE and E is alcohol. Surrounding glial cells also degrade DA by MAO and catechol-O methyl transferase (COMT) [74]. DA mainly degrades into homovanillic acid (HVA) that is produced from DOPAC after the action of COMT. Nigrostriatal dopaminergic neurons are devoid of COMT activity and it is only present in glial cells [114, 115]. Phase II conjugation reaction can take place before the excretion of DA and its metabolites. Thus, urine samples contain HVA, DOPAC, their sulfates, glucuronides and DA conjugates as the main DA excretion products [94].

DA Oxidation and Inflammation

The electron-rich catechol moiety present in DA and other CAs is more prone to oxidation reactions catalyzed by metals (Fe^{3+}) [116]. Cyclooxygenases, tyrosinase and other enzymes are involved in the enzymatic oxidation of CAs [116]. Superoxide radical anions generated after these reactions in the presence of oxygen act as an electron acceptor. The unspecific reaction of quinones and ROS with different cellular components may change their original function and can be fatal for neurons [117]. Inter and intramolecular reactions of DOPA-Q and DA-Q can take place with the nucleophiles. The central oxidation intermediate, CA quinones result in the formation of different products [118]. Neuromelanin, the neuronal pigment, is formed in presence of a precursor, 5,6-dihydroxyindole. Further 6-hydroxydopamine, a neurotoxin, is synthesized from DA-quinone in the presence of iron [119]. An endogenous neurotoxin like salsolinol is also formed from DA-quinone and is responsible for the onset of oxidative stress and mitochondrial dysfunction. Salsolinol can also inhibit the activity of TH, DA-β-hydroxylase, COMT and MAO and interfere with the CA metabolism [120].

Different 5-cysteinyl-catechol derivatives are formed after a CA-quinone reaction with thiol groups of amino acids. The derivatization of cysteinyl residues of proteins may hamper the normal protein function [74]. Stress caused due to DA also affects the function of DAT and TH. The capacity of the cell to cope with the condition of oxidative stress goes unwell after the conjugation of DA-quinone with glutathione [120 - 122]. The major component associated with PD is α-syn, a cytosolic inclusion body or LB. It is a small protein that is known to be present ubiquitously in the brain and by interacting with TH, it negatively regulates the biosynthesis of DA [123]. In the case of PD, the α-syn protofibrils cause cellular dysfunction by forming pores in the membrane for permeabilization. The derivatization of hydroxyl groups present on catechol moiety can prevent oxidation [120, 124].

Neuromelanin

A specific complex pigment present in the substantia nigra and locus coeruleus of the brain is named neuromelanin (NM). The building components of NM are DA-derivatives, covalently bound amino acids (15%) and absorbed lipids (20%) [124]. The formation of NM requires iron but the role of enzyme catalysis is still obscure. NM is somewhat similar to melanin but the exact structure is yet not clear [125]. Non-vesicular DA is responsible for the synthesis of NM and it is also found in the autophagic organelle such as lysosome present in cytoplasm. NM not only acts as a DA sink but it can also prevent the cell from oxidative stress by binding with iron to prevent Fenton reaction [125, 126]. As compared to the other cells, NM plays an important role in DAergic cells due to the abundance of ROS. Oxidative stress can worsen the situation by degradation of NM through peroxidation and causes the release of metal ions or toxins that were previously captured. Microglia activation and inflammation may start after the release of NM and neuronal cell death [126, 127].

Role of Oxidative Damage and Inflammation in PD

The underlying chemistry makes oxidative stress the key player in the metabolism of DA and is also associated with the process of neurodegeneration. Superoxide dismutase (SOD), catalase and glutathione peroxidase (GPX) are the antioxidant enzymes that work to clear the excessive ROS from the system [128]. Studies have shown that the level of SOD, catalase and GPX decreases significantly in substantia nigra and makes DAergic neurons more vulnerable to death [129]. Lipid oxidation, mitochondrial dysfunction and DNA damage occur due to the excessive ROS level [130]. Microglia activation and inflammatory reactions take place due to the signals related to the oxidative stress and chemoattractants released by dopamine neurons [131].

Microglia activation induces the production of intracellular NO, cytokine synthesis, inflammatory glycoproteins and cell adhesion molecules and results in the adhesion of microglial cells to neurons [132]. These processes are promoted from the releasing of chemoattractants from degrading neurons. Finally, the dopaminergic neurons are phagocytized by microglia cells [133]. NO can diffuse into DAergic neurons from activated microglia cells and forms a potent oxidizing agent, peroxynitrite (NO-3) from superoxide anions. Further, ROS level in neurons is increased due to the excess production of hydrogen peroxide. In addition, the activity of TH is interrupted after the production of tyrosine nitration from peroxynitrite. S-thiolation mediated by NO on cysteine residues can also suppress the activity of TH [74].

Biochemical Aspects of PD Pathogenesis

In PD patients, the most prominent reason for motor disorder is related to biochemical changes. The neurotransmitter, DA insufficiency in the striatal caudate putamen nuclei has been found in the autopsied brain of PD patients [134]. There are many limiting factors present at the presynaptic dopaminergic terminals, biochemical markers that are the metabolites of DA, such as homovanillic acid (HVA), tyrosine hydroxylase enzyme help in DA synthesis, L-dopa-decarboxylase, affects the whole nigrostriatal pathway [135]. The loss of DA level is more prominent in the putamen of the striatum as compared to caudate nuclei and it is approximately 80% in the advanced PD stages [74]. This difference in putamen-caudate is due to the loss of DA perikarya containing melanin in SNpc. However, on comparing the loss of nigral cell loss to DA loss in the striatum, the latter was found to be more than the former [45].

The loss of DA in the striatum follows two specific patterns and in idiopathic PD inter- and sub-regional depletion of DA occurs. In the inter depletion, the common pattern of loss occurs more in the putamen than the caudate while in subregional the DA depletion occurs within the putamen in two different portions in a different manner [74]. Within the putamen, DA depletion was found to be more pronounced in the caudal portion as compared to the rostral portion, whereas it is opposite for the caudate nuclei [136]. Until the loss of DA reaches more than 60%, the loss remains clinically silent. The clinical manifestation starts beyond the threshold value of 60-80% DA depletion and at this stage the symptoms of PD can be easily correlated with the depleted level of DA [137]. Thus PD is a motor disorder that essentially arises due to the putamen dysfunction caused by the loss of dopamine [138].

Biochemical Changes that Compensate for the Striatal DA Depletion

Higher levels of DA loss in the striatum may give rise to the clinical manifestation of PD but to a certain extent, the level of DA can be compensated by the other dopaminergic neurons, depending on the level of depletion [139]. These compensatory mechanisms work by increasing the process of DA metabolism in the rest of dopaminergic neurons, then the ratio of HVA:DA is found to be shifted towards the HVA in the striatum [140]. This is known as the early mechanism operated even when the loss of DA remains less than 60% while in advance cases, when the loss of DA is more than 90% the number of D2 receptors in the striatum present at postsynaptic terminals increases therapeutic efficacy to overcome the PD conditions clinically by DA substitutes [74].

Involvement of DA Transporters in PD Pathogenesis

DA transporters are physiologically known for their involvement in DA reuptake at synaptic terminals, which may also allow the entry of a few DA-toxic compounds such as the MPP^+ (product of MPTP) [141]. Thus, this toxic insult through the DA transporter may also contribute to the etiology of PD. This possibility is supported after the correspondence among the DA transporter and loss of cells present in nigra [142]. The uneven distribution of DA transporters may be responsible for allowing the entry of DA-toxic agents and causing cell death. The enriched presence of DA transporters at striatal terminals as compared to the substantia nigra cell bodies, supports the possibility of retrograde degeneration of dopaminergic neurons in PD [143].

CONVENTIONAL THERAPEUTIC TARGETS IN PD

PD is the 2nd most common neurodegenerative disorder that becomes more prevalent with advanced age [144]. To date, a number of animal models have been developed for gaining in-depth information about PD pathogenesis [145]. Presently both genetically and chemically induced models are being used in the research field. Chemical agents, such as MPTP, 6-hydroxydopamine (6-OHDA), rotenone, and paraquat *etc.* are used to develop PD-like phenotype in animal models [146]. However, none of the presently used models completely replicates the behavioural symptoms and neuropathology of the human PD. Limitations of animal models of PD like phenotypes, restrict the development of the novel therapeutic drug for PD, in humans [147]. Present medical management of PD gives symptomatic relief from PD symptoms along with some adverse effects [148]. For a long time, dopamine replacement therapies are well established and effective for the treatment of PD [149]. Although administration of levodopa is effective for the treatment of PD and gives symptomatic relief, over time the chronic treatment of levodopa induced dyskinesia [150]. Present therapies used in

PD are unable to sustain reversal of existing disabilities and disease progression. So, the development of effective treatment for PD requires an understanding of different conventional therapeutic targets (Fig. **4**) which could be involved in individual or combination therapy.

Therapeutic Targets to Treat PD

1. Targeting α-Synuclein
2. Targeting Oxidative Stress and Neuroinflammation
3. Mitochondrial dysfunction as a therapeutic target
4. Targeting Autophagy-Lysosome System
5. Targeting ubiquitin-proteasome system
6. Epigenome as a therapeutic target
7. Gut microbiota as a therapeutic target
8. Neurotrophic factors as a therapeutic target

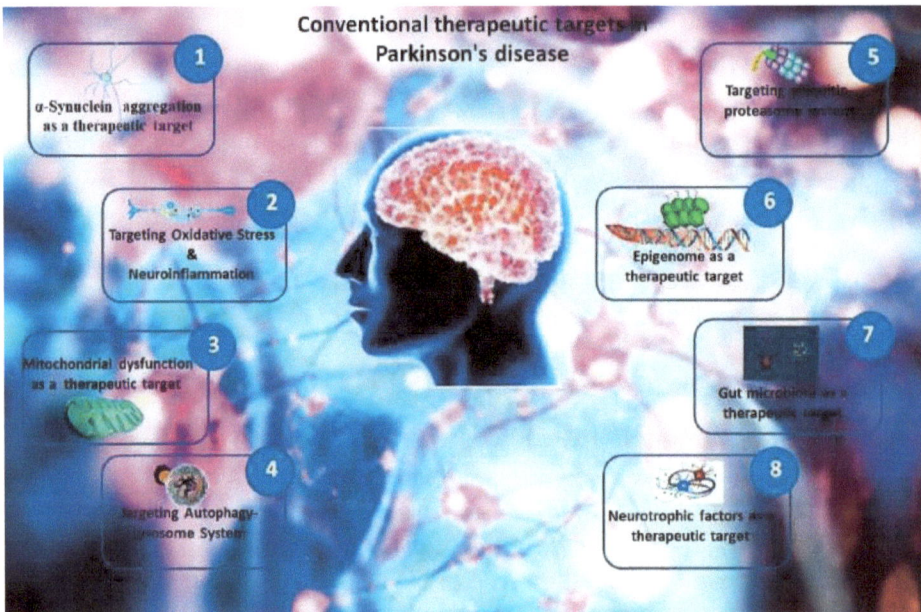

Fig. (4). Conventional therapeutic targets in PD.

Targeting α-Synuclein

Several lines of evidence support a pivotal role of α-syn in PD pathogenesis [151, 152], as α-syn was also found as a major component of Lewy-bodies (LBs), one of the hallmark pathological features of PD [153]. LBs were first described by Lewy in 1912 [154]. α-Syn is a protein consisting of 140 amino acids with a molecular weight of 14 kDa. There are three domains present in its architecture:

(1) an N-terminal lipid-binding alpha-helix; (2) a non-amyloid-component (NAC); and (3) an acidic C-terminal tail [155]. The presence of misfolded protein in the form of α-syn aggregation gives toxic effects on neurons and represents a potential common therapeutic target for PD treatment [156]. Although there is no confirmation about the toxic moiety of α-syn, some *in vitro* experiments indicate towards oligomers and protofibrils as the most toxic forms of α-Syn; however protofibrils are unstable intermediates and their link to neurotoxicity is presumed rather than conclusive [157, 158]. The significance between oligomers and protofibrils in terms of toxicity has not been clear, however it is predicted that together both these oligomeric and protofibrillar α-Syns, are more toxic than monomers [159]. Certain conditions such as low pH, molecular crowding, oxidative stress and interactions with other proteins and heavy metals *etc.* which induce the conversion of monomeric α-syn into toxic oligomeric and protofibrillar form are responsible for PD pathogenicity [160 - 162]. Declining the level of oligomeric and protofibrillar α-syn, is a targeted goal under the therapeutic strategy because neither the insoluble α-syn aggregates nor the intracytoplasmic misfolded proteinaceous inclusions, such as Lewy bodies, are considered toxic moieties in PD [163]. By preventing the formation of oligomeric and protofibrillar α-syn or stimulating their conversion to large less toxic mature aggregates, both the condition may result in a decrease in toxic oligomeric and protofibrillar α-syn, and could be useful in the treatment of PD pathogenesis. Therapeutic intervention in PD may be achieved by targeting the formation of massive insoluble aggregates, prevention of succeeding steps of misfolding of starting material after the oligomeric and protofibrillar α-syn formation [164]. Numerous *in vitro* studies show the role of small molecules that regulate the chaperone activity of αB-crystallin, were able to protect cells against α-syn aggregation [165 - 167]. The use of solid-state nuclear magnetic resonance (ssNMR) for the designing of N-methylated peptides as inhibitors of α-syn aggregation could prove as a novel method in this regard [168, 169]. In this line of work, another method known as Surface plasmon resonance was used by some researchers which resolve the α-syn binding affinities with screened peptide, however in this method evaluation of aggregation potential is also pursued by a Thioflavin T-based fibril formation assay [170]. Hence after the screening of hundreds of potential binding peptides, only a few promising candidates were taken for further *in vitro* validation.

Targeting Oxidative Stress and Neuroinflammation

Progressive degeneration and loss of dopaminergic neurons in the nigrostriatal region of the brain are considered the main pathological feature of PD, but the particular underlying cause(s) of death of a dopaminergic cell is still unclear [1]. A number of reports show the association of dopaminergic cell death with oxidative stress and neuroinflammation [171]. Oxidative stress and neuroinflammation combinedly promote a particular phenomenon in dopaminergic cells of substantia nigra, known as senescence-associated secretory phenotype (SASP) which is associated with various biomolecules such as proinflammatory cytokines, growth factors, and proteases *etc* [172]. In both sporadic and familial PD, an increased oxidative stress level has been observed in different brain regions [173]. Free radical production in nigrostriatal DAergic neurons affects dopamine metabolism, mitochondrial functioning and also induces neuroinflammatory conditions [174 - 176]. During PD pathogenesis unbalanced GSH/GSSG (reduced GSH / oxidized GSH (GSSG)) has been seen in the substantia nigra region of the brain which is probably due to the formation of free radicals which oxidize the reduced form of GSH, which is one of the crucial antioxidants in the cytoplasm [177]. Some reports also show leaking of GSH from cells during PD pathology; these leaked GSH molecules interact with the amino acids glutamate and cysteine in the extracellular matrix and transform these into cysteinyl peptides or glutamyl peptides. These compounds cross back into the cells and inhibit the complex I of mitochondria, which in turn leads to the production of ROS and oxidative stress. Cysteinyl and glutamyl peptides are toxic for dopaminergic cells where they are formed [178, 179]. In the brain, glial cells account for over 50% of the cells, among which astroglia and microglia cells show an indispensable role [180]. In PD brain, activated astrocytes induce ROS production whereas activated microglia mediate the overproduction of inflammatory cytokines, including TNF, IL-1β, IL-2, IL-6, and IFN-γ in the substantia nigra region [181, 182].

Recently researches reported the enrolment of many drugs in clinical trials which shows propitious outcomes in PD such as DPA-714-PET/MRI [NCT03457493], used to measure concentration and regional brain distribution of activated brain microglia/ macrophages using PET ligand [18F] DPA-714, this study was enrolled in the UAB Neuroinflammation in PD study. This study recruited participants in phase II/III in the trial on March 22nd, 2018. There is another study which targeted the activated microglia induced neuroinflammation by using memantine, an N-methyl-D-aspartate (NMDA) receptor antagonist [NCT039-18616] and showed the protective effect on the P2×7 purinergic receptors (P2 × 7R) of microglial cells, this particular receptor activates NLR Family Pyrin Domain Containing 3 (NLRP3) α-syn in the murine model, which induces neuroinflammation, that leads to degeneration of dopaminergic neurons. This particular study was completed on February 20th, 2017 (clinicaltrials.gov).

In summary drug agents that can inhibit both oxidative stress and inflammation, have shown a potent and promising role in the treatment of PD pathology, but to date, it has been observed that preclinical as well as clinical treatment of PD subjects, with such type of agents, could reduce neurodegeneration in PD just to a very limited extent, due to not reversing neurodegeneration but only slowing down the disease progression and preventing more damage [173]. Therefore, such types of agents can be used as supplemental and adjuvant medicines for the management of PD and as preventive medicines.

Targeting Mitochondrial Dysfunction

Among the several phenomena occurring during PD pathogenesis, the role of mitochondrial dysfunction has received strong attention [183, 184]. Because of the involvement of mitochondrial abnormality in PD pathogenesis, researchers have extensive interest in developing drug targets that can mitigate aberrant mitochondrial function [185, 186]. Both the preclinical and clinical data showed moderate glitches in Complex I (NADH dehydrogenase) [187] of mitochondrial respiratory electron transport system activity in the substantia nigra [188 - 190] as well as lymphocytes of PD subjects [191]. Mitochondrial dysfunction might lead to increased oxidative damage to lipids, proteins and DNA in brain tissue of PD subjects [192, 193]. Mitochondrial dynamics, especially fusion and fission steps are an essential process for cell survival, and aberrant mitochondrial dynamics are related to excessive free radicals and ROS production, which further lead to several abnormalities [194]. Involvement of common pathways such as mitochondria-associated cytopathology, altered mitochondrial dynamics and defective bioenergetics in the genetic mutation and neurotoxin-induced models of PD allocate possible targets for the extension of therapeutic approaches in PD.

Targeting Autophagy-Lysosome System

In cells, the process of autophagy is basically involved in the removal of unwanted or damaged cellular components or organelles. In mammalian cells, three different classes of autophagy are present: microautophagy, macroautophagy, and chaperone-mediated autophagy (CMA) [195]. Macroautophagy is further subdivided into two categories: nonselective (hereinafter demonstrated as autophagy) and selective autophagy.

Nonselective autophagy usually occurs at the time of cell starvation and inadequate nutrients for the cell. Unwanted damage components are removed by autophagosomes, which engulf these defective cellular components followed by lysosomal degradation. When the macroautophagy pathway mediates the degradation of damaged mitochondria this process is termed mitophagy. While selective autophagy is involved in the removal of damaged organelles and unwanted protein aggregates which is also referred to as aggrephagy [196]. Accumulation of abnormal and misfolded proteins which leads to impaired protein homeostasis is usually seen in PD pathology [197]. The CMA pathway is basically involved in lysosomal degradation of unfolded proteins with recognition of specific sequence (KFERQ-like peptide motif). Cytosolic chaperone proteins such as Hsp70 are engaged in the CMA pathway for the lysosomal degradation of targeted protein with the involvement of LAMP2A (Lysosomal Associated Membrane Protein 2) [198, 199]. Accumulation of α-syn protein occurs in the majority of sporadic PD brains and reports also showed autophagic degeneration in dopaminergic neurons of substantia nigra brain region in PD patients [200]. The autophagy system forms a vesicular continuum through connecting endosomal, lysosomal and other secretory pathways, all of these organelles work as central hub for routing the cargo of various steps like recycling, degradation or secretion [201, 202]. Initially, PD was considered a non genetic disease but after the discovery of different autosomal dominant forms of PD, many PD-related genes have been identified and several genes of PD pathogenesis were also found to be involved in the regulation of autophagy. PD-related mutations in SNCA such as A53T, A30P, and E46K could be targeted for reversing the PD pathogenesis [203 - 205]. Transgenic mice with A53T overexpression showed autophagy induction in the substantia nigra brain region through the upregulation of LC3-II (a lapidated form of LC3) and BECN1 (Beclin-1) levels [206, 207]. Leucine-rich repeat kinase 2 (LRRK2) mutations have been seen in the majority of autosomal dominant PD cases. LRRK2 regulates lysosomal homeostasis through its kinase activity *via* establishing its substrates such as RAB8A and RAB10 [208, 209].

Natively unfolded α-syn, the major component of LBs can be reduced through several clearance routes including CMA, autophagy, and proteasome pathways [210]. Hence there are numbers of evidence which suggests interconnection of PD pathways with synaptic equilibrium and autophagy could be used as therapeutic targets for PD management [211 - 213].

Targeting Ubiquitin-Proteasome System

A non-lysosomal proteolytic pathway termed ubiquitin-proteasome system (UPS) is a well-known proteolytic system crucial for the regulation of vital processes such as organelles biogenesis, cell cycle, differentiation and development [214]. As UPS also affects various molecular and cellular pathways so altered UPS could result in numerous pathological conditions [215]. UPS is responsible for the degradation of misfolded, mutant, and oxidatively damaged proteins [216]. Literature showed major involvement of UPS in the management of normal functioning of the nervous system such as neurotransmitter release, proper synaptic functioning, membrane receptor recycling and removal of defective and non-functional regulatory intracellular proteins [217, 218]. In PD pathological conditions cytoplasmic inclusions known as Lewy bodies with major component α-syn are present in the spared dopaminergic neurons of substantia nigra [219]. In addition to α-syn mutation in UPS-related genes such as ubiquitin carboxy-terminal hydrolase L1 (UCHL1) and parkin have also been linked to PD pathogenesis [220], the mutation in such genes suggests a role for UPS in PD pathogenesis. Defective UPS has been also reported in the case of sporadic PD [221]. Clinical data showed impaired proteasomal activity and diminished expression of a proteasomal activator such as 19S/PA700 whereas almost undetectable levels of PA28 in the substantia nigra region of the brain. Findings in the preclinical model of PD are also in support of clinical data which show a reduced activity of 20S proteasomal enzymes and increased ubiquitin-conjugated proteins, indicators of proteins marked for degradation, in ventral midbrain regions in rotenone infused rats. However, the occurrence of proteinaceous inclusions is not observed in most widely used animal models of PD such as 6-OHDA induced and conventional dosing of MPTP in rodent models [222], but in some reports αSyn upregulation has been shown to be linked with aggravated mitochondrial pathology in MPTP-challenged α-syn transgenic [223]. In the 6-OHDA induced model, UPS dysfunction has been observed due to the production of severe oxidative stress that could mediate proteasome failure that has been reversed by elevating proteasomes activity. 6-OHDA induced neurotoxicity also raises the inhibition of an increase in the proteasome function, apparently by suppressing UPS-related detoxification [224]. Another report also showed upregulation of ubiquitin with nigrostriatal degeneration in 6-OHDA-induced rat models of PD, demonstrating the participation of UPS in the nigrostriatal degeneration process [225]. The above reports indicate the role of UPS in PD pathology that could be a therapeutic target for the treatment of PD.

Epigenome as a Therapeutic Target

Although the standard drug such as levodopa used for the treatment of PD gives symptomatic relief, the occurrence of side effects after prolonged use demands new therapeutic drugs and drug targets to protect further degeneration of dopaminergic neurons in PD [226, 227]. Although Epigenetic factors play a critical role in embryonic development, early brain programming, brain plasticity and neurogenesis [228], the literature shows that epigenetic dysregulation is also essential for the onset of several neurodegenerative diseases, such as PD [229]. Recent evidence has highlighted epigenetic disturbances in PD patients such as an imbalance in the acetylation and deacetylation of the histone proteins, alterations in the methylome that might play a role in Parkinsonian pathology [230, 231]. Mechanism of acetylation and deacetylation of histone lysine residues control the gene expression, in CNS, and is an essential phenomenon for neuronal survival [232]. Histone acetylation is an extremely dynamic process which is controlled by two classes of enzyme: histone acetyltransferases (HATs) and histone deacetylase (HDACs) [233]. HATs are classified into three groups. **1.** p300/CREB binding proteins (CBP) HATs; **2.** Gcn5-related acetyltransferases (GNATs); **3.** the MOZ, Ybf2/Sas3, Sas2 and Tip60 (MYST)-related HATs [234]. To transfer an acetyl group to the ε-amino of lysine residues on the N-terminal tails of the four core histones, H2A, H2B, H3 and H4, all of the HATs use acetyl-coenzyme A as an acetyl group donor [235]. The acetylation process makes chromatin relaxed which is referred to as euchromatin that allows transcriptional factor access to the DNA [236]. The process of deacetylation removes the acetyl groups from ε-amino of lysine residues, is facilitated by HDACs, and causes chromatin condensation, referred to as heterochromatin [237]. Under the phylogenetic classification, HDACs have been categorised into 4 main classes: classes I, II (a and b), III and IV [238]. In the brain, different types of cells express different classes of HDACs [239]. A stable equilibrium exists between HAT and HDAC activity [240] and a tightly controlled equilibrium between HAT and HDAC activity is required for dynamic control of transcription of genes and is responsible for various cellular processes [241, 242]. In neuronal cells such harmonised balance subsequently facilitates appropriate neuronal health and homeostasis [242]. In PD pathogenesis HAT/HDAC imbalance occurs which is likely responsible for neuronal cell death and disease condition [242] [243]. In PD subject's histone hypoacetylation has been observed [244]. Previous reports showed that α-Syn accumulation in PD also assists histone H3 hypoacetylation in *in vitro* and α-Syn transgenic Drosophila models [245]. Therefore, it is also theorised that accumulated misfolded α-Syn promotes neurotoxicity in PD by 'masking' histone proteins through preventing histone acetylation that results in chromatin condensation, repressing gene expression and eventually leading to cell death. HDAC inhibitors are used to restore the imbalance between HAT/HDACs for the treatment of neurodegenerative disorders [245, 246]. In the neurodegeneration, HDAC inhibitor treatment reduces deacetylation of histones and allows the activation of multiple gene products such as anti-inflammation, brain-derived neurotrophic factor (BDNF), glial-derived neurotrophic factor (GDNF), heat shock protein 70 (HSP70) and antiapoptotic signaling related genes *etc* [246, 247]. Although several studies have shown the use of HDAC inhibitors for the treatment of PD, further work is required to validate the previous findings before implicating them for clinical trials. For the clinical significance and translatability there is a need to administer this drug after neurodegeneration, *i.e.*, in animal model treatment, HDAC inhibitor should be given after the toxin administration rather than prophylactically. In a similar way, PD is a progressive neurodegenerative disease so it is necessary to validate the neuroprotective effects of HDACIs in such a preclinical model of PD which reiterate the progressive aspect of disorder. Earlier, an extensive genomic study recognized numerous PD risk loci in the cerebellum and frontal cortex of PD brains, together with STX1B, GPNMB and PARK16 *etc.* those were related to differential DNA methylation at proximal CpG sites [248].

Depending on the particular site of methylation, histone methylation induces various alterations in gene expression [249]. In the MPTP-induced PD model, dopamine depletion is associated with a reduction in H3K4me3 in striatal histones [250], whereas another report shows the role of histone demethylase inhibitor for rescuing the dopaminergic neuron loss and motor defects in 6-OHDA-induced PD rats [251]. In the case of PD whether any gene will go for hyper- or hypomethylation, it varies according to the gene involved in pathology like peroxisome proliferator-activated receptor gamma coactivator-1 α (PGC-1α) found hypermethylated in substantia nigra region samples of sporadic PD subject [252]. Recently, PARK7 (DJ-1) DNA methylation was also reported in peripheral blood leukocytes of PD subjects [253]. Thus, epigenetic regulation may prove to be a crucial aspect in PD pathogenesis and targeting epigenetic variance in PD may open the scope for rational therapeutic intervention in PD treatment.

Gut Microbiota as a Therapeutic Target

Researches in the last two decades have drawn attention towards the presence of α-syn in the nuclei of lower cranial nerves which suggest the transportation of α-syn in the gut [254]. Clinical data have also verified the occurrence of aggregated α-syn in the gut of PD patients [255]. These observations advised the link between α-syn aggregation and infectious agents affecting the gut epithelia [255], and that gut microbiota may contribute to the aggregation of α-syn [256]. Recent reports show abnormal gut microbiota in PD patients which is termed dysbiosis [257]. The origin of such dysbiosis is unclear however certain factors such as food habits, genetic background, exposure to toxins, microbe's infection, chronic stress *etc.* have been suggested to play a role [258]. Histopathological experiments show evidence of hyperpermeability of the gut in PD patients which enhanced translocation of microbiota and their products across the gut. Altered gut permeability in PD patients also leads to a pathological condition known as "leaky gut" [259]. Increased gut permeability in PD pathology allows translocation of endotoxins (secreted by Gram negative bacteria residing in the intestines) such as LPS, in systemic circulation and CNS [260]. LPS binds to toll-like receptor type-4 of brain microglia and activates it [261]. Activated microglia leads to release of reactive oxygen species and inflammatory mediators responsible for damage of dopaminergic neurons in the substantia nigra [261]. Malfunctioning of gut microbiota causes abnormal synaptic transmission and affects plasticity in the striatum of PD patients [262]. A form of dysbiosis termed Intestinal bacterial overgrowth (IBO), is also a symptom found in PD patients [263]. Levodopa-induced dyskinesia is observed during the long-term treatment of PD [264]. However the microbiota is also involved in levodopa metabolism [265]. Supporting literature shows the role of *Enterococcus faecalis* with conserved tyrosine decarboxylase in the metabolism of levodopa [266]. If tyrosine decarboxylase is highly expressed in the dysbiotic gut, it causes abnormal levodopa function and would be expected to promote motor fluctuations during PD pathogenesis [267]. Evidence suggests that gut microbiota may affect the cardinal symptoms of PD [268 - 270], but still more research is required to explain these primary findings. Henceforth a therapeutic idea with the aim to re-establish a healthy microbiota would be required for PD treatment.

Neurotrophic Factors as a Therapeutic Target

In the developing brain, some secreted peptides are present that act as growth factors for the survival of brain cells. These growth factors combinedly termed Neurotrophic factors (NTFs) [271]. However, in the adult brain, the functions of NTFs are not well understood. According to the neurotrophic factor hypothesis, it is proposed that NTFs are important in both developing and the adult brain for phenotypic maintenance, survival and plasticity of growing and fully matured neurons [272]. Numerous studies demonstrated that adequate availability of NTFs may be crucial also in the pathogenesis of neurodegenerative disorders [273 - 275]. The neurotrophins family encompass several NTFs such as Nerve growth factor (NGF), Glial-derived neurotrophic factors (GDNF), Brain-derived neurotrophic factor (BDNF), Neurotrophin 3 (NT3), Neurotrophin 4 (NT4), cerebral dopamine neurotrophic factor (CDNF), mesencephalic astrocyte-derived neurotrophic factor (MANF), and Ciliary neurotrophic factors (CNTF). Several lines of evidence show an improper activity of NTFs which may induce loss of dopaminergic neurons in the case of PD pathogenesis [275, 276]. Recent research shows the effect of CDNF protein therapy in PD, according to this report Intraputamen administration of recombinant human CDNF has given measurable neuroprotective effect and reverse dopaminergic neurodegeneration in various animal models of PD. Besides its protective role, it also shows a good safety profile in preclinical toxicology studies. In phase I-II clinical study, intermittent monthly bilateral Intraputamen infusions of CDNF are also being examined in moderately advanced PD patients [277]. Growing evidence shows the probability of decreasing the PD pathogenesis through manipulating endogenous levels of GDNF, MANF, CDNF (Table **1**) [278]. Some literature favours this perspective and shows that dietary manipulations, physical exercise, and other lifestyle changes also potentiates the maintenance of the endogenous optimal level of these NTFs [279, 280]. Another recent research demonstrated that optimal production of endogenous CNTF, basically regulated by transient receptor potential vanilloid 1 in astrocytes, inhibits the degeneration of DAergic neurons and shows behavioral recovery in rat models of PD [281]. Moreover, CNTF has not been tested clinically in the case of PD so further studies are required before declaring it as a therapeutic drug in humans affected with PD pathogenesis. Undoubtedly endogenous NTFs promote neuronal survival and differentiation during development and even exogenous administration also restores the normal functioning and plasticity of degenerated neurons in animal models. However, under clinical application the same exogenous treatment strategy shows relatively weak effects [282]. Previous research explored the feasibility but not the potency of NTFs therapy for the treatment of neurodegenerative diseases [283]. Gene therapy is an emergent strategy to deliver NTFs in PD clinical trials [284].

Table 1. Therapeutic effect of NTFs on PD.

NTFs	Effect on PD	Ref.
BDNF	Regulates motor function, plasticity, and repair and promotes the survival of DAergic neurons	[285]
GDNF	Exerts neuroprotective and neurorestorative effects on DAergic neurons	[286]
CDNF	Reduces ER stress, attenuates the loss of DAergic neurons, and exhibits antioxidant, antiapoptotic, and anti-inflammatory properties as well as neurorestorative effects on DAergic neurons	[287]
CNTF	Prevents the degeneration of DAergic neurons	[288]
MANF	Reduces ER stress, inhibits oxidative stress, neuroinflammation, apoptotic cell death thus increases DAergic cells survival	[276]
NGF	Protects against oxidative stress and delays PD progression	[289]
NT4/5	Improve either motor or cognitive dysfunctions in Parkinsonian rats	[290]

MANAGEMENT AND TREATMENT OF PD

At present, there is no cure or treatment available for PD that can halt or slow down the disease progression [291]. PD is a very diversified disorder and several studies suggest early addition of drug treatment as early as the disease is diagnosed [292, 293]. Currently, levodopa (LD) is the gold standard and most widely used therapy for PD management. However, long-term use of LD may lead to levodopa-induced dyskinesia (LID) in nearly 40% of the patients having this therapy [294]. Before the commencement of the PD therapy, a proper diagnosis is required regarding the level of impairment. Every therapy should be individualized accordingly with the combination of therapy. Undoubtedly, various therapies are available targeting dopaminergic and non-dopaminergic pathways for slowing down or halting the PD progression.

There is a large number of good quality reviews that have discussed the clinical application of PD therapy with A2A receptor antagonist [295 - 297], muscarinic antagonists [298, 299], serotonergics and glutamatergics [82, 300, 301].

Current Scenario of Approved Drugs for PD

Currently, there is no disease-modifying therapy available for PD and available treatments only offer a little symptomatic relief, related to motor and non-motor complications in PD. Table **2** summarizes available treatment for PD treatment.

Levodopa

Since the 1960s, dopamine replacement therapy has become the predominant therapy in PD. LD, the gold standard for PD treatment, is a dopamine precursor that crosses the blood brain barrier and compensates for the DA loss [302]. The initial dose of levodopa (LD) is low and can be gradually increased as per patient's requirement. Dose ranges from 150-1000 mg daily in divided doses [303]. It is generally administered in combination with carbidopa to prevent the peripheral decarboxylation that leads to higher bioavailability of dopamine as well as decreased side effects. Long-term use of LD results in significant motor side effects such as in ADHD although these motor symptoms appear to be due to the dose variability [304]. Levodopa-induced dyskinesia can be classified into three different classes: peak-dose dyskinesia, biphasic dyskinesia and wearing-off dyskinesia [227, 305]

Dopamine Agonists

Dopamine receptor agonists were introduced in 1978 for PD (Table **2**). These drugs basically have ethanolamine moiety and can be classified as ergot and non-ergot alkaloid [306]. These act differently from LD by stimulating dopamine receptors (D1 like and D2 like). These are prescribed at an early stage of the disease especially in younger patients [307]. Dopamine agonists are preferred over LD as they have a longer duration of action and may delay the use of LD so that motor fluctuation can be reduced. However, the half-life of the dopamine agonist varies from person to person. The other benefit of using agonists is that they stimulate receptors and depress dopamine turnover and could be used as adjuvant therapy to LD for motor complications. The main side effect of dopamine agonists is nausea as this triggers area postrema in the medulla, however, these side effects can be treated by prescribing peripherally acting dopamine antagonists domperidone. Studies have reported antiparkinsonian effect could be achieved by activating D2 receptor activation and it has been observed that concomitant administration D1 and D2 agonists may produce better physiological and behavioral effects [308]. Many other studies have reported beneficial properties of dopamine agonists like antioxidant and antiapoptotic properties, although all these properties were based on *in vitro* experimentation and have not been clinically correlated yet [309]. Dopamine agonists are more preferred therapy over other treatments such as MAO-B inhibitors.

Table 2. FDA approved pharmacological management of PD.

S. No.	Drug Class	Drug Name	Mechanism of Action	Indication for Use
1.	Dopamine Precursor	Levodopa, Levodopa-Carbidopa	Converted to dopamine in brain	Used for both early and late-stage PD
2.	Dopamine Receptor Agonist	Pramipexole, Rotigotine, Dinoxyline, Dinapsoline, Dihydrexidine, Ropinirole, SKF-81297, Stepholidine, Cabergoline, SKF-38393, Pergolide	Enhances the level of dopamine *via* activation of dopamine receptor	Can be used alone and adjuvant therapy to levodopa, used to minimize the dose of levodopa
3.	COMT Inhibitor	Entacapone, Tolcapone	Blocks enzyme responsible for the breakdown of dopamine	Used to combat the motor fluctuation, levodopa dose reduction
4.	MAO-B Inhibitor	Selegiline, Rasagiline	Inhibits the activity of MAO-B, an enzyme responsible for breakdown and metabolism of dopamine	Used to mitigate motor problems in addition to levodopa
5.	Anticholinergics	Trihexyphenidyl, Biperiden, Orphenadrine, Procyclidine, Benztropine	Blocks action of Acetylcholine	Prescribed for young patients at an early stage for motor fluctuations
6.	Other	Amantadine	Triggers release of dopamine	Recommended for later phase of Parkinson disease

Catechol-O-Methyltransferase (COMT) Inhibitors

These are the enzymes that inhibit the degradation of dopamine, hence preserving the dopamine level [307]. These are generally not prescribed as monotherapy and mainly used as adjuvant therapy to levodopa, enhancing the onset of action by increasing the half-life of dopamine reaching the brain [310]. Notably, these could amplify the side effects of LD like dyskinesia and it is required to reduce the dose when needed. Currently, two COMT inhibitors have been widely used: tolcapone and entacapone. Tolcapone was recently withdrawn from the market due to its observed hepatic toxicity. However, the exact relationship between drugs and PD is still unknown.

Monoamine Oxidase - B (MAO-B) Inhibitors

Like the other therapy MAO-B inhibitors acts by inhibiting enzymes involved in the metabolism of dopamine, this results in enhanced dopamine level. A classical example of this class of drug is selegiline and rasagiline, which selectively and irreversibly blocks and reduces the conversion to/of dopamine to its metabolites and also inhibits dopaminergic reuptake to the synapse. When administered along with LD it reduces LD dose about 10%-15% [311]. Similar to COMT inhibitors, this class of drugs may also aggravate the LD side effects. The prescribed dose for selegiline is 5–10 mg daily, and for rasagiline is 0.5–1 mg once daily. MAO-B inhibitors are well tolerated, having minor gastrointestinal side effects and orthostatic hypotension [312].

Anticholinergics

All the above medications are supposed to enhance or regulate the dopamine level in the striatum. Few treatments come under the non-dopaminergic class and one of them is anticholinergics, which inhibit acetylcholine neurotransmitters, which are responsible for the tremors produced due to the loss of dopamine in PD. Generally, these are prescribed in young patients at an early phase of PD to treat the tremor and rigidity. It could be used as monotherapy as well as adjuvant therapy at an early stage of PD. Furthermore, cholinesterase inhibitors rivastigmine and donepezil have been evaluated clinically for efficacy in PD and researchers have found promising results [313, 314]. Apart from these beneficial effects, anticholinergics are less preferred in elderly patients because of cognitive problems [315]. Various anticholinergics are listed in Table 1. Common adverse effects of anticholinergics are blurred vision, dryness of mouth, urine retention, constipation, impaired memory, dizziness, dyskinetic movement [316].

Amantadine

It is also included in no dopaminergic class of PD treatment. Amantadine is the adamantine derivative and is basically an antiviral drug, administered to treat tremors, rigidity that leads to a temporary improvement in symptoms. Like other non-dopaminergic drugs, it is also used as adjuvant therapy to the LD and is helpful in reducing side effects of LD by reducing dose [317]. Its mechanism of action is still debatable but it is believed that it works by augmenting dopamine release from the storage site as well as antagonizing N-methyl-D-aspartate receptor (NMDA receptor) weakly. Similar to other antiparkinsonian drugs it is also prescribed at a lower dose (100 mg) in the initial phase and escalated (500-600 mg) accordingly. It is well tolerated and common side effects are hallucinations, confusion and impaired concentration, leg swelling, blurred vision, nausea and vomiting *etc* [318].

Phytotherapeutic Approach for PD

Recently, many studies have reported the effect of plant extracts on the inhibition of oligomerization and fibrillization of α-syn protein in PD [319 - 321]. The various plant origin compounds, which were found to be neuroprotective in different experimental PD models, are listed in Table **3**.

S-allylcysteine (SAC): a natural constituent of fresh garlic, is a well-established antioxidant molecule showing protection against 3-nitropropionic acid (3-NP) in an *in vivo* model. 3-NP is a specific inhibitor of mitochondrial complex II, inhibits mitochondrial complex I activity and also alters the Krebs cycle thereby causing mitochondrial dysfunction. Chronic administration of 3-NP causes a selective axonal loss in the striatum region but when 3-NP induced rat model is treated with SAC, axonal loss in the striatum is reversed followed by a blockade of complex I inhibition and mitochondrial dysfunction [322].

Idebenone and Menadione: (vitamin K3) is a quinone compound and well-reported for treating a complex I deficiency, implicated in various diseases such as neurodegenerative disease and often referred to as "complex I bypass factors". Recent studies have reported that four plant-derived 1, 4-naphthoquinone compounds isolated from *Stereospermum euphoroides* and structurally similar to menadione are used as complex I bypass factors. In these studies, they created a cellular model of complex I deficiency by using CRISPR genome editing to knock out Ndufa9 in mouse myoblasts, then checked the effect of 1, 4-naphthoquinone compounds in knock out Ndufa9 mouse myoblast, and found that compounds are showing a significant effect against complex I deficiency in Ndufa9 mouse myoblasts [323]. Currently, this compound is not explored in neurodegenerative diseases and might be useful as a tool compound for investigating complex I disease biology.

Tetramethylpyrazine (TMP): Some other naturally derived compounds are used as bypass factors of *etc* such as TMP derived from *Ligusticum wallichii Franchat* is used in cardiovascular and cerebrovascular disorder. TMP shows protection against kainate-induced excitotoxicity *in vitro* and *in vivo* and also preserving mitochondrial membrane potential, ATP production and complex I and III activities [324].

Ginseng: It is obtained from the dried rhizomes and roots of *Panax ginseng* (Araliaceae). Reports have been published demonstrating the *in vitro* application of an aqueous extract of *Ginseng* that constitutes the active constituent, ginsenoside, against MPTP induced cytotoxicity model of PD, by decreasing the oxidative stress and caspase-3 activation in human neuroblastoma cells [17] and is under clinical trial [325].

Scutellaria baicalensis: It is a Chinese herb that contains bioactive flavones known as baicalein, which gets oxidized to form baicalein quinone. This further downregulates and upregulates many pathways in the SNpc region of the striatum such as iNOS inhibition, decrease in α-syn aggregation, antagonizing the NMDA receptor by inhibiting excitotoxicity, upregulating BDNF expression and showing anti-apoptotic effect in 6-OHDA induced rodent model of PD [17, 326].

Curcuma longa: It is commonly known as turmeric, contains bioactive polyphenolic constituents such as curcuminoids. Administration of curcumin (150 mg/kg/day) showed a promising effect in PD by decreasing oxidative stress in the SNpc region of the striatum. It also showed antiapoptotic properties and down regulates astrocyte, NFκβ, TNFα, iNOS, in MPTP induced rodent model of PD, thus it exhibiting neuroprotective effect in PD [17, 327].

Gastrodia elata: It is an herbal plant and contains an active constituent known as gastrodin which has shown neuroprotection in PD. Some studies have reported that gastrodin 800/kg in association with L-dopa decreases L-Dopa induced dyskinesia. Gastrodin mainly down-regulates the expression of DRPP32, inhibits MAPK signaling, restricts ERK½ signaling, inhibits iNOS, down-regulates JNK signaling, reduces oxidative stress in the SNpc region of the striatum in 6-OHDA induced rodent model in PD, and hence became a novel strategy for the treatment of PD [328].

Ginkgo biloba: The dried leaves extract obtained from the plant *Ginkgo biloba* contains active pharmacological components such as flavonoids and terpenoids (ginkgolide) which have shown the potential benefit in PD. Recent studies have shown that *Ginkgo biloba* extract ameliorates locomotor activity in the A53T α-syn transgenic rat model of PD at different doses. *Ginkgo biloba* extract decreases oxidative stress, enhances the DA level in the SNpc region of the striatum, thus this study suggests that the *Ginkgo biloba* extract may become a therapeutic agent for the treatment of PD [9, 329].

Bacopa monnieri: It is an herbal plant, which contains saponins such as bacoside, bacopaside, bacopasaponin, bacosine, luteolin, and apigenin exhibiting promising effect in PD. Recent studies have exhibited that *Bacopa monnieri* extract shows potential antioxidant activity, which neutralizes the free radicals of MDA, ROS, increases the activity of Acetylcholinesterases (AChE), in the paraquat-induced toxicity in mice model of PD. Bacopa monnieri extract also decrease motor symptoms in PD [9, 330].

Mucuna pruriens: It is an herbal plant which contains naturally occurring L-Dopa up to 5%, genistein (peripheral DOPA decarboxylase inhibitor), genistein (dopamine precursor) which shows promising effect in PD and a higher safety profile than L-Dopa/carbidopa therapy. Some studies have shown that *Mucuna pruriens* also consists of coenzyme Q-10 which increase the mitochondrial functioning and shows antioxidant activity by inhibiting iNOS expression [23, 331].

Withania somnifera: It is commonly called Indian ginseng, it is an herbal plant, which contains withanolide A, withaferin A that show neuroprotection in PD. Recent studies have shown that *Withania somnifera* extract increases BCl-2 expression, downregulates BAX expression, inhibits iNOS and promotes motor function in MPTP induced rodent model of PD [332, 333].

Camellia sinensis: It is commonly called green tea which contains polyphenol such as Epigallocatechin-3-gallate (EGCG) that has shown potential effect in Parkinson's disease. Epigallocatechin-3-gallate (EGCG) inhibits iNOS, ROS, COX-2, MDA, enhances the expression of catalase enzyme, increases GSH formation, thus it shows potential antioxidant activity in *in vivo* as well as *in vitro* conditions. EGCG also upregulates mitochondrial function, inhibits α-syn aggregation, reduces neuroinflammation and increases motor functions *drosophila* model [334, 335].

Centella asiatica: It is an herbal plant containing Asiatic acid which shows neuroprotection in MPTP-induced rodent model of PD. Asiatic acid increases dopaminergic transmission in the striatum and improves motor function in MPTP-induced mice. Asiatic acid (100mg/ day) up-regulates the expression of ERK and PI3K/AKT/mTOR/GSK-3β in MPTP-induced mice. Recent studies have shown that *Centella asiatica* also contains madecassoside which exhibits anti-parkinsonian effect by up-regulating BDNF and BCl2 expression in MPTP induced rat model of PD [336, 337].

Table 3. Phytochemicals showing neuroprotective effect in PD.

S.No	Herbal Plants	Mechanism of Action Involved in PD	Clinical Trial	Reference
1	**Ginseng** (*Panax ginseng*)	↓Oxidative stress ↓Caspase-3 activation	Under clinical trial	[325, 338]
2	**Baikal skullcap** (*Scutellaria baicalensis*)	↓iNOS, ↓α-syn aggregation, ↓Excitotoxicity, ↓BDNF ↓Apoptosis	-	[326, 339]
3	**Turmeric** (*Curcuma longa*)	↓Astrocytes, ↓NFκβ, ↓TNFα, ↓iNOS, ↓Oxidative stress, ↑TH expression	-	[327, 340, 341]
4	**Tianma** (*Gastrodia elata*)	↓DRPP32, ↓MAPK signaling, ↓iNOS, ↓JNK signaling, ↓oxidative stress	-	[328, 342]

(Table 3) cont.....

S.No	Herbal Plants	Mechanism of Action Involved in PD	Clinical Trial	Reference
5	**Maidenhair tree** *(Ginkgo biloba)*	↑Dopamine level, ↓ oxidative stress	Phase II	[329]
6	**Brahmi** *(Bacopa monnieri)*	↑AChE ↓MDA, ↓ROS ↑ TH expression	-	[330, 343]
7	**Velvet bean** *(Mucuna pruriens)*	↑ safety profile of L-Dopa/carbidopa therapy. ↑coenzyme Q-10 which increase mitochondrial functioning ↓ iNOS expression	Completed phase II study	[331]
8	**Poisonous gooseberry** *(Withania somnifera)*	↑BCl-2 expression, ↓BAX expression, ↓iNOS ↑Promotes motor function	-	[332, 344]
9	**Tea plant** *(Camellia sinensis)*	↓iNOS, ↓ROS, ↓COX-2, ↓MDA, ↑Catalase enzyme, ↑GSH ↑Mitochondrial function, ↓α-syn aggregation, ↓Neuroinflammation and increase motor functions	Phase II	[335]
10	**Gotu kola** *(Centella asiatica)*	↑ERK, P13K/AKT/mTOR/GSK-3β, ↑ BDNF, ↑ BCl2 expression	-	[337]

Recent Advances in PD Therapy

On the basis of different pathological aspects of PD progression, few other agents could be a better possible target for PD treatment. Since the last 20 years, various newer therapies have been developed for PD treatment that can be categorized as follows: (1) those providing symptomatic relief from the motor and non-motor symptoms; and (2) targeting major causes of PD, they are listed below in Table **4** and Table **5**. To avoid the role of delayed gastric emptying in motor problems, different oral and parenteral formulations have been developed. Furthermore, a therapeutic approach has been made using a vaccine (PD01A) against α-syn protein, where immunization against the oligomeric form of α-syn has shown promising results in clinical trial phase 1 [345, 346].

Table 4. Recently approved drugs for improvement of motor and non-motor symptoms in PD.

Chemical Name / Brand Name	Mode of Action	Phase of Development	Reference
Istradefylline (Nourianz)	Selective A_{2A} receptor antagonist	Approved	Torti *et. al.* [347]

(Table 4) cont.....

Chemical Name / Brand Name	Mode of Action	Phase of Development	Reference
Safinamide (Xadago)	Monoamine oxidase type B (MAO-B) inhibitor	Approved	Bette *et. al.* [348]
Opicapone (Ongentys)	COMT-inhibitor, long-acting, an adjuvant to L-DOPA	Approved	Fabbri *et. al.* [349]
XP066 (Rytary)	Extended-release capsule of levodopa and carbidopa, Restore Dopamine level	Approved	Yao *et. al.* [350]
Melevodopa (Sirio)	Enhanced bioavailability of L-DOPA	Approved	Fasano *et. al.* [351]
Zonisamide (Zonegran)	blockage of sodium and T-type calcium channels, inhibition of carbonic anhydrase, inhibition of glutamate release and modulation of the GABA receptor	Approved	Bermejo *et. al.* [352]

Table 5. Drugs under clinical trial for improvement of motor and non-motor symptoms in PD.

Compound	Indication	Mode of action	Phase of Development	Reference
Nabilone (NCT03773796)	Non-motor symptoms	partial agonist of both Cannabinoid 1 (CB1) and Cannabinoid 2 (CB2)	Phase III	Peball *et. al.* [353]
Exenatide (NCT04232969)	Motor and cognitive improvement	GLP-1 Agonist	Phase III	Athauda *et. al.* [354]
LY03003 (Rotigotine)	Motor and cognitive improvement and sleep	non-ergolinic aminotetralin dopamine agonist	Phase I	Rosa-Grilo *et. al.* [355]
Isradipine (NCT02168842)	Slowed Progression of PD	Dihydropyridine calcium-channel blocker	Phase III	Ilijic *et. al.* [356]
Duloxetine (NCT00437125)	Depression in Parkinson disease	NERI and SSRI	Phase IV	Nishijima *et. al.* [357]
Amantadine NCT01538329	L-DOPA induced Dyskinesia in Early PD	non-competitive NMDA receptor antagonist of the	Phase II	Crosby *et. al.* [358]
Yohimbine HCL (NCT04346394)	Regulation of blood pressure, cognition and mood symptoms	α-2 Adrenergic Blocker	Early Phase I	Rechard *et. al.* [359]

Deep Brain Stimulation

Deep Brain Stimulation (DBS) was first approved in 1997 for tremors only, but later on, in 2002 it was approved for advanced stages of PD. Recently in 2016, it has got approval for the early stage of PD *i.e.*, patients having an onset of disease less than 4 years with motor symptoms.DBS is the most important advancement in the drug discovery of PD therapy after LD.

The principal mechanisms of action of DBS are synaptic inhibition and/or depression, depolarization blockade and stimulation of afferent axons projecting to SNpc [360 - 363]. DBS is a quite massive therapy that includes surgical implantation of electrodes into a fixed brain area mainly associated with movements like subthalamus and globus pallidus internus. The impulse generated from these electrodes improves motor dysfunction but it is limited to tremors [364 - 367]. The pronounced effects of DBS on motor symptoms have been well documented while it is less prominent in non-motor symptoms. There are also reports that inculcate that DBS may improve sleep quality, musculoskeletal pain, urinary symptoms, gastrointestinal symptoms (gastric emptying and constipation), and weight loss and odor discrimination. A recent study has reported a new region with DBS in the nucleus basalis of Meynert, as it is one of the important sources of the neurotransmitter acetylcholine, a crucial player for cognitive functioning [368]. Similar to other brain surgeries, the risk of infections, seizures, stroke and slurred speech has been seen in some cases.

Cell Transplantation

One of the major causes of PD progression is the loss of the dopaminergic neurons in the SNpc region. The idea of transplantation of neural stem cells, pluripotent cells and somatic cells that gets converted to dopaminergic neurons to replace to replenish the loss of dopaminergic neurons during PD is a newer treatment approach (Fig. **5**). For symptomatic relief, the cell must be of human origin, should have the capability of generation and survival of newly generated DAergic neurons and should have the capability of generation of SNpc neurons to recover the lost neuronal population [369 - 371]. Previous findings suggest that implantation of DAergic neurons derived from rodent and human embryonic stem cells have been shown to survive and function for a long time into the striatum of PD rats [372, 373]. Similarly, transplantation of embryonic stem cell-derived DAergic neurons improved survival of the neuronal population in 1-methyl-4-phenyl-1, 2, 3, 6-tetrahydropyridine (MPTP) induced monkey model.

Several open-label clinical trials have been done utilizing cell transplantation and it was found beneficial for the survival and recovery of DAergic neurons for many years [374]. The substantial information of human DAergic neuronal transplantation in the human striatum bilaterally gives a promising response towards improvement in disease progression even after 10 years of implantation [375, 376]. One of the major concerns about the cell transplantation technique is its immune rejection. A clinical study with bilateral putamen transplantation demonstrated a very limited effect on motor function even after 12 months follow up which could be due to decreased survival of the implanted neuronal population [377]. A similar observation was found by Ma *et. al.* and other groups after 4 years and 6-9 months of follow up where they did not find significant improvement in PD symptoms due to graft rejection [378, 379].

The major drawback with the use of the cell transplantation technique is the initiation of dyskinesia. Dyskinesia was first reported in 1996. About 15% of patients have developed dyskinesia after 1 year of transplantation. Similarly, Olanow and colleagues observed about 56% dyskinesia in patients transplanted with embryonic stem cells-derived DAergic neurons [377, 379]. The major factor behind the development of dyskinesia is the graft rejection and heterogeneous graft. This transplantation results in the abnormal production of DA and no relief in PD progression as well [380 - 382]. Therefore, to overcome these situations, transplantation of a sufficient number (≥100000) of cells having purely DAergic neuronal population at suitable regional insertion along with immunosuppressants suggested.

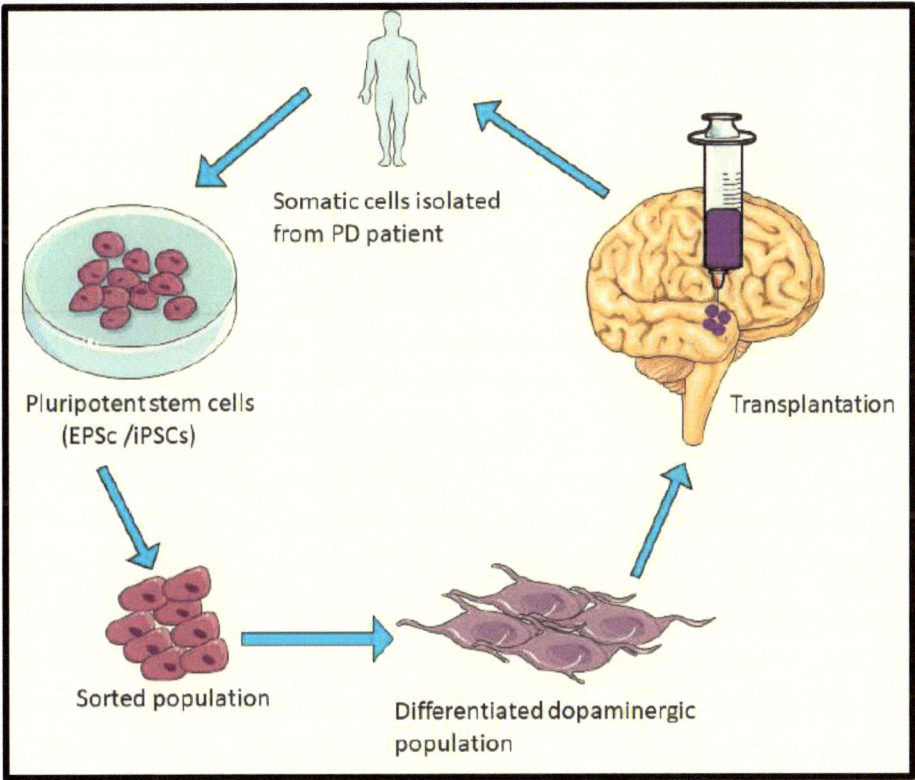

Fig. (5). Road map to induced pluripotent stem cell (iPSC) transplantation in the treatment of PD.

New Insight in Mitochondrial Abnormalities in PD

The role of mitochondria in PD was introduced in 1979 when the administration of MPTP (1-methyl-1,4-phenyl-1,2,3,6-tetrahydropyridine) a neurotoxin and a byproduct of illegitimate synthesis of a heroin analogue, resulted in the development of Parkinson's like phenotype in young drug addicts. MPTP is a lipophilic molecule that easily crosses the blood-brain barrier; it is oxidized by monoamine oxidase to yield MPP$^+$. The structure of MPP$^+$ is similar to dopamine and it binds to a dopamine transporter to be taken up into dopaminergic neurons. Previous studies have shown that the MPP$^+$ induces mitochondrial impairment and inactivates Complex I of the electron transport chain (*etc*) resulting in the development of the PD-like symptoms [320, 383]. Similarly, other neurotoxin and complex 1 inhibitors such as rotenone are used in animal studies to model PD. Besides, the dysfunction of the mitochondrial respiratory chain, altered expression of PD-related dominant and recessive genes such as Parkin, Pink1, the α-syn genes and LRRK2 has also been related to mitochondrial impairment in PD [384, 385]. Recent studies identified that the α-syn gene, a major component of cytoplasmic inclusions (LBs) in surviving DA neurons, is present in the mitochondrial associated membrane that connects mitochondria to the endoplasmic reticulum. Overexpression of α-syn protein (A53T or A30P) induced mitochondrial fragmentation through the cleavage of dynamin-like 120 kDa proteins (OPA1), a negative regulator of mitochondrial fragmentation [386]. Similarly, studies in transgenic mice have shown that pathogenic α-syn (A53T) inhibits the complex 1 activity and generates mitochondrial fragmentation and cell death [387]. Thus, these studies suggest that α-syn has an effect on mitochondria, and besides α-syn, other genes are also involved in mitochondrial impairment in PD. LRRK2 G2019S mutation decreased mitochondrial membrane potential and ATP level but mitochondrial elongation is increased in fibroblasts from PD patients [388]. LRRK2 G2019S mutations also impair mitochondrial fission [389] suggesting that LRRK2 is involved in the regulation of mitochondrial dynamics. Similarly, PINK1, a recessive PD gene product, is involved in mitochondrial functionality, including mitophagy [390], mitochondrial trafficking [391], mitochondrial depolarization and complex 1 activity [392]. PINK1 mutation decreases complex I activity, mitochondrial depolarization and increased sensitivity to apoptotic stress [393]. Interestingly the PINK1 knockout mice have impaired mitochondrial respiratory chain in the striatum and SNpc region but not in the cerebral cortex suggesting specific involvement of PINK1 in the dopaminergic system [394].

Electron Transport Chain (etc) Inhibition: A Determining Factor for PD Pathogenesis

The respiratory chain or the electron transport chain (*etc*) is located in the inner mitochondrial membrane. It comprises five protein complexes: complex I (NADH dehydrogenase) where two electrons are removed from NADH, complex II (succinate dehydrogenase) where electrons are received and ultimately transferred to coenzyme Q which subsequently passes electrons to complex III (cytochrome c reductase). These electrons are then transferred by cytochrome c to complex IV (cytochrome c oxidase), the final step where molecular oxygen is reduced to water. In the *etc*, each complex is more electronegative than the previous one. Electrons are transferred along with these complexes and finally to oxygen molecules that act as the terminal electron acceptor in most aerobic organisms (Fig. **6**). Being a very important pathway, even small defects in any one of the complexes can have devastating and deleterious effects on the cell.

Mitochondrial respiratory chain impairment has been involved in the pathogenesis of PD and other neurodegenerative disorders. Several studies have indicated that mitochondrial complex 1 activity is selectively decreased in the substantia nigra of post-mortem brains of PD patients. Overexpression of α-syn (A53T) protein inhibits the complex 1 activity in SNpc of PD patients. Inhibition of complex 1 activity leads to the degeneration of dopaminergic neurons through different types of factors such as oxidative stress, free radical generation and excitotoxicity. Inhibition of these complex activities is also found in lymphocytes and platelets of PD patients. Human platelets contain a high level of MAO B enzyme and have a number of properties for amine neurotransmitters. Platelets are more easily available for research than PD brain samples or muscle biopsy, so for an early stage of PD, platelets are a first choice to measure the status of pathophysiology such as complex activity, oxidative stress, neuroinflammation. Several studies have reported that in platelets and mitochondrial samples of PD patients, a decreased level of complex I activity is present even if the degree of inhibition is varied in a wide range [395]. Similarly, other respiratory complexes such as complex II [191], Complex III [396] and Complex IV [397] were also altered in PD platelets but a few studies have shown no difference in the complex activity in platelets samples of PD patients [398, 399]. In contrast, an altered expression of complex I, II, III and IV was found in PD muscle [400, 401], whereas other studies have shown no difference of complex activity in muscles [402, 403] as well as in PD lymphocytes [404]. These studies have suggested that other tissues of PD patients have a contradictory result regarding the functioning of the respiratory chain, so there is a further need to explore the mechanism behind the dysfunction of complex activity.

Several epidemiological studies have reported that pesticides and other toxins from the environment that impair mitochondrial respiratory complex activity are involved in the pathogenesis of sporadic PD. MPTP and rotenone both are neurotoxins that inhibit complex I, complex II, complex III and complex IV activity and also decrease ATP activity, increase the ROS production, alter mitochondrial metabolic enzyme, oxidative phosphorylation-related proteins and inner and outer mitochondrial proteins. Some pesticides like paraquat and maneb act as potent redox cyclers which convert the free radicals into superoxide and reactive oxygen species and inhibits the complex III activity (Fig. **6**). Additionally, the specific inhibitor of complex I mimics the biochemical, anatomical and behavioural characteristics of PD in the animal model. Besides ROS generation, reactive nitrogen species (RNS) lead to nitrosative stress resulting in neuronal loss in the substantia nigra region in PD [405]. Reactive nitrogen species are also involved in the inhibition of the mitochondrial respiratory chain. Inhibited complexes I and IV of the mitochondrial electron transport chain generate ROS in the PD model [406]. Several studies have shown that the expression of iNOS and eNOS is increased in the striatum and SNpc regions of the postmortem brain of PD patients [407]. A similar result is also found in the mouse MPTP model of PD [408]. These findings suggest that RNS are also involved in the pathogenesis of PD.

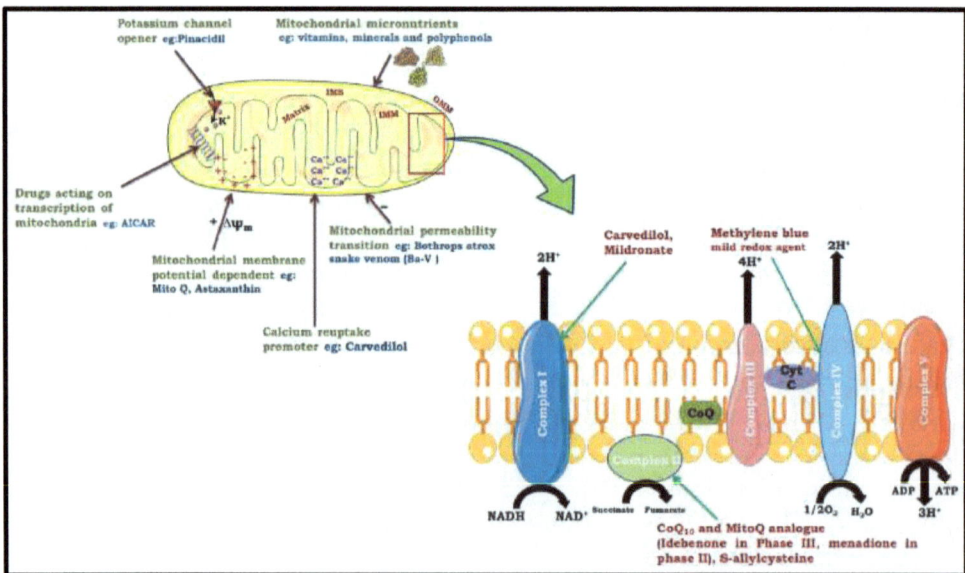

Fig. (6). Drugs targeting mitochondrial respiratory chain complex.

Mitochondria have their own DNA that adds to functional complexity. In Parkinson's disease, mtDNA is also defective and alters the mitochondrial respiratory chain. Defect of mtDNA which encodes 7 of the 49 protein subunits of the complex I enzyme and mutation of POLG1 *i.e.* nuclear-encoded protein imported into mitochondria cause dysfunction of complex I activity, complex III, complex II and complex IV activity [409 - 411] and lead to defective ROS production in PD platelets that are transferable into mitochondrial deficient cybrid cell lines [412, 413]. These defects also increase ROS production and impair mitochondrial calcium buffering [414]. These results suggest that complex I defects in PD are caused by alteration in the mitochondrial genome or in the somatic mtDNA, although mtDNA mutation in PD is rare and maybe a coincidence [415, 416]. However, the abnormalities of mtDNA might contribute to the pathophysiology of PD.

Mitochondria-Targeted Drugs in PD Pathogenesis: From the evidence presented above, it is clear that mitochondrial dysfunction is present and could be one of the main reasons for neuronal cell death in PD patients. Thus, drugs targeting mitochondrial dysfunction and therapies involving the same were tested for treating PD. Electron transport chain bypass strategy is current therapy to treat mitochondrial dysfunction in PD pathogenesis (Table **6**) Several natural compounds, mitochondrial metabolites, bioactive molecules, and pharmaceuticals were reported to bypass the electron transport chain and provide partial protection to mitochondria.

Table 6. Compound targeting Mitochondria Not explored in PD.

Antioxidant	Mechanism of Action	Disease	Clinical Trial
SS31(NCT01572909)	Antioxidant	neuroprotection [417]; insulin resistance; immobilization-induced muscle atrophy [418]; skeletal muscle burn injury [419]	Phase II 1. Advanced Medical Research Center Port Orange, Florida, United States. 2. Henry Ford Hospital Detroit, Michigan, United States. 3. Creighton Cardiac Center Omaha, Nebraska, United States.

(Table 6) cont.....

Antioxidant	Mechanism of Action	Disease	Clinical Trial
			4. Universitätsmedizin Berlin, Charité Campus Benjamin Franklin Berlin, Germany. **5.** Staedtische Kliniken Bielefeld Bielefeld, Germany and +25 more **Sponsored by 1.** Stealth BioTherapeutics Inc. **2.** ICON Clinical Research
MitoTEMPO	Antioxidant	Hypertension [420]	-
MitoSNO	S-Nitrosation	Cardiac IR injury [421]	-
RP-103(NCT02473445)	Increases intracellular glutathione levels by increasing cysteine availability; Leigh Syndrome	Inherited Mitochondrial Disease; Leigh Syndrome	Phase II **1.** University of California at San Diego (UCSD) **2.** University of California at San Diego (UCSD) **3.** Akron Children's Hospital **4.** Baylor College of Medicine University of Utah, Division of Medical Genetics **Sponsored by** Horizon Pharma USA, Inc
Permeability Transition Pore Complex Acting	**Mechanism of Action**	**Disease**	**Clinical Trial**
Sanglifehrin A	Binds on IMPDH2	Ischemia [422]	-

(Table 6) cont.....

Antioxidant	Mechanism of Action	Disease	Clinical Trial
Elamipretide (MTP-131) (NCT02976038)	Prevent oxidation of cardiolipin	Mitochondrial Myopathy [423]	Phase I **1.** University of California **2.** Massachusetts General Hospital **3.** Akron Children's Hospital **4.** Children's Hospital of Pittsburg of UPMC **Sponsored by** Stealth BioTherapeutics Inc.
omaveloxolone (RTA 408) (NCT02255422)	Activates Nrf 2	Mitochondrial Myopathy	Phase II **1.** UCLA California USA **2.** Mass General Hospital Massachusetts USA **3.** Akron Children's Hospital Ohio, USA **4.** The Children's Hospital of Philadelphia Pennsylvania, USA **5.** University of Pittsburgh Pennsylvania USA and +4 more **Sponsored By-1.**Reata Pharmaceuticals, Inc.
Involved in Fatty Acid Oxidation	**Mechanism of Action**	**Disease**	**Clinical Trial**
Ranolazine	Beta oxidation	Angina; *In vitro* study in Astrocytes culture [424, 425] Clinical trial for diabetic neuropathic pain	**2.** AbbVie Phase IV trial terminated **1.** Cardiology Associates, Alabama, USA. **2.** Cardiovascular Institute of South Louisiana, USA. Cardiovascular Institute of South Louisiana, USA **Sponsored By-** **1.** Horizons International **2.** Peripheral Group Gilead Sciences
Aminobutyric acid	Beta oxidation	Diabetes [426]	-
Involved in Glucose Metabolism	**Mechanism of Action**	**Disease**	**Clinical Trial**

(Table 6) cont.....

Antioxidant	Mechanism of Action	Disease	Clinical Trial
Dichloroacetate (NCT03356457)	stimulates pyruvate dehydrogenase	Cardiac myopathy [427, 428] Type I Diabetes	Early Phase I trial Yale New Haven Hospital **Sponsor**-Yale University
ELQ-300 derivative of Atovaquone	Cytochrome *bc*1	Malaria [429]	-
Acipimox (NCT02796950)	Elevate levels of NAD⁺	Type II Diabetes	Phase III University Hospital of Aarhus, Denmark

CONCLUSION

The currently available therapies are focused mainly on restoring the dopamine level in the striatum region, subsequently reducing the motor impairment in PD. A battery of evidence is there which suggests the lack of suitable treatment for non-motor abnormality. All these available treatments, including exogenous and endogenous drugs treatments, have an only a moderate effect and produce side effects if used for a longer duration. Similarly, there is no standard dosage regimen for PD treatment and all the treatment schedule is decided accordingly.

Here, we have compiled the information on some of the newer drugs and approaches towards the development of PD treatment along with earlier available therapies. After exhaustive research, we could identify Safinamide, Istradefylline and XP066 *etc.* as recently approved drugs for the treatment of PD, while others such as Nabilone, Exenatide, and Isradipine *etc.* are under different phases of clinical trials. They all act either by selectively antagonizing the A2A receptor, Monoamine oxidase type B (MAO-B) inhibition or activating both Cannabinoid 1 (CB1) and Cannabinoid 2 (CB2) and GLP-1. Further, cell transplantation and DBS have been the recently approved technique for PD therapy. Although various therapies and techniques are being developed, this still necessitates finding novel CNS active lead molecules for the treatment of PD.

CONSENT FOR PUBLICATION

Not applicable.

CONFLICT OF INTEREST

The authors declare no conflict of interest, financial or otherwise.

ACKNOWLEDGEMENTS

The authors are thankful to the Director, CSIR-CDRI, for continuous encouragement and support. Virendra Tiwari, Akanksha Mishra and Parul acknowledge CSIR, Govt. of India, for providing Doctoral research fellowships for adequate furtherance of work. Pratibha Tripathi is supported by a Post-Doctoral Fellowship from SERB-DST, Govt. of India. Sonu Singh is supported by a Doctoral research fellowship from ICMR, Govt. of India. CDRI communication number for this article is 10295.

REFERENCES

[1]   Kalia LV, Lang AE. Parkinson's disease. Lancet 2015; 386(9996): 896-912.
       [http://dx.doi.org/10.1016/S0140-6736(14)61393-3] [PMID: 25904081]

[2]   Tysnes OB, Storstein A. Epidemiology of Parkinson's disease. J Neural Transm (Vienna) 2017; 124(8): 901-5.
       [http://dx.doi.org/10.1007/s00702-017-1686-y] [PMID: 28150045]

[3]   Zhu MY, Raza MU, Zhan Y, Fan Y. Norepinephrine upregulates the expression of tyrosine hydroxylase and protects dopaminegic neurons against 6-hydrodopamine toxicity. Neurochem Int 2019; 131: 104549.
       [http://dx.doi.org/10.1016/j.neuint.2019.104549] [PMID: 31539561]

[4]   Schapira AH, Bezard E, Brotchie J, *et al.* Novel pharmacological targets for the treatment of Parkinson's disease. Nat Rev Drug Discov 2006; 5(10): 845-54.
       [http://dx.doi.org/10.1038/nrd2087] [PMID: 17016425]

[5]   Schneider RB, Iourinets J, Richard IH. Parkinson's disease psychosis: presentation, diagnosis and management. Neurodegener Dis Manag 2017; 7(6): 365-76.
       [http://dx.doi.org/10.2217/nmt-2017-0028] [PMID: 29160144]

[6]   Singh S, Dikshit M. Apoptotic neuronal death in Parkinson's disease: involvement of nitric oxide. Brain Res Brain Res Rev 2007; 54(2): 233-50.
       [http://dx.doi.org/10.1016/j.brainresrev.2007.02.001] [PMID: 17408564]

[7]   Maiti P, Manna J, Dunbar GL. Current understanding of the molecular mechanisms in Parkinson's disease: Targets for potential treatments. Transl Neurodegener 2017; 6: 28.
       [http://dx.doi.org/10.1186/s40035-017-0099-z] [PMID: 29090092]

[8]   Cannon JR, Greenamyre JT. Neurotoxic *in vivo* models of Parkinson's disease recent advances. Prog Brain Res 2010; 184: 17-33.
       [http://dx.doi.org/10.1016/S0079-6123(10)84002-6] [PMID: 20887868]

[9]   Srivastav S, Fatima M, Mondal AC. Important medicinal herbs in Parkinson's disease pharmacotherapy. Biomed Pharmacother 2017; 92: 856-63.
       [http://dx.doi.org/10.1016/j.biopha.2017.05.137] [PMID: 28599249]

[10]   Celesia GG, Wanamaker WM. L-dopa-carbidopa: combined therapy for the treatment of Parkinson's disease. Dis Nerv Syst 1976; 37(3): 123-5.
       [PMID: 1253661]

[11]   Yuan H, Zhang ZW, Liang LW, *et al.* Treatment strategies for Parkinson's disease. Neurosci Bull 2010; 26(1): 66-76.
       [http://dx.doi.org/10.1007/s12264-010-0302-z] [PMID: 20101274]

[12]   Waters C. Catechol-O-methyltransferase (COMT) inhibitors in Parkinson's disease. J Am Geriatr Soc 2000; 48(6): 692-8.

[http://dx.doi.org/10.1111/j.1532-5415.2000.tb04732.x] [PMID: 10855610]

[13]   Finberg JPM. Inhibitors of MAO-B and COMT: their effects on brain dopamine levels and uses in Parkinson's disease. J Neural Transm (Vienna) 2019; 126(4): 433-48.
[http://dx.doi.org/10.1007/s00702-018-1952-7] [PMID: 30386930]

[14]   Factor SA. Dopamine agonists. Med Clin North Am 1999; 83(2): 415-443, vi-vii.
[http://dx.doi.org/10.1016/S0025-7125(05)70112-7] [PMID: 10093586]

[15]   Raz A, Lev N, Orbach-Zinger S, Djaldetti R. Safety of perioperative treatment with intravenous amantadine in patients with Parkinson disease. Clin Neuropharmacol 2013; 36(5): 166-9.
[http://dx.doi.org/10.1097/WNF.0b013e31829bd066] [PMID: 24045608]

[16]   Thomas A, Bonanni L, Gambi F, Di Iorio A, Onofrj M. Pathological gambling in Parkinson disease is reduced by amantadine. Ann Neurol 2010; 68(3): 400-4.
[http://dx.doi.org/10.1002/ana.22029] [PMID: 20687121]

[17]   More SV, Kumar H, Kang SM, Song SY, Lee K, Choi DK. Advances in neuroprotective ingredients of medicinal herbs by using cellular and animal models of Parkinson's disease. Evid Based Complement Alternat Med 2013; 2013: 957875.
[http://dx.doi.org/10.1155/2013/957875] [PMID: 24073012]

[18]   Nam SM, Choi JH, Yoo DY, *et al.* Effects of curcumin (Curcuma longa) on learning and spatial memory as well as cell proliferation and neuroblast differentiation in adult and aged mice by upregulating brain-derived neurotrophic factor and CREB signaling. J Med Food 2014; 17(6): 641-9.
[http://dx.doi.org/10.1089/jmf.2013.2965] [PMID: 24712702]

[19]   Tanaka K, Galduróz RF, Gobbi LT, Galduróz JC. Ginkgo biloba extract in an animal model of Parkinson's disease: a systematic review. Curr Neuropharmacol 2013; 11(4): 430-5.
[http://dx.doi.org/10.2174/1570159X11311040006] [PMID: 24381532]

[20]   Ittiyavirah SP, R R. Effect of hydro-alcoholic root extract of Plumbago zeylanica l alone and its combination with aqueous leaf extract of Camellia sinensis on haloperidol induced parkinsonism in wistar rats. Ann Neurosci 2014; 21(2): 47-50.
[http://dx.doi.org/10.5214/ans.0972.7531.210204] [PMID: 25206060]

[21]   Pervin M, Unno K, Ohishi T, Tanabe H, Miyoshi N, Nakamura Y. Beneficial Effects of Green Tea Catechins on Neurodegenerative Diseases. Molecules 2018; 23(6): E1297.
[http://dx.doi.org/10.3390/molecules23061297] [PMID: 29843466]

[22]   Song JX, Sze SC, Ng TB, *et al.* Anti-Parkinsonian drug discovery from herbal medicines: what have we got from neurotoxic models? J Ethnopharmacol 2012; 139(3): 698-711.
[http://dx.doi.org/10.1016/j.jep.2011.12.030] [PMID: 22212501]

[23]   Manyam BV, Dhanasekaran M, Hare TA. Neuroprotective effects of the antiparkinson drug Mucuna pruriens. Phytother Res 2004; 18(9): 706-12.
[http://dx.doi.org/10.1002/ptr.1514] [PMID: 15478206]

[24]   Choi HS, Kim HJ, Oh JH, *et al.* Therapeutic potentials of human adipose-derived stem cells on the mouse model of Parkinson's disease. Neurobiol Aging 2015; 36(10): 2885-92.
[http://dx.doi.org/10.1016/j.neurobiolaging.2015.06.022] [PMID: 26242706]

[25]   Blanchard V, Raisman-Vozari R, Vyas S, *et al.* Differential expression of tyrosine hydroxylase and membrane dopamine transporter genes in subpopulations of dopaminergic neurons of the rat mesencephalon. Brain Res Mol Brain Res 1994; 22(1-4): 29-38.
[http://dx.doi.org/10.1016/0169-328X(94)90029-9] [PMID: 7912404]

[26]   Hare DJ, Lei P, Ayton S, Roberts BR, Grimm R, George JL, *et al.* An iron–dopamine index predicts risk of parkinsonian neurodegeneration in the substantia nigra pars compacta. Chem Sci (Camb) 2014; 5(6): 2160-9.
[http://dx.doi.org/10.1039/C3SC53461H]

[27]   Burkett JP, Young LJ. The behavioral, anatomical and pharmacological parallels between social

attachment, love and addiction. Psychopharmacology (Berl) 2012; 224(1): 1-26.
[http://dx.doi.org/10.1007/s00213-012-2794-x] [PMID: 22885871]

[28]   Jankovic J, Chen S, Le WD. The role of Nurr1 in the development of dopaminergic neurons and Parkinson's disease. Prog Neurobiol 2005; 77(1-2): 128-38.
[http://dx.doi.org/10.1016/j.pneurobio.2005.09.001] [PMID: 16243425]

[29]   Kadkhodaei B, Ito T, Joodmardi E, *et al.* Nurr1 is required for maintenance of maturing and adult midbrain dopamine neurons. J Neurosci 2009; 29(50): 15923-32.
[http://dx.doi.org/10.1523/JNEUROSCI.3910-09.2009] [PMID: 20016108]

[30]   Martinat C, Bacci J-J, Leete T, *et al.* Cooperative transcription activation by Nurr1 and Pitx3 induces embryonic stem cell maturation to the midbrain dopamine neuron phenotype. Proc Natl Acad Sci USA 2006; 103(8): 2874-9.
[http://dx.doi.org/10.1073/pnas.0511153103] [PMID: 16477036]

[31]   Blesa J, Trigo-Damas I, Quiroga-Varela A, Jackson-Lewis VR. Oxidative stress and Parkinson's disease. Front Neuroanat 2015; 9: 91.
[http://dx.doi.org/10.3389/fnana.2015.00091] [PMID: 26217195]

[32]   McNaught KSP, Olanow CW. Proteolytic stress: a unifying concept for the etiopathogenesis of Parkinson's disease. Ann Neurol 2003; 53(S3) (Suppl. 3): S73-84.
[http://dx.doi.org/10.1002/ana.10512] [PMID: 12666100]

[33]   Sulzer D. Multiple hit hypotheses for dopamine neuron loss in Parkinson's disease. Trends Neurosci 2007; 30(5): 244-50.
[http://dx.doi.org/10.1016/j.tins.2007.03.009] [PMID: 17418429]

[34]   Chinta SJ, Andersen JK. Dopaminergic neurons. Int J Biochem Cell Biol 2005; 37(5): 942-6.
[http://dx.doi.org/10.1016/j.biocel.2004.09.009] [PMID: 15743669]

[35]   Sillitoe RV, Vogel MW. Desire, disease, and the origins of the dopaminergic system. Schizophr Bull 2008; 34(2): 212-9.
[http://dx.doi.org/10.1093/schbul/sbm170] [PMID: 18283047]

[36]   Henny P, Brown MTC, Northrop A, *et al.* Structural correlates of heterogeneous *in vivo* activity of midbrain dopaminergic neurons. Nat Neurosci 2012; 15(4): 613-9.
[http://dx.doi.org/10.1038/nn.3048] [PMID: 22327472]

[37]   Margolis EB, Lock H, Hjelmstad GO, Fields HL. The ventral tegmental area revisited: is there an electrophysiological marker for dopaminergic neurons? J Physiol 2006; 577(Pt 3): 907-24.
[http://dx.doi.org/10.1113/jphysiol.2006.117069] [PMID: 16959856]

[38]   Roberts KM, Fitzpatrick PF. Mechanisms of tryptophan and tyrosine hydroxylase. IUBMB Life 2013; 65(4): 350-7.
[http://dx.doi.org/10.1002/iub.1144] [PMID: 23441081]

[39]   Thuret S, Bhatt L, O'Leary DDM, Simon HH. Identification and developmental analysis of genes expressed by dopaminergic neurons of the substantia nigra pars compacta. Mol Cell Neurosci 2004; 25(3): 394-405.
[http://dx.doi.org/10.1016/j.mcn.2003.11.004] [PMID: 15033168]

[40]   Meredith GE, Kang UJ. Behavioral models of Parkinson's disease in rodents: a new look at an old problem. Mov Disord 2006; 21(10): 1595-606.
[http://dx.doi.org/10.1002/mds.21010] [PMID: 16830310]

[41]   Hosp JA, Pekanovic A, Rioult-Pedotti MS, Luft AR. Dopaminergic projections from midbrain to primary motor cortex mediate motor skill learning. J Neurosci 2011; 31(7): 2481-7.
[http://dx.doi.org/10.1523/JNEUROSCI.5411-10.2011] [PMID: 21325515]

[42]   Molina-Luna K, Pekanovic A, Röhrich S, *et al.* Dopamine in motor cortex is necessary for skill learning and synaptic plasticity. PLoS One 2009; 4(9): e7082.
[http://dx.doi.org/10.1371/journal.pone.0007082] [PMID: 19759902]

[43]   Grace AA, Bunney BS, Moore H, Todd CL. Dopamine-cell depolarization block as a model for the therapeutic actions of antipsychotic drugs. Trends Neurosci 1997; 20(1): 31-7.
[http://dx.doi.org/10.1016/S0166-2236(96)10064-3] [PMID: 9004417]

[44]   Ljungberg T, Ungerstedt U. A rapid and simple behavioural screening method for simultaneous assessment of limbic and striatal blocking effects of neuroleptic drugs. Pharmacol Biochem Behav 1985; 23(3): 479-85.
[http://dx.doi.org/10.1016/0091-3057(85)90025-5] [PMID: 2864704]

[45]   Beckstead RM, Domesick VB, Nauta WJH. Efferent connections of the substantia nigra and ventral tegmental area in the rat. Brain Res 1979; 175(2): 191-217.
[http://dx.doi.org/10.1016/0006-8993(79)91001-1] [PMID: 314832]

[46]   Uhl GR, Hedreen JC, Price DL. Parkinson's disease: loss of neurons from the ventral tegmental area contralateral to therapeutic surgical lesions. Neurology 1985; 35(8): 1215-8.
[http://dx.doi.org/10.1212/WNL.35.8.1215] [PMID: 4022359]

[47]   Ikemoto S, Panksepp J. The role of nucleus accumbens dopamine in motivated behavior: a unifying interpretation with special reference to reward-seeking. Brain Res Brain Res Rev 1999; 31(1): 6-41.
[http://dx.doi.org/10.1016/S0165-0173(99)00023-5] [PMID: 10611493]

[48]   Yim CY, Mogenson GJ. Response of nucleus accumbens neurons to amygdala stimulation and its modification by dopamine. Brain Res 1982; 239(2): 401-15.
[http://dx.doi.org/10.1016/0006-8993(82)90518-2] [PMID: 6284305]

[49]   Ayano G. Dopamine: receptors, functions, synthesis, pathways, locations and mental disorders: review of literatures. J Ment Disord Treat 2016; 2(120): 2.
[http://dx.doi.org/10.4172/2471-271X.1000120]

[50]   Salimpoor VN, Benovoy M, Larcher K, Dagher A, Zatorre RJ. Anatomically distinct dopamine release during anticipation and experience of peak emotion to music. Nat Neurosci 2011; 14(2): 257-62.
[http://dx.doi.org/10.1038/nn.2726] [PMID: 21217764]

[51]   Salamone JD, Correa M, Mingote SM, Weber SM. Beyond the reward hypothesis: alternative functions of nucleus accumbens dopamine. Curr Opin Pharmacol 2005; 5(1): 34-41.
[http://dx.doi.org/10.1016/j.coph.2004.09.004] [PMID: 15661623]

[52]   Gurden H, Tassin JP, Jay TM. Integrity of the mesocortical dopaminergic system is necessary for complete expression of *in vivo* hippocampal-prefrontal cortex long-term potentiation. Neuroscience 1999; 94(4): 1019-27.
[http://dx.doi.org/10.1016/S0306-4522(99)00395-4] [PMID: 10625044]

[53]   Stam CJ, de Bruin JPC, van Haelst AM, van der Gugten J, Kalsbeek A. Influence of the mesocortical dopaminergic system on activity, food hoarding, social-agonistic behavior, and spatial delayed alternation in male rats. Behav Neurosci 1989; 103(1): 24-35.
[http://dx.doi.org/10.1037/0735-7044.103.1.24] [PMID: 2923675]

[54]   Lozoff B. Early iron deficiency has brain and behavior effects consistent with dopaminergic dysfunction. J Nutr 2011; 141(4): 740S-6S.
[http://dx.doi.org/10.3945/jn.110.131169] [PMID: 21346104]

[55]   Costa A, Peppe A, Dell'Agnello G, Caltagirone C, Carlesimo GA. Dopamine and cognitive functioning in de novo subjects with Parkinson's disease: effects of pramipexole and pergolide on working memory. Neuropsychologia 2009; 47(5): 1374-81.
[http://dx.doi.org/10.1016/j.neuropsychologia.2009.01.039] [PMID: 19428401]

[56]   Duka T, Lupp A. The effects of incentive on antisaccades: is a dopaminergic mechanism involved? Behav Pharmacol 1997; 8(5): 373-82.
[http://dx.doi.org/10.1097/00008877-199710000-00001] [PMID: 9832976]

[57]   Annunziato L. Regulation of the tuberoinfundibular and nigrostriatal systems. Evidence for different kinds of dopaminergic neurons in the brain. Neuroendocrinology 1979; 29(1): 66-76.

[http://dx.doi.org/10.1159/000122906] [PMID: 471198]

[58]   Gudelsky GA. Tuberoinfundibular dopamine neurons and the regulation of prolactin secretion. Psychoneuroendocrinology 1981; 6(1): 3-16.
[http://dx.doi.org/10.1016/0306-4530(81)90044-5] [PMID: 7017786]

[59]   Boecker H, Jankowski J, Ditter P, Scheef L. A role of the basal ganglia and midbrain nuclei for initiation of motor sequences. Neuroimage 2008; 39(3): 1356-69.
[http://dx.doi.org/10.1016/j.neuroimage.2007.09.069] [PMID: 18024158]

[60]   Mink JW. The basal ganglia: focused selection and inhibition of competing motor programs. Prog Neurobiol 1996; 50(4): 381-425.
[http://dx.doi.org/10.1016/S0301-0082(96)00042-1] [PMID: 9004351]

[61]   Calabresi P, Picconi B, Tozzi A, Ghiglieri V, Di Filippo M. Direct and indirect pathways of basal ganglia: a critical reappraisal. Nat Neurosci 2014; 17(8): 1022-30.
[http://dx.doi.org/10.1038/nn.3743] [PMID: 25065439]

[62]   Smith Y, Bevan MD, Shink E, Bolam JP. Microcircuitry of the direct and indirect pathways of the basal ganglia. Neuroscience 1998; 86(2): 353-87.
[PMID: 9881853]

[63]   Reiner A, Medina L, Haber SN. The distribution of dynorphinergic terminals in striatal target regions in comparison to the distribution of substance P-containing and enkephalinergic terminals in monkeys and humans. Neuroscience 1999; 88(3): 775-93.
[http://dx.doi.org/10.1016/S0306-4522(98)00254-1] [PMID: 10363817]

[64]   Van Bockstaele EJ, Gracy KN, Pickel VM. Dynorphin-immunoreactive neurons in the rat nucleus accumbens: ultrastructure and synaptic input from terminals containing substance P and/or dynorphin. J Comp Neurol 1995; 351(1): 117-33.
[http://dx.doi.org/10.1002/cne.903510111] [PMID: 7534773]

[65]   Filion M, Tremblay L. Abnormal spontaneous activity of globus pallidus neurons in monkeys with MPTP-induced parkinsonism. Brain Res 1991; 547(1): 142-51.
[http://dx.doi.org/10.1016/0006-8993(91)90585-J] [PMID: 1677607]

[66]   Filion M, Tremblay L, Bédard PJ. Effects of dopamine agonists on the spontaneous activity of globus pallidus neurons in monkeys with MPTP-induced parkinsonism. Brain Res 1991; 547(1): 152-61.
[http://dx.doi.org/10.1016/0006-8993(91)90586-K] [PMID: 1677608]

[67]   DeLong MR. Primate models of movement disorders of basal ganglia origin. Trends Neurosci 1990; 13(7): 281-5.
[http://dx.doi.org/10.1016/0166-2236(90)90110-V] [PMID: 1695404]

[68]   Deutch AY. Prefrontal cortical dopamine systems and the elaboration of functional corticostriatal circuits: implications for schizophrenia and Parkinson's disease 1993; 91(2-3): 197-221.
[http://dx.doi.org/10.1007/BF01245232]

[69]   Tarazi FI, Campbell A, Yeghiayan SK, Baldessarini RJ. Localization of dopamine receptor subtypes in corpus striatum and nucleus accumbens septi of rat brain: comparison of D1-, D2-, and D4-like receptors. Neuroscience 1998; 83(1): 169-76.
[http://dx.doi.org/10.1016/S0306-4522(97)00386-2] [PMID: 9466407]

[70]   Mesulam MM, Mash D, Hersh L, Bothwell M, Geula C. Cholinergic innervation of the human striatum, globus pallidus, subthalamic nucleus, substantia nigra, and red nucleus. J Comp Neurol 1992; 323(2): 252-68.
[http://dx.doi.org/10.1002/cne.903230209] [PMID: 1401259]

[71]   Van Domburg PHMF, ten Donkelaar HJ. The human substantia nigra and ventral tegmental area. The Human Substantia Nigra and Ventral Tegmental Area. Springer 1991; pp. 32-69.
[http://dx.doi.org/10.1007/978-3-642-75846-1_4]

[72]   Haber SN. The place of dopamine in the cortico-basal ganglia circuit. Neuroscience 2014; 282: 248-

57.
[http://dx.doi.org/10.1016/j.neuroscience.2014.10.008] [PMID: 25445194]

[73]   Pérez-Neri I, Méndez-Sánchez I, Montes S, Ríos C. Acute dehydroepiandrosterone treatment exerts different effects on dopamine and serotonin turnover ratios in the rat corpus striatum and nucleus accumbens. Prog Neuropsychopharmacol Biol Psychiatry 2008; 32(6): 1584-9.
[http://dx.doi.org/10.1016/j.pnpbp.2008.06.002] [PMID: 18585426]

[74]   Hornykiewicz O. Biochemical aspects of Parkinson's disease. Neurology 1998; 51(2) (Suppl. 2): S2-9.
[http://dx.doi.org/10.1212/WNL.51.2_Suppl_2.S2] [PMID: 9711973]

[75]   Hurley MJ, Mash DC, Jenner P. Markers for dopaminergic neurotransmission in the cerebellum in normal individuals and patients with Parkinson's disease examined by RT-PCR. Eur J Neurosci 2003; 18(9): 2668-72.
[http://dx.doi.org/10.1046/j.1460-9568.2003.02963.x] [PMID: 14622169]

[76]   O'Donnell P, Greene J, Pabello N, Lewis BL, Grace AA. Modulation of cell firing in the nucleus accumbens. Ann N Y Acad Sci 1999; 877(1): 157-75.
[http://dx.doi.org/10.1111/j.1749-6632.1999.tb09267.x] [PMID: 10415649]

[77]   Bartels AL, Leenders KL. Parkinson's disease: the syndrome, the pathogenesis and pathophysiology. Cortex 2009; 45(8): 915-21.
[http://dx.doi.org/10.1016/j.cortex.2008.11.010] [PMID: 19095226]

[78]   Toulorge D, Schapira AHV, Hajj R. Molecular changes in the postmortem parkinsonian brain. J Neurochem 2016; 139 (Suppl. 1): 27-58.
[http://dx.doi.org/10.1111/jnc.13696] [PMID: 27381749]

[79]   Aarsland D, Creese B, Politis M, *et al.* Cognitive decline in Parkinson disease. Nat Rev Neurol 2017; 13(4): 217-31.
[http://dx.doi.org/10.1038/nrneurol.2017.27] [PMID: 28257128]

[80]   Conte A, Khan N, Defazio G, Rothwell JC, Berardelli A. Pathophysiology of somatosensory abnormalities in Parkinson disease. Nat Rev Neurol 2013; 9(12): 687-97.
[http://dx.doi.org/10.1038/nrneurol.2013.224] [PMID: 24217516]

[81]   Huot P, Fox SH. The serotonergic system in motor and non-motor manifestations of Parkinson's disease. Exp Brain Res 2013; 230(4): 463-76.
[http://dx.doi.org/10.1007/s00221-013-3621-2] [PMID: 23811734]

[82]   Politis M, Niccolini F. Serotonin in Parkinson's disease. Behav Brain Res 2015; 277: 136-45.
[http://dx.doi.org/10.1016/j.bbr.2014.07.037] [PMID: 25086269]

[83]   Pagano G, Politis M. Molecular imaging of the serotonergic system in Parkinson's disease. International review of neurobiology 141. Elsevier 2018; pp. 173-210.

[84]   Nayyar T, Bubser M, Ferguson MC, *et al.* Cortical serotonin and norepinephrine denervation in parkinsonism: preferential loss of the beaded serotonin innervation. Eur J Neurosci 2009; 30(2): 207-16.
[http://dx.doi.org/10.1111/j.1460-9568.2009.06806.x] [PMID: 19659923]

[85]   Birtwistle J, Baldwin D. Role of dopamine in schizophrenia and Parkinson's disease. Br J Nurs 1998; 7(14): 832-834, 836, 838-841.
[http://dx.doi.org/10.12968/bjon.1998.7.14.5636] [PMID: 9849144]

[86]   Sharp T, Cowen PJ. 5-HT and depression: is the glass half-full? Curr Opin Pharmacol 2011; 11(1): 45-51.
[http://dx.doi.org/10.1016/j.coph.2011.02.003] [PMID: 21377932]

[87]   Asbreuk CHJ, van Schaick HSA, Cox JJ, Kromkamp M, Smidt MP, Burbach JPH. The homeobox genes Lhx7 and Gbx1 are expressed in the basal forebrain cholinergic system. Neuroscience 2002; 109(2): 287-98.
[http://dx.doi.org/10.1016/S0306-4522(01)00466-3] [PMID: 11801365]

[88]   Bohnen NI, Albin RL. The cholinergic system and Parkinson disease. Behav Brain Res 2011; 221(2): 564-73.
       [http://dx.doi.org/10.1016/j.bbr.2009.12.048] [PMID: 20060022]

[89]   Hurley MJ, Brandon B, Gentleman SM, Dexter DT. Parkinson's disease is associated with altered expression of CaV1 channels and calcium-binding proteins. Brain 2013; 136(Pt 7): 2077-97.
       [http://dx.doi.org/10.1093/brain/awt134] [PMID: 23771339]

[90]   Amenta F, Ricci A, Tayebati SK, Zaccheo D. The peripheral dopaminergic system: morphological analysis, functional and clinical applications. Ital J Anat Embryol 2002; 107(3): 145-67.
       [PMID: 12437142]

[91]   Missale C, Nash SR, Robinson SW, Jaber M, Caron MG. Dopamine receptors: from structure to function. Physiol Rev 1998; 78(1): 189-225.
       [http://dx.doi.org/10.1152/physrev.1998.78.1.189] [PMID: 9457173]

[92]   Sibley DR, Monsma FJ Jr. Molecular biology of dopamine receptors. Trends Pharmacol Sci 1992; 13(2): 61-9.
       [http://dx.doi.org/10.1016/0165-6147(92)90025-2] [PMID: 1561715]

[93]   Marsden CA. Dopamine: the rewarding years. Br J Pharmacol 2006; 147(S1) (Suppl. 1): S136-44.
       [http://dx.doi.org/10.1038/sj.bjp.0706473] [PMID: 16402097]

[94]   Meiser J, Weindl D, Hiller K. Complexity of dopamine metabolism. Cell Commun Signal 2013; 11(1): 34.
       [http://dx.doi.org/10.1186/1478-811X-11-34] [PMID: 23683503]

[95]   Kopin IJ, Eisenhofer G, Goldstein D. Sympathoadrenal medullary system and stress. Mechanisms of physical and emotional stress. Springer 1988; pp. 11-23.
       [http://dx.doi.org/10.1007/978-1-4899-2064-5_2]

[96]   Flatmark T. Catecholamine biosynthesis and physiological regulation in neuroendocrine cells. Acta Physiol Scand 2000; 168(1): 1-17.
       [http://dx.doi.org/10.1046/j.1365-201x.2000.00596.x] [PMID: 10691773]

[97]   Hyland K, Gunasekara RS, Munk-Martin TL, Arnold LA, Engle T. The hph-1 mouse: a model for dominantly inherited GTP-cyclohydrolase deficiency. Ann Neurol 2003; 54(S6) (Suppl. 6): S46-8.
       [http://dx.doi.org/10.1002/ana.10695] [PMID: 12891653]

[98]   Lentz SI, Kapatos G. Tetrahydrobiopterin biosynthesis in the rat brain: heterogeneity of GTP cyclohydrolase I mRNA expression in monoamine-containing neurons. Neurochem Int 1996; 28(5-6): 569-82.
       [http://dx.doi.org/10.1016/0197-0186(95)00124-7] [PMID: 8792338]

[99]   Bromek E, Haduch A, Gołembiowska K, Daniel WA. Cytochrome P450 mediates dopamine formation in the brain *in vivo*. J Neurochem 2011; 118(5): 806-15.
       [http://dx.doi.org/10.1111/j.1471-4159.2011.07339.x] [PMID: 21651557]

[100] Brennenstuhl H, Jung-Klawitter S, Assmann B, Opladen T. 2019; 50(01): 002-14.

[101] Weingarten P, Zhou QY. Protection of intracellular dopamine cytotoxicity by dopamine disposition and metabolism factors. J Neurochem 2001; 77(3): 776-85.
       [http://dx.doi.org/10.1046/j.1471-4159.2001.00263.x] [PMID: 11331406]

[102] Dix TA, Kuhn DM, Benkovic SJ. Mechanism of oxygen activation by tyrosine hydroxylase. Biochemistry 1987; 26(12): 3354-61.
       [http://dx.doi.org/10.1021/bi00386a016] [PMID: 2888478]

[103] Urano F, Hayashi N, Arisaka F, Kurita H, Murata S, Ichinose H. Molecular mechanism for pterin-mediated inactivation of tyrosine hydroxylase: formation of insoluble aggregates of tyrosine hydroxylase. J Biochem 2006; 139(4): 625-35.
       [http://dx.doi.org/10.1093/jb/mvj073] [PMID: 16672262]

[104]   Zhu Y, Zhang J, Zeng Y. Overview of tyrosine hydroxylase in Parkinson's disease 2012; 11(4): 350-8.
[http://dx.doi.org/10.2174/187152712800792901]

[105]   O'Neil W. Dopamine Homeostasis and Environmental Risk Factors in a Parkinson's Disease Model 2011.

[106]   Thöny B, Auerbach G, Blau N. Tetrahydrobiopterin biosynthesis, regeneration and functions. Biochem J 2000; 347(Pt 1): 1-16.
[http://dx.doi.org/10.1042/bj3470001] [PMID: 10727395]

[107]   Gershanik OS. Improving L-dopa therapy: the development of enzyme inhibitors. Mov Disord 2015; 30(1): 103-13.
[http://dx.doi.org/10.1002/mds.26050] [PMID: 25335824]

[108]   Chagraoui A, Boulain M, Juvin L, Anouar Y, Barrière G, Deurwaerdère P. L-DOPA in Parkinson's Disease: Looking at the "False" Neurotransmitters and Their Meaning. Int J Mol Sci 2019; 21(1): E294.
[http://dx.doi.org/10.3390/ijms21010294] [PMID: 31906250]

[109]   Opacka-Juffry J, Brooks DJ. L-dihydroxyphenylalanine and its decarboxylase: new ideas on their neuroregulatory roles. Mov Disord 1995; 10(3): 241-9.
[http://dx.doi.org/10.1002/mds.870100302] [PMID: 7651438]

[110]   Monzani E, Nicolis S, Dell'Acqua S, et al. Dopamine, Oxidative Stress and Protein-Quinone Modifications in Parkinson's and Other Neurodegenerative Diseases. Angew Chem Int Ed Engl 2019; 58(20): 6512-27.
[http://dx.doi.org/10.1002/anie.201811122] [PMID: 30536578]

[111]   Xu W, Zhu JP, Angulo JA. Induction of striatal pre- and postsynaptic damage by methamphetamine requires the dopamine receptors. Synapse 2005; 58(2): 110-21.
[http://dx.doi.org/10.1002/syn.20185] [PMID: 16088948]

[112]   Florang VR, Rees JN, Brogden NK, Anderson DG, Hurley TD, Doorn JA. Inhibition of the oxidative metabolism of 3,4-dihydroxyphenylacetaldehyde, a reactive intermediate of dopamine metabolism, by 4-hydroxy-2-nonenal. Neurotoxicology 2007; 28(1): 76-82.
[http://dx.doi.org/10.1016/j.neuro.2006.07.018] [PMID: 16956664]

[113]   Huang SY, Lin WW, Ko HC, et al. Possible interaction of alcohol dehydrogenase and aldehyde dehydrogenase genes with the dopamine D2 receptor gene in anxiety-depressive alcohol dependence. Alcohol Clin Exp Res 2004; 28(3): 374-84.
[http://dx.doi.org/10.1097/01.ALC.0000117832.62901.61] [PMID: 15084894]

[114]   Westerink BH, Korf J. Turnover of acid dopamine metabolites in striatal and mesolimbic tissue of the rat brain. Eur J Pharmacol 1976; 37(2): 249-55.
[http://dx.doi.org/10.1016/0014-2999(76)90032-7] [PMID: 954808]

[115]   Kastner A, Anglade P, Bounaix C, et al. Immunohistochemical study of catechol-O-methyltransferase in the human mesostriatal system. Neuroscience 1994; 62(2): 449-57.
[http://dx.doi.org/10.1016/0306-4522(94)90379-4] [PMID: 7830891]

[116]   Sotomatsu A, Nakano M, Hirai S. Phospholipid peroxidation induced by the catechol-Fe3+(Cu2+) complex: a possible mechanism of nigrostriatal cell damage. Arch Biochem Biophys 1990; 283(2): 334-41.
[http://dx.doi.org/10.1016/0003-9861(90)90651-E] [PMID: 2125819]

[117]   Berman SB, Zigmond MJ, Hastings TG. Modification of dopamine transporter function: effect of reactive oxygen species and dopamine. J Neurochem 1996; 67(2): 593-600.
[http://dx.doi.org/10.1046/j.1471-4159.1996.67020593.x] [PMID: 8764584]

[118]   Siopa F, Pereira AS, Ferreira LM, Matilde Marques M, Branco PS. Synthesis of catecholamine conjugates with nitrogen-centered bionucleophiles. Bioorg Chem 2012; 44: 19-24.
[http://dx.doi.org/10.1016/j.bioorg.2012.05.002] [PMID: 22784829]

[119] Fedorow H, Tribl F, Halliday G, Gerlach M, Riederer P, Double KL. Neuromelanin in human dopamine neurons: comparison with peripheral melanins and relevance to Parkinson's disease. Prog Neurobiol 2005; 75(2): 109-24.
[http://dx.doi.org/10.1016/j.pneurobio.2005.02.001] [PMID: 15784302]

[120] Napolitano A, Manini P, d'Ischia M. Oxidation chemistry of catecholamines and neuronal degeneration: an update. Curr Med Chem 2011; 18(12): 1832-45.
[http://dx.doi.org/10.2174/092986711795496863] [PMID: 21466469]

[121] Zhou ZD, Lim TM. Roles of glutathione (GSH) in dopamine (DA) oxidation studied by improved tandem HPLC plus ESI-MS. Neurochem Res 2009; 34(2): 316-26.
[http://dx.doi.org/10.1007/s11064-008-9778-6] [PMID: 18600447]

[122] Zhou ZD, Lim TM. Glutathione conjugates with dopamine-derived quinones to form reactive or non-reactive glutathione-conjugates. Neurochem Res 2010; 35(11): 1805-18.
[http://dx.doi.org/10.1007/s11064-010-0247-7] [PMID: 20721623]

[123] Lotharius J, Brundin P. Impaired dopamine storage resulting from alpha-synuclein mutations may contribute to the pathogenesis of Parkinson's disease. Hum Mol Genet 2002; 11(20): 2395-407.
[http://dx.doi.org/10.1093/hmg/11.20.2395] [PMID: 12351575]

[124] Ding TT, Lee SJ, Rochet JC, Lansbury PT Jr. Annular alpha-synuclein protofibrils are produced when spherical protofibrils are incubated in solution or bound to brain-derived membranes. Biochemistry 2002; 41(32): 10209-17.
[http://dx.doi.org/10.1021/bi020139h] [PMID: 12162735]

[125] Zecca L, Shima T, Stroppolo A, *et al.* Interaction of neuromelanin and iron in substantia nigra and other areas of human brain. Neuroscience 1996; 73(2): 407-15.
[http://dx.doi.org/10.1016/0306-4522(96)00047-4] [PMID: 8783258]

[126] Zecca L, Wilms H, Geick S, *et al.* Human neuromelanin induces neuroinflammation and neurodegeneration in the rat substantia nigra: implications for Parkinson's disease. Acta Neuropathol 2008; 116(1): 47-55.
[http://dx.doi.org/10.1007/s00401-008-0361-7] [PMID: 18343932]

[127] Viceconte N, Burguillos MA, Herrera AJ, De Pablos RM, Joseph B, Venero JL. Neuromelanin activates proinflammatory microglia through a caspase-8-dependent mechanism. J Neuroinflammation 2015; 12: 5.
[http://dx.doi.org/10.1186/s12974-014-0228-x] [PMID: 25586882]

[128] Jenner P. Oxidative stress in Parkinson's disease. Ann Neurol 2003; 53 (Suppl. 3): S26-36.
[http://dx.doi.org/10.1002/ana.10483] [PMID: 12666096]

[129] Koppula S, Kumar H, More SV, Lim HW, Hong SM, Choi DK. Recent updates in redox regulation and free radical scavenging effects by herbal products in experimental models of Parkinson's disease. Molecules 2012; 17(10): 11391-420.
[http://dx.doi.org/10.3390/molecules171011391] [PMID: 23014498]

[130] Bhat AH, Dar KB, Anees S, *et al.* Oxidative stress, mitochondrial dysfunction and neurodegenerative diseases; a mechanistic insight. Biomed Pharmacother 2015; 74: 101-10.
[http://dx.doi.org/10.1016/j.biopha.2015.07.025] [PMID: 26349970]

[131] de Pablos RM, Herrera AJ, Espinosa-Oliva AM, *et al.* Chronic stress enhances microglia activation and exacerbates death of nigral dopaminergic neurons under conditions of inflammation. J Neuroinflammation 2014; 11: 34.
[http://dx.doi.org/10.1186/1742-2094-11-34] [PMID: 24565378]

[132] Aquilano K, Baldelli S, Rotilio G, Ciriolo MR. Role of nitric oxide synthases in Parkinson's disease: a review on the antioxidant and anti-inflammatory activity of polyphenols. Neurochem Res 2008; 33(12): 2416-26.
[http://dx.doi.org/10.1007/s11064-008-9697-6] [PMID: 18415676]

[133] Zhang W, Phillips K, Wielgus AR, *et al.* Neuromelanin activates microglia and induces degeneration of dopaminergic neurons: implications for progression of Parkinson's disease. Neurotox Res 2011; 19(1): 63-72.
[http://dx.doi.org/10.1007/s12640-009-9140-z] [PMID: 19957214]

[134] Kish SJ, Morito C, Hornykiewicz O. Glutathione peroxidase activity in Parkinson's disease brain. Neurosci Lett 1985; 58(3): 343-6.
[http://dx.doi.org/10.1016/0304-3940(85)90078-3] [PMID: 4047494]

[135] Leenders KL, Palmer AJ, Quinn N, *et al.* Brain dopamine metabolism in patients with Parkinson's disease measured with positron emission tomography. J Neurol Neurosurg Psychiatry 1986; 49(8): 853-60.
[http://dx.doi.org/10.1136/jnnp.49.8.853] [PMID: 3091770]

[136] Kish SJ, Shannak K, Hornykiewicz O. Uneven pattern of dopamine loss in the striatum of patients with idiopathic Parkinson's disease. Pathophysiologic and clinical implications. N Engl J Med 1988; 318(14): 876-80.
[http://dx.doi.org/10.1056/NEJM198804073181402] [PMID: 3352672]

[137] Bernheimer H, Birkmayer W, Hornykiewicz O, Jellinger K, Seitelberger F. Brain dopamine and the syndromes of Parkinson and Huntington. Clinical, morphological and neurochemical correlations. J Neurol Sci 1973; 20(4): 415-55.
[http://dx.doi.org/10.1016/0022-510X(73)90175-5] [PMID: 4272516]

[138] Hünerli D, Emek-Savaş DD, Çavuşoğlu B, Dönmez Çolakoğlu B, Ada E, Yener GG. Mild cognitive impairment in Parkinson's disease is associated with decreased P300 amplitude and reduced putamen volume. Clin Neurophysiol 2019; 130(8): 1208-17.
[http://dx.doi.org/10.1016/j.clinph.2019.04.314] [PMID: 31163365]

[139] Zigmond MJ, Abercrombie ED, Berger TW, Grace AA, Stricker EM. Compensations after lesions of central dopaminergic neurons: some clinical and basic implications. Trends Neurosci 1990; 13(7): 290-6.
[http://dx.doi.org/10.1016/0166-2236(90)90112-N] [PMID: 1695406]

[140] Hornykiewicz O. Brain neurotransmitter changes in Parkinson's disease. Movement disorders. Elsevier 1981; pp. 41-58.

[141] Kitayama S, Wang JB, Uhl GR. Dopamine transporter mutants selectively enhance MPP$^+$ transport. Synapse 1993; 15(1): 58-62.
[http://dx.doi.org/10.1002/syn.890150107] [PMID: 8310426]

[142] Miller GW, Gainetdinov RR, Levey AI, Caron MG. Dopamine transporters and neuronal injury. Trends Pharmacol Sci 1999; 20(10): 424-9.
[http://dx.doi.org/10.1016/S0165-6147(99)01379-6] [PMID: 10498956]

[143] Morales I, Sanchez A, Rodriguez-Sabate C, Rodriguez M. The degeneration of dopaminergic synapses in Parkinson's disease: A selective animal model. Behav Brain Res 2015; 289: 19-28.
[http://dx.doi.org/10.1016/j.bbr.2015.04.019] [PMID: 25907749]

[144] Reeve A, Simcox E, Turnbull D. Ageing and Parkinson's disease: why is advancing age the biggest risk factor? Ageing Res Rev 2014; 14: 19-30.
[http://dx.doi.org/10.1016/j.arr.2014.01.004] [PMID: 24503004]

[145] Marras C, Chaudhuri KR. Nonmotor features of Parkinson's disease subtypes. Mov Disord 2016; 31(8): 1095-102.
[http://dx.doi.org/10.1002/mds.26510] [PMID: 26861861]

[146] Zeng XS, Geng WS, Jia JJ. Neurotoxin-Induced Animal Models of Parkinson Disease: Pathogenic Mechanism and Assessment. ASN Neuro 2018; 10: 1759091418777438.
[http://dx.doi.org/10.1177/1759091418777438] [PMID: 29809058]

[147] Bezard E, Yue Z, Kirik D, Spillantini MG. Animal models of Parkinson's disease: limits and relevance

to neuroprotection studies. Mov Disord 2013; 28(1): 61-70.
[http://dx.doi.org/10.1002/mds.25108] [PMID: 22753348]

[148] Poewe W, Seppi K, Tanner CM, *et al.* Parkinson disease. Nat Rev Dis Primers 2017; 3: 17013.
[http://dx.doi.org/10.1038/nrdp.2017.13] [PMID: 28332488]

[149] Ferrazzoli D, Carter A, Ustun FS, *et al.* Dopamine Replacement Therapy, Learning and Reward Prediction in Parkinson's Disease: Implications for Rehabilitation. Front Behav Neurosci 2016; 10: 121.
[http://dx.doi.org/10.3389/fnbeh.2016.00121] [PMID: 27378872]

[150] Thanvi B, Lo N, Robinson T. Levodopa-induced dyskinesia in Parkinson's disease: clinical features, pathogenesis, prevention and treatment. Postgrad Med J 2007; 83(980): 384-8.
[http://dx.doi.org/10.1136/pgmj.2006.054759] [PMID: 17551069]

[151] Zhang G, Xia Y, Wan F, *et al.* New Perspectives on Roles of Alpha-Synuclein in Parkinson's Disease. Front Aging Neurosci 2018; 10: 370.
[http://dx.doi.org/10.3389/fnagi.2018.00370] [PMID: 30524265]

[152] Fields CR, Bengoa-Vergniory N, Wade-Martins R. Targeting Alpha-Synuclein as a Therapy for Parkinson's Disease. Front Mol Neurosci 2019; 12: 299.
[http://dx.doi.org/10.3389/fnmol.2019.00299] [PMID: 31866823]

[153] Conway KA, Harper JD, Lansbury PT Jr. Fibrils formed *in vitro* from alpha-synuclein and two mutant forms linked to Parkinson's disease are typical amyloid. Biochemistry 2000; 39(10): 2552-63.
[http://dx.doi.org/10.1021/bi991447r] [PMID: 10704204]

[154] Engelhardt E, Gomes MDM. Lewy and his inclusion bodies: Discovery and rejection. Dement Neuropsychol 2017; 11(2): 198-201.
[http://dx.doi.org/10.1590/1980-57642016dn11-020012] [PMID: 29213511]

[155] Lashuel HA, Overk CR, Oueslati A, Masliah E. The many faces of α-synuclein: from structure and toxicity to therapeutic target. Nat Rev Neurosci 2013; 14(1): 38-48.
[http://dx.doi.org/10.1038/nrn3406] [PMID: 23254192]

[156] Ross CA, Pickart CM. The ubiquitin-proteasome pathway in Parkinson's disease and other neurodegenerative diseases. Trends Cell Biol 2004; 14(12): 703-11.
[http://dx.doi.org/10.1016/j.tcb.2004.10.006] [PMID: 15564047]

[157] Cookson MR, van der Brug M. Cell systems and the toxic mechanism(s) of alpha-synuclein. Exp Neurol 2008; 209(1): 5-11.
[http://dx.doi.org/10.1016/j.expneurol.2007.05.022] [PMID: 17603039]

[158] Meredith SC. Protein denaturation and aggregation: Cellular responses to denatured and aggregated proteins. Ann N Y Acad Sci 2005; 1066: 181-221.
[http://dx.doi.org/10.1196/annals.1363.030] [PMID: 16533927]

[159] Bengoa-Vergniory N, Roberts RF, Wade-Martins R, Alegre-Abarrategui J. Alpha-synuclein oligomers: a new hope. Acta Neuropathol 2017; 134(6): 819-38.
[http://dx.doi.org/10.1007/s00401-017-1755-1] [PMID: 28803412]

[160] Conway KA, Rochet JC, Bieganski RM, Lansbury PT Jr. Kinetic stabilization of the alpha-synuclein protofibril by a dopamine-alpha-synuclein adduct. Science 2001; 294(5545): 1346-9.
[http://dx.doi.org/10.1126/science.1063522] [PMID: 11701929]

[161] Lee HJ, Shin SY, Choi C, Lee YH, Lee SJ. Formation and removal of alpha-synuclein aggregates in cells exposed to mitochondrial inhibitors. J Biol Chem 2002; 277(7): 5411-7.
[http://dx.doi.org/10.1074/jbc.M105326200] [PMID: 11724769]

[162] Ischiropoulos H. Oxidative modifications of alpha-synuclein. Ann N Y Acad Sci 2003; 991: 93-100.
[http://dx.doi.org/10.1111/j.1749-6632.2003.tb07466.x] [PMID: 12846977]

[163] Tanaka M, Kim YM, Lee G, Junn E, Iwatsubo T, Mouradian MM. Aggresomes formed by alpha-

synuclein and synphilin-1 are cytoprotective. J Biol Chem 2004; 279(6): 4625-31.
[http://dx.doi.org/10.1074/jbc.M310994200] [PMID: 14627698]

[164] Murphy RM. Peptide aggregation in neurodegenerative disease. Annu Rev Biomed Eng 2002; 4: 155-74.
[http://dx.doi.org/10.1146/annurev.bioeng.4.092801.094202] [PMID: 12117755]

[165] Sharma SK, Priya S. Expanding role of molecular chaperones in regulating α-synuclein misfolding; implications in Parkinson's disease. Cell Mol Life Sci 2017; 74(4): 617-29.
[http://dx.doi.org/10.1007/s00018-016-2340-9] [PMID: 27522545]

[166] Cox D, Carver JA, Ecroyd H. Preventing α-synuclein aggregation: the role of the small heat-shock molecular chaperone proteins. Biochim Biophys Acta 2014; 1842(9): 1830-43.
[http://dx.doi.org/10.1016/j.bbadis.2014.06.024] [PMID: 24973551]

[167] Ecroyd H, Carver JA. The effect of small molecules in modulating the chaperone activity of alphaB-crystallin against ordered and disordered protein aggregation. FEBS J 2008; 275(5): 935-47.
[http://dx.doi.org/10.1111/j.1742-4658.2008.06257.x] [PMID: 18218039]

[168] Madine J, Doig AJ, Middleton DA. Design of an N-methylated peptide inhibitor of alpha-synuclein aggregation guided by solid-state NMR. J Am Chem Soc 2008; 130(25): 7873-81.
[http://dx.doi.org/10.1021/ja075356q] [PMID: 18510319]

[169] Sciarretta KL, Gordon DJ, Meredith SC. Peptide-based inhibitors of amyloid assembly. Methods Enzymol 2006; 413: 273-312.
[http://dx.doi.org/10.1016/S0076-6879(06)13015-3] [PMID: 17046402]

[170] Abe K, Kobayashi N, Sode K, Ikebukuro K. Peptide ligand screening of alpha-synuclein aggregation modulators by *in silico* panning. BMC Bioinformatics 2007; 8: 451.
[http://dx.doi.org/10.1186/1471-2105-8-451] [PMID: 18005454]

[171] Puspita L, Chung SY, Shim JW. Oxidative stress and cellular pathologies in Parkinson's disease. Mol Brain 2017; 10(1): 53.
[http://dx.doi.org/10.1186/s13041-017-0340-9] [PMID: 29183391]

[172] Chinta SJ, Woods G, Demaria M, *et al.* Cellular Senescence Is Induced by the Environmental Neurotoxin Paraquat and Contributes to Neuropathology Linked to Parkinson's Disease. Cell Rep 2018; 22(4): 930-40.
[http://dx.doi.org/10.1016/j.celrep.2017.12.092] [PMID: 29386135]

[173] Giordano S, Darley-Usmar V, Zhang J. Autophagy as an essential cellular antioxidant pathway in neurodegenerative disease. Redox Biol 2013; 2: 82-90.
[http://dx.doi.org/10.1016/j.redox.2013.12.013] [PMID: 24494187]

[174] Segura-Aguilar J, Paris I, Muñoz P, Ferrari E, Zecca L, Zucca FA. Protective and toxic roles of dopamine in Parkinson's disease. J Neurochem 2014; 129(6): 898-915.
[http://dx.doi.org/10.1111/jnc.12686] [PMID: 24548101]

[175] Franco-Iborra S, Vila M, Perier C. The Parkinson Disease Mitochondrial Hypothesis: Where Are We at? Neuroscientist 2016; 22(3): 266-77.
[http://dx.doi.org/10.1177/1073858415574600] [PMID: 25761946]

[176] Surace MJ, Block ML. Targeting microglia-mediated neurotoxicity: the potential of NOX2 inhibitors. Cell Mol Life Sci 2012; 69(14): 2409-27.
[http://dx.doi.org/10.1007/s00018-012-1015-4] [PMID: 22581365]

[177] Schulz JB, Lindenau J, Seyfried J, Dichgans J. Glutathione, oxidative stress and neurodegeneration. Eur J Biochem 2000; 267(16): 4904-11.
[http://dx.doi.org/10.1046/j.1432-1327.2000.01595.x] [PMID: 10931172]

[178] Conrad M, Sato H. The oxidative stress-inducible cystine/glutamate antiporter, system x (c) (-) : cystine supplier and beyond. Amino Acids 2012; 42(1): 231-46.
[http://dx.doi.org/10.1007/s00726-011-0867-5] [PMID: 21409388]

[179] Olanow CW, Schapira AH, LeWitt PA, *et al.* TCH346 as a neuroprotective drug in Parkinson's disease: a double-blind, randomised, controlled trial. Lancet Neurol 2006; 5(12): 1013-20.
[http://dx.doi.org/10.1016/S1474-4422(06)70602-0] [PMID: 17110281]

[180] Sheeler C, Rosa JG, Ferro A, McAdams B, Borgenheimer E, Cvetanovic M. Glia in Neurodegeneration: The Housekeeper, the Defender and the Perpetrator. Int J Mol Sci 2020; 21(23): E9188.
[http://dx.doi.org/10.3390/ijms21239188] [PMID: 33276471]

[181] Sidoryk-Wegrzynowicz M, Strużyńska L. Dysfunctional glia: contributors to neurodegenerative disorders. Neural Regen Res 2021; 16(2): 218-22.
[http://dx.doi.org/10.4103/1673-5374.290877] [PMID: 32859767]

[182] He J, Zhu G, Wang G, Zhang F. Oxidative Stress and Neuroinflammation Potentiate Each Other to Promote Progression of Dopamine Neurodegeneration. Oxid Med Cell Longev 2020; 2020: 6137521.
[http://dx.doi.org/10.1155/2020/6137521] [PMID: 32714488]

[183] Luo Y, Hoffer A, Hoffer B, Qi X. Mitochondria: A Therapeutic Target for Parkinson's Disease? Int J Mol Sci 2015; 16(9): 20704-30.
[http://dx.doi.org/10.3390/ijms160920704] [PMID: 26340618]

[184] Thomas B, Beal MF. Mitochondrial therapies for Parkinson's disease. Mov Disord 2010; 25 (Suppl. 1): S155-60.
[http://dx.doi.org/10.1002/mds.22781] [PMID: 20187246]

[185] Grünewald A, Kumar KR, Sue CM. New insights into the complex role of mitochondria in Parkinson's disease. Prog Neurobiol 2019; 177: 73-93.
[http://dx.doi.org/10.1016/j.pneurobio.2018.09.003] [PMID: 30219247]

[186] Macdonald R, Barnes K, Hastings C, Mortiboys H. Mitochondrial abnormalities in Parkinson's disease and Alzheimer's disease: can mitochondria be targeted therapeutically? Biochem Soc Trans 2018; 46(4): 891-909.
[http://dx.doi.org/10.1042/BST20170501] [PMID: 30026371]

[187] Greenamyre JT, Sherer TB, Betarbet R, Panov AV. Complex I and Parkinson's disease. IUBMB Life 2001; 52(3-5): 135-41.
[http://dx.doi.org/10.1080/15216540152845939] [PMID: 11798025]

[188] Reeve AK, Grady JP, Cosgrave EM, *et al.* Mitochondrial dysfunction within the synapses of substantia nigra neurons in Parkinson's disease. NPJ Parkinsons Dis 2018; 4: 9.
[http://dx.doi.org/10.1038/s41531-018-0044-6] [PMID: 29872690]

[189] Chen C, Turnbull DM, Reeve AK. Mitochondrial Dysfunction in Parkinson's Disease-Cause or Consequence? Biology (Basel) 2019; 8(2): E38.
[http://dx.doi.org/10.3390/biology8020038] [PMID: 31083583]

[190] Blandini F, Nappi G, Greenamyre JT. Quantitative study of mitochondrial complex I in platelets of parkinsonian patients. Mov Disord 1998; 13(1): 11-5.
[http://dx.doi.org/10.1002/mds.870130106] [PMID: 9452319]

[191] Yoshino H, Nakagawa-Hattori Y, Kondo T, Mizuno Y. Mitochondrial complex I and II activities of lymphocytes and platelets in Parkinson's disease. J Neural Transm Park Dis Dement Sect 1992; 4(1): 27-34.
[http://dx.doi.org/10.1007/BF02257619] [PMID: 1347219]

[192] Dias V, Junn E, Mouradian MM. The role of oxidative stress in Parkinson's disease. J Parkinsons Dis 2013; 3(4): 461-91.
[http://dx.doi.org/10.3233/JPD-130230] [PMID: 24252804]

[193] Trist BG, Hare DJ, Double KL. Oxidative stress in the aging substantia nigra and the etiology of Parkinson's disease. Aging Cell 2019; 18(6): e13031.
[http://dx.doi.org/10.1111/acel.13031] [PMID: 31432604]

[194] Gao J, Wang L, Liu J, Xie F, Su B, Wang X. Abnormalities of Mitochondrial Dynamics in Neurodegenerative Diseases. Antioxidants 2017; 6(2): E25.
[http://dx.doi.org/10.3390/antiox6020025] [PMID: 28379197]

[195] Mizushima N, Komatsu M. Autophagy: renovation of cells and tissues. Cell 2011; 147(4): 728-41.
[http://dx.doi.org/10.1016/j.cell.2011.10.026] [PMID: 22078875]

[196] Reggiori F, Komatsu M, Finley K, Simonsen A. Autophagy: more than a nonselective pathway. Int J Cell Biol 2012; 2012: 219625.
[http://dx.doi.org/10.1155/2012/219625] [PMID: 22666256]

[197] Cook C, Stetler C, Petrucelli L. Disruption of protein quality control in Parkinson's disease. Cold Spring Harb Perspect Med 2012; 2(5): a009423.
[http://dx.doi.org/10.1101/cshperspect.a009423] [PMID: 22553500]

[198] Yang Q, Wang R, Zhu L. Chaperone-Mediated Autophagy. Adv Exp Med Biol 2019; 1206: 435-52.
[http://dx.doi.org/10.1007/978-981-15-0602-4_20] [PMID: 31776997]

[199] Cuervo AM, Stefanis L, Fredenburg R, Lansbury PT, Sulzer D. Impaired degradation of mutant alpha-synuclein by chaperone-mediated autophagy. Science 2004; 305(5688): 1292-5.
[http://dx.doi.org/10.1126/science.1101738] [PMID: 15333840]

[200] Anglade P, Vyas S, Javoy-Agid F, *et al.* Apoptosis and autophagy in nigral neurons of patients with Parkinson's disease. Histol Histopathol 1997; 12(1): 25-31.
[PMID: 9046040]

[201] Davis S, Wang J, Ferro-Novick S. Crosstalk between the Secretory and Autophagy Pathways Regulates Autophagosome Formation. Dev Cell 2017; 41(1): 23-32.
[http://dx.doi.org/10.1016/j.devcel.2017.03.015] [PMID: 28399396]

[202] Ponpuak M, Mandell MA, Kimura T, Chauhan S, Cleyrat C, Deretic V. Secretory autophagy. Curr Opin Cell Biol 2015; 35: 106-16.
[http://dx.doi.org/10.1016/j.ceb.2015.04.016] [PMID: 25988755]

[203] Polymeropoulos MH, Lavedan C, Leroy E, *et al.* Mutation in the alpha-synuclein gene identified in families with Parkinson's disease. Science 1997; 276(5321): 2045-7.
[http://dx.doi.org/10.1126/science.276.5321.2045] [PMID: 9197268]

[204] Krüger R, Kuhn W, Müller T, *et al.* Ala30Pro mutation in the gene encoding alpha-synuclein in Parkinson's disease. Nat Genet 1998; 18(2): 106-8.
[http://dx.doi.org/10.1038/ng0298-106] [PMID: 9462735]

[205] Zarranz JJ, Alegre J, Gómez-Esteban JC, *et al.* The new mutation, E46K, of alpha-synuclein causes Parkinson and Lewy body dementia. Ann Neurol 2004; 55(2): 164-73.
[http://dx.doi.org/10.1002/ana.10795] [PMID: 14755719]

[206] Song JX, Lu JH, Liu LF, *et al.* HMGB1 is involved in autophagy inhibition caused by SNCA/α-synuclein overexpression: a process modulated by the natural autophagy inducer corynoxine B. Autophagy 2014; 10(1): 144-54.
[http://dx.doi.org/10.4161/auto.26751] [PMID: 24178442]

[207] Yu WH, Dorado B, Figueroa HY, *et al.* Metabolic activity determines efficacy of macroautophagic clearance of pathological oligomeric alpha-synuclein. Am J Pathol 2009; 175(2): 736-47.
[http://dx.doi.org/10.2353/ajpath.2009.080928] [PMID: 19628769]

[208] Li JQ, Tan L, Yu JT. The role of the LRRK2 gene in Parkinsonism. Mol Neurodegener 2014; 9: 47.
[http://dx.doi.org/10.1186/1750-1326-9-47] [PMID: 25391693]

[209] Eguchi T, Kuwahara T, Sakurai M, *et al.* LRRK2 and its substrate Rab GTPases are sequentially targeted onto stressed lysosomes and maintain their homeostasis. Proc Natl Acad Sci USA 2018; 115(39): E9115-24.
[http://dx.doi.org/10.1073/pnas.1812196115] [PMID: 30209220]

[210] Lopes da Fonseca T, Villar-Piqué A, Outeiro TF. The Interplay between Alpha-Synuclein Clearance and Spreading. Biomolecules 2015; 5(2): 435-71.
[http://dx.doi.org/10.3390/biom5020435] [PMID: 25874605]

[211] Soukup SF, Vanhauwaert R, Verstreken P. Parkinson's disease: convergence on synaptic homeostasis. EMBO J 2018; 37(18): e98960.
[http://dx.doi.org/10.15252/embj.201898960] [PMID: 30065071]

[212] Sheehan P, Yue Z. Deregulation of autophagy and vesicle trafficking in Parkinson's disease. Neurosci Lett 2019; 697: 59-65.
[http://dx.doi.org/10.1016/j.neulet.2018.04.013] [PMID: 29627340]

[213] Nguyen M, Wong YC, Ysselstein D, Severino A, Krainc D. Synaptic, Mitochondrial, and Lysosomal Dysfunction in Parkinson's Disease. Trends Neurosci 2019; 42(2): 140-9.
[http://dx.doi.org/10.1016/j.tins.2018.11.001] [PMID: 30509690]

[214] Zheng Q, Huang T, Zhang L, *et al.* Dysregulation of Ubiquitin-Proteasome System in Neurodegenerative Diseases. Front Aging Neurosci 2016; 8: 303.
[http://dx.doi.org/10.3389/fnagi.2016.00303] [PMID: 28018215]

[215] Leestemaker Y, Ovaa H. Tools to investigate the ubiquitin proteasome system. Drug Discov Today Technol 2017; 26: 25-31.
[http://dx.doi.org/10.1016/j.ddtec.2017.11.006] [PMID: 29249239]

[216] Ciechanover A, Kwon YT. Degradation of misfolded proteins in neurodegenerative diseases: therapeutic targets and strategies. Exp Mol Med 2015; 47: e147.
[http://dx.doi.org/10.1038/emm.2014.117] [PMID: 25766616]

[217] Hegde AN, Upadhya SC. Role of ubiquitin-proteasome-mediated proteolysis in nervous system disease. Biochim Biophys Acta 2011; 1809(2): 128-40.
[http://dx.doi.org/10.1016/j.bbagrm.2010.07.006] [PMID: 20674814]

[218] Momtaz S, Memariani Z, El-Senduny FF, *et al.* Targeting Ubiquitin-Proteasome Pathway by Natural Products: Novel Therapeutic Strategy for Treatment of Neurodegenerative Diseases. Front Physiol 2020; 11: 361.
[http://dx.doi.org/10.3389/fphys.2020.00361] [PMID: 32411012]

[219] Braak H, Braak E, Yilmazer D, Schultz C, de Vos RA, Jansen EN. Nigral and extranigral pathology in Parkinson's disease. J Neural Transm Suppl 1995; 46: 15-31.
[PMID: 8821039]

[220] Dawson TM, Dawson VL. Rare genetic mutations shed light on the pathogenesis of Parkinson disease. J Clin Invest 2003; 111(2): 145-51.
[http://dx.doi.org/10.1172/JCI200317575] [PMID: 12531866]

[221] Lim KL, Tan JM. Role of the ubiquitin proteasome system in Parkinson's disease. BMC Biochem 2007; 8 (Suppl. 1): S13.
[http://dx.doi.org/10.1186/1471-2091-8-S1-S13] [PMID: 18047737]

[222] McNaught KS, Belizaire R, Isacson O, Jenner P, Olanow CW. Altered proteasomal function in sporadic Parkinson's disease. Exp Neurol 2003; 179(1): 38-46.
[http://dx.doi.org/10.1006/exnr.2002.8050] [PMID: 12504866]

[223] Song DD, Shults CW, Sisk A, Rockenstein E, Masliah E. Enhanced substantia nigra mitochondrial pathology in human alpha-synuclein transgenic mice after treatment with MPTP. Exp Neurol 2004; 186(2): 158-72.
[http://dx.doi.org/10.1016/S0014-4886(03)00342-X] [PMID: 15026254]

[224] Höglinger GU, Carrard G, Michel PP, *et al.* Dysfunction of mitochondrial complex I and the proteasome: interactions between two biochemical deficits in a cellular model of Parkinson's disease. J Neurochem 2003; 86(5): 1297-307.
[http://dx.doi.org/10.1046/j.1471-4159.2003.01952.x] [PMID: 12911637]

[225] Pierson J, Svenningsson P, Caprioli RM, Andren PE. Increased levels of ubiquitin in the 6-OHD-
        -lesioned striatum of rats. J Proteome Res 2005; 4(2): 223-6.
        [http://dx.doi.org/10.1021/pr049836h] [PMID: 15822896]

[226] Tambasco N, Romoli M, Calabresi P. Levodopa in Parkinson's Disease: Current Status and Future
        Developments. Curr Neuropharmacol 2018; 16(8): 1239-52.
        [http://dx.doi.org/10.2174/1570159X15666170510143821] [PMID: 28494719]

[227] Fahn S. The spectrum of levodopa-induced dyskinesias. Ann Neurol 2000; 47(4) (Suppl. 1): S2-9.
        [PMID: 10762127]

[228] Yao B, Christian KM, He C, Jin P, Ming GL, Song H. Epigenetic mechanisms in neurogenesis. Nat
        Rev Neurosci 2016; 17(9): 537-49.
        [http://dx.doi.org/10.1038/nrn.2016.70] [PMID: 27334043]

[229] Ammal Kaidery N, Tarannum S, Thomas B. Epigenetic landscape of Parkinson's disease: emerging
        role in disease mechanisms and therapeutic modalities. Neurotherapeutics 2013; 10(4): 698-708.
        [http://dx.doi.org/10.1007/s13311-013-0211-8] [PMID: 24030213]

[230] Hegarty SV, Sullivan AM, O'Keeffe GW. The Epigenome as a therapeutic target for Parkinson's
        disease. Neural Regen Res 2016; 11(11): 1735-8.
        [http://dx.doi.org/10.4103/1673-5374.194803] [PMID: 28123403]

[231] Labbé C, Lorenzo-Betancor O, Ross OA. Epigenetic regulation in Parkinson's disease. Acta
        Neuropathol 2016; 132(4): 515-30.
        [http://dx.doi.org/10.1007/s00401-016-1590-9] [PMID: 27358065]

[232] Konsoula Z, Barile FA. Epigenetic histone acetylation and deacetylation mechanisms in experimental
        models of neurodegenerative disorders. J Pharmacol Toxicol Methods 2012; 66(3): 215-20.
        [http://dx.doi.org/10.1016/j.vascn.2012.08.001] [PMID: 22902970]

[233] Ito K, J Barnes P, M Adcock I. Histone acetylation and deacetylation. Methods Mol Med 2000; 44:
        309-19.
        [PMID: 21312138]

[234] Yang XJ, Seto E. HATs and HDACs: from structure, function and regulation to novel strategies for
        therapy and prevention. Oncogene 2007; 26(37): 5310-8.
        [http://dx.doi.org/10.1038/sj.onc.1210599] [PMID: 17694074]

[235] Roth SY, Denu JM, Allis CD. Histone acetyltransferases. Annu Rev Biochem 2001; 70: 81-120.
        [http://dx.doi.org/10.1146/annurev.biochem.70.1.81] [PMID: 11395403]

[236] Bannister AJ, Kouzarides T. Regulation of chromatin by histone modifications. Cell Res 2011; 21(3):
        381-95.
        [http://dx.doi.org/10.1038/cr.2011.22] [PMID: 21321607]

[237] Ziemka-Nalecz M, Jaworska J, Sypecka J, Zalewska T. Histone Deacetylase Inhibitors: A Therapeutic
        Key in Neurological Disorders? J Neuropathol Exp Neurol 2018; 77(10): 855-70.
        [http://dx.doi.org/10.1093/jnen/nly073] [PMID: 30165682]

[238] Xu WS, Parmigiani RB, Marks PA. Histone deacetylase inhibitors: molecular mechanisms of action.
        Oncogene 2007; 26(37): 5541-52.
        [http://dx.doi.org/10.1038/sj.onc.1210620] [PMID: 17694093]

[239] Broide RS, Redwine JM, Aftahi N, Young W, Bloom FE, Winrow CJ. Distribution of histone
        deacetylases 1-11 in the rat brain. J Mol Neurosci 2007; 31(1): 47-58.
        [http://dx.doi.org/10.1007/BF02686117] [PMID: 17416969]

[240] Yamagoe S, Kanno T, Kanno Y, *et al.* Interaction of histone acetylases and deacetylases *in vivo*. Mol
        Cell Biol 2003; 23(3): 1025-33.
        [http://dx.doi.org/10.1128/MCB.23.3.1025-1033.2003] [PMID: 12529406]

[241] Dietz KC, Casaccia P. HDAC inhibitors and neurodegeneration: at the edge between protection and

damage. Pharmacol Res 2010; 62(1): 11-7.
[http://dx.doi.org/10.1016/j.phrs.2010.01.011] [PMID: 20123018]

[242] Saha RN, Pahan K. HATs and HDACs in neurodegeneration: a tale of disconcerted acetylation homeostasis. Cell Death Differ 2006; 13(4): 539-50.
[http://dx.doi.org/10.1038/sj.cdd.4401769] [PMID: 16167067]

[243] Kazantsev AG, Thompson LM. Therapeutic application of histone deacetylase inhibitors for central nervous system disorders. Nat Rev Drug Discov 2008; 7(10): 854-68.
[http://dx.doi.org/10.1038/nrd2681] [PMID: 18827828]

[244] Harrison IF, Smith AD, Dexter DT. Pathological histone acetylation in Parkinson's disease: Neuroprotection and inhibition of microglial activation through SIRT 2 inhibition. Neurosci Lett 2018; 666: 48-57.
[http://dx.doi.org/10.1016/j.neulet.2017.12.037] [PMID: 29273397]

[245] Kontopoulos E, Parvin JD, Feany MB. Alpha-synuclein acts in the nucleus to inhibit histone acetylation and promote neurotoxicity. Hum Mol Genet 2006; 15(20): 3012-23.
[http://dx.doi.org/10.1093/hmg/ddl243] [PMID: 16959795]

[246] Chuang DM, Leng Y, Marinova Z, Kim HJ, Chiu CT. Multiple roles of HDAC inhibition in neurodegenerative conditions. Trends Neurosci 2009; 32(11): 591-601.
[http://dx.doi.org/10.1016/j.tins.2009.06.002] [PMID: 19775759]

[247] de Ruijter AJ, van Gennip AH, Caron HN, Kemp S, van Kuilenburg AB. Histone deacetylases (HDACs): characterization of the classical HDAC family. Biochem J 2003; 370(Pt 3): 737-49.
[http://dx.doi.org/10.1042/bj20021321] [PMID: 12429021]

[248] A two-stage meta-analysis identifies several new loci for Parkinson's disease. PLoS Genet 2011; 7(6): e1002142.
[http://dx.doi.org/10.1371/journal.pgen.1002142] [PMID: 21738488]

[249] Gibney ER, Nolan CM. Epigenetics and gene expression. Heredity 2010; 105(1): 4-13.
[http://dx.doi.org/10.1038/hdy.2010.54] [PMID: 20461105]

[250] Nicholas AP, Lubin FD, Hallett PJ, *et al.* Striatal histone modifications in models of levodopa-induced dyskinesia. J Neurochem 2008; 106(1): 486-94.
[http://dx.doi.org/10.1111/j.1471-4159.2008.05417.x] [PMID: 18410512]

[251] Mu MD, Qian ZM, Yang SX, Rong KL, Yung WH, Ke Y. Therapeutic effect of a histone demethylase inhibitor in Parkinson's disease. Cell Death Dis 2020; 11(10): 927.
[http://dx.doi.org/10.1038/s41419-020-03105-5] [PMID: 33116116]

[252] Su X, Chu Y, Kordower JH, *et al.* PGC-1α Promoter Methylation in Parkinson's Disease. PLoS One 2015; 10(8): e0134087.
[http://dx.doi.org/10.1371/journal.pone.0134087] [PMID: 26317511]

[253] Miranda-Morales E, Meier K, Sandoval-Carrillo A, Salas-Pacheco J, Vázquez-Cárdenas P, Arias-Carrión O. Implications of DNA Methylation in Parkinson's Disease. Front Mol Neurosci 2017; 10: 225.
[http://dx.doi.org/10.3389/fnmol.2017.00225] [PMID: 28769760]

[254] Baizabal-Carvallo JF. Gut microbiota: a potential therapeutic target for Parkinson's disease. Neural Regen Res 2021; 16(2): 287-8.
[http://dx.doi.org/10.4103/1673-5374.290896] [PMID: 32859778]

[255] Schaeffer E, Kluge A, Böttner M, *et al.* Alpha Synuclein Connects the Gut-Brain Axis in Parkinson's Disease Patients - A View on Clinical Aspects, Cellular Pathology and Analytical Methodology. Front Cell Dev Biol 2020; 8: 573696.
[http://dx.doi.org/10.3389/fcell.2020.573696] [PMID: 33015066]

[256] Fitzgerald E, Murphy S, Martinson HA. Alpha-Synuclein Pathology and the Role of the Microbiota in Parkinson's Disease. Front Neurosci 2019; 13: 369.

[http://dx.doi.org/10.3389/fnins.2019.00369] [PMID: 31068777]

[257] Dutta SK, Verma S, Jain V, *et al.* Parkinson's Disease: The Emerging Role of Gut Dysbiosis, Antibiotics, Probiotics, and Fecal Microbiota Transplantation. J Neurogastroenterol Motil 2019; 25(3): 363-76.
[http://dx.doi.org/10.5056/jnm19044] [PMID: 31327219]

[258] Karl JP, Hatch AM, Arcidiacono SM, *et al.* Effects of Psychological, Environmental and Physical Stressors on the Gut Microbiota. Front Microbiol 2018; 9: 2013.
[http://dx.doi.org/10.3389/fmicb.2018.02013] [PMID: 30258412]

[259] Baizabal-Carvallo JF, Alonso-Juarez M. The Link between Gut Dysbiosis and Neuroinflammation in Parkinson's Disease. Neuroscience 2020; 432: 160-73.
[http://dx.doi.org/10.1016/j.neuroscience.2020.02.030] [PMID: 32112917]

[260] Forsyth CB, Shannon KM, Kordower JH, *et al.* Increased intestinal permeability correlates with sigmoid mucosa alpha-synuclein staining and endotoxin exposure markers in early Parkinson's disease. PLoS One 2011; 6(12): e28032.
[http://dx.doi.org/10.1371/journal.pone.0028032] [PMID: 22145021]

[261] Stefanova N, Fellner L, Reindl M, Masliah E, Poewe W, Wenning GK. Toll-like receptor 4 promotes α-synuclein clearance and survival of nigral dopaminergic neurons. Am J Pathol 2011; 179(2): 954-63.
[http://dx.doi.org/10.1016/j.ajpath.2011.04.013] [PMID: 21801874]

[262] Santos SF, de Oliveira HL, Yamada ES, Neves BC, Pereira A Jr. The Gut and Parkinson's Disease-A Bidirectional Pathway. Front Neurol 2019; 10: 574.
[http://dx.doi.org/10.3389/fneur.2019.00574] [PMID: 31214110]

[263] Uyar GO, Yildiran H. A nutritional approach to microbiota in Parkinson's disease. Biosci Microbiota Food Health 2019; 38(4): 115-27.
[http://dx.doi.org/10.12938/bmfh.19-002] [PMID: 31763115]

[264] Pandey S, Srivanitchapoom P. Levodopa-induced Dyskinesia: Clinical Features, Pathophysiology, and Medical Management. Ann Indian Acad Neurol 2017; 20(3): 190-8.
[PMID: 28904447]

[265] Jameson KG, Hsiao EY. A novel pathway for microbial metabolism of levodopa. Nat Med 2019; 25(8): 1195-7.
[http://dx.doi.org/10.1038/s41591-019-0544-x] [PMID: 31388180]

[266] Hegelmaier T, Lebbing M, Duscha A, *et al.* Interventional Influence of the Intestinal Microbiome Through Dietary Intervention and Bowel Cleansing Might Improve Motor Symptoms in Parkinson's Disease. Cells 2020; 9(2): E376.
[http://dx.doi.org/10.3390/cells9020376] [PMID: 32041265]

[267] van Kessel SP, Frye AK, El-Gendy AO, *et al.* Gut bacterial tyrosine decarboxylases restrict levels of levodopa in the treatment of Parkinson's disease. Nat Commun 2019; 10(1): 310.
[http://dx.doi.org/10.1038/s41467-019-08294-y] [PMID: 30659181]

[268] Yang D, Zhao D, Ali Shah SZ, *et al.* The Role of the Gut Microbiota in the Pathogenesis of Parkinson's Disease. Front Neurol 2019; 10: 1155.
[http://dx.doi.org/10.3389/fneur.2019.01155] [PMID: 31781020]

[269] Bhattarai Y, Kashyap PC. Parkinson's disease: Are gut microbes involved? Am J Physiol Gastrointest Liver Physiol 2020; 319(5): G529-40.
[http://dx.doi.org/10.1152/ajpgi.00058.2020] [PMID: 32877215]

[270] Ilie OD, Ciobica A, McKenna J, Doroftei B, Mavroudis I. Minireview on the Relations between Gut Microflora and Parkinson's Disease: Further Biochemical (Oxidative Stress), Inflammatory, and Neurological Particularities. Oxid Med Cell Longev 2020; 2020: 4518023.
[http://dx.doi.org/10.1155/2020/4518023] [PMID: 32089768]

[271] Huang EJ, Reichardt LF. Neurotrophins: roles in neuronal development and function. Annu Rev

Neurosci 2001; 24: 677-736.
[http://dx.doi.org/10.1146/annurev.neuro.24.1.677] [PMID: 11520916]

[272]   Sofroniew MV, Howe CL, Mobley WC. Nerve growth factor signaling, neuroprotection, and neural repair. Annu Rev Neurosci 2001; 24: 1217-81.
[http://dx.doi.org/10.1146/annurev.neuro.24.1.1217] [PMID: 11520933]

[273]   Jellinger KA. Basic mechanisms of neurodegeneration: a critical update. J Cell Mol Med 2010; 14(3): 457-87.
[http://dx.doi.org/10.1111/j.1582-4934.2010.01010.x] [PMID: 20070435]

[274]   Bathina S, Das UN. Brain-derived neurotrophic factor and its clinical implications. Arch Med Sci 2015; 11(6): 1164-78.
[http://dx.doi.org/10.5114/aoms.2015.56342] [PMID: 26788077]

[275]   Weissmiller AM, Wu C. Current advances in using neurotrophic factors to treat neurodegenerative disorders. Transl Neurodegener 2012; 1(1): 14.
[http://dx.doi.org/10.1186/2047-9158-1-14] [PMID: 23210531]

[276]   Nasrolahi A, Mahmoudi J, Akbarzadeh A, *et al.* Neurotrophic factors hold promise for the future of Parkinson's disease treatment: is there a light at the end of the tunnel? Rev Neurosci 2018; 29(5): 475-89.
[http://dx.doi.org/10.1515/revneuro-2017-0040] [PMID: 29305570]

[277]   Huttunen HJ, Saarma M. CDNF Protein Therapy in Parkinson's Disease. Cell Transplant 2019; 28(4): 349-66.
[http://dx.doi.org/10.1177/0963689719840290] [PMID: 30947516]

[278]   Rodrigues TM, Jerónimo-Santos A, Outeiro TF, Sebastião AM, Diógenes MJ. Challenges and promises in the development of neurotrophic factor-based therapies for Parkinson's disease. Drugs Aging 2014; 31(4): 239-61.
[http://dx.doi.org/10.1007/s40266-014-0160-x] [PMID: 24610720]

[279]   Oliveira de Carvalho A, Filho ASS, Murillo-Rodriguez E, Rocha NB, Carta MG, Machado S. Physical Exercise For Parkinson's Disease: Clinical And Experimental Evidence. Clin Pract Epidemiol Ment Health 2018; 14: 89-98.
[http://dx.doi.org/10.2174/1745017901814010089] [PMID: 29785199]

[280]   Lister T. Nutrition and Lifestyle Interventions for Managing Parkinson's Disease: A Narrative Review. J Mov Disord 2020; 13(2): 97-104.
[http://dx.doi.org/10.14802/jmd.20006] [PMID: 32498495]

[281]   Jeong KH, Nam JH, Jin BK, Kim SR. Activation of CNTF/CNTFRα signaling pathway by hRheb(S16H) transduction of dopaminergic neurons *in vivo.* PLoS One 2015; 10(3): e0121803.
[http://dx.doi.org/10.1371/journal.pone.0121803] [PMID: 25799580]

[282]   Chmielarz P, Saarma M. Neurotrophic factors for disease-modifying treatments of Parkinson's disease: gaps between basic science and clinical studies. Pharmacol Rep 2020; 72(5): 1195-217.
[http://dx.doi.org/10.1007/s43440-020-00120-3] [PMID: 32700249]

[283]   Levy YS, Gilgun-Sherki Y, Melamed E, Offen D. Therapeutic potential of neurotrophic factors in neurodegenerative diseases. BioDrugs 2005; 19(2): 97-127.
[http://dx.doi.org/10.2165/00063030-200519020-00003] [PMID: 15807629]

[284]   Hitti FL, Yang AI, Gonzalez-Alegre P, Baltuch GH. Human gene therapy approaches for the treatment of Parkinson's disease: An overview of current and completed clinical trials. Parkinsonism Relat Disord 2019; 66: 16-24.
[http://dx.doi.org/10.1016/j.parkreldis.2019.07.018] [PMID: 31324556]

[285]   Palasz E, Wysocka A, Gasiorowska A, Chalimoniuk M, Niewiadomski W, Niewiadomska G. BDNF as a Promising Therapeutic Agent in Parkinson's Disease. Int J Mol Sci 2020; 21(3): E1170.
[http://dx.doi.org/10.3390/ijms21031170] [PMID: 32050617]

[286] Grondin R, Gash DM. Glial cell line-derived neurotrophic factor (GDNF): a drug candidate for the treatment of Parkinson's disease. J Neurol 1998; 245(11) (Suppl. 3): 35-42.
[http://dx.doi.org/10.1007/PL00007744] [PMID: 9808338]

[287] Miyazaki I, Asanuma M. Neuron-Astrocyte Interactions in Parkinson's Disease. Cells 2020; 9(12): E2623.
[http://dx.doi.org/10.3390/cells9122623] [PMID: 33297340]

[288] Hagg T, Varon S. Ciliary neurotrophic factor prevents degeneration of adult rat substantia nigra dopaminergic neurons *in vivo*. Proc Natl Acad Sci USA 1993; 90(13): 6315-9.
[http://dx.doi.org/10.1073/pnas.90.13.6315] [PMID: 8101002]

[289] Salinas M, Diaz R, Abraham NG, Ruiz de Galarreta CM, Cuadrado A. Nerve growth factor protects against 6-hydroxydopamine-induced oxidative stress by increasing expression of heme oxygenase-1 in a phosphatidylinositol 3-kinase-dependent manner. J Biol Chem 2003; 278(16): 13898-904.
[http://dx.doi.org/10.1074/jbc.M209164200] [PMID: 12578834]

[290] Haque NS, Hlavin ML, Fawcett JW, Dunnett SB. The neurotrophin NT4/5, but not NT3, enhances the efficacy of nigral grafts in a rat model of Parkinson's disease. Brain Res 1996; 712(1): 45-52.
[http://dx.doi.org/10.1016/0006-8993(95)01427-6] [PMID: 8705306]

[291] Cheong SL, Federico S, Spalluto G, Klotz KN, Pastorin G. The current status of pharmacotherapy for the treatment of Parkinson's disease: transition from single-target to multitarget therapy. Drug Discov Today 2019; 24(9): 1769-83.
[http://dx.doi.org/10.1016/j.drudis.2019.05.003] [PMID: 31102728]

[292] Warren Olanow C, Kieburtz K, Rascol O, *et al.* Factors predictive of the development of Levodopa-induced dyskinesia and wearing-off in Parkinson's disease. Mov Disord 2013; 28(8): 1064-71.
[http://dx.doi.org/10.1002/mds.25364] [PMID: 23630119]

[293] Löhle M, Ramberg CJ, Reichmann H, Schapira AH. Early *versus* delayed initiation of pharmacotherapy in Parkinson's disease. Drugs 2014; 74(6): 645-57.
[http://dx.doi.org/10.1007/s40265-014-0209-5] [PMID: 24756431]

[294] Ahlskog JE, Muenter MD. Frequency of levodopa-related dyskinesias and motor fluctuations as estimated from the cumulative literature. Mov Disord 2001; 16(3): 448-58.
[http://dx.doi.org/10.1002/mds.1090] [PMID: 11391738]

[295] Zheng J, Zhang X, Zhen X. Development of Adenosine A_{2A} Receptor Antagonists for the Treatment of Parkinson's Disease: A Recent Update and Challenge. ACS Chem Neurosci 2019; 10(2): 783-91.
[http://dx.doi.org/10.1021/acschemneuro.8b00313] [PMID: 30199223]

[296] Pinna A. Adenosine A2A receptor antagonists in Parkinson's disease: progress in clinical trials from the newly approved istradefylline to drugs in early development and those already discontinued. CNS Drugs 2014; 28(5): 455-74.
[http://dx.doi.org/10.1007/s40263-014-0161-7] [PMID: 24687255]

[297] Cacciari B, Spalluto G, Federico S. A2A Adenosine Receptor Antagonists as Therapeutic Candidates: Are They Still an Interesting Challenge? Mini Rev Med Chem 2018; 18(14): 1168-74.
[http://dx.doi.org/10.2174/1389557518666180423113051] [PMID: 29692248]

[298] Katzenschlager R, Sampaio C, Costa J, Lees A. Anticholinergics for symptomatic management of Parkinson's disease. Cochrane Database Syst Rev 2003; (2): CD003735.
[PMID: 12804486]

[299] Langmead CJ, Watson J, Reavill C. Muscarinic acetylcholine receptors as CNS drug targets. Pharmacol Ther 2008; 117(2): 232-43.
[http://dx.doi.org/10.1016/j.pharmthera.2007.09.009] [PMID: 18082893]

[300] Huot P, Sgambato-Faure V, Fox SH, McCreary AC. Serotonergic Approaches in Parkinson's Disease: Translational Perspectives, an Update. ACS Chem Neurosci 2017; 8(5): 973-86.
[http://dx.doi.org/10.1021/acschemneuro.6b00440] [PMID: 28460160]

[301] Mao Q, Qin WZ, Zhang A, Ye N. Recent advances in dopaminergic strategies for the treatment of Parkinson's disease. Acta Pharmacol Sin 2020; 41(4): 471-82.
[http://dx.doi.org/10.1038/s41401-020-0365-y] [PMID: 32112042]

[302] Samii A, Nutt JG, Ransom BR. Parkinson's disease. Lancet 2004; 363(9423): 1783-93.
[http://dx.doi.org/10.1016/S0140-6736(04)16305-8] [PMID: 15172778]

[303] MayoClinic. Levodopa (Oral Route) 2020.https://www.mayoclinic.org/drugs-supplements/levodop-
-oral-route/proper-use/drg-20064498

[304] Schrag A, Quinn N. Dyskinesias and motor fluctuations in Parkinson's disease. A community-based study. Brain 2000; 123(Pt 11): 2297-305.
[http://dx.doi.org/10.1093/brain/123.11.2297] [PMID: 11050029]

[305] Jankovic J. Levodopa strengths and weaknesses. Neurology 2002; 58(4) (Suppl. 1): S19-32.
[http://dx.doi.org/10.1212/WNL.58.suppl_1.S19] [PMID: 11909982]

[306] Deleu D, Northway MG, Hanssens Y. Clinical pharmacokinetic and pharmacodynamic properties of drugs used in the treatment of Parkinson's disease. Clin Pharmacokinet 2002; 41(4): 261-309.
[http://dx.doi.org/10.2165/00003088-200241040-00003] [PMID: 11978145]

[307] Goldenberg MM. Medical management of Parkinson's disease. P&T 2008; 33(10): 590-606.
[PMID: 19750042]

[308] Brooks DJ. Dopamine agonists: their role in the treatment of Parkinson's disease. J Neurol Neurosurg Psychiatry 2000; 68(6): 685-9.
[http://dx.doi.org/10.1136/jnnp.68.6.685] [PMID: 10811688]

[309] Calne DB, Teychenne PF, Claveria LE, Eastman R, Greenacre JK, Petrie A. Bromocriptine in Parkinsonism. BMJ 1974; 4(5942): 442-4.
[http://dx.doi.org/10.1136/bmj.4.5942.442] [PMID: 4425916]

[310] Connolly BS, Lang AE. Pharmacological treatment of Parkinson disease: a review. JAMA 2014; 311(16): 1670-83.
[http://dx.doi.org/10.1001/jama.2014.3654] [PMID: 24756517]

[311] Ives NJ, Stowe RL, Marro J, *et al.* Monoamine oxidase type B inhibitors in early Parkinson's disease: meta-analysis of 17 randomised trials involving 3525 patients. BMJ 2004; 329(7466): 593.
[http://dx.doi.org/10.1136/bmj.38184.606169.AE] [PMID: 15310558]

[312] Cereda E, Cilia R, Canesi M, *et al.* Efficacy of rasagiline and selegiline in Parkinson's disease: a head-to-head 3-year retrospective case-control study. J Neurol 2017; 264(6): 1254-63.
[http://dx.doi.org/10.1007/s00415-017-8523-y] [PMID: 28550482]

[313] Henderson EJ, Lord SR, Brodie MA, *et al.* Rivastigmine for gait stability in patients with Parkinson's disease (ReSPonD): a randomised, double-blind, placebo-controlled, phase 2 trial. Lancet Neurol 2016; 15(3): 249-58.
[http://dx.doi.org/10.1016/S1474-4422(15)00389-0] [PMID: 26795874]

[314] Chung KA, Lobb BM, Nutt JG, Horak FB. Effects of a central cholinesterase inhibitor on reducing falls in Parkinson disease. Neurology 2010; 75(14): 1263-9.
[http://dx.doi.org/10.1212/WNL.0b013e3181f6128c] [PMID: 20810998]

[315] Cooper JA, Sagar HJ, Doherty SM, Jordan N, Tidswell P, Sullivan EV. Different effects of dopaminergic and anticholinergic therapies on cognitive and motor function in Parkinson's disease. A follow-up study of untreated patients. Brain 1992; 115(Pt 6): 1701-25.
[http://dx.doi.org/10.1093/brain/115.6.1701] [PMID: 1486457]

[316] Ahlskog JE. Slowing Parkinson's disease progression: recent dopamine agonist trials. Neurology 2003; 60(3): 381-9.
[http://dx.doi.org/10.1212/01.WNL.0000044047.58984.2F] [PMID: 12580184]

[317] Sawada H, Oeda T, Kuno S, *et al.* Amantadine for dyskinesias in Parkinson's disease: a randomized

controlled trial. PLoS One 2010; 5(12): e15298.
[http://dx.doi.org/10.1371/journal.pone.0015298] [PMID: 21217832]

[318] Zahoor I, Shafi A, Haq E. Pharmacological Treatment of Parkinson's Disease. 2018.
[http://dx.doi.org/10.15586/codonpublications.parkinsonsdisease.2018.ch7]

[319] Lobbens ES, Breydo L, Skamris T, *et al.* Mechanistic study of the inhibitory activity of Geum
urbanum extract against α-Synuclein fibrillation. Biochim Biophys Acta 2016; 1864(9): 1160-9.
[http://dx.doi.org/10.1016/j.bbapap.2016.06.009] [PMID: 27353564]

[320] Ren R, Shi C, Cao J, *et al.* Neuroprotective Effects of A Standardized Flavonoid Extract of Safflower
Against Neurotoxin-Induced Cellular and Animal Models of Parkinson's Disease. Sci Rep 2016; 6:
22135.
[http://dx.doi.org/10.1038/srep22135] [PMID: 26906725]

[321] Briffa M, Ghio S, Neuner J, *et al.* Extracts from two ubiquitous Mediterranean plants ameliorate
cellular and animal models of neurodegenerative proteinopathies. Neurosci Lett 2017; 638: 12-20.
[http://dx.doi.org/10.1016/j.neulet.2016.11.058] [PMID: 27919712]

[322] Elinos-Calderón D, Robledo-Arratia Y, Pérez-De La Cruz V, *et al.* Antioxidant strategy to rescue
synaptosomes from oxidative damage and energy failure in neurotoxic models in rats: protective role
of S-allylcysteine. J Neural Transm (Vienna) 2010; 117(1): 35-44.
[http://dx.doi.org/10.1007/s00702-009-0299-5] [PMID: 19866339]

[323] Vafai SB, Mevers E, Higgins KW, *et al.* Natural Product Screening Reveals Naphthoquinone Complex
I Bypass Factors. PLoS One 2016; 11(9): e0162686.
[http://dx.doi.org/10.1371/journal.pone.0162686] [PMID: 27622560]

[324] Li SY, Jia YH, Sun WG, *et al.* Stabilization of mitochondrial function by tetramethylpyrazine protects
against kainate-induced oxidative lesions in the rat hippocampus. Free Radic Biol Med 2010; 48(4):
597-608.
[http://dx.doi.org/10.1016/j.freeradbiomed.2009.12.004] [PMID: 20006702]

[325] Song L, Xu MB, Zhou XL, Zhang DP, Zhang SL, Zheng GQ. A Preclinical Systematic Review of
Ginsenoside-Rg1 in Experimental Parkinson's Disease. Oxid Med Cell Longev 2017; 2017: 2163053.
[http://dx.doi.org/10.1155/2017/2163053] [PMID: 28386306]

[326] Li Y, Zhao J, Hölscher C. Therapeutic Potential of Baicalein in Alzheimer's Disease and Parkinson's
Disease. CNS Drugs 2017; 31(8): 639-52.
[http://dx.doi.org/10.1007/s40263-017-0451-y] [PMID: 28634902]

[327] Wang XS, Zhang ZR, Zhang MM, Sun MX, Wang WW, Xie CL. Neuroprotective properties of
curcumin in toxin-base animal models of Parkinson's disease: a systematic experiment literatures
review. BMC Complement Altern Med 2017; 17(1): 412.
[http://dx.doi.org/10.1186/s12906-017-1922-x] [PMID: 28818104]

[328] Doo AR, Kim SN, Hahm DH, *et al.* Gastrodia elata Blume alleviates L-DOPA-induced dyskinesia by
normalizing FosB and ERK activation in a 6-OHDA-lesioned Parkinson's disease mouse model. BMC
Complement Altern Med 2014; 14: 107.
[http://dx.doi.org/10.1186/1472-6882-14-107] [PMID: 24650244]

[329] Kuang S, Yang L, Rao Z, *et al.* Effects of Ginkgo Biloba Extract on A53T α-Synuclein Transgenic
Mouse Models of Parkinson's Disease. Can J Neurol Sci 2018; 45(2): 182-7.
[http://dx.doi.org/10.1017/cjn.2017.268] [PMID: 29506601]

[330] Hosamani R, Krishna G, Muralidhara . Standardized Bacopa monnieri extract ameliorates acute
paraquat-induced oxidative stress, and neurotoxicity in prepubertal mice brain. Nutr Neurosci 2016;
19(10): 434-46.
[http://dx.doi.org/10.1179/1476830514Y.0000000149] [PMID: 25153704]

[331] Yadav SK, Rai SN, Singh SP. Mucuna pruriens reduces inducible nitric oxide synthase expression in
Parkinsonian mice model. J Chem Neuroanat 2017; 80: 1-10.

[http://dx.doi.org/10.1016/j.jchemneu.2016.11.009] [PMID: 27919828]

[332] Prakash J, Chouhan S, Yadav SK, Westfall S, Rai SN, Singh SP. Withania somnifera alleviates parkinsonian phenotypes by inhibiting apoptotic pathways in dopaminergic neurons. Neurochem Res 2014; 39(12): 2527-36.
[http://dx.doi.org/10.1007/s11064-014-1443-7] [PMID: 25403619]

[333] Dar NJ, Hamid A, Ahmad M. Pharmacologic overview of Withania somnifera, the Indian Ginseng. Cell Mol Life Sci 2015; 72(23): 4445-60.
[http://dx.doi.org/10.1007/s00018-015-2012-1] [PMID: 26306935]

[334] Renaud J, Nabavi SF, Daglia M, Nabavi SM, Martinoli MG. Epigallocatechin-3-Gallate, a Promising Molecule for Parkinson's Disease? Rejuvenation Res 2015; 18(3): 257-69.
[http://dx.doi.org/10.1089/rej.2014.1639] [PMID: 25625827]

[335] Ye Q, Ye L, Xu X, *et al.* Epigallocatechin-3-gallate suppresses 1-methyl-4-phenyl-pyridine-induced oxidative stress in PC12 cells via the SIRT1/PGC-1α signaling pathway. BMC Complement Altern Med 2012; 12: 82.
[http://dx.doi.org/10.1186/1472-6882-12-82] [PMID: 22742579]

[336] Nataraj J, Manivasagam T, Justin Thenmozhi A, Essa MM. Neurotrophic Effect of Asiatic acid, a Triterpene of Centella asiatica Against Chronic 1-Methyl 4-Phenyl 1, 2, 3, 6-Tetrahydropyridine Hydrochloride/Probenecid Mouse Model of Parkinson's disease: The Role of MAPK, PI3K-Ak--GSK3β and mTOR Signalling Pathways. Neurochem Res 2017; 42(5): 1354-65.
[http://dx.doi.org/10.1007/s11064-017-2183-2] [PMID: 28181071]

[337] Xu CL, Qu R, Zhang J, Li LF, Ma SP. Neuroprotective effects of madecassoside in early stage of Parkinson's disease induced by MPTP in rats. Fitoterapia 2013; 90: 112-8.
[http://dx.doi.org/10.1016/j.fitote.2013.07.009] [PMID: 23876367]

[338] Rajabian A, Rameshrad M, Hosseinzadeh H. Therapeutic potential of Panax ginseng and its constituents, ginsenosides and gintonin, in neurological and neurodegenerative disorders: a patent review 2019; 29(1): 55-72.
[http://dx.doi.org/10.1080/13543776.2019.1556258]

[339] Mu X, He G, Cheng Y, Li X, Xu B, Du G. Baicalein exerts neuroprotective effects in 6-hydroxydopamine-induced experimental parkinsonism *in vivo* and *in vitro.* Pharmacol Biochem Behav 2009; 92(4): 642-8.
[http://dx.doi.org/10.1016/j.pbb.2009.03.008] [PMID: 19327378]

[340] Mythri RB, Bharath MM. Curcumin: a potential neuroprotective agent in Parkinson's disease. Curr Pharm Des 2012; 18(1): 91-9.
[http://dx.doi.org/10.2174/138161212798918995] [PMID: 22211691]

[341] Abbaoui A, Chatoui H, El Hiba O, Gamrani H. Neuroprotective effect of curcumin-I in copper-induced dopaminergic neurotoxicity in rats: A possible link with Parkinson's disease. Neurosci Lett 2017; 660: 103-8.
[http://dx.doi.org/10.1016/j.neulet.2017.09.032] [PMID: 28919537]

[342] Jiang G, Hu Y, Liu L, Cai J, Peng C, Li Q. Gastrodin protects against MPP(+)-induced oxidative stress by up regulates heme oxygenase-1 expression through p38 MAPK/Nrf2 pathway in human dopaminergic cells. Neurochem Int 2014; 75: 79-88.
[http://dx.doi.org/10.1016/j.neuint.2014.06.003] [PMID: 24932697]

[343] Singh B, Pandey S, Verma R, Ansari JA, Mahdi AA. Comparative evaluation of extract of Bacopa monnieri and Mucuna pruriens as neuroprotectant in MPTP model of Parkinson's disease. Indian J Exp Biol 2016; 54(11): 758-66.
[PMID: 30179419]

[344] Ahmad M, Saleem S, Ahmad AS, *et al.* Neuroprotective effects of Withania somnifera on 6-hydroxydopamine induced Parkinsonism in rats. Hum Exp Toxicol 2005; 24(3): 137-47.
[http://dx.doi.org/10.1191/0960327105ht509oa] [PMID: 15901053]

[345] Freitas ME, Ruiz-Lopez M, Fox SH. Novel Levodopa Formulations for Parkinson's Disease. CNS Drugs 2016; 30(11): 1079-95.
       [http://dx.doi.org/10.1007/s40263-016-0386-8] [PMID: 27743318]

[346] Jankovic J, Goodman I, Safirstein B, *et al.* Safety and Tolerability of Multiple Ascending Doses of PRX002/RG7935, an Anti-α-Synuclein Monoclonal Antibody, in Patients With Parkinson Disease: A Randomized Clinical Trial. JAMA Neurol 2018; 75(10): 1206-14.
       [http://dx.doi.org/10.1001/jamaneurol.2018.1487] [PMID: 29913017]

[347] Torti M, Vacca L, Stocchi F. Istradefylline for the treatment of Parkinson's disease: is it a promising strategy? Expert Opin Pharmacother 2018; 19(16): 1821-8.
       [http://dx.doi.org/10.1080/14656566.2018.1524876] [PMID: 30232916]

[348] Bette S, Shpiner DS, Singer C, Moore H. Safinamide in the management of patients with Parkinson's disease not stabilized on levodopa: a review of the current clinical evidence. Ther Clin Risk Manag 2018; 14: 1737-45.
       [http://dx.doi.org/10.2147/TCRM.S139545] [PMID: 30271159]

[349] Fabbri M, Rosa MM, Ferreira JJ. Clinical pharmacology review of opicapone for the treatment of Parkinson's disease. Neurodegener Dis Manag 2016; 6(5): 349-62.
       [http://dx.doi.org/10.2217/nmt-2016-0022] [PMID: 27599671]

[350] Yao HM, Hsu A, Gupta S, Modi NB. Clinical Pharmacokinetics of IPX066: Evaluation of Dose Proportionality and Effect of Food in Healthy Volunteers. Clin Neuropharmacol 2016; 39(1): 10-7.
       [http://dx.doi.org/10.1097/WNF.0000000000000126] [PMID: 26626430]

[351] Fasano A, Bove F, Gabrielli M, *et al.* Liquid melevodopa *versus* standard levodopa in patients with Parkinson disease and small intestinal bacterial overgrowth. Clin Neuropharmacol 2014; 37(4): 91-5.
       [http://dx.doi.org/10.1097/WNF.0000000000000034] [PMID: 24992085]

[352] Bermejo PE, Anciones B. A review of the use of zonisamide in Parkinson's disease. Ther Adv Neurol Disord 2009; 2(5): 313-7.
       [http://dx.doi.org/10.1177/1756285609338501] [PMID: 21180621]

[353] Peball M, Werkmann M, Ellmerer P, *et al.* Nabilone for non-motor symptoms of Parkinson's disease: a randomized placebo-controlled, double-blind, parallel-group, enriched enrolment randomized withdrawal study (The NMS-Nab Study). J Neural Transm (Vienna) 2019; 126(8): 1061-72.
       [http://dx.doi.org/10.1007/s00702-019-02021-z] [PMID: 31129719]

[354] Athauda D, Maclagan K, Skene SS, *et al.* Exenatide once weekly *versus* placebo in Parkinson's disease: a randomised, double-blind, placebo-controlled trial. Lancet 2017; 390(10103): 1664-75.
       [http://dx.doi.org/10.1016/S0140-6736(17)31585-4] [PMID: 28781108]

[355] Rosa-Grilo M, Qamar MA, Taddei RN, *et al.* Rotigotine transdermal patch and sleep in Parkinson's disease: where are we now? NPJ Parkinsons Dis 2017; 3: 28.
       [http://dx.doi.org/10.1038/s41531-017-0030-4] [PMID: 28890931]

[356] Ilijic E, Guzman JN, Surmeier DJ. The L-type channel antagonist isradipine is neuroprotective in a mouse model of Parkinson's disease. Neurobiol Dis 2011; 43(2): 364-71.
       [http://dx.doi.org/10.1016/j.nbd.2011.04.007] [PMID: 21515375]

[357] Nishijima H, Ueno T, Kon T, *et al.* Effects of duloxetine on motor and mood symptoms in Parkinson's disease: An open-label clinical experience. J Neurol Sci 2017; 375: 186-9.
       [http://dx.doi.org/10.1016/j.jns.2017.01.066] [PMID: 28320128]

[358] Crosby N, Deane KH, Clarke CE. Amantadine in Parkinson's disease. Cochrane Database Syst Rev 2003; (1): CD003468.
       [PMID: 12535476]

[359] Richard IH, Szegethy E, Lichter D, Schiffer RB, Kurlan R. Parkinson's disease: a preliminary study of yohimbine challenge in patients with anxiety. Clin Neuropharmacol 1999; 22(3): 172-5.
       [PMID: 10367182]

[360] Groiss SJ, Wojtecki L, Südmeyer M, Schnitzler A. Deep brain stimulation in Parkinson's disease. Ther Adv Neurol Disord 2009; 2(6): 20-8.
[http://dx.doi.org/10.1177/1756285609339382] [PMID: 21180627]

[361] Beurrier C, Bioulac B, Audin J, Hammond C. High-frequency stimulation produces a transient blockade of voltage-gated currents in subthalamic neurons. J Neurophysiol 2001; 85(4): 1351-6.
[http://dx.doi.org/10.1152/jn.2001.85.4.1351] [PMID: 11287459]

[362] Dostrovsky JO, Levy R, Wu JP, Hutchison WD, Tasker RR, Lozano AM. Microstimulation-induced inhibition of neuronal firing in human globus pallidus. J Neurophysiol 2000; 84(1): 570-4.
[http://dx.doi.org/10.1152/jn.2000.84.1.570] [PMID: 10899228]

[363] Gradinaru V, Mogri M, Thompson KR, Henderson JM, Deisseroth K. Optical deconstruction of parkinsonian neural circuitry. Science 2009; 324(5925): 354-9.
[http://dx.doi.org/10.1126/science.1167093] [PMID: 19299587]

[364] Kalia SK, Sankar T, Lozano AM. Deep brain stimulation for Parkinson's disease and other movement disorders. Curr Opin Neurol 2013; 26(4): 374-80.
[http://dx.doi.org/10.1097/WCO.0b013e3283632d08] [PMID: 23817213]

[365] Okun MS. Deep-brain stimulation--entering the era of human neural-network modulation. N Engl J Med 2014; 371(15): 1369-73.
[http://dx.doi.org/10.1056/NEJMp1408779] [PMID: 25197963]

[366] Follett KA, Weaver FM, Stern M, *et al.* Pallidal *versus* subthalamic deep-brain stimulation for Parkinson's disease. N Engl J Med 2010; 362(22): 2077-91.
[http://dx.doi.org/10.1056/NEJMoa0907083] [PMID: 20519680]

[367] Grabli D, Karachi C, Folgoas E, *et al.* Gait disorders in parkinsonian monkeys with pedunculopontine nucleus lesions: a tale of two systems. J Neurosci 2013; 33(29): 11986-93.
[http://dx.doi.org/10.1523/JNEUROSCI.1568-13.2013] [PMID: 23864685]

[368] Gratwicke J, Zrinzo L, Kahan J, *et al.* Bilateral Deep Brain Stimulation of the Nucleus Basalis of Meynert for Parkinson Disease Dementia: A Randomized Clinical Trial. JAMA Neurol 2018; 75(2): 169-78.
[http://dx.doi.org/10.1001/jamaneurol.2017.3762] [PMID: 29255885]

[369] Lindvall O, Barker RA, Brüstle O, Isacson O, Svendsen CN. Clinical translation of stem cells in neurodegenerative disorders. Cell Stem Cell 2012; 10(2): 151-5.
[http://dx.doi.org/10.1016/j.stem.2012.01.009] [PMID: 22305565]

[370] Lindvall O, Kokaia Z. Stem cells for the treatment of neurological disorders. Nature 2006; 441(7097): 1094-6.
[http://dx.doi.org/10.1038/nature04960] [PMID: 16810245]

[371] Lindvall O, Kokaia Z, Martinez-Serrano A. Stem cell therapy for human neurodegenerative disorders-how to make it work. Nat Med 2004; 10 (Suppl.): S42-50.
[http://dx.doi.org/10.1038/nm1064] [PMID: 15272269]

[372] Kim JH, Auerbach JM, Rodríguez-Gómez JA, *et al.* Dopamine neurons derived from embryonic stem cells function in an animal model of Parkinson's disease. Nature 2002; 418(6893): 50-6.
[http://dx.doi.org/10.1038/nature00900] [PMID: 12077607]

[373] Yang D, Zhang ZJ, Oldenburg M, Ayala M, Zhang SC. Human embryonic stem cell-derived dopaminergic neurons reverse functional deficit in parkinsonian rats. Stem Cells 2008; 26(1): 55-63.
[http://dx.doi.org/10.1634/stemcells.2007-0494] [PMID: 17951220]

[374] Lindvall O. Developing dopaminergic cell therapy for Parkinson's disease--give up or move forward? Mov Disord 2013; 28(3): 268-73.
[http://dx.doi.org/10.1002/mds.25378] [PMID: 23401015]

[375] Politis M, Wu K, Loane C, *et al.* Serotonergic neurons mediate dyskinesia side effects in Parkinson's

patients with neural transplants. Sci Transl Med 2010; 2(38): 38ra46.
[http://dx.doi.org/10.1126/scitranslmed.3000976] [PMID: 20592420]

[376] Kefalopoulou Z, Politis M, Piccini P, *et al.* Long-term clinical outcome of fetal cell transplantation for Parkinson disease: two case reports. JAMA Neurol 2014; 71(1): 83-7.
[http://dx.doi.org/10.1001/jamaneurol.2013.4749] [PMID: 24217017]

[377] Freed CR, Greene PE, Breeze RE, *et al.* Transplantation of embryonic dopamine neurons for severe Parkinson's disease. N Engl J Med 2001; 344(10): 710-9.
[http://dx.doi.org/10.1056/NEJM200103083441002] [PMID: 11236774]

[378] Ma Y, Tang C, Chaly T, *et al.* Dopamine cell implantation in Parkinson's disease: long-term clinical and (18)F-FDOPA PET outcomes. J Nucl Med 2010; 51(1): 7-15.
[http://dx.doi.org/10.2967/jnumed.109.066811] [PMID: 20008998]

[379] Olanow CW, Goetz CG, Kordower JH, *et al.* A double-blind controlled trial of bilateral fetal nigral transplantation in Parkinson's disease. Ann Neurol 2003; 54(3): 403-14.
[http://dx.doi.org/10.1002/ana.10720] [PMID: 12953276]

[380] Piccini P, Pavese N, Hagell P, *et al.* Factors affecting the clinical outcome after neural transplantation in Parkinson's disease. Brain 2005; 128(Pt 12): 2977-86.
[http://dx.doi.org/10.1093/brain/awh649] [PMID: 16246865]

[381] Lane EL, Soulet D, Vercammen L, Cenci MA, Brundin P. Neuroinflammation in the generation of post-transplantation dyskinesia in Parkinson's disease. Neurobiol Dis 2008; 32(2): 220-8.
[http://dx.doi.org/10.1016/j.nbd.2008.06.011] [PMID: 18675359]

[382] Carlsson T, Carta M, Muñoz A, *et al.* Impact of grafted serotonin and dopamine neurons on development of L-DOPA-induced dyskinesias in parkinsonian rats is determined by the extent of dopamine neuron degeneration. Brain 2009; 132(Pt 2): 319-35.
[PMID: 19039008]

[383] Davis GC, Williams AC, Markey SP, *et al.* Chronic Parkinsonism secondary to intravenous injection of meperidine analogues. Psychiatry Res 1979; 1(3): 249-54.
[http://dx.doi.org/10.1016/0165-1781(79)90006-4] [PMID: 298352]

[384] Hsu LJ, Sagara Y, Arroyo A, *et al.* alpha-synuclein promotes mitochondrial deficit and oxidative stress. Am J Pathol 2000; 157(2): 401-10.
[http://dx.doi.org/10.1016/S0002-9440(10)64553-1] [PMID: 10934145]

[385] Müftüoglu M, Elibol B, Dalmizrak O, *et al.* Mitochondrial complex I and IV activities in leukocytes from patients with parkin mutations. Mov Disord 2004; 19(5): 544-8.
[http://dx.doi.org/10.1002/mds.10695] [PMID: 15133818]

[386] Guardia-Laguarta C, Area-Gomez E, Rüb C, *et al.* α-Synuclein is localized to mitochondria-associated ER membranes. J Neurosci 2014; 34(1): 249-59.
[http://dx.doi.org/10.1523/JNEUROSCI.2507-13.2014] [PMID: 24381286]

[387] Martin LJ, Pan Y, Price AC, *et al.* Parkinson's disease alpha-synuclein transgenic mice develop neuronal mitochondrial degeneration and cell death. J Neurosci 2006; 26(1): 41-50.
[http://dx.doi.org/10.1523/JNEUROSCI.4308-05.2006] [PMID: 16399671]

[388] Mortiboys H, Johansen KK, Aasly JO, Bandmann O. Mitochondrial impairment in patients with Parkinson disease with the G2019S mutation in LRRK2. Neurology 2010; 75(22): 2017-20.
[http://dx.doi.org/10.1212/WNL.0b013e3181ff9685] [PMID: 21115957]

[389] Saez-Atienzar S, Bonet-Ponce L, Blesa JR, *et al.* The LRRK2 inhibitor GSK2578215A induces protective autophagy in SH-SY5Y cells: involvement of Drp-1-mediated mitochondrial fission and mitochondrial-derived ROS signaling. Cell Death Dis 2014; 5: e1368.
[http://dx.doi.org/10.1038/cddis.2014.320] [PMID: 25118928]

[390] Narendra DP, Jin SM, Tanaka A, *et al.* PINK1 is selectively stabilized on impaired mitochondria to activate Parkin. PLoS Biol 2010; 8(1): e1000298.

[http://dx.doi.org/10.1371/journal.pbio.1000298] [PMID: 20126261]

[391] Wang X, Winter D, Ashrafi G, *et al.* PINK1 and Parkin target Miro for phosphorylation and degradation to arrest mitochondrial motility. Cell 2011; 147(4): 893-906.
[http://dx.doi.org/10.1016/j.cell.2011.10.018] [PMID: 22078885]

[392] Itoh K, Nakamura K, Iijima M, Sesaki H. Mitochondrial dynamics in neurodegeneration. Trends Cell Biol 2013; 23(2): 64-71.
[http://dx.doi.org/10.1016/j.tcb.2012.10.006] [PMID: 23159640]

[393] Morais VA, Verstreken P, Roethig A, *et al.* Parkinson's disease mutations in PINK1 result in decreased Complex I activity and deficient synaptic function. EMBO Mol Med 2009; 1(2): 99-111.
[http://dx.doi.org/10.1002/emmm.200900006] [PMID: 20049710]

[394] Gautier CA, Kitada T, Shen J. Loss of PINK1 causes mitochondrial functional defects and increased sensitivity to oxidative stress. Proc Natl Acad Sci USA 2008; 105(32): 11364-9.
[http://dx.doi.org/10.1073/pnas.0802076105] [PMID: 18687901]

[395] Buckman TD, Chang R, Sutphin MS, Eiduson S. Interaction of 1-methyl-4-phenylpyridinium ion with human platelets. Biochem Biophys Res Commun 1988; 151(2): 897-904.
[http://dx.doi.org/10.1016/S0006-291X(88)80366-8] [PMID: 3258155]

[396] Haas RH, Nasirian F, Nakano K, *et al.* Low platelet mitochondrial complex I and complex II/III activity in early untreated Parkinson's disease. Ann Neurol 1995; 37(6): 714-22.
[http://dx.doi.org/10.1002/ana.410370604] [PMID: 7778844]

[397] Benecke R, Strümper P, Weiss H. Electron transfer complexes I and IV of platelets are abnormal in Parkinson's disease but normal in Parkinson-plus syndromes. Brain 1993; 116(Pt 6): 1451-63.
[http://dx.doi.org/10.1093/brain/116.6.1451] [PMID: 8293280]

[398] Swerdlow RH, Parks JK, Cassarino DS, *et al.* Biochemical analysis of cybrids expressing mitochondrial DNA from Contursi kindred Parkinson's subjects. Exp Neurol 2001; 169(2): 479-85.
[http://dx.doi.org/10.1006/exnr.2001.7674] [PMID: 11358461]

[399] Aomi Y, Chen CS, Nakada K, *et al.* Cytoplasmic transfer of platelet mtDNA from elderly patients with Parkinson's disease to mtDNA-less HeLa cells restores complete mitochondrial respiratory function. Biochem Biophys Res Commun 2001; 280(1): 265-73.
[http://dx.doi.org/10.1006/bbrc.2000.4113] [PMID: 11162509]

[400] Cardellach F, Martí MJ, Fernández-Solá J, *et al.* Mitochondrial respiratory chain activity in skeletal muscle from patients with Parkinson's disease. Neurology 1993; 43(11): 2258-62.
[http://dx.doi.org/10.1212/WNL.43.11.2258] [PMID: 8232939]

[401] Blin O, Desnuelle C, Rascol O, *et al.* Mitochondrial respiratory failure in skeletal muscle from patients with Parkinson's disease and multiple system atrophy. J Neurol Sci 1994; 125(1): 95-101.
[http://dx.doi.org/10.1016/0022-510X(94)90248-8] [PMID: 7964895]

[402] DiDonato S, Zeviani M, Giovannini P, *et al.* Respiratory chain and mitochondrial DNA in muscle and brain in Parkinson's disease patients. Neurology 1993; 43(11): 2262-8.
[http://dx.doi.org/10.1212/WNL.43.11.2262] [PMID: 8232940]

[403] Reichmann H, Janetzky B, Bischof F, *et al.* Unaltered respiratory chain enzyme activity and mitochondrial DNA in skeletal muscle from patients with idiopathic Parkinson's syndrome. Eur Neurol 1994; 34(5): 263-7.
[http://dx.doi.org/10.1159/000117053] [PMID: 7995300]

[404] Shinde S, Pasupathy K. Respiratory-chain enzyme activities in isolated mitochondria of lymphocytes from patients with Parkinson's disease: preliminary study. Neurol India 2006; 54(4): 390-3.
[http://dx.doi.org/10.4103/0028-3886.28112] [PMID: 17114849]

[405] Nakamura T, Prikhodko OA, Pirie E, *et al.* Aberrant protein S-nitrosylation contributes to the pathophysiology of neurodegenerative diseases. Neurobiol Dis 2015; 84: 99-108.
[http://dx.doi.org/10.1016/j.nbd.2015.03.017] [PMID: 25796565]

[406] van Muiswinkel FL, Steinbusch HW, Drukarch B, de Vente J. Identification of NO-producing and -receptive cells in mesencephalic transplants in a rat model of Parkinson's disease: a study using NADPH-d enzyme- and NOSc/cGMP immunocytochemistry. Ann N Y Acad Sci 1994; 738: 289-304.
[http://dx.doi.org/10.1111/j.1749-6632.1994.tb21815.x] [PMID: 7530418]

[407] Levecque C, Elbaz A, Clavel J, *et al.* Association between Parkinson's disease and polymorphisms in the nNOS and iNOS genes in a community-based case-control study. Hum Mol Genet 2003; 12(1): 79-86.
[http://dx.doi.org/10.1093/hmg/ddg009] [PMID: 12490535]

[408] Joniec I, Ciesielska A, Kurkowska-Jastrzebska I, Przybylkowski A, Czlonkowska A, Czlonkowski A. Age- and sex-differences in the nitric oxide synthase expression and dopamine concentration in the murine model of Parkinson's disease induced by 1-methyl-4-phenyl-1,2,3,6-tetrahydropyridine. Brain Res 2009; 1261: 7-19.
[http://dx.doi.org/10.1016/j.brainres.2008.12.081] [PMID: 19401171]

[409] Luoma PT, Eerola J, Ahola S, *et al.* Mitochondrial DNA polymerase gamma variants in idiopathic sporadic Parkinson disease. Neurology 2007; 69(11): 1152-9.
[http://dx.doi.org/10.1212/01.wnl.0000276955.23735.eb] [PMID: 17846414]

[410] Mancuso M, Filosto M, Bellan M, *et al.* POLG mutations causing ophthalmoplegia, sensorimotor polyneuropathy, ataxia, and deafness. Neurology 2004; 62(2): 316-8.
[http://dx.doi.org/10.1212/WNL.62.2.316] [PMID: 14745080]

[411] Taanman JW, Schapira AH. Analysis of the trinucleotide CAG repeat from the DNA polymerase gamma gene (POLG) in patients with Parkinson's disease. Neurosci Lett 2005; 376(1): 56-9.
[http://dx.doi.org/10.1016/j.neulet.2004.11.023] [PMID: 15694274]

[412] Swerdlow RH, Parks JK, Miller SW, *et al.* Origin and functional consequences of the complex I defect in Parkinson's disease. Ann Neurol 1996; 40(4): 663-71.
[http://dx.doi.org/10.1002/ana.410400417] [PMID: 8871587]

[413] Gu M, Cooper JM, Taanman JW, Schapira AH. Mitochondrial DNA transmission of the mitochondrial defect in Parkinson's disease. Ann Neurol 1998; 44(2): 177-86.
[http://dx.doi.org/10.1002/ana.410440207] [PMID: 9708539]

[414] Sheehan JP, Swerdlow RH, Parker WD, Miller SW, Davis RE, Tuttle JB. Altered calcium homeostasis in cells transformed by mitochondria from individuals with Parkinson's disease. J Neurochem 1997; 68(3): 1221-33.
[http://dx.doi.org/10.1046/j.1471-4159.1997.68031221.x] [PMID: 9048769]

[415] Swerdlow RH, Parks JK, Davis JN II, *et al.* Matrilineal inheritance of complex I dysfunction in a multigenerational Parkinson's disease family. Ann Neurol 1998; 44(6): 873-81.
[http://dx.doi.org/10.1002/ana.410440605] [PMID: 9851431]

[416] Thyagarajan D, Bressman S, Bruno C, *et al.* A novel mitochondrial 12SrRNA point mutation in parkinsonism, deafness, and neuropathy. Ann Neurol 2000; 48(5): 730-6.
[http://dx.doi.org/10.1002/1531-8249(200011)48:5<730::AID-ANA6>3.0.CO;2-0] [PMID: 11079536]

[417] Pang Y, Wang C, Yu L. Mitochondria-Targeted Antioxidant SS-31 is a Potential Novel Ophthalmic Medication for Neuroprotection in Glaucoma. Med Hypothesis Discov Innov Ophthalmol 2015; 4(3): 120-6.
[PMID: 27350953]

[418] Cho S, Szeto HH, Kim E, Kim H, Tolhurst AT, Pinto JT. A novel cell-permeable antioxidant peptide, SS31, attenuates ischemic brain injury by down-regulating CD36. J Biol Chem 2007; 282(7): 4634-42.
[http://dx.doi.org/10.1074/jbc.M609388200] [PMID: 17178711]

[419] Lee HY, Kaneki M, Andreas J, Tompkins RG, Martyn JA. Novel mitochondria-targeted antioxidant peptide ameliorates burn-induced apoptosis and endoplasmic reticulum stress in the skeletal muscle of mice. Shock 2011; 36(6): 580-5.

[http://dx.doi.org/10.1097/SHK.0b013e3182366872] [PMID: 21937949]

[420] Dikalova AE, Bikineyeva AT, Budzyn K, *et al.* Therapeutic targeting of mitochondrial superoxide in hypertension. Circ Res 2010; 107(1): 106-16.
[http://dx.doi.org/10.1161/CIRCRESAHA.109.214601] [PMID: 20448215]

[421] Wilson RJ, Drake JC, Cui D, *et al.* Mitochondrial protein S-nitrosation protects against ischemia reperfusion-induced denervation at neuromuscular junction in skeletal muscle. Free Radic Biol Med 2018; 117: 180-90.
[http://dx.doi.org/10.1016/j.freeradbiomed.2018.02.006] [PMID: 29432799]

[422] Pua KH, Stiles DT, Sowa ME, Verdine GL. IMPDH2 Is an Intracellular Target of the Cyclophilin A and Sanglifehrin A Complex. Cell Rep 2017; 18(2): 432-42.
[http://dx.doi.org/10.1016/j.celrep.2016.12.030] [PMID: 28076787]

[423] Sabbah HN, Gupta RC, Kohli S, Wang M, Hachem S, Zhang K. Chronic Therapy With Elamipretide (MTP-131), a Novel Mitochondria-Targeting Peptide, Improves Left Ventricular and Mitochondrial Function in Dogs With Advanced Heart Failure. Circ Heart Fail 2016; 9(2): e002206.
[http://dx.doi.org/10.1161/CIRCHEARTFAILURE.115.002206] [PMID: 26839394]

[424] Chaitman BR. Ranolazine for the treatment of chronic angina and potential use in other cardiovascular conditions. Circulation 2006; 113(20): 2462-72.
[http://dx.doi.org/10.1161/CIRCULATIONAHA.105.597500] [PMID: 16717165]

[425] Aldasoro M, Guerra-Ojeda S, Aguirre-Rueda D, *et al.* Effects of Ranolazine on Astrocytes and Neurons in Primary Culture. PLoS One 2016; 11(3): e0150619.
[http://dx.doi.org/10.1371/journal.pone.0150619] [PMID: 26950436]

[426] Shi CX, Zhao MX, Shu XD, *et al.* β-aminoisobutyric acid attenuates hepatic endoplasmic reticulum stress and glucose/lipid metabolic disturbance in mice with type 2 diabetes. Sci Rep 2016; 6: 21924.
[http://dx.doi.org/10.1038/srep21924] [PMID: 26907958]

[427] Le Page LM, Rider OJ, Lewis AJ, *et al.* Increasing Pyruvate Dehydrogenase Flux as a Treatment for Diabetic Cardiomyopathy: A Combined 13C Hyperpolarized Magnetic Resonance and Echocardiography Study. Diabetes 2015; 64(8): 2735-43.
[http://dx.doi.org/10.2337/db14-1560] [PMID: 25795215]

[428] Lewis JF, DaCosta M, Wargowich T, Stacpoole P. Effects of dichloroacetate in patients with congestive heart failure. Clin Cardiol 1998; 21(12): 888-92.
[http://dx.doi.org/10.1002/clc.4960211206] [PMID: 9853180]

[429] Miley GP, Pou S, Winter R, *et al.* ELQ-300 prodrugs for enhanced delivery and single-dose cure of malaria. Antimicrob Agents Chemother 2015; 59(9): 5555-60.
[http://dx.doi.org/10.1128/AAC.01183-15] [PMID: 26124159]

CHAPTER 3

Neurotrophic Factors to Combat Neurodegeneration

Yulia A. Sidorova[1,*]

[1] Institute of Biotechnology, HiLIFE, Viikinkaari 5D, FI-00014 University of Helsinki, Helsinki, Finland

Abstract: Conditions caused by the lesion and progressive death of neuronal cells in the organism include neurodegenerative disorders and neuropathic pain. They represent the major causes of disability in Western countries. These conditions are more common in elderly people, and their prevalence, therefore, is expected to grow in the future because of the aging population. Currently, curative therapies against neurodegenerative disorders and neuropathic pain are not available. Existing treatments may provide temporary symptomatic relief to some patients but fail to stop neuronal degeneration, protect and restore damaged neurons. Neurotrophic factors are small secretory proteins whose main function is to support the survival of neurons. Therefore, they hold considerable promise for disease-modifying treatment of neurodegenerative disorders and neuropathic pain. However, despite promising results in preclinical studies, clinical translation of neurotrophic factors has so far achieved limited success. Neurotrophic factors are different from traditional chemical compounds used as drugs in the majority of cases, and this complicates their clinical use. Biology of neurotrophic factors and their absorption, distribution, metabolism, excretion, and pharmacokinetics properties dictate special requirements to clinical trials design. Patients taking part in clinical trials, delivery system, delivery paradigm, and the dose of neurotrophic factor should be carefully considered in trial design in order to ensure that the treatment will improve the condition of patients.

In the present chapter, the author summarizes the available literature regarding signaling of neurotrophic factors, provides the data about their preclinical evaluation in animal models of neurodegenerative disorders and neuropathic pain, describes the results of clinical trials conducted with neurotrophic factors in patients, and discusses the limitations of these trials and translational problems faced by researchers and clinicians in this field. The author will further discuss emerging alternatives to neurotrophic factor proteins with improved translational perspectives, such as mutant proteins, small molecules, and peptides targeting the receptors of neurotrophic factors. The author will review attempts of clinical translation of glial cell line-derived neurotrophic factor family ligands for the treatment of Parkinson's disease and neuropathic pain. The author will briefly describe the non-conventional cerebral dopamine neurotrophic factor tested in Phase I/II clinical trial in patients with Parkin-

---

* Corresponding author Yulia A. Sidorova: Institute of Biotechnology, HiLIFE, Viikinkaari 5D, FI-00014 University of Helsinki, Helsinki, Finland; Tel: +358400267815; E-mail: yulia.sidorova@helsinki.fi

Atta-ur-Rahman & Zareen Amtul (Eds.)

son's disease. The author will also describe the data concerning the clinical evaluation of other neurotrophic factors in the above-mentioned conditions.

Keywords: Amyotrophic lateral sclerosis, Alzheimer's Disease, Brain-Derived Neurotrophic Factor (BDNF), Cerebral Dopamine Neurotrophic Factor (CDNF), Ciliary Neurotrophic Factor (CNTF), Clinical trials, Clinical translation, Glial Cell Line-Derived Neurotrophic Factor (GDNF), Glial Cell Line-Derived Neurotrophic Factor Family Ligands (GFLs), Leukemia Inhibitory Factor (LIF), Neurodegenerative Disorders, Neuropathic Pain, Neurotrophic Factors, Neurotrophins, Nerve Growth Factor (NGF), Neurokines, Neurorestoration, Neuroprotection, Neurodegeneration, Parkinson's Disease, Small Molecules Targeting Neurotrophic Factor Receptors.

NEUROTROPHIC FACTORS

What are Neurotrophic Factors?

Neurotrophic factors (NTFs) are small secreted proteins that support the development, survival, and functioning of neurons. While many growth factors are able to positively influence certain aspects of neuronal cell life, the distinctive feature of NTFs is the ability to promote neuronal survival. The relations between neuronal survival and neurotrophic factors can be described by the target field concept, according to which, during the development, a large number of neurons are born. However, only neurons reaching their target tissues are able to receive neurotrophic support and survive, while those unable to access NTFs produced by target tissue will die [1]. A recent study published by Wang and co-authors provided new insights on the mechanisms of neuronal deaths in development. Authors showed that intrinsic factors, such as the expression of NTFs receptors in neurons, are important for the prediction of neuronal fate: sensory neuron precursors expressing a low level of Tropomyosin receptor kinase (Trk) C, a receptor for neurotrophin 3 (NT-3), were more prone to death compared to the neurons expressing high levels of TrkC [2]. The target field hypothesis is generally accepted for the peripheral nervous system. Available data for the central nervous system is less straightforward. However, general principles seem to be the same. In the organism, NTFs are synthesized by target tissues, neuronal, glial, and endothelial cells [3].

Classification of Neurotrophic Factors

Currently, NTFs include 4 protein families: neurotrophins, neuropoietic cytokines, glial cell line-derived (GDNF) family ligands (GFLs), and a family of cerebral dopamine neurotrophic factor (CDNF) and mesencephalic astrocyte-derived neurotrophic factor (MANF) (Table **1**). Mammalian neurotrophins include nerve growth factor (NGF), which was the first discovered NTF [4, 5], brain-derived neurotrophic factor (BDNF), neurotrophin-3 (NT-3), and neurotrophin-4/5 (NT-4/5) [6]. Neuropoietic cytokines or neurokines are a group of diverse proteins with pleiotropic effects in the brain, including Interleukin-6 (IL-6) as a founding member, leukemia inhibitory factor (LIF), Ciliary Neurotrophic Factor (CNTF), oncostatin M, cardiotrophin-1, neuropoietin, and cytokine cardiotrophin-like (also known as new neurotrophin 1 and B cell stimulatory factor-3) [3, 7]. GFLs include 4 conventional members which are GDNF, neurturin (NRTN), artemin (ARTN) (also known as enovin and neuroblastin), and persephin (PSPN) [8, 9], as well as a distantly related protein Growth Differentiation Factor 15 (GDF15) [10 - 14]. MANF and CDNF are two structurally related proteins which are different from classic target-derived NTFs. They reside in the endoplasmic reticulum and may play a role in controlling protein homeostasis [15, 16].

NTFs are synthesized as preproproteins and are enzymatically cleaved during the processing in the endoplasmic reticulum. ProNTFs and mature NTFs may have opposite functions in the cells; for instance, mature neurotrophins promote neuronal survival while proneurotrophins activate apoptotic pathways in the cells [17]. Moreover, aminoacid composition of pro region of NTF can affect biology; for example, secretion and cellular localization of NTF [18].

Neurotrophic Factor Receptors and Signaling

Neurotrophins, GFLs, and neuropoietic cytokines signal *via* kinase receptors while the signaling of MANF/CDNF is less thoroughly understood [19]. The receptors of NTFs (Table **1**) and the biological aspects of neurotrophic factors signaling which are important for clinical translation are described below.

Table 1. Neurotrophic factors in their receptors.

Neurotrophic Factor Family	Neurotrophic Factor	Receptors and Co-Receptors	Comments
Glial cell line-derived neurotrophic factor (GDNF) family ligands	GDNF	GFRα1, RET, NCAM, Syndecan-3	Can also interact with GFRα2
	Neurturin	GFRα2, RET, NCAM, Syndecan-3	All can signal also *via* GFRa1
	Artemin	GFRα3, RET, Syndecan-3	
	Persephin	GFRα4, RET	
	GDF15	GFRAL, RET	-

(Table 1) cont.....

Neurotrophic Factor Family	Neurotrophic Factor	Receptors and Co-Receptors	Comments
Neurotrophins	NGF	p75[NTR], TrkA	Immature proneurotrophins can bind to p75[NTR] and stimulate apoptosis
-	BDNF	p75[NTR], TrkB	
-	NT-3	p75[NTR], TrkC	
-	NT-4/5	p75[NTR], TrkB	
Neurokines	CNTF	LIF receptor, CNTF receptor and gp130	-
	LIF	gp130, LIF receptor	-
	Others	gp130 as a dimer or monomer in combination with different co-receptors	-
CDNF/MANF family	CDNF and MANF	Unknown, uptake is probably mediated *via* interactions with lipids	Unconventional, ER-resident proteins

Neurotrophins

Neurotrophins signal *via* transmembrane receptor tyrosine kinases Trks and p75[NTR] receptor, which lacks intrinsic catalytic activity but can form complexes with other proteins. There are three Trk receptors that are selective for individual neurotrophins: NGF preferentially signals *via* TrkA, BDNF, and NT-4/5- through TrkB, NT-3 - through TrkC [20]. However, cross-talks between these NTFs and receptors are also possible [21]. Trks are classical receptor tyrosine kinases, ligand binding leads to their dimerization which is followed by autophosphorylation, binding of adaptor proteins and activation of intracellular signaling cascades such as mitogen-activated protein kinase (MAPK) and phosphatidylinositol 3-kinase (PI3K)/Akt [20]. Mature and proneurotrophins can also interact with low-affinity receptor p75[NTR]. p75[NTR] can form a complex with Trk, which has a higher affinity to mature neurotrophins compared to Trk alone [22]. Interaction of proneurotrophins with p75[NTR] stimulates the formation of a complex with sortilin and activation of apoptotic pathways resulting in cell death [17, 23]. In the cells expressing p75[NTR] in the absence of Trk, mature neurotrophins can also induce cell death [24].

Glial Cell Line-derived Neurotrophic Factor Family Ligands

The receptor complex for GFLs consists of two structurally different subunits: a transmembrane receptor tyrosine kinase RET which transmits signals from all five members of the family and the glycosylphosphatidylinositol (GPI)-anchored ligand-binding subunit called GDNF family receptor alpha 1-4 (GFRα1-4) or transmembrane GDNF family receptor alpha-like (GFRAL) selective for individual proteins. GDNF has a high affinity to GFRa1, NRTN – to GFRa2, ARTN – to GFRa3 and PSPN – to GFRa4. In addition, all these four proteins can signal *via* GFRa1 [25], and GDNF can signal *via* GFRa2 [8]. GDF15 requires GFRAL to elicit cellular response [10 - 13]; it seems to be unable to interact with GFRa1-4. Also, other GFLs do not signal *via* GFRAL. GFRa co-receptors locate in so-called lipid rafts, membrane domains enriched in cholesterol and sphingolipids [26]. Ligand binding to GFRa promotes the recruitment of RET into raft, RET autophosphorylation and activation of intracellular signaling cascades such as MAPK and PI3K/Akt [27, 28]. GFRa co-receptors can also stimulate the formation of signaling complex in a soluble form, when they are cleaved from the membrane. GFRAL is structurally different from GFRa and has an intracellular domain which seems to be important for its biological function; however, it also stimulates RET phosphorylation and subsequent downstream signaling events. While all five GFLs activate RET and similar downstream signaling cascades, the kinetics of these events can be different for individual proteins. The structural features, in particular, the angle between RET monomers vary for the receptor complexes formed by individual GFLs and their co-receptors. This seems to affect the duration and speed of downstream cascades activation allowing fine-tuning the cellular response to individual proteins [29, 30].

Apart from RET, signals from GDNF, NRTN and ARTN can be transmitted by proteoglycan Syndecan-3. Binding of GFLs to this receptor does not require the presence of co-receptor [31]. Another alternative receptor for GFLs is neural cell adhesion molecule (NCAM). GFLs bind to NCAM in complex with GFRa coreceptor [32]. Both syndecan-3 and NCAM are more widely expressed in the nervous system compared to RET.

Neuropoietic Cytokines

The signals from neurokines are transmitted by a receptor complex which includes a transmembrane glycoprotein gp130 as a dimer or monomer and, in many cases, also a low-affinity ligand-binding receptor(s) selective for individual proteins. The exact composition of the signaling complex for every individual neurokine varies pronouncedly. For instance, the signalling complex for IL-6 includes two molecules of gp130 and one molecule of IL-6 receptor; the signaling complexes for LIF or oncostatin M consist of monomeric gp130 and monomeric LIF receptor or oncostatin M receptor, respectively. CNTF signals *via* a trimeric complex which includes LIF receptor, CNTF receptor, and gp130, all as monomers [3, 33, 34]. Ligand-binding subunits of neurokine receptors may have different structural organizations, while IL-6 and LIF receptors are transmembrane, CNTF receptor is GPI-anchored and can also transmit the signal in the soluble form [34]. Components of receptor complexes for neuropoietic cytokines lack intrinsic catalytic activity. Instead, they are associated with Janus kinases (JAKs) and activate upon interaction with ligands, mainly JAK-Stat, but also MAPK, PI3K, and insulin receptor substrate (IRS) signaling pathways [3].

MANF/CDNF

MANF and CDNF signal through very different mechanisms compared to those activated by other neurotrophic factors. In fact, at least MANF has a prominent role outside the nervous system, controlling the proliferation and survival of insulin-producing β-cells in the pancreas and mice lacking *MANF* gene develop diabetes mellitus [35]. MANF and CDNF are expressed in the majority if not all tissues in the organism. MANF expression is highest in the tissues characterized by the high level of protein secretion and in endocrine cells. High levels of CDNF mRNA and protein were detected in the heart and muscle [36]. MANF and CDNF reside in ER. MANF expression and secretion are increased by unfolded protein response (UPR) and, therefore, these NTFs can potentially play a role in UPR and ER stress regulation [16]. ER stress is a process occurring in the cells as a result of the accumulation of unfolded proteins. At the beginning, ER stress has a protective role in cells: it reduces misfolded proteins load by decreasing the overall rate of protein synthesis, stimulation of chaperone production, and activation of protein degradation pathways. However, prolonged ER stress is detrimental and may lead to the death of the affected cells [37]. As described below, protein aggregation is often observed in neurodegenerative disorders. Perhaps, the neurotrophic activity of MANF and CDNF is a concomitant effect of their function in the regulation of unfolded protein load in the cells.

Although traditional transmembrane protein receptors for MANF and CDNF have not been identified, these proteins, when delivered exogenously to extracellular space, are able to protect and repair neurons. This implies the presence of either a cell membrane receptor or an internalization mechanism for these proteins in the cells. MANF and CDNF contain N-terminal lipid-binding domains [38]. Moreover, recently, it was demonstrated that MANF interacts with the lipid called sulfatide or 3-O-sulfogalactosylceramide, which mediates the cellular uptake of this NTF. Therefore, it is possible that the effects of exogenously delivered MANF and CDNF are mediated *via* their interactions with lipids [39]. Suggested mechanisms of MANF and CDNF's positive effect on cell survival include downregulation of ER stress pathways, regulation of calcium homeostasis and/or activation of antiapoptotic pathways in the cells [16].

Biology of Neurotrophic Factor Signaling: Aspects Important for Clinical Translation

In this section, the overall sequence of events happening in the cells upon stimulation with NTFs will be described. Not all described mechanisms were directly demonstrated experimentally using NTFs. However, the similarity of NTF receptors and signalling mechanisms with that seen for other growth factors allows approximating findings made for the signaling elicited by other receptor tyrosine kinases also to the NTF receptor-dependent signaling. At the moment, it is difficult to say whether all principles described below apply to MANF and CDNF signaling as they are very different from classic NTFs such as neurokines, GFLs, and neurotrophins and have a distinct mechanism of action. Since little is known about MANF and CDNF receptors, the following text will be mainly focused on the discussion of cellular signaling elicited for conventional NTFs. However, homeostatic mechanisms, regulated synthesis, and release to extracellular space are likely applicable to a wide variety of biological processes and targets. Therefore, basic principles may be similar for MANF and CDNF as well.

NTFs are released by the cells intermittently. Upon binding to cell surface receptors, NTFs are internalized, and receptors may be degraded or recycled. The half-life of endogenous or exogenously delivered neurotrophic factors is rather short. Exogenously delivered growth factors are either internalized and degraded inside the cells or cleaved by proteases in intracellular space within several hours. However, the cellular changes elicited by these proteins can outlast the presence of neurotrophic factor itself for many days or even months.

After binding to receptors, at least classic neurotrophic factors stimulate a cascade of intracellular events that shape the cell response to the stimulation. This cascade includes rapid and delayed events. Within minutes after the interaction, the ligand stimulates the phosphorylation of the receptor, which in turn triggers phosphorylation of a number of secondary messengers in signaling cascades. Signalling cascade effectors finally activate transcription factors which initiate the expression of responsive genes. This process takes longer to start and usually occurs within hours after the introduction of the stimulus, and may last for several days. Newly synthesized proteins influence the fate and function of affected cells stimulating even more long-lasting changes, such as, for instance, the growth of neurites, functional adaptations, better survival, *etc.*, that can persist for weeks or even months.

Stimulation with neurotrophic factors also activates mechanisms of negative feedback that inactivate receptors and signaling cascade effectors in order to limit the extent and duration of stimulation and prevent uncontrollable signal amplification. Ligand-bound receptors are internalized and can be either degraded or delivered back to the plasma membrane during the recycling process. An increase in the concentration of ligand results in a decrease in the number of available receptor molecules on the plasma membrane (Fig. **1**). In experimental settings, it was demonstrated for a receptor tyrosine kinase transmitting signals from epidermal growth factor (EGF). In the presence of a low concentration of EGF, internalized receptors are mainly recycled, while in the presence of high EGF concentration, the majority of internalized receptors are degraded, rendering cells irresponsive to further stimulation [40].

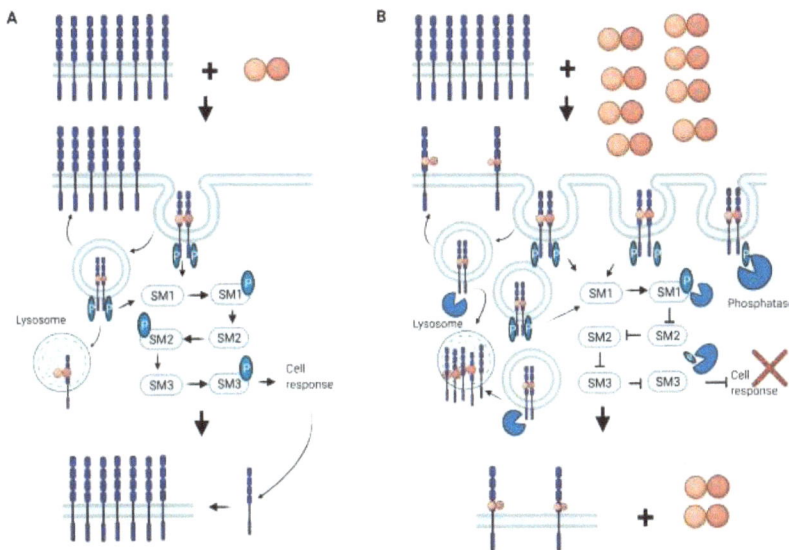

Fig. (1). Cellular events elicited by neurotrophic factors in physiological (A) and supraphysiological concentrations (B). **A**. In physiological concentration neurotrophic factor binds to the receptor and stimulates its dimerization (or oligomerization). Receptor becomes phosphorylated and activates secondary messengers (SM) in signaling cascades which in turn triggers cellular response *via* activation of gene expression. Receptor itself can be recycled or degraded. If necessary additional copies of the receptor can be synthesized and the cell is fully prepared to the next cycle. **B**. When neurotrophic factor is added to the cell in excessive concentration it stimulates internalization of several receptor-ligand complexes and bind remaining receptor molecules on cell membrane in 1:1 ratio rendering them unable to dimerise. Excessive signaling activates phosphatases and other negative feedback mechanisms in signaling cascades and stimulates degradation of the receptor. These events reduce or even block cellular response. The new receptor molecules to replace degraded copies are not synthesized. Cell is not equipped to respond to the neurotrophic factor in the next cycle even if the ligand is still available.

Another event happening in the cells upon stimulation with the ligand signaling *via* receptor tyrosine kinase or kinase-associated receptors is the activation of phosphatases which dephosphorylate receptors [41] and effectors [42] in signaling cascades, thus limiting signal propagation. In this manner, cells maintain homeostasis. Experimental evidence indicates that the response of cultured neurons to the neurotrophic factors is bell-shaped [43]. An initial increase in neurotrophic factor concentration is accompanied by the proportional increase in neuronal survival until a certain limit, after which the survival of neurons is decreasing despite a continuous increase in neurotrophic factor concentration [43, 44]. An elegant model to explain this phenomenon apart from the options presented above was proposed by Shlee and co-authors: dimerization or oligodimerization of receptors around a single molecule of a ligand is an important step in the initiation of neurotrophic factor signaling. It is possible that in the presence of a high concentrations of neurotrophic factor, every monomeric receptor is bound to a ligand (Fig. **1**), making the formation of functional dimeric/multimeric signaling complex impossible [45].

Another noteworthy aspect to mention in this context is the possibility to induce excessive sprouting and incorrect wiring of neurons upon the treatment with supraphysiological concentrations of growth factors. In the physiological context, the neurons are surrounded by a complex network of stimuli, some of which serve as chemoattractants, and the others are repulsive for forming neurites. This complex network of attractive and aversive stimuli allows growing neuronal axons to innervate correct targets. The application of high doses of NTFs may distort the balance of positive and negative stimuli and result in the growth of axons into incorrect regions. This was demonstrated in the dorsal root crush model in rats upon application of NGF. In the dorsal root crush model, the axons of sensory neurons projecting into the spinal cord are damaged. This leads to the formation of a highly repulsive region, which prevents the entry of regenerating axons into the spinal cord. NGF is a potent survival- and neurite outgrowth-promoting factor for a subset of sensory neurons. Upon injection of adenovirus-encoded NGF into the spinal cord of animals with crushed dorsal routes, regenerating axons became able to overcome repulsive region and enter the spinal cord. However, they failed to innervate topographically correct regions of the spinal cord. Instead, these axons projected massively to incorrect regions of the spinal cord [46, 47], which is an undesirable effect that may lead to negative consequences, *e.g.*, pain.

In essence, it is important to understand that an excessive and/or long-term stimulation of cells with neurotrophic factors will not increase the neuronal capacity to survive or function above a certain threshold. Instead, the activation of negative feedback mechanisms will result in the loss of cellular responsiveness to the stimulation and may even have a detrimental effect on cell survival [45]. Excessive stimulation with neurotrophic factors can also result in undesirable excessive sprouting caused by disruption of repulsive and attractive stimuli network in the tissue.

These aspects make neurotrophic factors stay apart from the majority of clinically-used small molecule drugs, which inhibit certain enzymes in the cells, and are characterized by linear dose-response relationships. These issues should be considered in the context of the clinical translation of NTFs.

NEURODEGENERATIVE DISORDERS

Neurodegenerative disorders are a group of diseases that are characterized by the progressive death of specific neuronal populations in the nervous system [48]. The two most common neurodegenerative disorders are Alzheimer's disease (AD) and Parkinson's disease (PD), which affect more than 25 mln of people [49] and over 10 mln people in the world, respectively [50, 51]. Huntington's disease (HD) and Amyotrophic lateral sclerosis (ALS) are less common, with prevalence estimates between 4 and 14 per 100000 persons per year [52, 53]. Apart from listed diseases, a few other rare conditions such as spinocerebellar ataxia, spinal muscular atrophy, and multiple system atrophy, *etc.*, are considered neurodegenerative disorders [54 - 56].

A common molecular feature of neurodegenerative disorders seems to be the presence of insoluble protein aggregates in the central nervous system. AD is characterized by the presence of amyloid plaques containing accumulated β-amyloid peptides Aβ-40 and Aβ-42 and neurofibrillary tangles composed of filamentous tau proteins [57]. In the brains of PD patients, Lewy bodies consisting of aggregated α-synuclein as a major component along with more than 90 other proteins [58] and, according to the recent data, also organelles and lipid membrane structure [20], are present. In ALS, aggregates of various proteins were identified in the spinal cords, including superoxide dismutase (SOD), notably fused in sarcoma (FUS), TAR DNA-binding protein 43 (TDP-43), and others [59]. In HD, intracellular inclusion bodies consisting of the aggregated mutated huntingtin are found in the brains [60].

The role of these aggregates in neurodegeneration is somewhat unclear. Although based on preclinical research prevailing concept considers these aggregates as causative agents for the development of neurodegenerative disorders, some data from clinical studies do not fit this hypothesis. In particular, the severity of behavioral pathology in PD and AD does not always correlate with the severity of histopathological changes related to protein aggregation [61]. At the early stage, the aggregates are mainly seen in the cerebral cortex and striatal interneurons in HD patients, while degeneration occurs predominantly in medium spiny projection neurons [60]. In addition, multiple agents targeting amyloid plaques or preventing their formation failed to demonstrate efficacy in AD patients [61]. Thus, while protein aggregation in neurodegeneration may exacerbate neuronal death by, *e.g.*, interfering with cellular transport [62], it can also be a neutral event (co-existing with or appearing as a result of cellular malfunction) or even a neuroprotective mechanism reducing the toxicity of soluble forms of above-indicated proteins or sequestering toxic misfolded proteins [58, 60, 61].

Multiple attempts were made to target protein aggregation in patients with neurodegeneration. Neither of them resulted in unequivocal success. One of the arguments explaining the lack of efficacy of such treatments in patients with neurodegenerative disorders is an unsuitable time for the intervention. At the moment, reliable methods of early diagnostics of neurodegenerative disorders are missing. Therefore, when the disease is diagnosed, a significant portion of neurons have already degenerated. Thus, even if the treatment targeting protein aggregation has had a pharmacodynamic effect and reduced the number of aggregates, it may not lead to cognitive or motor improvement due to the insufficient number of remaining target neurons to elicit a clinically relevant functional response. In addition, it is unclear whether elimination of protein aggregates from the cells is sufficient to stimulate at least functional recovery of affected neurons or additional neurorestorative cues are required in order to achieve this goal. The development of early diagnostic criteria for neurodegenerative disorders can help to address these matters in the future.

Neuronal cells are postmitotic and do not divide. The birth of new neuronal cells, while reported in model organisms, seems to occur very rarely in the adult human brain [63]. Therefore, strategies focused on the protection and restoration of remaining cells are especially attractive in the context of finding a cure against neurodegenerative disorders. Due to their ability to support neuronal cells, neurotrophic factors attracted significant attention as potential disease-modifying treatments for neurodegeneration. They were extensively tested in preclinical models of these diseases and in clinical trials in patients with neurodegeneration, as described in the subsequent sections.

Alzheimer's Disease

AD is the most common, mainly sporadic, neurodegenerative disorder characterized by progressive cognitive decline [64]. Inherited forms are associated with early-onset and occur according to different estimates in less than 5% or even less than 1% of patients [65, 66]. The most common symptom of AD in early-stage patients is short-term memory impairment [67]. In the majority of patients, it is accompanied by a variety of neuropsychiatric symptoms ranging from apathy to aggression and even euphoria [68].

AD is characterized by progressive brain atrophy predominantly in the hippocampus and cortex, which accelerates when the disease progresses [69, 70]. Cholinergic neurons are among the most affected [71]. Currently, there are two classes of approved drugs in most countries to treat cognitive decline in AD, cholinesterase inhibitors for early-stage patients and an N-methyl-D-aspartate (NMDA) receptor inhibitor memantine for moderate and late-stage AD patients [65, 72]. Improvement seen as a result of either of these treatments is modest to moderate. The use of cholinesterase inhibitors is associated with multiple side effects, which result in drug intolerance in up to 30% of patients. Memantine is usually well tolerated [72]. An acetylcholine precursor, choline conjugated with phospholipid to form choline alfoscerate, is used in some countries to treat memory impairment. According to the results of the interim evaluation in AD patients treated with a combination of a cholinesterase inhibitor donepezil and choline alfoscerate, the grey matter loss was temporarily reduced compared to patients receiving only donepezil. This was accompanied by the reduced rate of cognitive decline [73].

Parkinson's Disease

Parkinson's disease is diagnosed based on the appearance of characteristic motor symptoms, which include resting tremor, slowness of movement, rigidity, postural instability and may co-exist with some other motor abnormalities [74]. These symptoms are caused by the reduction of dopamine in basal ganglia as a result of progressive neurodegeneration of nigrostriatal dopamine neurons, which cell bodies locate in Substantia nigra pars compacta (SNpc) and axons project to caudate nucleus and putamen. The majority of Parkinson's disease cases are sporadic (> 90%) [75].

Current treatments for Parkinson's disease include dopamine replacement therapy with Levodopa, dopamine receptors agonists, compounds preventing the breakdown of dopamine, anticholinergics, and amantadine [76, 77]. Neither of these treatments is able to halt disease progressions. Drugs generally lose efficacy when the disease progresses, and the emergence of side effects prevents the use of effective doses of these therapeutics by PD patients. Deep brain stimulation is an efficient method to alleviate some motor symptoms but requires complicated and expensive brain surgery and can produce undesirable side effects (*e.g.*, cognitive problems) [78].

A number of non-motor symptoms, including gastrointestinal disturbances, autonomic dysfunction, cognitive, neurobehavioral, and sensory abnormalities [74, 79 - 81], may precede the appearance of motor symptoms. They are caused by dysfunction and/or degeneration of various neuronal populations in the CNS or peripheral nervous system and impair the health-related quality of life of the patient even more than motor symptoms. Non-motor symptoms of PD are managed poorly by currently available therapeutics [79, 80].

Amyotrophic Lateral Sclerosis

ALS is characterized by the degeneration of motor neurons in the spinal cord and brain, which results in muscle weakness and paralysis. Apart from motor symptoms, many patients experience cognitive decline and behavioral impairment [82]. The typical age of onset is 51–66 years, and the prognosis is poor, with the death of most patients occurring within 5 years after diagnosis, predominantly from respiratory tract infections and respiratory failure [52, 83, 84]. The vast majority of cases are sporadic, while 5-10% are hereditary and represent familial ALS [85, 86]. Among patients with familial ALS, 20% have a gain of function mutation in the *SOD1* gene.

Currently, two drugs which are considered disease-modifying for ALS are available: an antiglutamatergic drug, riluzole [87, 88] and an antioxidant, edavarone (the latter is approved in the USA and Japan) [87, 89, 90]. These drugs demonstrated higher efficacy in early-stage ALS patients in clinical trials [87, 90, 91]. According to the conclusions of the Cochrane systematic review, the increase in median survival of ALS patients treated with riluzole compared to patients receiving placebo was equal to 3 months. An increase in the probability of surviving for 1 year constituted 10% [92]. Edavarone reduced the speed of motor function decline and slowed down deterioration of quality of life in early-stage ALS patients. The ability of the drug to prolong survival was not evaluated [90].

Huntington's Disease

In contrast to many other common neurodegenerative disorders, which are mainly sporadic, HD is caused by an autosomal dominant mutation (trinucleotide CAG repeat expansion) in the huntingtin gene [93]. The disease penetrance depends on the repeat length and is equal to 100% in carriers with more than 39 CAG [94]. HD usually manifests between 20 and 65 years, and the median disease duration is about 20 years [93, 95]. HD patients experience motor, cognitive and neuropsychiatric symptoms [93, 96]. The most recognizable symptoms of HD are chorea and dystonia, involuntary movements that at first appear in distal extremities and when the disease progresses also affect more proximal muscles [93, 96]. In patients with late-stage HD, hyperkinetic movements are replaced by akinetic symptoms such as rigidity and bradykinesia [96]. The cognitive decline initially seen as slowness in information processing, and memory retrieval, progresses into dementia in late-stage patients [93, 96]. Psychiatric symptoms are diverse: depression, apathy, irritability and suicidal behavior are often reported in individuals affected by HD [96, 97].

HD is characterized by progressive degeneration of caudate and putamen and loss of cortical neurons at the late stage of the disease, accompanied by an enlargement of ventricles and overall brain size reduction [96, 98]. The most affected are striatal medium spiny neurons, while interneurons remain relatively intact [96]. Available drugs relieve HD symptoms but do not stop the disease progression. To control hyperkinetic movements, vesicular monoamine transporter type 2 (VMAT2) inhibitors tetrabenazine and deutetrabenazine are used [99]. Also, antipsychotics, anticonvulsants, and tranquilizers can be prescribed to patients to treat chorea [100]. To relieve rigidity, levodopa can be used. Antidepressants, mood stabilizers, and antipsychotics are used to control neuropsychiatric symptoms [100].

Neuropathic Pain

Although neuropathic pain (NP) caused by the lesion or dysfunction in the sensory system is not generally considered to be a neurodegenerative disorder, it is also characterized by neuronal lesions and degeneration. Therefore, it is included into this book chapter. NP is widespread and a difficult to treat condition affecting up to 10% of adults [101, 102]. NP may appear as a result of another systemic disease (*e.g.*, diabetes) or infection (*e.g.*, COVID-19, HIV), trauma, and treatment (*e.g.*, with chemotherapeutic agents) [103 - 105].

Currently, the first-line treatments for NP management include pregabalin and gabapentin as well as antidepressants (serotonin and noradrenaline reuptake inhibitors along with tricyclic antidepressants). Local treatment with lidocaine and capsaicin patches and weak opioid tramadol (systemically) can be used as a second-line medication against NP. Strong opioids are prescribed as the third-line drugs due to side effects [102, 106]. Neither of the available treatments for NP is considered disease-modifying. Existing drugs show efficacy only in a subset of patients [106, 107], and in more than 50% of patients, effective pain management is not achieved with any combination of available analgesics [102]. Many of the listed drugs produce serious adverse effects. In particular, the use of strong opioids and gabapentinoids is associated with the risk of abuse and misuse [106, 108].

To summarize, neurodegenerative disorders are currently incurable. While there are drugs that may alleviate symptoms, available therapeutics are mainly unable to stop underlying neuronal degeneration, protect and restore remaining neurons. Existing drugs also cause multiple side-effects, cause tolerance, dependence, and lose efficacy over time because of disease progression.

Neurodegenerative Disorders in Translational Context

Several aspects complicate clinical translation of drug-candidates in the neurodegenerative disorders field. Neurodegenerative disorders are diagnosed based on the manifestation of clearly distinguishable symptoms when a significant portion of affected neurons has already disappeared. Differential diagnosis at the early stages of the disease is hard, and confirmatory tests require time to perform, analyze the data, and conclude the diagnosis and treatment. Diagnostic methods are expensive, and the availability of samples for direct analysis is rather limited. Neurodegenerative disorders are progressive and neuronal loss is aggravated with the time during the diagnostic period and beyond. After the diagnosis and initiation of therapy, the disease progresses even if the treatment is effective as available drugs are symptomatic and are unsuitable for disease modification. At the early stage of the disease, affected neurons may stay alive but they loose connection to target tissues, as it is shown in PD. PD is diagnosed when approximately 70% of dopamine neuron axons in the striatum are lost, but the majority of neuronal cell bodies in SNpc are still alive [109]. These neurons are an attractive target for neurorestorative treatments. However, in 3-5 years after the diagnosis, these neurons that could have been rescued at the early stages of the disease will almost completely disappear from the brain [110]. Being among the few molecules able to protect and restore lesioned neurons, NTFs bear a high potential for disease-modifying treatment of neurodegenerative disorders. However, currently available treatments with NTFs, which are currently under development, are highly invasive and are, therefore, used in patients with advanced disease, which have little chance to benefit from such treatments (see below). Advances in disease diagnostics and the development of less invasive methods for NTFs delivery can transform the world for people suffering from neurodegenerative disorders.

Preclinical Models of Neurodegenerative Disorders

One of the major problems in the clinical translation of medications against neurodegenerative disorders is the lack of preclinical models based on the disease mechanism that recapitulate all disease features. The author has reviewed the common types of existing models of Parkinson's disease in a recent publication, but general principles can also be applied to animal models of other neurodegenerative disorders [111]. Most of the available models of neurodegenerative disorders can be divided into three groups: models where neurodegeneration is caused by neurotoxin treatment, genetic models, and models where the protein aggregation process is affected. Toxin models are relatively cheap and suitable for quick collection of the data, and at least some of them were used to develop currently available drugs against neurodegenerative disorders.

Thus, their translational value has been confirmed. However, they are not disease mechanism-based. The main problem with genetic models of neurodegenerative disorders is that these diseases except HD are mainly sporadic; therefore, genetic models are representative only for a small cohort of patients. Besides, while behavioral changes can be present, neuronal degeneration in many of these models is mild or absent. An overexpression of a prone to aggregation protein or introduction into the organism of a seed consisting of preformed aggregated protein oligomers seems to reflect the pathological process in patients as protein aggregation is seen in the majority of patients with neurodegenerative disorders. Such models are useful to understand the mechanisms of neurodegenerative processes. However, they are relatively novel; therefore, their further evaluation for reproducibility and translational potential is needed before final conclusions can be formed. As it was discussed above, it is unclear whether protein aggregation is a cause of neurodegeneration, neutral, or even protective mechanism. In addition, the pathology in the models from the last 2 groups usually develops slowly, which makes the long and expensive process of drug development even lengthier and more costly.

To model AD, aged animals are used sometimes [112, 113]. Indeed during the normal process of aging, some loss of cortical and hippocampal neurons occurs. However, these processes in the brains of AD patients are accelerated many times, and the reason for this is not fully understood [69]. It is unclear how well age-based models of neuronal loss recapitulate pathological processes in the brains of AD patients.

Pharmacokinetics of Neurotrophic Factors in the Context of Clinical Translation

Neurotrophic factors do not cross the blood-brain barrier and poorly penetrate and spread into tissues. This dictates the necessity to deliver them locally in close proximity to affected neurons. For classic neurodegenerative disorders such as PD and AD, this means the necessity to use invasive, risky, and complicated stereotactic brain surgery. Besides intrinsic complications, such delivery methods also raise ethical concerns that prevent the use of NTFs in early-stage patients. Therefore, NTF-based treatments are mainly considered for patients with advanced neurodegeneration who are not the primary target group for neurorestorative treatments because of a small number of remaining target neurons [109, 110]

Even when delivered directly into the brain, neurotrophic factors spread in tissues poorly compared to small molecule drugs. For example, GFLs, except for PSPN, bind to the extracellular matrix [31], which limits their diffusion. This issue is less obvious in experimental animals whose brains are relatively small. However, in the large brains of human beings, the limited spreading of NTFs constitutes a significant problem. According to some estimates, the coverage of target fields of dopamine neurons in the brain in clinical trials in PD patients treated with GFLs did not exceed 22% [114, 115]. This means that only a portion of nigrostriatal dopamine neurons received trophic support. In AD, ALS, and NP, in which several brain or CNS structures are affected, this problem will be even worse than in PD.

Expression of Neurotrophic Factor Receptors in the Healthy and Diseased Nervous System

An important aspect to discuss in the context of clinical translation of NTF is the expression of their receptors in the healthy and diseased nervous system. Literature data regarding this topic are often controversial. Several aspects are important to consider when interpreting available data regarding NTF receptors expression in the tissues of patients with neurodegeneration.

On one hand, since NTF receptors are expressed in neurons that degenerate, the overall expression level of these proteins can decrease, especially in the brains of late-stage patients with neurodegenerative disorders. Many NTF receptors are expressed not only in neuronal but also in non-neuronal, in particular glial cells. Neurodegenerative disorders are accompanied by neuroinflammation, astro- and microgliosis. Thus, the overall expression of NTF receptors in certain brain regions can increase. Therefore, when analyzing the expression of NTF receptors, it is important to consider cell type, tissue, and disease stage. The development of novel techniques allowing sorting and cultivating specific cell types or reprogramming other cells in the organism into neurons may allow addressing these points more precisely in the future. Many NTFs may signal through several receptors/co-receptors, which can be differentially regulated in diseased tissues. For example, Konishi *et al.* (2014) reported that the expression of GFRα1, but not other GFRα co-receptors, was downregulated in cultured cortical neurons from the brains of AD patients compared to that seen in neurons from healthy people. In line with these data, the neurons from AD patients were also irresponsive to GDNF stimulation. However, the infection of these neurons with adenoviral vector encoding GFRα1 restored responsiveness to GDNF. These results taken together indicated that the expression of the signal transducing module of GFL receptor complex (in this case, likely NCAM) was not affected by the disease [116]. This is in line with the data of Gillian *et al.* (1994), showing an unaltered NCAM expression in the brains of AD patients. Ginsberg *et al.* (2006), using single-cell expression profiling, demonstrated a decrease in TRKs, but not p75[NTR] receptor mRNA levels in cholinergic basal forebrain neurons from patients with AD [117].

Several years ago, based on preclinical findings in rats overexpressing α−synuclein, it was proposed that protein aggregation downregulates expression of GFL receptors in nigrostriatal dopamine neurons [118]. Indeed protein aggregation may interfere with axonal transport, which is important for the trafficking of NTF receptors. However, these results were related to the extremely high level of transgene expression and were not reproduced in animals with mild α−synuclein overexpression. Moreover, in the brains of PD patients, mRNA levels of α−synuclein are not elevated, and downregulation of GFL receptors is not reported [119]. The results of postmortem analysis of striata of PD patients treated with an adeno-associated viral (AAV) vector-encoded NRTN are also in line with these data: there was an increase in the density of dopamine neuron fibers in the regions where NRTN overexpression was detected. This indicates that NRTN produced a pharmacodynamics effect and thus implies the presence of its receptors in striata of PD patients [115].

In summary, receptors of NTFs in diseases maybe be regulated differently: while the expression of some of them is downregulated, the expression of others remains unchanged. This provides the basis for the rational selection of NTFs for the treatment of a particular condition. Differential regulation of NTF receptors in disease can also be used in translational research by developing agonists selectively activating the receptor complex or its part, which is not affected by the disease.

EFFICACY OF NEUROTROPHIC FACTORS IN PRECLINICAL MODELS OF NEURODEGENERATIVE DISORDERS AND NEUROPATHIC PAIN

GFLs in Preclinical Models of Neurodegenerative Disorders and Neuropathic Pain

GDNF was discovered as a survival factor for dopamine neurons [120]. Therefore, GFLs (mainly GDNF and NRTN) have been extensively studied in animal models of PD. A detailed description of GFL effects in preclinical models of PD can be found elsewhere [122 - 130]. In brief, GFLs, when delivered into the striatum, demonstrated the ability to alleviate motor deficits and protect/restore dopamine neurons in the brains of experimental animals [121 - 129]. Importantly GFLs were inefficient in animal models of advanced PD [121, 130].

Since GFL's main receptor RET is not expressed in the hippocampus in normal conditions, the data on biological effect of these proteins in preclinical models of

AD are limited. It was shown that GDNF encoded by lentiviral vector might alleviate memory deficit in the genetic model of AD [131].

Several studies in which GDNF encoded by viral or naked DNA vectors were conducted in animal models of ALS. In most of them, intramuscularly delivered GDNF delayed the disease onset, improved motor deficits, and prolonged the survival of experimental animals [132 - 134]. Moreover, GDNF encoded by the AAV vector prevented the loss of motoneurons in mice with ALS [134]. At the same time, GDNF overexpressed in the central nervous system by genetic method or delivered into the spinal cord with the help of lentiviral vector failed to demonstrate efficacy in rodent models of ALS [135, 136]. In ALS rats, intravenously delivered AAV-GDNF alleviated motor deficits, protected motor neurons in the spinal cord, but failed to influence the lifespan of animals. In this study, adverse effects of GDNF on weight gain, locomotor activity, and memory were reported in both healthy and ALS rats [137]. Approaches with the delivery of cells overexpressing GDNF into the spinal cord or muscles of ALS animals also demonstrated protective efficacy in preclinical studies [138, 139]. AAV-NRTN injected into the spinal cord protected motoneurons and reduced the speed of motor function deterioration but did not influence the survival of ALS mice [140].

In neurotoxin models of HD recombinant or AAV-GDNF alleviated motor deficits and preserved functional or morphological integrity of striatum [141 - 143]. The most pronounced effect was observed for calbindin-positive neurons [143, 144]. In the transgenic model of HD, GDNF and NRTN encoded by AAV vector attenuated functional deficits and prevented the degeneration of striatal neurons [145 - 147]. GDNF also reduced the number of striatal neurons with aggregated Htt inclusions [145]. At the same time, GDNF overexpressed by lentiviral vector failed to demonstrate efficacy in animal models of HD [148]. Similarly, GDNF overexpressed by regulated AAV failed to alleviate motor deficit or neuronal loss in animal models of HD, but prevented excessive weight gain [149].

GDNF and ARTN were studied in animal models of NP. The author recently reviewed available data regarding the efficacy of GFLs in preclinical models of NP in detail [14]. In brief, effects of GFLs on animals with NP depended on the delivery paradigm, dose, model type, and model organism. In rats with surgery-induced NP, GFLs mainly alleviated pain. In mice and inflammatory pain models, GFLs seem to induce pain [9].

Neurotrophins in Preclinical Models of Neurodegenerative Disorders and Neuropathic Pain

The efficacy of BDNF and NGF in preclinical models of AD was described in recent reviews [150, 151]. In brief, in animal models of AD, these neurotrophins, at least in some studies, attenuated cognitive decline and showed neuroprotective properties. NGF being a survival factor for basal forebrain cholinergic neurons, attracted special attention in the context of AD [152]. A recent research topic of Frontiers in Neuroscience provides a detailed review of NGF's role in AD pathology in 7 excellent articles [153].

The author reviewed preclinical data regarding the biological activity of neurotrophins in PD in a previous review [19]. The effects of these proteins in the dopamine system are mainly limited to neuroprotection rather than to neurorestoration. Therefore, they were not chosen for translational research.

In animal models of HD, BDNF reduced the striatal lesion and protected striatal neurons, including medium spiny neurons from neurotoxin lesions [143, 154]. In transgenic models of HD, this neurotrophin ameliorated depressive behavior [155] and improved motor and cognitive function [156, 157], which was also accompanied in some cases by reduction of striatal atrophy and neuronal loss [155, 157]. Recombinant BDNF protein-infused subcutaneously improved motor function and reduced neuronal loss in the striatum of transgenic animals with HD. This was accompanied by the restoration of BDNF levels in the brains of experimental animals [158].

Neurotrophins play an important role in the pathogenesis of NP. Both pronociceptive and antinociceptive effects are reported for the neurotrophic factors from this family. A review of Khan and Smith (2015) provides a good summary of available data on the biological effects of neurotrophins in preclinical models of NP. In brief, NGF and BDNF are mostly pronociceptive, NT-4/5 seems to be mainly neutral in regard to nociception, and the majority of reports indicate that NT-3 is antinociceptive [159].

Neurokines in Preclinical Models of Neurodegenerative Disorders and Neuropathic Pain

Lee *et al.* (2016) demonstrated that in the drosophila model of AD, LIF fed to flies reduced the expression of autophagy marker. According to the authors, the level of LIF in the brain of flies treated with this neurokine was qualitatively higher compared to that seen in control insects [160].

In neurotoxin models of HD, CNTF encoded by lentiviral or adenoviral vector and CNTF recombinant protein protected striatal neurons from the death [161 - 164], increased weight gain, reduced the volume of ventricles and extent of the striatal lesion [165]. IL6 encoded by lentiviral vector was also neuroprotective in the neurotoxin model of HD [166]. However, in the transgenic model of HD, AAV-CNTF constitutively expressed in the brain augmented motor deficit, induced abnormal behavior, and reduced the expression level of striatal neuron markers [167].

The data concerning the biological activity of LIF in the genetic model of ALS are contradictory. Feeney *et al.* (2003) reported no benefits to transgenic animals from LIF delivered intramuscularly, subcutaneously, or intrathecally. Moreover, LIF delivered to ALS mice intrathecally was rather detrimental, exacerbating motor deficit and reducing survival of experimental animals [168]. However, in another study, intraperitoneally injected recombinant LIF reduced grip strength impairment in transgenic mice but was unable to rescue motoneurons [169].

In neurotoxin models of PD, LIF infused into the cerebrospinal fluid or injected intraperitoneally reduced motor deficit and increased neurogenesis in the brain [170, 171]. In the surgical model of PD LIF prevented the loss of nigrostriatal neurons but failed to influence the reduction of tyrosine hydroxylase (a key enzyme of dopamine synthesis) expression [172].

Neurokines influence the inflammation process that plays an important part in the development and maintenance of neuropathic pain. They are pronociceptive at least for some pain modalities, and anti-neurokine antibodies or genetic ablation of CNTF gene produced analgesic effects in experimental animals [173, 174].

CDNF and MANF in Animal Models of Neurodegenerative Disorders and Neuropathic Pain

In the transgenic model of AD, recombinant or AAV-encoded GDNF improved memory deficit but did not influence neurogenesis, dopamine innervation in the hippocampus, or synaptic marker expression [175].

Multiple studies were conducted to evaluate the efficacy of CDNF and MANF in preclinical models of PD. Recombinant and AAV-encoded MANF alleviated motor deficit and protected dopamine neuron bodies in rodent neurotoxin models of PD [176 - 178]. In α-synuclein model of PD in *C. elegans*, MANF slowed down neurodegeneration, attenuated motor impairment, and reduced protein aggregation [179]. Interestingly, recombinant MANF produced no positive effect in experimental PD when it was continuously infused into the brain [180].

Recombinant CDNF alleviated motor deficit and protected and restored dopamine neurons in rodent neurotoxin models of PD [180 - 182]. Similar effects were observed for AAV-encoded CDNF [183, 184], especially in early-stage PD model [185]. In nonhuman primates with lesions caused by the neurotoxin, CDNF improved the functional status of the nigrostriatal dopamine system [186]. Interestingly, according to the data collected by Cordero-Llana, neither MANF nor CDNF overexpressed in striatum from lentiviral vectors produced beneficial effects in animal models of PD. When delivered into SNpc, lentiviral vector-encoded MANF preserved dopamine fibers but did not influence motor deficit. Overexpression of both proteins in SNpc led to reduced motor deficit and prevention of neurodegeneration in the nigrostriatal dopamine system [187].

In animal models of ALS, recombinant CDNF prevented the development of motor deficit, prolonged the survival of experimental animals, and protected motoneurons in the spinal cord from degeneration [188]. In the neurotoxin-based model of HD CDNF alleviated motor function deterioration and prevented the neuronal loss in the striatum [189].

At the moment, the information regarding the effects of CDNF and MANF in neuropathic pain is limited. Cheng *et al.* (2013) demonstrate the ability of CDNF encoded by the lentiviral vector to stimulate the regeneration of the sciatic nerve after surgical injury [190].

It is noteworthy that there are a number of reports in the literature about the effects of neurotrophic factors in animal models of neurodegeneration when these proteins are delivered using genetically engineered cells synthesizing neurotrophic factors. In the majority of these studies, beneficial effects of grafted cells are reported. However, these cells might secrete a cocktail of proteins, which can act in a synergistic manner. It complicates the understanding of the effects of individual growth factors. Therefore, these studies are not reviewed here.

EFFICACY OF NEUROTROPHIC FACTORS IN CLINICAL TRIALS IN PATIENTS WITH NEURODEGENERATIVE DISORDERS AND NEUROPATHIC PAIN

Neurotrophins in a Clinical Trials in Alzheimer's Disease Patients

Intraventrically delivered NGF was tested in 3 patients with AD. NGF-treated patients with AD had an increase in nicotine binding in several brain areas, as shown by positron emission tomography (PET) scanning [150, 191, 192]. Some improvements in certain neuropsychological tests were also seen. However, all 3 patients reported back pain and 2 – significant weight loss resulting in premature termination of the study [150, 192]. Gene therapy-based approach to deliver NGF to the brain of AD patients was also used. Although some signs of efficacy of this approach were reported in Phase I clinical trial [193] along with signs of improvement on a cellular level according to the data of postmortem analysis [194], in Phase 2 randomized placebo-controlled clinical trial, AAV2-encoded NGF failed to demonstrate efficacy in AD patients [195].

Neurotrophic Factors in Clinical Trials in Parkinson's Disease Patients

GFLs in Clinical Trials in Parkinson's Disease Patients

Recombinant GDNF protein, AAV2-GDNF, and AAV2-NRTN (CERE-120) have been tested in several Phase I/II and Phase II clinical trials in PD patients (Table 2). In the first clinical trial initiated in 1996, GDNF protein was infused into the ventricles. In this trial, no improvement in the condition of patients was seen. Moreover, GDNF produced significant side effects [196]. Later it became obvious that due to the inability to cross the tissue barriers, intraventrically delivered GDNF has never reached target dopamine neurons.

Therefore, in all subsequent clinical trials in PD patients, GFLs were delivered into the putamen. In all studies conducted with intraputaminally delivered GFLs, the treatments were found safe and well-tolerated by the patients. In two small scale, open-label clinical trials, intraputaminally delivered recombinant GDNF protein increased ^{18}F-DOPA tracer signal in PET scans, indicative of increased dopamine transporter (DAT) function and possible neuroprotective and neurorestorative effect of the treatment. These changes were also accompanied by long-lasting improvement in motor symptoms evaluated using the Unified Parkinson's Disease Rating Scale (UPDRS) [197 - 199]. However, the subsequent Phase II randomized, double-blinded, placebo-control clinical trial failed to reach its primary end-points. No statistically significant differences in UPDRS score were found between patients treated with GDNF and patients treated with placebo [200]. The possible reasons for the lack of GDNF efficacy in this clinical trial are insufficient coverage of the putamen with GDNF [114], insufficient power of the study [201], or failure to deliver GDNF protein into the brain due to poor connection of delivery pump and brain catheter because in this study the formation of function-blocking antibodies against GDNF was reported, implying the leakage of GDNF the periphery [200].

A series of Phase II clinical trials with AAV2-NRTN delivered into the nigrostriatal system also achieved only limited success [201]. Initially, the treatment was injected only into the putamen [202]. However, it was noted that the transport of NRTN into SNpc in humans was significantly impaired compared to the results collected for nonhuman primates [203]. Therefore, in a subsequent clinical trial, CERE-120 was delivered into both putamen and SNpc [204].

Neither of these clinical trials reached its primary efficacy endpoints. Nevertheless, postmortem analysis revealed an increase in the density of TH-positive fibers in the areas of putamina, also positive for NRTN, which supports the idea about neuroprotective and neurorestorative effects of GFLs in the human brain [115, 205]. However, only a small portion of putamina (< 25%) turned out to be covered with NRTN in these samples [115, 205]. This can be one of the reasons explaining the lack of significant changes in UPDRS scores in treated patients in CERE-120 clinical trials.

A better understanding of GFLs's biology and tissue distribution resulted in a substantial change in the GDNF delivery paradigm in clinical trials. The most recent Phase II clinical trial was conducted in patients with PD of moderate severity, and low doses of recombinant GDNF were delivered into putamina intermittently using a convection-enhanced delivery system. This delivery paradigm better resembles the physiological situation when GDNF is released in a pulsatile manner. While this trial also did not reach primary efficacy end-point and statistically significant change in UPDRS score in GDNF treated patients compared to placebo-treated patients was not seen, clinically significant improvement was seen only in patients treated with GDNF. Also, according to the results of PET scanning in GDNF-treated patients, an improvement in DAT function was demonstrated [206, 207]. Recently also AAV2-encoded GDNF was tested upon convection-enhanced delivery into the putamina of PD patients. The treatment was found safe and enhanced ^{18}F-DOPA signal in PET scans implying an improvement in DAT function [208].

One of the major problems in clinical trials with GFLs in PD patients is related to the inability of recombinant or viral vector-encoded proteins to penetrate the blood-brain barrier. This necessitates the delivery of these treatments directly into the brain by means of complicated and risky brain surgery. Due to ethical reasons, brain surgery is not recommended for early-stage PD patients. Therefore, participants in clinical trials with GFL-based treatments had moderate or advanced PD with an average disease duration equal to 10 years. However, while early-stage PD patients have approximately 70% of alive but atrophic dopamine neuron bodies in SNpc and approximately 30% of dopamine neuron fibers in putamina, a few (3-5) years after the diagnosis, almost all fibers and the majority of dopamine neuron bodies disappear from the brain [109, 110]. This negatively affects the efficacy of GFLs and may explain the inability of many trials to reach their primary efficacy end-points.

Table 2. List of Phase I/II and Phase II clinical trials with GFLs in Parkinson's disease patients. PET – positron emission tomography, UPDRS - Unified Parkinson's Disease Rating Scale, GDNF - glial cell line-derived neurotrophic factor, GFLs – GDNF family ligands, SNpc – substantia nigra pars compacta.

Growth factor and Delivery System Used in the Trial	Main Results	Main Problems	Reference
Recombinant GDNF, intracerebroventricular delivery	No motor improvement, a number of adverse effects, weight loss, appetite disturbances, paresthesias	Inability to reach target neurons due to the delivery to the wrong place	[196]
Recombinant GDNF, intraputaminal delivery	Improvement in nigrostriatal dopamine system function shown by PET scanning, motor improvement, well-tolerated	Small scale, no placebo control	[197, 198]
Recombinant GDNF, intraputaminal delivery	Improvement in nigrostriatal dopamine system function shown by PET scanning, the motor improvement seen by UPDRs score, quality of life improvement, well-tolerated.	Small scale, no placebo control	[199]
Recombinant GDNF, intraputaminal delivery	No statistically significant improvement in motor performance of patients, side effects, the formation of anti-GDNF antibodies	Poor spreading in the brain parenchyma and extremely low putamen coverage, leakage of the protein on the periphery, insufficient number of patients in the treatment groups, late-stage patients	[200]
AAV2-encoded NRTN (CERE-120), intraputaminal delivery	No statistically significant improvement in motor performance of patients, improvement in the quality of life (self-assessment), well-tolerated.	Poor transport of the protein from putamina to SNpc, late-stage patients	[202]
AAV2-encoded NRTN (CERE-120), intraputaminal and intranigral delivery	No statistically significant improvement in motor performance of patients, improvement in the quality of life (self-assessment), well tolerated.	Rather low coverage of the putamen, late-stage patients	[204]

(Table 2) cont.....

Growth factor and Delivery System Used in the Trial	Main Results	Main Problems	Reference
Recombinant GDNF, intermittent convection-enhanced delivery of low dose into putamina	Improvement in nigrostriatal dopamine system function shown by PET scanning, no statistically significant motor improvement shown by UPDRs, but clinically significant improvement by ≥ 10 points seen only in GDNF-treated patients, well-tolerated	Probably low dose and biological activity of the treatment, incomplete putamen coverage, late-stage patients	[206, 207]
AAV2-encoded GDNF	Improvement in nigrostriatal dopamine system function shown by PET scanning, well-tolerated	Incomplete putamen coverage, late-stage patients, small scale	[208]

CDNF in Clinical Trials in Parkinson's Disease Patients

CDNF was tested in a randomized placebo-controlled double-blind multicenter Phase 1/2 clinical trial in 17 PD patients with an average duration of the disease equal to 10 years. The trial was completed in August 2020, but the results of the study are yet to be published in a peer-reviewed journal. However, preliminary data are available on the TreatER project webpage. The main period of the trial lasted for 6 months, during which 11 patients received monthly intraputaminal infusions of recombinant CDNF (in one of 2 selected doses, mid-, and high dose) and 6 patients – placebo. During the 6-month extension study, the therapy regimen remained the same for patients from active treatment groups, and patients from the placebo group were also infused with CDNF. The main outcome of the study is the established safety of CDNF for patients. After a 6-month period of the main study, the treatment seemed to also increase the signal from DAT-tracer in PET scans which may indicate neuroprotective and/or neurorestorative effect of CDNF on dopamine fibers in the putamen. Improvement in motor performance was seen in patients receiving a mid dose of CDNF [209].

Neurotrophic Factors in a Clinical Trials in Patients with Amyotrophic Lateral Sclerosis

BDNF was administered intrathecally to patients with ALS (n = 25) in a randomized dose-escalation study. The treatment was found safe, but due to the small number of participants, conclusions on efficacy were not made [210]. A Phase III clinical trial failed to demonstrate the ability of BDNF to promote the survival of ALS patients. However, according to post-hoc analysis results in subgroups of patients with early respiratory function deterioration and especially in those with bowel dysfunction developing as a side effect of the treatment, BDNF significantly increased survival probability [211].

Recombinant CNTF delivered subcutaneously was tested in a randomized double-blinded placebo-controlled phase 2/3 clinical trial in patients with ALS (n = 730). No statistically significant difference in the rate of disease progression and mortality was seen between patients from placebo and active treatment groups. Dose-limiting side-effects, such as weight loss, anorexia, and cough, were observed in many patients [212].

Several clinical trials with the cells expressing neurotrophic factors have been conducted in ALS patients. In the single-center blinded Phase 1/2a clinical trial, neural progenitor cells genetically modified to express GDNF were surgically implanted into the spinal cords of ALS patients. The treatment was found to be safe for the patients [213]. Also, bone marrow-derived mesenchymal stem cells secreting GDNF, BDNF, Vascular Endothelial Growth factor, and Hepatocyte Growth factor [214] were studied in several phase 1 and phase 2 clinical trials. In several open-label phase 1 (n = 12) and phase 2a (n = 14) clinical trials in ALS patients (n = 26), this treatment, when injected intrathecally and/or intramuscularly, was found safe. In addition, in the majority of the patients who received the NTF-expressing stem cells in phase 2a clinical trial either intrathecally and/or intramuscularly, the rate of disease progression was lower compared to the pretreatment period [215]. In follow-up phase 2, randomized placebo-controlled clinical trial conducted in 48 ALS patients (randomization 3:1) safety end-points were also met. In the total population, the rate of disease progression among placebo- and the NTF-expressing stem cells – treated patients differed insignificantly, but the number of responders among rapid progressors treated with NTF-expressing stem cells was significantly higher during the first few weeks of therapy compared to placebo-treated patients [216].

Neurotrophic Factors in Clinical Trials in Patients with Neuropathy

GFLs in Clinical Trials in Patients with Neuropathic Pain

ARTN was studied in several randomized, blinded, placebo-controlled clinical trials in patients with neuropathic pain [217 - 219]. ARTN was rather well-tolerated, and the main treatment-related adverse effects included headache, changes in temperature perception, and itching [218]. In two of these studies, somewhat efficacy of ARTN was demonstrated: clear dose-response was not observed in either of these studies, but in self-assessment and both self-assessment and objective scoring in Phase II clinical trials, pain relief was noted [217, 219]. Importantly, ARTN was able to reduce pain in painful lumbosacral radiculopathy patients resistant to at least 2 standard analgesics. Pain relief lasted for up to 5 weeks after intravenous administration of 3 doses of the drug (every other day) [217]. These data are consistent with ARTN pharmacodynamics as the treatment is expected to restore the function and integrity of lesioned neurons, and thus, it should provide long-lasting pain relief [217]. Up to 50% of patients treated with ARTN compared to 21% of those in the placebo-treated group also reported big or very big improvement based on self-assessment using the Patient Global Impression of Change scale. Improvement in quality of sleep was also registered [217]. However, in both clinical trials the dose-response relations curve was not linear [217, 219].

Neurotrophins in Patients with Neuropathy

Recombinant NGF and BDNF delivered subcutaneously were tested in several trials in patients with neuropathy. Although in phase II clinical trials in patients with human immunodeficiency virus or diabetes-induced neuropathy, encouraging results for NGF were obtained, the Phase III clinical trial in patients with diabetic polyneuropathy failed to reach primary efficacy end-points [220]. BDNF delivered subcutaneously also failed to demonstrate efficacy in patients with diabetic neuropathy (n = 21 treated with BDNF and n = 9 treated with placebo) [221].

ALTERNATIVE APPROACHES TO UTILIZE THE POTENTIAL OF NEUROTROPHIC FACTORS FOR THE TREATMENT OF NEURODEGENERATION

Although NTFs hold considerable promise as potential treatments for neurodegenerative disorders, their clinical translation turned out to be difficult. Neurotrophic factors are unable to cross blood-brain barrier and require invasive delivery in patients with neurodegeneration. They also poorly diffuse in the tissues and spread in the body, and may signal *via* different receptors in different cell types resulting in difficult to predict effects on the organism. Their receptors can be differentially regulated in the diseased nervous system. Therefore, alternatives with better drug-like characteristics such as fusion proteins, peptides, peptidomimetics, and small molecules targeting receptors of neurotrophic factors are being developed to improve perspectives of NTF's clinical translation. In the recent reviews of the author, the approaches to the development of compounds with biological activity similar to that of GFLs are reviewed in detail [111, 222]. Therefore, only a short summary of these developments will be presented here. More attention will be given to an update on the status of compounds mimicking the biological activity of neurotrophins and neurokines. Advances in the development of CDNF/MANF mimetics are limited by insufficient knowledge on the mechanisms of their action and unknown receptors. Thus, only ligand-based drug design approaches, which will be briefly mentioned at the end of this chapter, are possible in this case.

GFLs

Mutated NRTN and GDNF variants with improved tissue diffusion were developed. GDNF mutant had reduced biological activity [223], while mutated NRTN showed higher efficacy in an animal model of PD [224] compared to the wild-type protein.

To improve the delivery to motor neurons, a fusion protein containing mature GDNF and C-fragment of tetanus toxin was generated. Conjugation improved delivery of GDNF into spinal motoneurons [225]. Intramuscular injection of this fusion to symptomatic animals with ALS prolonged the survival of mice and increased spontaneous muscular activity [226].

To address the issue of blood-brain-barrier penetration, fusion variants of GDNF protein with monoclonal antibody to transferrin receptor or with a cell-penetrating peptide of human immunodeficiency virus were developed. They were able to get into the brain after systemic delivery [227 - 229]. GDNF fused with monoclonal antibody to transferrin also alleviated motor deficit in the 6-OHDA model of Parkinson's disease in mice after intravenous injection [230].

Peptide from the GDNF pro region, DNSP-11, also showed potent neuroprotective activity in the 6-OHDA model of PD in rats when it was delivered directly into the brain [130] or intranasally [231]. Importantly this peptide showed efficacy in rats with a severe lesion of the nigrostriatal dopamine pathway in which GDNF itself failed to alleviate the motor deficit. According to the results of the in vitro studies, DNSP-11 signals via different mechanisms compared to GDNF. Interestingly, DNSP-11 was also shown to influence the functional activity of hippocampal neurons; therefore, it might be interesting to test its biological activity in animal models of AD [232].

Tetrameric peptides produced from heel regions of GDNF and ARTN were shown to activate NCAM and RET-dependent signaling in the cells. These peptides possessed neuroprotective properties in cultured neurons [233, 234]. However, these compounds are yet to be tested in vivo, they are rather bulky, and it is unclear if they can reach the target neurons if injected systemically into experimental animals.

Three structurally different chemical scaffolds targeting RET directly were discovered via high-throughput screening of several chemical libraries [9, 235 - 239]. These compounds activate RET and RET downstream signaling in vitro and in vivo, and were shown to support the survival of cultured dopamine neurons [240], promote the neurite outgrowth from sensory neurons [237, 238], and protect retinal cells from the death [235, 236]. RET agonists alleviated pain in animal models of NP, and also showed neuroprotective and neurorestorative properties towards sensory neurons *in vivo* [237, 238, 241]. In the 6-OHDA model of PD, RET agonists alleviated motor deficits [239]. A compound increasing biological activity of GDNF, XIB4035, was discovered in 2003 and later shown to be able to reduce signs of diabetic neuropathy in experimental animals [242, 243]. Using rational drug design methods, the author with colleagues also developed a compound targeting GDNF co-receptor GFRa1, which activated RET and RET-downstream signaling in immortalized cells. However, its biological activity was very weak, which prevented us from testing this compound in neurons and animal models of neurodegeneration [244].

Neurotrophins

Dimeric cyclic peptides from loop 1 and loop 4 of NGF promoted the survival and neurite outgrowth from DRG neurons *via* activation of MAPK signaling cascade [245, 246]. Cyclic peptides of various lengths comprising loop 1 and loop 4 regions of NGF activated TrkA, but not TrkB in PC12 cells, promoted neurite outgrowth from DRGs, reduced mechanical and thermal sensitivity in chronic constriction injury model of neuropathic pain, and promoted the regrowth of sensory neuron axons into the spinal cord [247]. A peptide comprising the first 14 amino acids of human NGF, especially in acetylated form, promoted TrkA activation, TrkA-dependent signaling, increased functional activity in cholinergic neurons, and also stimulated the survival and neurite outgrowth from DRG neurons [248]. A dipeptide blood-brain barrier-penetrating NGF mimetic activating TrkA and TrkA-dependent AKT signaling showed neuroprotective properties and reduced motor deficit in several toxin-based models of PD. It also alleviated memory loss in animal models of AD. Importantly, in experimental animals, this dipeptide NGF mimetic did not produce hyperalgesic and anorexic effects [249]. Non-peptidic NGF mimetic was described in 2012. The compound activated TrkA and TrkA-dependent signaling in immortalized cell and promoted neurite outgrowth from sensory neurons [250].

An NGF peptidomimetic (MIM-D3) was shown to stimulate TrkA-dependent survival of fetal DRG neurons in the absence but also synergistically in the presence of NGF. In addition, it stimulated the differentiation of fetal DRGs and septal neurons [251]. The compound was tested in Phase II clinical trial in patients with dry eye syndrome (n = 150) as NGF was shown to play a role in the regulation of lacrimal function. MIM-D3 reduced symptoms according to both subjective and objective estimates [252]. Currently, MIM-D3 is being tested in Phase III clinical trial in patients with dry eye syndrome (n = 600) [253].

Using the pharmacophore model of NGF hairpin loop 1, Massa *et al.* (2006) designed a library of chemical compounds and identified several non-peptide small molecules (named LM11) targeting neurotrophin receptor p75^{NTR}. These compounds interacted with p75^{NTR}, blocked proNGF-induced death of oligodendrocytes, and promoted the survival of embryonic hippocampal neurons p75^{NTR}-dependently [254]. One of these molecules, LM11-31, was chosen for further preclinical development based on predicted and tested pharmacological characteristics. LM11A-31 was tested in several animal models of neurodegenerative diseases and neuropathic pain. In an animal model of HD, LM11A-31 reduced aggregation of hungtingtin in the brain, inhibited degeneration of dendrites of interneurons in the striatum, extended the survival and motor performance of mice [255]. In animal models of AD this compound penetrated blood-brain barrier, reduced cognitive deficits, prevented the loss of neurites of basal forebrain cholinergic neurons, decreased neuroinflammation [256 - 259], suppressed phosphorylation and clustering of tau protein [259]. The compound also prevented age-related degeneration and loss of basal forebrain cholinergic neurons [260]. Recently a Phase IIa clinical trials in patients with mild-to-moderate AD receiving two different doses of LM11A-31 has been completed [261], but the results of this study are yet to be published.

Several approaches were used to target the BDNF receptor TrkB. The developed compounds are mainly applied in the field of depression as this neurotrophin receptor seems to be strongly associated with this condition. A number of small molecules acting as TrkB agonists were reported. The first described of those, 7,8-dihydroxyflavone [262, 263], was shown to alle*via*te motor deficits and protect nigrostriatal dopamine neurons in animal models of PD [264, 265]. These effects were also accompanied by the reduction of α-synuclein overexpression [264]. In an animal model of HD, 7,8-dihydroxyflavone ameliorated motor deficits, brain atrophy, impairment in neuritogenesis, and prolonged the survival of experimental animals [266]. In the animal model of AD 7,8-dihydroxyflavone reduced memory deficits, prevented synaptic loss, and amyloid β deposition [267 - 270]. In addition, in a model of peripheral nerve injury, this compound promoted axonal regeneration [271]. Recently, in order to improve pharmacokinetics of 7,8-dihydroxyflavone, a prodrug which retained the biological activity of a parent compound in AD model was developed [272]. Mimetic of loop II of BDNF, compound LM22A-4 which binds and activates TrkB and shows neuroprotective properties in *in vitro* experiments [273] was shown to reduce neuropathology and improve motor performance in an animal model of HD [274]. A compound with a chalcone-coumarin-related structure, LMDS-1, was shown to activate TrkB, alle*via*te cognitive deficit and neuronal loss in an animal model of AD, [275]. The ability of deoxygedunin [276], amitriptyline [277], and isocoumarin 1 [278] to activate TrkB has also been reported. Interestingly, while in *in vivo* conditions described compounds seem to indeed activate TrkB [265, 269], the reports on their effects on TrkB phosphorylation in cultured cells are inconsistent. In the recent independent studies, 7,8-dihydroxyflavone, LM22A-4, deoxygedunin, and amitriptyline failed to increase the level of TrkB phosphorylation or activate intracellular signaling in cultured neurons [279, 280].

Therapeutic modalities other than small molecules targeting TrkB were also discovered. Several groups reported TrkB targeting peptides with neuroprotective properties in cultured neurons [281, 282]. In addition, several other TrkB agonistic peptides whose neuroprotective properties are yet to be established were discovered [283 - 285]. An RNA aptamer binding TrkB with high affinity demonstrated the ability to reduce the death of cultured cortical neurons [286]. Several agonistic TrkB antibodies [280, 287, 288] have been identified. Two of them demonstrated neuroprotective properties in näive cortical [288] and Htt-treated striatal neurons [280]. Further evaluation and optimization of these compounds may reveal their potential for the treatment of neurodegeneration.

Neurokines

A peptide comprising four 14 amino acid residues derived from the CD-loop-D helix region of human CNTF coupled to lysine backbone named Cintrofin was synthesized and shown to interact with high affinity to LIF receptor (but not to CNTF receptor) and gp130. This peptide promoted the survival and neurite outgrowth in cultured cerebellar granule neurons. The removal of 6 amino acids from the N-terminus of 14-mer peptide did not influence the ability of the resulting peptide to influence neurite outgrowth, while C-terminal amino acid residues were critical for the retention of biological activity [289]. Cintrofin was able to penetrate through the blood-brain barrier and alleviated memory deficits after seizures [290]. In an animal model of ALS, implanted into the spinal cord mesoporous silica particles loaded with Cintrofin, GDNF, and vascular endothelial growth factor mimicking peptide, delayed the onset, promoted survival, and alleviated motor deficits. However, in this experiment, unloaded mesoporous silica particles induced very similar changes. Therefore, it is difficult to conclude on the efficacy of neurotrophic factor mimicking peptides in this study [291].

An 11-mer peptide derived from CNTF aminoacid resides 146-156 (named P6) promoted proliferation/survival of adult hippocampal neurons progenitors *in vitro* and *in vivo*, increased neurogenesis in the dentate gyrus, promoted the migration and maturation of neuronal precursors. In normal mice, P6 improved spatial memory. The peptide was able to penetrate through the blood brain barrier and cleared from the brain less rapidly than from plasma. However, it seemed to signal *via* different mechanisms than neurokines, because it was unable to activate JAK/STAT signaling in neuronal progenitor cells in contrast to CNTF and failed to stimulate haptoglobin secretion in contrast to LIF. At the same time, P6 concentration-dependently blocked the effects of both CNTF and LIF on intracellular signaling cascades and protein secretion, probably interacting with the same receptors as these NTFs [292]. A shorter version of this peptide, a tetramer P6c, correspondent to amino acid residues 147-150 of CNTF demonstrated similar biological activity to P6 *in vitro* and *in vivo* [293].

An optimization of P6c *via* the addition of a C-terminal adamantylated group resulted in a peptide P021 with enhanced lipophilicity and improved stability. In aged rats, P021 reduced memory impairment, increased expression of dendritic and synaptic markers in the cortex and hippocampus and stimulated neurogenesis. These effects were accompanied by the increase in BDNF expression; therefore, taking into account the inability of the parental compound to activate intracellular signaling in neuronal cells, neuroprotective effects of P021 might be indirect and mediated by other neurotrophic factors [294]. P021 was detected in the brain 10 and 30 min after intraperitoneal injection, indicative of its ability to cross through the blood-brain barrier. In an animal model of AD P021 reduced tau hyperphosphorylation in the hippocampus, diminished concentration of soluble $A\beta_{1-4o}$ и $A\beta_{1-42}$ in the cortex, alle*via*ted cognitive deficit, enhanced neurogenesis in the dentate gyrus, reduced expression of neuroinflammation markers, and ameliorated the loss of dendrites and synapses in the hippocampus [295, 296].

In addition to peptides targeting neurokine receptor mentioned above, agonistic antibodies to components of LIF/CNTF receptor complex are described [297, 298]. However, their neurotrophic effects have not yet been established. Mutants of cardiotrophin-like cytokine factor 1 with enhanced activity towards neurokine receptors were also developed. This mutant more efficiently promoted the survival of E18 cortical neurons than cardiotrophin-like cytokine factor 1. However, the superiority of its biological activity compared to that of CNTF was not demonstrated [299].

While developed peptides seem to have promising biological activity, it is unclear what they target and how specific they are. While inhibitory effects of P6 towards the events induced by LIF/CNTF in the cells may suggest its affinity towards respective receptors, it can not be excluded that it targets some other proteins in the signaling cascades. Besides, its described mechanism of action related to stimulation of neurogenesis may not be translatable to humans. While the presence of neurogenesis in the brain of experimental animals seems to be quite well documented, it is unclear if it occurs in the adult human brain on a reasonable level. According to the most of available data, only a few neurons are born in the brains of humans older than 18 years old [63]. These data indicate that the process of neurogenesis may be regulated differently in humans and experimental animals. Therefore, a translational perspective of therapeutics influencing neurogenesis may not be very bright. For example, the efficacy of platelet growth factor acting *via* this mechanism in PD was not demonstrated in clinical trials despite promising preclinical data [300].

CDNF/MANF

The lack of detailed understanding of CDNF/MANF signaling and uncertainties regarding their receptors in the cells preclude the implication of a target-based methods to the development of alternative compounds with similar biological activity for these NTFs. However, ligand-based approaches can be used. MANF and CDNF have in their structure two domains connected by a polypeptide linker. It was shown that the C-terminal domain of MANF supports the survival of superior cervical ganglion neurons even more prominently than full-length protein [301]. Therefore, it is possible to reduce the size of these neurotrophic factors in order to improve their pharmacokinetics and tissue distribution without the loss of biological activity. Here, it is worth noting that even full-length CDNF and MANF have a higher volume of distribution in the brain parenchyma compared to GFLs [180]. The use of their domains instead of full-length proteins may improve their ability to diffuse in tissues even further and thus reach more neurons in the brain. Since limited spreading of GFLs in, *e.g.*, PD patients could have contributed to the lack of statistically significant improvement of motor symptoms, better diffusing NTFs may help to overcome this limitation.

Targeting Multiple Neuronal Populations in the Organism

Neurodegeneration or functional dysfunction can occur in patients with neurodegenerative disorders in different neuronal populations. As described above PD patients, apart from motor symptoms experience number of non-motor symptoms. In ALS patients both brain and spinal cord motor neurons are affected. In AD and late-stage HD pathology is seen in both hippocampal and cortical brain regions. In NP patients both local and central mechanisms seem to contribute to the development of allodynia and hyperalgesia. The inability of native NTFs to spread in the body restricts their effects only to certain neuronal populations to which they can be delivered in a sufficient dose. Alternatives such as small molecules, peptides, or antibodies with improved ability to cross tissue barriers and diffuse in tissues may serve as good starting points to develop therapies maintaining the well-being of different malfunctioning neuronal populations simultaneously. This will allow generating a single drug to treat multiple symptoms of neurodegenerative disorders and acting on different levels in patients with NP. Such an approach will be useful to improve the compliance of patients and to avoid undesirable drug-drug interactions. Neurodegenerative disorders mainly affect elderly people, often with multiple comorbidities, who have to take multiple drugs. Development of a single drug to address multiple symptoms of neurodegenerative disease is an important step forward in the therapeutic management of these patients.

CONCLUSIONS AND FUTURE PERSPECTIVES

Neurotrophic factors hold considerable potential for disease-modification in neurodegeneration and neuropathic pain because, in contrast to available treatments, they are able to protect and restore the integrity and function of damaged neurons. However, their clinical translation turned out to be complicated. Differently from traditional CNS drugs, neurotrophic factors are unable to cross tissue barriers and reach the brain when delivered systemically. Besides, they do not spread well in tissues. Therefore, they have to be delivered to the proximity of target neurons which implies, in the case of traditional neurodegenerative disorders, the need to use complicated brain surgery. Even in this delivery paradigm, NTF can concentrate near the injection site and fail to reach a sufficient number of target cells in the quantity necessary to elicit a biological response on the behavioral level. Neurotrophic factors interact with multiple receptors in the organism, which are differently expressed in different tissues and in various diseases, making it difficult to predict the overall effect the NTF will produce in a patient. The signaling pathways activated by neurotrophic factors are characterized by multiple negative feedback mechanisms, necessary to limit the amplification of the signal, resulting in non-linear dose-response curves in *in vitro* systems, animal models, and clinical trials. These factors made the scientific community try to leverage neurotrophic factors' potential into other therapeutic modalities in order to find a cure for neurodegenerative disorders. Several peptides, peptidomimetics, and chemical compounds with improved pharmacokinetics were developed recently and some of them have been tested in clinical trials. Further research in this field is warranted, and hopefully, these attempts will at the end, result in establishing novel disease-modifying drugs to combat neurodegeneration.

CONSENT FOR PUBLICATION

Not applicable.

CONFLICT OF INTEREST

The author is a minor shareholder in GeneCode Ltd, a company developing small molecular weight compounds targeting receptors of glial cell line-derived neurotrophic factor family ligands for disease-modifying treatment of Parkinson's disease .

ACKNOWLEDGEMENTS

The author is receiving funding from the Academy of Finland, project number 1325555. The illustration was created with Biorender.com..

REFERENCES

[1]   Davies AM. The neurotrophic hypothesis: where does it stand? Philos Trans R Soc Lond B Biol Sci 1996; 351(1338): 389-94.
[http://dx.doi.org/10.1098/rstb.1996.0033] [PMID: 8730776]

[2]   Wang Y, Wu H, Fontanet P, *et al.* A cell fitness selection model for neuronal survival during development. Nat Commun 2019; 10(1): 4137.
[http://dx.doi.org/10.1038/s41467-019-12119-3] [PMID: 31515492]

[3]   Erta M, Quintana A, Hidalgo J. Interleukin-6, a major cytokine in the central nervous system. Int J Biol Sci 2012; 8(9): 1254-66.
[http://dx.doi.org/10.7150/ijbs.4679] [PMID: 23136554]

[4]   Cohen S, Levi-Montalcini R. Purification and properties of a nerve growth-promoting factor isolated from mouse sarcoma 180. Cancer Res 1957; 17(1): 15-20.
[PMID: 13413830]

[5]   Cohen S, Levi-Montalcini R. A Nerve Growth-stimulating factor isolated from snake venom*. Proc Natl Acad Sci USA 1956; 42(9): 571-4.
[http://dx.doi.org/10.1073/pnas.42.9.571] [PMID: 16589907]

[6]   Huang EJ, Reichardt LF. Neurotrophins: roles in neuronal development and function. Annu Rev Neurosci 2001; 24: 677-736.
[http://dx.doi.org/10.1146/annurev.neuro.24.1.677] [PMID: 11520916]

[7]   Stolp HB. Neuropoietic cytokines in normal brain development and neurodevelopmental disorders. Mol Cell Neurosci 2013; 53: 63-8.
[http://dx.doi.org/10.1016/j.mcn.2012.08.009] [PMID: 22926235]

[8]   Airaksinen MS, Saarma M. The GDNF family: signalling, biological functions and therapeutic value. Nat Rev Neurosci 2002; 3(5): 383-94.
[http://dx.doi.org/10.1038/nrn812] [PMID: 11988777]

[9]   Mahato AK, Sidorova YA. Glial cell line-derived neurotrophic factors (GFLs) and small molecules targeting RET receptor for the treatment of pain and Parkinson's disease. Cell Tissue Res 2020; 382(1): 147-60.
[http://dx.doi.org/10.1007/s00441-020-03227-4] [PMID: 32556722]

[10]   Hsu J-Y, Crawley S, Chen M, *et al.* Non-homeostatic body weight regulation through a brainstem-restricted receptor for GDF15. Nature 2017; 550(7675): 255-9.
[http://dx.doi.org/10.1038/nature24042] [PMID: 28953886]

[11]   Mullican SE, Lin-Schmidt X, Chin C-N, *et al.* GFRAL is the receptor for GDF15 and the ligand promotes weight loss in mice and nonhuman primates. Nat Med 2017; 23(10): 1150-7.
[http://dx.doi.org/10.1038/nm.4392] [PMID: 28846097]

[12]   Yang L, Chang C-C, Sun Z, *et al.* GFRAL is the receptor for GDF15 and is required for the anti-obesity effects of the ligand. Nat Med 2017; 23(10): 1158-66.
[http://dx.doi.org/10.1038/nm.4394] [PMID: 28846099]

[13]   Emmerson PJ, Wang F, Du Y, *et al.* The metabolic effects of GDF15 are mediated by the orphan

receptor GFRAL. Nat Med 2017; 23(10): 1215-9.
[http://dx.doi.org/10.1038/nm.4393] [PMID: 28846098]

[14]   Mahato AK, Sidorova YA. RET Receptor Tyrosine Kinase: Role in Neurodegeneration, Obesity, and Cancer. Int J Mol Sci 2020; 21(19): 7108.
[http://dx.doi.org/10.3390/ijms21197108] [PMID: 32993133]

[15]   Lindholm P, Saarma M. Novel CDNF/MANF family of neurotrophic factors. Dev Neurobiol 2010; 70(5): 360-71.
[http://dx.doi.org/10.1002/dneu.20760] [PMID: 20186704]

[16]   Lindahl M, Saarma M, Lindholm P. Unconventional neurotrophic factors CDNF and MANF: Structure, physiological functions and therapeutic potential. Neurobiol Dis 2017; 97(Pt B): 90-102.

[17]   Meeker RB, Williams KS. The p75 neurotrophin receptor: at the crossroad of neural repair and death. Neural Regen Res 2015; 10(5): 721-5.
[http://dx.doi.org/10.4103/1673-5374.156967] [PMID: 26109945]

[18]   Lonka-Nevalaita L, Lume M, Leppänen S, Jokitalo E, Peränen J, Saarma M. Characterization of the intracellular localization, processing, and secretion of two glial cell line-derived neurotrophic factor splice isoforms. J Neurosci 2010; 30(34): 11403-13.
[http://dx.doi.org/10.1523/JNEUROSCI.5888-09.2010] [PMID: 20739562]

[19]   Sidorova YA, Volcho KP, Salakhutdinov NF. Neuroregeneration in Parkinson's Disease: From Proteins to Small Molecules. Curr Neuropharmacol 2019; 17(3): 268-87.
[http://dx.doi.org/10.2174/1570159X16666180905094123] [PMID: 30182859]

[20]   Shahmoradian SH, Lewis AJ, Genoud C, *et al.* Lewy pathology in Parkinson's disease consists of crowded organelles and lipid membranes. Nat Neurosci 2019; 22(7): 1099-109.
[http://dx.doi.org/10.1038/s41593-019-0423-2] [PMID: 31235907]

[21]   Mitre M, Mariga A, Chao MV. Neurotrophin signalling: novel insights into mechanisms and pathophysiology. Clin Sci (Lond) 2017; 131(1): 13-23.
[http://dx.doi.org/10.1042/CS20160044] [PMID: 27908981]

[22]   Hempstead BL, Martin-Zanca D, Kaplan DR, Parada LF, Chao MV. High-affinity NGF binding requires coexpression of the trk proto-oncogene and the low-affinity NGF receptor. Nature 1991; 350(6320): 678-83.
[http://dx.doi.org/10.1038/350678a0] [PMID: 1850821]

[23]   Nykjaer A, Lee R, Teng KK, *et al.* Sortilin is essential for proNGF-induced neuronal cell death. Nature 2004; 427(6977): 843-8.
[http://dx.doi.org/10.1038/nature02319] [PMID: 14985763]

[24]   Friedman WJ. Neurotrophins induce death of hippocampal neurons *via* the p75 receptor. J Neurosci 2000; 20(17): 6340-6.
[http://dx.doi.org/10.1523/JNEUROSCI.20-17-06340.2000] [PMID: 10964939]

[25]   Sidorova YA, Mätlik K, Paveliev M, *et al.* Persephin signaling through GFRalpha1: the potential for the treatment of Parkinson's disease. Mol Cell Neurosci 2010; 44(3): 223-32.
[http://dx.doi.org/10.1016/j.mcn.2010.03.009] [PMID: 20350599]

[26]   Sezgin E, Levental I, Mayor S, Eggeling C. The mystery of membrane organization: composition, regulation and roles of lipid rafts. Nat Rev Mol Cell Biol 2017; 18(6): 361-74.
[http://dx.doi.org/10.1038/nrm.2017.16] [PMID: 28356571]

[27]   Pierchala BA, Milbrandt J, Johnson EM Jr. Glial cell line-derived neurotrophic factor-dependent recruitment of Ret into lipid rafts enhances signaling by partitioning Ret from proteasome-dependent degradation. J Neurosci 2006; 26(10): 2777-87.
[http://dx.doi.org/10.1523/JNEUROSCI.3420-05.2006] [PMID: 16525057]

[28]   Tsui CC, Gabreski NA, Hein SJ, Pierchala BA. Lipid Rafts Are Physiologic Membrane Microdomains Necessary for the Morphogenic and Developmental Functions of Glial Cell Line-Derived

Neurotrophic Factor In Vivo. J Neurosci 2015; 35(38): 13233-43.
[http://dx.doi.org/10.1523/JNEUROSCI.2935-14.2015] [PMID: 26400951]

[29]   Parkash V, Leppänen V-M, Virtanen H, *et al.* The structure of the glial cell line-derived neurotrophic factor-coreceptor complex: insights into RET signaling and heparin binding. J Biol Chem 2008; 283(50): 35164-72.
[http://dx.doi.org/10.1074/jbc.M802543200] [PMID: 18845535]

[30]   Saarma M, Goldman A. Obesity: Receptors identified for a weight regulator. Nature 2017; 550(7675): 195-7.
[http://dx.doi.org/10.1038/nature24143] [PMID: 28953879]

[31]   Bespalov MM, Sidorova YA, Tumova S, *et al.* Heparan sulfate proteoglycan syndecan-3 is a novel receptor for GDNF, neurturin, and artemin. J Cell Biol 2011; 192(1): 153-69.
[http://dx.doi.org/10.1083/jcb.201009136] [PMID: 21200028]

[32]   Paratcha G, Ledda F, Ibáñez CF. The neural cell adhesion molecule NCAM is an alternative signaling receptor for GDNF family ligands. Cell 2003; 113(7): 867-79.
[http://dx.doi.org/10.1016/S0092-8674(03)00435-5] [PMID: 12837245]

[33]   Bauer S, Kerr BJ, Patterson PH. The neuropoietic cytokine family in development, plasticity, disease and injury. Nat Rev Neurosci 2007; 8(3): 221-32.
[http://dx.doi.org/10.1038/nrn2054] [PMID: 17311007]

[34]   Nathanson NM. Regulation of neurokine receptor signaling and trafficking. Neurochem Int 2012; 61(6): 874-8.
[http://dx.doi.org/10.1016/j.neuint.2012.01.018] [PMID: 22306348]

[35]   Lindahl M, Danilova T, Palm E, *et al.* MANF is indispensable for the proliferation and survival of pancreatic β cells. Cell Rep 2014; 7(2): 366-75.
[http://dx.doi.org/10.1016/j.celrep.2014.03.023] [PMID: 24726366]

[36]   Danilova T, Galli E, Pakarinen E, *et al.* Mesencephalic Astrocyte-Derived Neurotrophic Factor (MANF) Is Highly Expressed in Mouse Tissues With Metabolic Function. Front Endocrinol (Lausanne) 2019; 10: 765.
[http://dx.doi.org/10.3389/fendo.2019.00765] [PMID: 31781038]

[37]   Pilla E, Schneider K, Bertolotti A. Coping with Protein Quality Control Failure. Annu Rev Cell Dev Biol 2017; 33: 439-65.
[http://dx.doi.org/10.1146/annurev-cellbio-111315-125334] [PMID: 28992440]

[38]   Parkash V, Lindholm P, Peränen J, *et al.* The structure of the conserved neurotrophic factors MANF and CDNF explains why they are bifunctional. Protein Eng Des Sel 2009; 22(4): 233-41.
[http://dx.doi.org/10.1093/protein/gzn080] [PMID: 19258449]

[39]   Bai M, Vozdek R, Hnízda A, *et al.* Conserved roles of C. elegans and human MANFs in sulfatide binding and cytoprotection. Nat Commun 2018; 9(1): 897.
[http://dx.doi.org/10.1038/s41467-018-03355-0] [PMID: 29497057]

[40]   Tanaka T, Zhou Y, Ozawa T, *et al.* Ligand-activated epidermal growth factor receptor (EGFR) signaling governs endocytic trafficking of unliganded receptor monomers by non-canonical phosphorylation. J Biol Chem 2018; 293(7): 2288-301.
[http://dx.doi.org/10.1074/jbc.M117.811299] [PMID: 29255092]

[41]   Yadav L, Pietilä E, Öhman T, *et al.* PTPRA Phosphatase Regulates GDNF-Dependent RET Signaling and Inhibits the RET Mutant MEN2A Oncogenic Potential. iScience 2020; 23(2): 100871.
[http://dx.doi.org/10.1016/j.isci.2020.100871] [PMID: 32062451]

[42]   Arkun Y, Yasemi M. Dynamics and control of the ERK signaling pathway: Sensitivity, bistability, and oscillations. PLoS One 2018; 13(4): e0195513.
[http://dx.doi.org/10.1371/journal.pone.0195513] [PMID: 29630631]

[43]   Hou JG, Lin LF, Mytilineou C. Glial cell line-derived neurotrophic factor exerts neurotrophic effects

on dopaminergic neurons *in vitro* and promotes their survival and regrowth after damage by 1-methy-
-4-phenylpyridinium. J Neurochem 1996; 66(1): 74-82.
[http://dx.doi.org/10.1046/j.1471-4159.1996.66010074.x] [PMID: 8522992]

[44]   Saarenpää T, Kogan K, Sidorova Y, *et al.* Zebrafish GDNF and its co-receptor GFRα1 activate the
human RET receptor and promote the survival of dopaminergic neurons *in vitro.* PLoS One 2017;
12(5): e0176166.
[http://dx.doi.org/10.1371/journal.pone.0176166] [PMID: 28467503]

[45]   Schlee S, Carmillo P, Whitty A. Quantitative analysis of the activation mechanism of the
multicomponent growth-factor receptor Ret. Nat Chem Biol 2006; 2(11): 636-44.
[http://dx.doi.org/10.1038/nchembio823] [PMID: 17013378]

[46]   Romero MI, Rangappa N, Garry MG, Smith GM. Functional regeneration of chronically injured
sensory afferents into adult spinal cord after neurotrophin gene therapy. J Neurosci 2001; 21(21):
8408-16.
[http://dx.doi.org/10.1523/JNEUROSCI.21-21-08408.2001] [PMID: 11606629]

[47]   Smith GM, Falone AE, Frank E. Sensory axon regeneration: rebuilding functional connections in the
spinal cord. Trends Neurosci 2012; 35(3): 156-63.
[http://dx.doi.org/10.1016/j.tins.2011.10.006] [PMID: 22137336]

[48]   Dugger BN, Dickson DW. Pathology of Neurodegenerative Diseases. Cold Spring Harb Perspect Biol
2017; 9(7): a028035.
[http://dx.doi.org/10.1101/cshperspect.a028035] [PMID: 28062563]

[49]   Qiu C, Kivipelto M, von Strauss E. Epidemiology of Alzheimer's disease: occurrence, determinants,
and strategies toward intervention. Dialogues Clin Neurosci 2009; 11(2): 111-28.
[http://dx.doi.org/10.31887/DCNS.2009.11.2/cqiu] [PMID: 19585947]

[50]   Naqvi E. https://parkinsonsnewstoday.com/parkinsons-disease-statistics/

[51]   Han S, Kim S, Kim H, Shin H-W, Na K-S, Suh HS. Prevalence and incidence of Parkinson's disease
and drug-induced parkinsonism in Korea. BMC Public Health 2019; 19(1): 1328.
[http://dx.doi.org/10.1186/s12889-019-7664-6] [PMID: 31640652]

[52]   Longinetti E, Fang F. Epidemiology of amyotrophic lateral sclerosis: an update of recent literature.
Curr Opin Neurol 2019; 32(5): 771-6.
[http://dx.doi.org/10.1097/WCO.0000000000000730] [PMID: 31361627]

[53]   Baig SS, Strong M, Quarrell OW. The global prevalence of Huntington's disease: a systematic review
and discussion. Neurodegener Dis Manag 2016; 6(4): 331-43.
[http://dx.doi.org/10.2217/nmt-2016-0008] [PMID: 27507223]

[54]   https://www.ncbi.nlm.nih.gov/books/NBK557816/

[55]   Monzio Compagnoni G, Di Fonzo A. Understanding the pathogenesis of multiple system atrophy:
state of the art and future perspectives. Acta Neuropathol Commun 2019; 7(1): 113.
[http://dx.doi.org/10.1186/s40478-019-0730-6] [PMID: 31300049]

[56]   D'Amico A, Mercuri E, Tiziano FD, Bertini E. Spinal muscular atrophy. Orphanet J Rare Dis 2011;
6(1): 71.
[http://dx.doi.org/10.1186/1750-1172-6-71] [PMID: 22047105]

[57]   DeTure MA, Dickson DW. The neuropathological diagnosis of Alzheimer's disease. Mol
Neurodegener 2019; 14(1): 32.
[http://dx.doi.org/10.1186/s13024-019-0333-5] [PMID: 31375134]

[58]   Wakabayashi K, Tanji K, Odagiri S, Miki Y, Mori F, Takahashi H. The Lewy body in Parkinson's
disease and related neurodegenerative disorders. Mol Neurobiol 2013; 47(2): 495-508.
[http://dx.doi.org/10.1007/s12035-012-8280-y] [PMID: 22622968]

[59]   Ramesh N, Pandey UB. Autophagy Dysregulation in ALS: When Protein Aggregates Get Out of

Hand. Front Mol Neurosci 2017; 10: 263.
[http://dx.doi.org/10.3389/fnmol.2017.00263] [PMID: 28878620]

[60] Arrasate M, Finkbeiner S. Protein aggregates in Huntington's disease. Exp Neurol 2012; 238(1): 1-11.
[http://dx.doi.org/10.1016/j.expneurol.2011.12.013] [PMID: 22200539]

[61] Espay AJ, Vizcarra JA, Marsili L, *et al.* Revisiting protein aggregation as pathogenic in sporadic
Parkinson and Alzheimer diseases. Neurology 2019; 92(7): 329-37.
[http://dx.doi.org/10.1212/WNL.0000000000006926] [PMID: 30745444]

[62] Volpicelli-Daley LA. Effects of α-synuclein on axonal transport. Neurobiol Dis 2017; 105: 321-7.
[http://dx.doi.org/10.1016/j.nbd.2016.12.008] [PMID: 27956085]

[63] Kumar A, Pareek V, Faiq MA, Ghosh SK, Kumari C. Adult neurogenesis in humans: a review of basic
concepts, history, current research, and clinical implications. Innov Clin Neurosci 2019; 16(5-6): 30-7.
[PMID: 31440399]

[64] Eratne D, Loi SM, Farrand S, Kelso W, Velakoulis D, Looi JC. Alzheimer's disease: clinical update
on epidemiology, pathophysiology and diagnosis. Australas Psychiatry 2018; 26(4): 347-57.
[http://dx.doi.org/10.1177/1039856218762308] [PMID: 29614878]

[65] Mucke L. Neuroscience: Alzheimer's disease. Nature 2009; 461(7266): 895-7.
[http://dx.doi.org/10.1038/461895a] [PMID: 19829367]

[66] Imbimbo BP, Lombard J, Pomara N. Pathophysiology of Alzheimer's disease. Neuroimaging Clin N
Am 2005; 15(4): 727-753, ix.
[http://dx.doi.org/10.1016/j.nic.2005.09.009] [PMID: 16443487]

[67] 2016 Alzheimer's disease facts and figures. Alzheimers Dement 2016; 12(4): 459-509.
[http://dx.doi.org/10.1016/j.jalz.2016.03.001] [PMID: 27570871]

[68] Wolinsky D, Drake K, Bostwick J. Diagnosis and Management of Neuropsychiatric Symptoms in
Alzheimer's Disease. Curr Psychiatry Rep 2018; 20(12): 117.
[http://dx.doi.org/10.1007/s11920-018-0978-8] [PMID: 30367272]

[69] Toepper M. Dissociating Normal Aging from Alzheimer's Disease: A View from Cognitive
Neuroscience. J Alzheimers Dis 2017; 57(2): 331-52.
[http://dx.doi.org/10.3233/JAD-161099] [PMID: 28269778]

[70] Bozzali M, Serra L, Cercignani M. Quantitative MRI to understand Alzheimer's disease
pathophysiology. Curr Opin Neurol 2016; 29(4): 437-44.
[http://dx.doi.org/10.1097/WCO.0000000000000345] [PMID: 27228309]

[71] Sultzer DL. Cognitive ageing and Alzheimer's disease: the cholinergic system redux. Brain 2018;
141(3): 626-8.
[http://dx.doi.org/10.1093/brain/awy040] [PMID: 30753416]

[72] Briggs R, Kennelly SP, O'Neill D. Drug treatments in Alzheimer's disease. Clin Med (Lond) 2016;
16(3): 247-53.
[http://dx.doi.org/10.7861/clinmedicine.16-3-247] [PMID: 27251914]

[73] Traini E, Carotenuto A, Fasanaro AM, Amenta F. Volume Analysis of Brain Cognitive Areas in
Alzheimer's Disease: Interim 3-Year Results from the ASCOMALVA Trial. J Alzheimers Dis 2020;
76(1): 317-29.
[http://dx.doi.org/10.3233/JAD-190623] [PMID: 32508323]

[74] Jankovic J. Parkinson's disease: clinical features and diagnosis. J Neurol Neurosurg Psychiatry 2008;
79(4): 368-76.
[http://dx.doi.org/10.1136/jnnp.2007.131045] [PMID: 18344392]

[75] Dawson TM, Dawson VL. The role of parkin in familial and sporadic Parkinson's disease. Mov
Disord 2010; 25(S1) (Suppl. 1): S32-9.
[http://dx.doi.org/10.1002/mds.22798] [PMID: 20187240]

[76] Armstrong MJ, Okun MS. Diagnosis and Treatment of Parkinson Disease: A Review. JAMA 2020; 323(6): 548-60.
[http://dx.doi.org/10.1001/jama.2019.22360] [PMID: 32044947]

[77] Aryun K, Young Eun K, Ji Young Y. Amantadine and the Risk of Dyskinesia in Patients with Early Parkinson's Disease An Open-Label, Pragmatic Trial JMD

[78] Mehanna R, Fernandez HH, Wagle Shukla A, Bajwa JA. Deep Brain Stimulation in Parkinson's Disease J Mov Disord 2018; 11(2): 65-71.https://www.hindawi.com/journals/pd/2018/9625291/

[79] Pfeiffer RF. Non-motor symptoms in Parkinson's disease. Parkinsonism Relat Disord 2016; 22 (Suppl. 1): S119-22.
[http://dx.doi.org/10.1016/j.parkreldis.2015.09.004] [PMID: 26372623]

[80] Todorova A, Jenner P, Ray Chaudhuri K. Non-motor Parkinson's: integral to motor Parkinson's, yet often neglected. Pract Neurol 2014; 14(5): 310-22.
[http://dx.doi.org/10.1136/practneurol-2013-000741] [PMID: 24699931]

[81] Tibar H, El Bayad K, Bouhouche A, et al. Non-Motor Symptoms of Parkinson's Disease and Their Impact on Quality of Life in a Cohort of Moroccan Patients. Front Neurol 2018; 9: 170.
[http://dx.doi.org/10.3389/fneur.2018.00170] [PMID: 29670566]

[82] Hardiman O, Al-Chalabi A, Chio A, Corr EM, Logroscino G, Robberecht W, et al. Amyotrophic lateral sclerosis. Nat Rev Dis Primers 2017; 3.

[83] Mehta P, Kaye W, Raymond J, et al. Prevalence of Amyotrophic Lateral Sclerosis - United States, 2014. MMWR Morb Mortal Wkly Rep 2018; 67(7): 216-8.
[http://dx.doi.org/10.15585/mmwr.mm6707a3] [PMID: 29470458]

[84] Corcia P, Pradat P-F, Salachas F, et al. Causes of death in a post-mortem series of ALS patients. Amyotroph Lateral Scler 2008; 9(1): 59-62.
[http://dx.doi.org/10.1080/17482960701656940] [PMID: 17924236]

[85] Talbott EO, Malek AM, Lacomis D. The epidemiology of amyotrophic lateral sclerosis.Handbook of Clinical Neurology. Elsevier 2016; pp. 225-38.

[86] Valadi N. https://www.clinicalkey.com/#!/content/playContent/1-s2.0-S0095454315000202?returnurl =https:%2F%2Flinkinghub.elsevier.com%2Fretrieve%2Fpii%2FS0095454315000202%3Fshowall%3 Dtrue&referrer=https:%2F%2Fpubmed.ncbi.nlm.nih.gov%2F

[87] Jaiswal MK. Riluzole and edaravone: A tale of two amyotrophic lateral sclerosis drugs. Med Res Rev 2019; 39(2): 733-48.
[http://dx.doi.org/10.1002/med.21528] [PMID: 30101496]

[88] Lazarevic V, Yang Y, Ivanova D, Fejtova A, Svenningsson P. Riluzole attenuates the efficacy of glutamatergic transmission by interfering with the size of the readily releasable neurotransmitter pool. Neuropharmacology 2018; 143: 38-48.
[http://dx.doi.org/10.1016/j.neuropharm.2018.09.021] [PMID: 30222983]

[89] https://reader.elsevier.com/reader/sd/pii/S0092867417311984?token=20CDACBF6F9C6FBAB127F2 B97A62D8DE6EF6B2D563DFCFA710F5F5D00C0BDD07FF7B89D9650461833E97997009CB8960

[90] Safety and efficacy of edaravone in well defined patients with amyotrophic lateral sclerosis: a randomised, double-blind, placebo-controlled trial. Lancet Neurol 2017; 16(7): 505-12.
[http://dx.doi.org/10.1016/S1474-4422(17)30115-1] [PMID: 28522181]

[91] Dharmadasa T, Kiernan MC. Riluzole, disease stage and survival in ALS. Lancet Neurol 2018; 17(5): 385-6.
[http://dx.doi.org/10.1016/S1474-4422(18)30091-7] [PMID: 29525493]

[92] Miller RG, Mitchell JD, Moore DH. Riluzole for amyotrophic lateral sclerosis (ALS)/motor neuron disease (MND). Cochrane Database Syst Rev 2012; 2012(3): CD001447.
[http://dx.doi.org/10.1002/14651858.CD001447.pub3] [PMID: 22419278]

[93]   Wyant KJ, Ridder AJ, Dayalu P. Huntington's Disease-Update on Treatments. Curr Neurol Neurosci Rep 2017; 17(4): 33.
[http://dx.doi.org/10.1007/s11910-017-0739-9] [PMID: 28324302]

[94]   McColgan P, Tabrizi SJ. Huntington's disease: a clinical review. Eur J Neurol 2018; 25(1): 24-34.
[http://dx.doi.org/10.1111/ene.13413] [PMID: 28817209]

[95]   Foroud T, Gray J, Ivashina J, Conneally PM. Differences in duration of Huntington's disease based on age at onset. J Neurol Neurosurg Psychiatry 1999; 66(1): 52-6.
[http://dx.doi.org/10.1136/jnnp.66.1.52] [PMID: 9886451]

[96]   Pandey M, Rajamma U. Huntington's disease: the coming of age. J Genet 2018; 97(3): 649-64.
[http://dx.doi.org/10.1007/s12041-018-0957-1] [PMID: 30027901]

[97]   Roman OC, Stovall J, Claassen DO. Perseveration and Suicide in Huntington's Disease. J Huntingtons Dis 2018; 7(2): 185-7.
[http://dx.doi.org/10.3233/JHD-170249] [PMID: 29614688]

[98]   Reiner A, Dragatsis I, Dietrich P. Genetics and Neuropathology of Huntington's Disease.International Review of Neurobiology. Academic Press 2011; pp. 325-72.

[99]   Heo Y-A, Scott LJ. Deutetrabenazine: A Review in Chorea Associated with Huntington's Disease. Drugs 2017; 77(17): 1857-64.
[http://dx.doi.org/10.1007/s40265-017-0831-0] [PMID: 29080203]

[100]   Ross CA, Tabrizi SJ. Huntington's disease: from molecular pathogenesis to clinical treatment. Lancet Neurol 2011; 10(1): 83-98.
[http://dx.doi.org/10.1016/S1474-4422(10)70245-3] [PMID: 21163446]

[101]   Yawn BP, Wollan PC, Weingarten TN, Watson JC, Hooten WM, Melton LJ III. The prevalence of neuropathic pain: clinical evaluation compared with screening tools in a community population. Pain Med 2009; 10(3): 586-93.
[http://dx.doi.org/10.1111/j.1526-4637.2009.00588.x] [PMID: 20849570]

[102]   Colloca L, Ludman T, Bouhassira D, *et al.* Neuropathic pain. Nat Rev Dis Primers 2017; 3: 17002.
[http://dx.doi.org/10.1038/nrdp.2017.2] [PMID: 28205574]

[103]   Odriozola A, Ortega L, Martinez L, *et al.* Widespread sensory neuropathy in diabetic patients hospitalized with severe COVID-19 infection. Diabetes Res Clin Pract 2021; 172: 108631.
[http://dx.doi.org/10.1016/j.diabres.2020.108631] [PMID: 33346072]

[104]   Özdağ Acarli AN, Samanci B, Ekizoğlu E, *et al.* Coronavirus Disease 2019 (COVID-19) From the Point of View of Neurologists: Observation of Neurological Findings and Symptoms During the Combat Against a Pandemic. Noro Psikiyatri Arsivi 2020; 57(2): 154-9.
[http://dx.doi.org/10.29399/npa.26148] [PMID: 32550783]

[105]   Ossipov MH. Growth factors and neuropathic pain. Curr Pain Headache Rep 2011; 15(3): 185-92.
[http://dx.doi.org/10.1007/s11916-011-0183-5] [PMID: 21327569]

[106]   Finnerup NB, Attal N, Haroutounian S, *et al.* Pharmacotherapy for neuropathic pain in adults: a systematic review and meta-analysis. Lancet Neurol 2015; 14(2): 162-73.
[http://dx.doi.org/10.1016/S1474-4422(14)70251-0] [PMID: 25575710]

[107]   Finnerup NB, Otto M, Jensen TS, Sindrup SH. An evidence-based algorithm for the treatment of neuropathic pain. MedGenMed 2007; 9(2): 36.
[PMID: 17955091]

[108]   Mathieson S, Lin CC, Underwood M, Eldabe S. Pregabalin and gabapentin for pain. BMJ 2020; 369: m1315.
[http://dx.doi.org/10.1136/bmj.m1315] [PMID: 32345589]

[109]   Cheng H-C, Ulane CM, Burke RE. Clinical progression in Parkinson disease and the neurobiology of axons. Ann Neurol 2010; 67(6): 715-25.

[http://dx.doi.org/10.1002/ana.21995] [PMID: 20517933]

[110]   Kordower JH, Olanow CW, Dodiya HB, *et al.* Disease duration and the integrity of the nigrostriatal system in Parkinson's disease. Brain 2013; 136(Pt 8): 2419-31.
[http://dx.doi.org/10.1093/brain/awt192] [PMID: 23884810]

[111]   Sidorova YA, Saarma M. Can Growth Factors Cure Parkinson's Disease? Trends Pharmacol Sci 2020; 41(12): 909-22.
[http://dx.doi.org/10.1016/j.tips.2020.09.010] [PMID: 33198924]

[112]   Latimer CS, Shively CA, Keene CD, *et al.* A nonhuman primate model of early Alzheimer's disease pathologic change: Implications for disease pathogenesis. Alzheimers Dement 2019; 15(1): 93-105.
[http://dx.doi.org/10.1016/j.jalz.2018.06.3057] [PMID: 30467082]

[113]   Bronzetti E, Felici L, Zaccheo D, Amenta F. Age-related anatomical changes in the rat hippocampus: retardation by choline alfoscerate treatment. Arch Gerontol Geriatr 1991; 13(2): 167-78.
[http://dx.doi.org/10.1016/0167-4943(91)90059-Y] [PMID: 15374427]

[114]   Salvatore MF, Ai Y, Fischer B, *et al.* Point source concentration of GDNF may explain failure of phase II clinical trial. Exp Neurol 2006; 202(2): 497-505.
[http://dx.doi.org/10.1016/j.expneurol.2006.07.015] [PMID: 16962582]

[115]   Chu Y, Bartus RT, Manfredsson FP, Olanow CW, Kordower JH. Long-term post-mortem studies following neurturin gene therapy in patients with advanced Parkinson's disease. Brain 2020; 143(3): 960-75.
[http://dx.doi.org/10.1093/brain/awaa020] [PMID: 32203581]

[116]   Konishi Y, Yang L-B, He P, *et al.* Deficiency of GDNF Receptor GFRα1 in Alzheimer's Neurons Results in Neuronal Death. J Neurosci 2014; 34(39): 13127-38.
[http://dx.doi.org/10.1523/JNEUROSCI.2582-13.2014] [PMID: 25253858]

[117]   Ginsberg SD, Che S, Wuu J, Counts SE, Mufson EJ. Down regulation of trk but not p75NTR gene expression in single cholinergic basal forebrain neurons mark the progression of Alzheimer's disease. J Neurochem 2006; 97(2): 475-87.
[http://dx.doi.org/10.1111/j.1471-4159.2006.03764.x] [PMID: 16539663]

[118]   Decressac M, Kadkhodaei B, Mattsson B, Laguna A, Perlmann T, Björklund A. α-Synuclein-induced down-regulation of Nurr1 disrupts GDNF signaling in nigral dopamine neurons. Sci Transl Med 2012; 4(163): 163ra156.
[http://dx.doi.org/10.1126/scitranslmed.3004676] [PMID: 23220632]

[119]   Su X, Fischer DL, Li X, Bankiewicz K, Sortwell CE, Federoff HJ. Alpha-Synuclein mRNA Is Not Increased in Sporadic PD and Alpha-Synuclein Accumulation Does Not Block GDNF Signaling in Parkinson's Disease and Disease Models. Mol Ther 2017; 25(10): 2231-5.
[http://dx.doi.org/10.1016/j.ymthe.2017.04.018] [PMID: 28522034]

[120]   Lin LF, Doherty DH, Lile JD, Bektesh S, Collins F. GDNF: a glial cell line-derived neurotrophic factor for midbrain dopaminergic neurons. Science 1993; 260(5111): 1130-2.
[http://dx.doi.org/10.1126/science.8493557] [PMID: 8493557]

[121]   Hoffer BJ, Hoffman A, Bowenkamp K, *et al.* Glial cell line-derived neurotrophic factor reverses toxin-induced injury to midbrain dopaminergic neurons *in vivo.* Neurosci Lett 1994; 182(1): 107-11.
[http://dx.doi.org/10.1016/0304-3940(94)90218-6] [PMID: 7891873]

[122]   Bowenkamp KE, Hoffman AF, Gerhardt GA, *et al.* Glial cell line-derived neurotrophic factor supports survival of injured midbrain dopaminergic neurons. J Comp Neurol 1995; 355(4): 479-89.
[http://dx.doi.org/10.1002/cne.903550402] [PMID: 7636027]

[123]   Gash DM, Zhang Z, Ovadia A, *et al.* Functional recovery in parkinsonian monkeys treated with GDNF. Nature 1996; 380(6571): 252-5.
[http://dx.doi.org/10.1038/380252a0] [PMID: 8637574]

[124]   Tomac A, Lindqvist E, Lin L-FH, *et al.* Protection and repair of the nigrostriatal dopaminergic system

by GDNF in vivo. Nature 1995; 373(6512): 335-9.
[http://dx.doi.org/10.1038/373335a0] [PMID: 7830766]

[125] Kirik D, Georgievska B, Rosenblad C, Björklund A. Delayed infusion of GDNF promotes recovery of motor function in the partial lesion model of Parkinson's disease. Eur J Neurosci 2001; 13(8): 1589-99.
[http://dx.doi.org/10.1046/j.0953-816x.2001.01534.x] [PMID: 11328352]

[126] Horger BA, Nishimura MC, Armanini MP, *et al.* Neurturin exerts potent actions on survival and function of midbrain dopaminergic neurons. J Neurosci 1998; 18(13): 4929-37.
[http://dx.doi.org/10.1523/JNEUROSCI.18-13-04929.1998] [PMID: 9634558]

[127] Oiwa Y, Yoshimura R, Nakai K, Itakura T. Dopaminergic neuroprotection and regeneration by neurturin assessed by using behavioral, biochemical and histochemical measurements in a model of progressive Parkinson's disease. Brain Res 2002; 947(2): 271-83.
[http://dx.doi.org/10.1016/S0006-8993(02)02934-7] [PMID: 12176170]

[128] Reyes-Corona D, Vázquez-Hernández N, Escobedo L, *et al.* Neurturin overexpression in dopaminergic neurons induces presynaptic and postsynaptic structural changes in rats with chronic 6-hydroxydopamine lesion. PLoS One 2017; 12(11): e0188239.
[http://dx.doi.org/10.1371/journal.pone.0188239] [PMID: 29176874]

[129] Herzog CD, Brown L, Kruegel BR, *et al.* Enhanced neurotrophic distribution, cell signaling and neuroprotection following substantia nigral *versus* striatal delivery of AAV2-NRTN (CERE-120). Neurobiol Dis 2013; 58: 38-48.
[http://dx.doi.org/10.1016/j.nbd.2013.04.011] [PMID: 23631873]

[130] Bradley LH, Fuqua J, Richardson A, *et al.* Dopamine neuron stimulating actions of a GDNF propeptide. PLoS One 2010; 5(3): e9752.
[http://dx.doi.org/10.1371/journal.pone.0009752] [PMID: 20305789]

[131] Revilla S, Ursulet S, Álvarez-López MJ, *et al.* Lenti-GDNF gene therapy protects against Alzheimer's disease-like neuropathology in 3xTg-AD mice and MC65 cells. CNS Neurosci Ther 2014; 20(11): 961-72.
[http://dx.doi.org/10.1111/cns.12312] [PMID: 25119316]

[132] Moreno-Igoa M, Calvo AC, Ciriza J, Muñoz MJ, Zaragoza P, Osta R. Non-viral gene delivery of the GDNF, either alone or fused to the C-fragment of tetanus toxin protein, prolongs survival in a mouse ALS model. Restor Neurol Neurosci 2012; 30(1): 69-80.
[http://dx.doi.org/10.3233/RNN-2011-0621] [PMID: 22124037]

[133] Acsadi G, Anguelov RA, Yang H, *et al.* Increased survival and function of SOD1 mice after glial cell-derived neurotrophic factor gene therapy. Hum Gene Ther 2002; 13(9): 1047-59.
[http://dx.doi.org/10.1089/104303402753812458] [PMID: 12067438]

[134] Wang L-J, Lu Y-Y, Muramatsu S, *et al.* Neuroprotective effects of glial cell line-derived neurotrophic factor mediated by an adeno-associated virus vector in a transgenic animal model of amyotrophic lateral sclerosis. J Neurosci 2002; 22(16): 6920-8.
[http://dx.doi.org/10.1523/JNEUROSCI.22-16-06920.2002] [PMID: 12177190]

[135] Guillot S, Azzouz M, Déglon N, Zurn A, Aebischer P. Local GDNF expression mediated by lentiviral vector protects facial nerve motoneurons but not spinal motoneurons in SOD1(G93A) transgenic mice. Neurobiol Dis 2004; 16(1): 139-49.
[http://dx.doi.org/10.1016/j.nbd.2004.01.017] [PMID: 15207271]

[136] Li W, Brakefield D, Pan Y, Hunter D, Myckatyn TM, Parsadanian A. Muscle-derived but not centrally derived transgene GDNF is neuroprotective in G93A-SOD1 mouse model of ALS. Exp Neurol 2007; 203(2): 457-71.
[http://dx.doi.org/10.1016/j.expneurol.2006.08.028] [PMID: 17034790]

[137] Thomsen GM, Alkaslasi M, Vit J-P, *et al.* Systemic injection of AAV9-GDNF provides modest functional improvements in the SOD1^{G93A} ALS rat but has adverse side effects. Gene Ther 2017; 24(4):

245-52.
[http://dx.doi.org/10.1038/gt.2017.9] [PMID: 28276446]

[138] Klein SM, Behrstock S, McHugh J, *et al.* GDNF delivery using human neural progenitor cells in a rat model of ALS. Hum Gene Ther 2005; 16(4): 509-21.
[http://dx.doi.org/10.1089/hum.2005.16.509] [PMID: 15871682]

[139] Mohajeri MH, Figlewicz DA, Bohn MC. Intramuscular grafts of myoblasts genetically modified to secrete glial cell line-derived neurotrophic factor prevent motoneuron loss and disease progression in a mouse model of familial amyotrophic lateral sclerosis. Hum Gene Ther 1999; 10(11): 1853-66.
[http://dx.doi.org/10.1089/10430349950017536] [PMID: 10446925]

[140] Gross SK, Shim BS, Bartus RT, *et al.* Focal and dose-dependent neuroprotection in ALS mice following AAV2-neurturin delivery. Exp Neurol 2020; 323: 113091.
[http://dx.doi.org/10.1016/j.expneurol.2019.113091] [PMID: 31678350]

[141] Araujo DM, Hilt DC. Glial cell line-derived neurotrophic factor attenuates the excitotoxin-induced behavioral and neurochemical deficits in a rodent model of Huntington's disease. Neuroscience 1997; 81(4): 1099-110.
[http://dx.doi.org/10.1016/S0306-4522(97)00079-1] [PMID: 9330371]

[142] McBride JL, During MJ, Wuu J, Chen EY, Leurgans SE, Kordower JH. Structural and functional neuroprotection in a rat model of Huntington's disease by viral gene transfer of GDNF. Exp Neurol 2003; 181(2): 213-23.
[http://dx.doi.org/10.1016/S0014-4886(03)00044-X] [PMID: 12781994]

[143] Kells AP, Fong DM, Dragunow M, During MJ, Young D, Connor B. AAV-mediated gene delivery of BDNF or GDNF is neuroprotective in a model of Huntington disease. Mol Ther 2004; 9(5): 682-8.
[http://dx.doi.org/10.1016/j.ymthe.2004.02.016] [PMID: 15120329]

[144] Pérez-Navarro E, Arenas E, Reiriz J, Calvo N, Alberch J. Glial cell line-derived neurotrophic factor protects striatal calbindin-immunoreactive neurons from excitotoxic damage. Neuroscience 1996; 75(2): 345-52.
[http://dx.doi.org/10.1016/0306-4522(96)00336-3] [PMID: 8931001]

[145] McBride JL, Ramaswamy S, Gasmi M, *et al.* Viral delivery of glial cell line-derived neurotrophic factor improves behavior and protects striatal neurons in a mouse model of Huntington's disease. Proc Natl Acad Sci USA 2006; 103(24): 9345-50.
[http://dx.doi.org/10.1073/pnas.0508875103] [PMID: 16751280]

[146] Ramaswamy S, McBride JL, Han I, *et al.* Intrastriatal CERE-120 (AAV-Neurturin) protects striatal and cortical neurons and delays motor deficits in a transgenic mouse model of Huntington's disease. Neurobiol Dis 2009; 34(1): 40-50.
[http://dx.doi.org/10.1016/j.nbd.2008.12.005] [PMID: 19150499]

[147] Ramaswamy S, McBride JL, Herzog CD, *et al.* Neurturin gene therapy improves motor function and prevents death of striatal neurons in a 3-nitropropionic acid rat model of Huntington's disease. Neurobiol Dis 2007; 26(2): 375-84.
[http://dx.doi.org/10.1016/j.nbd.2007.01.003] [PMID: 17336076]

[148] Popovic N, Maingay M, Kirik D, Brundin P. Lentiviral gene delivery of GDNF into the striatum of R6/2 Huntington mice fails to attenuate behavioral and neuropathological changes. Exp Neurol 2005; 193(1): 65-74.
[http://dx.doi.org/10.1016/j.expneurol.2004.12.009] [PMID: 15817265]

[149] Cheng S, Tereshchenko J, Zimmer V, *et al.* Therapeutic efficacy of regulable GDNF expression for Huntington's and Parkinson's disease by a high-induction, background-free "GeneSwitch" vector. Exp Neurol 2018; 309: 79-90.
[http://dx.doi.org/10.1016/j.expneurol.2018.07.017] [PMID: 30076831]

[150] Mitra S, Behbahani H, Eriksdotter M. Innovative Therapy for Alzheimer's Disease-With Focus on Biodelivery of NGF. Front Neurosci 2019; 13: 38.

[http://dx.doi.org/10.3389/fnins.2019.00038] [PMID: 30804738]

[151]   Tanila H. The role of BDNF in Alzheimer's disease. Neurobiol Dis 97(Pt B): 114-8.2017;
        [http://dx.doi.org/10.1016/j.nbd.2016.05.008]

[152]   Cuello AC, Pentz R, Hall H. The Brain NGF Metabolic Pathway in Health and in Alzheimer's
        Pathology. Front Neurosci 2019; 13: 62.
        [http://dx.doi.org/10.3389/fnins.2019.00062] [PMID: 30809111]

[153]   Cuello AC. Editorial: The Involvement of NGF in the Alzheimer's Pathology. Front Neurosci 2019;
        13: 872.
        [http://dx.doi.org/10.3389/fnins.2019.00872] [PMID: 31551671]

[154]   Bemelmans AP, Horellou P, Pradier L, Brunet I, Colin P, Mallet J. Brain-derived neurotrophic factor-
        mediated protection of striatal neurons in an excitotoxic rat model of Huntington's disease, as
        demonstrated by adenoviral gene transfer. Hum Gene Ther 1999; 10(18): 2987-97.
        [http://dx.doi.org/10.1089/10430349950016393] [PMID: 10609659]

[155]   da Fonsêca VS, da Silva Colla AR, de Paula Nascimento-Castro C, *et al.* Brain-Derived Neurotrophic
        Factor Prevents Depressive-Like Behaviors in Early-Symptomatic YAC128 Huntington's Disease
        Mice. Mol Neurobiol 2018; 55(9): 7201-15.
        [http://dx.doi.org/10.1007/s12035-018-0890-6] [PMID: 29388082]

[156]   Connor B, Sun Y, von Hieber D, Tang SK, Jones KS, Maucksch C. AAV1/2-mediated BDNF gene
        therapy in a transgenic rat model of Huntington's disease. Gene Ther 2016; 23(3): 283-95.
        [http://dx.doi.org/10.1038/gt.2015.113] [PMID: 26704721]

[157]   Xie Y, Hayden MR, Xu B. BDNF overexpression in the forebrain rescues Huntington's disease
        phenotypes in YAC128 mice. J Neurosci 2010; 30(44): 14708-18.
        [http://dx.doi.org/10.1523/JNEUROSCI.1637-10.2010] [PMID: 21048129]

[158]   Giampà C, Montagna E, Dato C, Melone MAB, Bernardi G, Fusco FR. Systemic delivery of
        recombinant brain derived neurotrophic factor (BDNF) in the R6/2 mouse model of Huntington's
        disease. PLoS One 2013; 8(5): e64037.
        [http://dx.doi.org/10.1371/journal.pone.0064037] [PMID: 23700454]

[159]   Khan N, Smith MT. Neurotrophins and Neuropathic Pain: Role in Pathobiology. Molecules 2015;
        20(6): 10657-88.
        [http://dx.doi.org/10.3390/molecules200610657] [PMID: 26065639]

[160]   Lee HJ, Lee JO, Lee YW, *et al.* LIF, a Novel Myokine, Protects Against Amyloid-Beta-Induced
        Neurotoxicity *via* Akt-Mediated Autophagy Signaling in Hippocampal Cells. Int J
        Neuropsychopharmacol 2019; 22(6): 402-14.
        [http://dx.doi.org/10.1093/ijnp/pyz016] [PMID: 31125414]

[161]   de Almeida LP, Zala D, Aebischer P, Déglon N. Neuroprotective effect of a CNTF-expressing
        lentiviral vector in the quinolinic acid rat model of Huntington's disease. Neurobiol Dis 2001; 8(3):
        433-46.
        [http://dx.doi.org/10.1006/nbdi.2001.0388] [PMID: 11442352]

[162]   Régulier E, Pereira de Almeida L, Sommer B, Aebischer P, Déglon N. Dose-dependent
        neuroprotective effect of ciliary neurotrophic factor delivered *via* tetracycline-regulated lentiviral
        vectors in the quinolinic acid rat model of Huntington's disease. Hum Gene Ther 2002; 13(16): 1981-
        90.
        [http://dx.doi.org/10.1089/10430340260355383] [PMID: 12427308]

[163]   Mittoux V, Ouary S, Monville C, *et al.* Corticostriatopallidal neuroprotection by adenovirus-mediated
        ciliary neurotrophic factor gene transfer in a rat model of progressive striatal degeneration. J Neurosci
        2002; 22(11): 4478-86.
        [http://dx.doi.org/10.1523/JNEUROSCI.22-11-04478.2002] [PMID: 12040055]

[164]   Anderson KD, Panayotatos N, Corcoran TL, Lindsay RM, Wiegand SJ. Ciliary neurotrophic factor

protects striatal output neurons in an animal model of Huntington disease. Proc Natl Acad Sci USA 1996; 93(14): 7346-51.
[http://dx.doi.org/10.1073/pnas.93.14.7346] [PMID: 8692996]

[165] Jørgensen JR, Emerich DF, Thanos C, *et al.* Lentiviral delivery of meteorin protects striatal neurons against excitotoxicity and reverses motor deficits in the quinolinic acid rat model. Neurobiol Dis 2011; 41(1): 160-8.
[http://dx.doi.org/10.1016/j.nbd.2010.09.003] [PMID: 20840868]

[166] Bensadoun J-C, de Almeida LP, Dréano M, Aebischer P, Déglon N. Neuroprotective effect of interleukin-6 and IL6/IL6R chimera in the quinolinic acid rat model of Huntington's syndrome. Eur J Neurosci 2001; 14(11): 1753-61.
[http://dx.doi.org/10.1046/j.0953-816x.2001.01802.x] [PMID: 11860469]

[167] Denovan-Wright EM, Attis M, Rodriguez-Lebron E, Mandel RJ. Sustained striatal ciliary neurotrophic factor expression negatively affects behavior and gene expression in normal and R6/1 mice. J Neurosci Res 2008; 86(8): 1748-57.
[http://dx.doi.org/10.1002/jnr.21636] [PMID: 18293418]

[168] Feeney SJ, Austin L, Bennett TM, *et al.* The effect of leukaemia inhibitory factor on SOD1 G93A murine amyotrophic lateral sclerosis. Cytokine 2003; 23(4-5): 108-18.
[http://dx.doi.org/10.1016/S1043-4666(03)00217-5] [PMID: 12967646]

[169] Azari MF, Lopes EC, Stubna C, *et al.* Behavioural and anatomical effects of systemically administered leukemia inhibitory factor in the SOD1(G93A G1H) mouse model of familial amyotrophic lateral sclerosis. Brain Res 2003; 982(1): 92-7.
[http://dx.doi.org/10.1016/S0006-8993(03)02989-5] [PMID: 12915243]

[170] Liu Y, Peng M, Zang D, Zhang B. Leukemia inhibitory factor promotes nestin-positive cells, and increases gp130 levels in the Parkinson disease mouse model of 6-hydroxydopamine. Neurosciences (Riyadh) 2013; 18(4): 363-70.
[PMID: 24141460]

[171] Liu J, Zang D. Response of neural precursor cells in the brain of Parkinson's disease mouse model after LIF administration. Neurol Res 2009; 31(7): 681-6.
[http://dx.doi.org/10.1179/174313209X382368] [PMID: 19108756]

[172] Hagg T, Varon S. Ciliary neurotrophic factor prevents degeneration of adult rat substantia nigra dopaminergic neurons in vivo. Proc Natl Acad Sci USA 1993; 90(13): 6315-9.
[http://dx.doi.org/10.1073/pnas.90.13.6315] [PMID: 8101002]

[173] Thompson SW, Dray A, Urban L. Leukemia inhibitory factor induces mechanical allodynia but not thermal hyperalgesia in the juvenile rat. Neuroscience 1996; 71(4): 1091-4.
[http://dx.doi.org/10.1016/0306-4522(95)00537-4] [PMID: 8684613]

[174] Hu Z, Deng N, Liu K, Zhou N, Sun Y, Zeng W. CNTF-STAT3-IL-6 Axis Mediates Neuroinflammatory Cascade across Schwann Cell-Neuron-Microglia. Cell Rep 2020; 31(7): 107657.
[http://dx.doi.org/10.1016/j.celrep.2020.107657] [PMID: 32433966]

[175] Kemppainen S, Lindholm P, Galli E, *et al.* Cerebral dopamine neurotrophic factor improves long-term memory in APP/PS1 transgenic mice modeling Alzheimer's disease as well as in wild-type mice. Behav Brain Res 2015; 291: 1-11.
[http://dx.doi.org/10.1016/j.bbr.2015.05.002] [PMID: 25975173]

[176] Voutilainen MH, Bäck S, Pörsti E, *et al.* Mesencephalic astrocyte-derived neurotrophic factor is neurorestorative in rat model of Parkinson's disease. J Neurosci 2009; 29(30): 9651-9.
[http://dx.doi.org/10.1523/JNEUROSCI.0833-09.2009] [PMID: 19641128]

[177] Hao F, Yang C, Chen S-S, *et al.* Long-term protective effects of AAV9-mesencephalic astrocyte-derived neurotrophic factor gene transfer in parkinsonian rats. Exp Neurol 2017; 291: 120-33.
[http://dx.doi.org/10.1016/j.expneurol.2017.01.008] [PMID: 28131727]

[178] Liu Y, Zhang J, Jiang M, Cai Q, Fang J, Jin L. MANF improves the MPP⁺/MPTP-induced Parkinson's disease via improvement of mitochondrial function and inhibition of oxidative stress. Am J Transl Res 2018; 10(5): 1284-94.
[PMID: 29887945]

[179] Zhang Z, Shen Y, Luo H, *et al.* MANF protects dopamine neurons and locomotion defects from a human α-synuclein induced Parkinson's disease model in C. elegans by regulating ER stress and autophagy pathways. Exp Neurol 2018; 308: 59-71.
[http://dx.doi.org/10.1016/j.expneurol.2018.06.016] [PMID: 29959908]

[180] Voutilainen MH, Bäck S, Peränen J, *et al.* Chronic infusion of CDNF prevents 6-OHDA-induced deficits in a rat model of Parkinson's disease. Exp Neurol 2011; 228(1): 99-108.
[http://dx.doi.org/10.1016/j.expneurol.2010.12.013] [PMID: 21185834]

[181] Lindholm P, Voutilainen MH, Laurén J, *et al.* Novel neurotrophic factor CDNF protects and rescues midbrain dopamine neurons in vivo. Nature 2007; 448(7149): 73-7.
[http://dx.doi.org/10.1038/nature05957] [PMID: 17611540]

[182] Airavaara M, Harvey BK, Voutilainen MH, *et al.* CDNF protects the nigrostriatal dopamine system and promotes recovery after MPTP treatment in mice. Cell Transplant 2012; 21(6): 1213-23.
[http://dx.doi.org/10.3727/096368911X600948] [PMID: 21943517]

[183] Ren X, Zhang T, Gong X, Hu G, Ding W, Wang X. AAV2-mediated striatum delivery of human CDNF prevents the deterioration of midbrain dopamine neurons in a 6-hydroxydopamine induced parkinsonian rat model. Exp Neurol 2013; 248: 148-56.
[http://dx.doi.org/10.1016/j.expneurol.2013.06.002] [PMID: 23764500]

[184] Bäck S, Peränen J, Galli E, *et al.* Gene therapy with AAV2-CDNF provides functional benefits in a rat model of Parkinson's disease. Brain Behav 2013; 3(2): 75-88.
[http://dx.doi.org/10.1002/brb3.117] [PMID: 23532969]

[185] Wang L, Wang Z, Zhu R, *et al.* Therapeutic efficacy of AAV8-mediated intrastriatal delivery of human cerebral dopamine neurotrophic factor in 6-OHDA-induced parkinsonian rat models with different disease progression. PLoS One 2017; 12(6): e0179476.
[http://dx.doi.org/10.1371/journal.pone.0179476] [PMID: 28622392]

[186] Garea-Rodríguez E, Eesmaa A, Lindholm P, *et al.* Comparative Analysis of the Effects of Neurotrophic Factors CDNF and GDNF in a Nonhuman Primate Model of Parkinson's Disease. PLoS One 2016; 11(2): e0149776.
[http://dx.doi.org/10.1371/journal.pone.0149776] [PMID: 26901822]

[187] Cordero-Llana Ó, Houghton BC, Rinaldi F, *et al.* Enhanced efficacy of the CDNF/MANF family by combined intranigral overexpression in the 6-OHDA rat model of Parkinson's disease. Mol Ther 2015; 23(2): 244-54.
[http://dx.doi.org/10.1038/mt.2014.206] [PMID: 25369767]

[188] Lorenzo FD, Lüningschrör P, Nam J, Pilotto F, Galli E, Lindholm P, *et al.* CDNF rescues motor neurons in three animal models of ALS by targeting ER stress. bioRxiv 2020.

[189] Stepanova P, Srinivasan V, Lindholm D, Voutilainen MH. Cerebral dopamine neurotrophic factor (CDNF) protects against quinolinic acid-induced toxicity in *in vitro* and *in vivo* models of Huntington's disease. Sci Rep 2020; 10(1): 19045.
[http://dx.doi.org/10.1038/s41598-020-75439-1] [PMID: 33154393]

[190] Cheng L, Liu Y, Zhao H, Zhang W, Guo Y-J, Nie L. Lentiviral-mediated transfer of CDNF promotes nerve regeneration and functional recovery after sciatic nerve injury in adult rats. Biochem Biophys Res Commun 2013; 440(2): 330-5.
[http://dx.doi.org/10.1016/j.bbrc.2013.09.084] [PMID: 24076387]

[191] Olson L, Nordberg A, von Holst H, *et al.* Nerve growth factor affects 11C-nicotine binding, blood flow, EEG, and verbal episodic memory in an Alzheimer patient (case report). J Neural Transm Park

Dis Dement Sect 1992; 4(1): 79-95.
[http://dx.doi.org/10.1007/BF02257624] [PMID: 1540306]

[192] Eriksdotter Jönhagen M, Nordberg A, Amberla K, *et al.* Intracerebroventricular infusion of nerve growth factor in three patients with Alzheimer's disease. Dement Geriatr Cogn Disord 1998; 9(5): 246-57.
[http://dx.doi.org/10.1159/000017069] [PMID: 9701676]

[193] Tuszynski MH, Thal L, Pay M, *et al.* A phase 1 clinical trial of nerve growth factor gene therapy for Alzheimer disease. Nat Med 2005; 11(5): 551-5.
[http://dx.doi.org/10.1038/nm1239] [PMID: 15852017]

[194] Tuszynski MH, Yang JH, Barba D, *et al.* Nerve Growth Factor Gene Therapy: Activation of Neuronal Responses in Alzheimer Disease. JAMA Neurol 2015; 72(10): 1139-47.
[http://dx.doi.org/10.1001/jamaneurol.2015.1807] [PMID: 26302439]

[195] Rafii MS, Tuszynski MH, Thomas RG, *et al.* Adeno-Associated Viral Vector (Serotype 2)-Nerve Growth Factor for Patients With Alzheimer Disease: A Randomized Clinical Trial. JAMA Neurol 2018; 75(7): 834-41.
[http://dx.doi.org/10.1001/jamaneurol.2018.0233] [PMID: 29582053]

[196] Nutt JG, Burchiel KJ, Comella CL, *et al.* Randomized, double-blind trial of glial cell line-derived neurotrophic factor (GDNF) in PD. Neurology 2003; 60(1): 69-73.
[http://dx.doi.org/10.1212/WNL.60.1.69] [PMID: 12525720]

[197] Gill SS, Patel NK, Hotton GR, *et al.* Direct brain infusion of glial cell line-derived neurotrophic factor in Parkinson disease. Nat Med 2003; 9(5): 589-95.
[http://dx.doi.org/10.1038/nm850] [PMID: 12669033]

[198] Patel NK, Bunnage M, Plaha P, Svendsen CN, Heywood P, Gill SS. Intraputamenal infusion of glial cell line-derived neurotrophic factor in PD: a two-year outcome study. Ann Neurol 2005; 57(2): 298-302.
[http://dx.doi.org/10.1002/ana.20374] [PMID: 15668979]

[199] Slevin JT, Gerhardt GA, Smith CD, Gash DM, Kryscio R, Young B. Improvement of bilateral motor functions in patients with Parkinson disease through the unilateral intraputaminal infusion of glial cell line-derived neurotrophic factor. J Neurosurg 2005; 102(2): 216-22.
[http://dx.doi.org/10.3171/jns.2005.102.2.0216] [PMID: 15739547]

[200] Lang AE, Gill S, Patel NK, *et al.* Randomized controlled trial of intraputamenal glial cell line-derived neurotrophic factor infusion in Parkinson disease. Ann Neurol 2006; 59(3): 459-66.
[http://dx.doi.org/10.1002/ana.20737] [PMID: 16429411]

[201] Hutchinson M, Gurney S, Newson R. GDNF in Parkinson disease: an object lesson in the tyranny of type II. J Neurosci Methods 2007; 163(2): 190-2.
[http://dx.doi.org/10.1016/j.jneumeth.2006.06.015] [PMID: 16876872]

[202] Marks WJ Jr, Bartus RT, Siffert J, *et al.* Gene delivery of AAV2-neurturin for Parkinson's disease: a double-blind, randomised, controlled trial. Lancet Neurol 2010; 9(12): 1164-72.
[http://dx.doi.org/10.1016/S1474-4422(10)70254-4] [PMID: 20970382]

[203] Bartus RT, Herzog CD, Chu Y, *et al.* Bioactivity of AAV2-neurturin gene therapy (CERE-120): differences between Parkinson's disease and nonhuman primate brains. Mov Disord 2011; 26(1): 27-36.
[http://dx.doi.org/10.1002/mds.23442] [PMID: 21322017]

[204] Warren Olanow C, Bartus RT, Baumann TL, *et al.* Gene delivery of neurturin to putamen and substantia nigra in Parkinson disease: A double-blind, randomized, controlled trial. Ann Neurol 2015; 78(2): 248-57.
[http://dx.doi.org/10.1002/ana.24436] [PMID: 26061140]

[205] Bartus RT, Kordower JH, Johnson EM Jr, *et al.* Post-mortem assessment of the short and long-term

effects of the trophic factor neurturin in patients with α-synucleinopathies. Neurobiol Dis 2015; 78: 162-71.
[http://dx.doi.org/10.1016/j.nbd.2015.03.023] [PMID: 25841760]

[206] Whone A, Luz M, Boca M, *et al.* Randomized trial of intermittent intraputamenal glial cell line-derived neurotrophic factor in Parkinson's disease. Brain 2019; 142(3): 512-25.
[http://dx.doi.org/10.1093/brain/awz023] [PMID: 30808022]

[207] Whone AL, Boca M, Luz M, *et al.* Extended Treatment with Glial Cell Line-Derived Neurotrophic Factor in Parkinson's Disease. J Parkinsons Dis 2019; 9(2): 301-13.
[http://dx.doi.org/10.3233/JPD-191576] [PMID: 30829619]

[208] Heiss JD, Lungu C, Hammoud DA, *et al.* Trial of magnetic resonance-guided putaminal gene therapy for advanced Parkinson's disease. Mov Disord 2019; 34(7): 1073-8.
[http://dx.doi.org/10.1002/mds.27724] [PMID: 31145831]

[209] https://treater.eu/treater-webcast-seminar-worlds-first-clinical-trial-with-cdnf-a-ming-for-a-breakthrough-in-parkinsons/

[210] Ochs G, Penn RD, York M, *et al.* A phase I/II trial of recombinant methionyl human brain derived neurotrophic factor administered by intrathecal infusion to patients with amyotrophic lateral sclerosis. Amyotroph Lateral Scler Other Motor Neuron Disord 2000; 1(3): 201-6.
[http://dx.doi.org/10.1080/14660820050515197] [PMID: 11464953]

[211] A controlled trial of recombinant methionyl human BDNF in ALS: The BDNF Study Group (Phase III). Neurology 1999; 52(7): 1427-33.
[http://dx.doi.org/10.1212/WNL.52.7.1427] [PMID: 10227630]

[212] A double-blind placebo-controlled clinical trial of subcutaneous recombinant human ciliary neurotrophic factor (rHCNTF) in amyotrophic lateral sclerosis. Neurology 1996; 46(5): 1244-9.
[http://dx.doi.org/10.1212/WNL.46.5.1244] [PMID: 8628460]

[213] Baloh RH, Glass JD, Svendsen CN. Stem cell transplantation for amyotrophic lateral sclerosis. Curr Opin Neurol 2018; 31(5): 655-61.
[http://dx.doi.org/10.1097/WCO.0000000000000598] [PMID: 30080719]

[214] Gothelf Y, Abramov N, Harel A, Offen D. Safety of repeated transplantations of neurotrophic factors-secreting human mesenchymal stromal stem cells. Clin Transl Med 2014; 3: 21.
[http://dx.doi.org/10.1186/2001-1326-3-21] [PMID: 25097724]

[215] Petrou P, Gothelf Y, Argov Z, *et al.* Safety and Clinical Effects of Mesenchymal Stem Cells Secreting Neurotrophic Factor Transplantation in Patients With Amyotrophic Lateral Sclerosis: Results of Phase 1/2 and 2a Clinical Trials. JAMA Neurol 2016; 73(3): 337-44.
[http://dx.doi.org/10.1001/jamaneurol.2015.4321] [PMID: 26751635]

[216] Berry JD, Cudkowicz ME, Windebank AJ, *et al.* NurOwn, phase 2, randomized, clinical trial in patients with ALS: Safety, clinical, and biomarker results. Neurology 2019; 93(24): e2294-305.
[http://dx.doi.org/10.1212/WNL.0000000000008620] [PMID: 31740545]

[217] Backonja M, Williams L, Miao X, Katz N, Chen C. Safety and efficacy of neublastin in painful lumbosacral radiculopathy: a randomized, double-blinded, placebo-controlled phase 2 trial using Bayesian adaptive design (the SPRINT trial). Pain 2017; 158(9): 1802-12.
[http://dx.doi.org/10.1097/j.pain.0000000000000983] [PMID: 28746076]

[218] Rolan PE, O'Neill G, Versage E, *et al.* First-In-Human, Double-Blind, Placebo-Controlled, Randomized, Dose-Escalation Study of BG00010, a Glial Cell Line-Derived Neurotrophic Factor Family Member, in Subjects with Unilateral Sciatica. PLoS One 2015; 10(5): e0125034.
[http://dx.doi.org/10.1371/journal.pone.0125034] [PMID: 25962165]

[219] Okkerse P, Hay JL, Versage E, *et al.* Pharmacokinetics and pharmacodynamics of multiple doses of BG00010, a neurotrophic factor with anti-hyperalgesic effects, in patients with sciatica. Br J Clin Pharmacol 2016; 82(1): 108-17.

[http://dx.doi.org/10.1111/bcp.12941] [PMID: 27016000]

[220] Apfel SC, Schwartz S, Adornato BT, *et al.* Efficacy and safety of recombinant human nerve growth factor in patients with diabetic polyneuropathy: A randomized controlled trial. rhNGF Clinical Investigator Group. JAMA 2000; 284(17): 2215-21.
[http://dx.doi.org/10.1001/jama.284.17.2215] [PMID: 11056593]

[221] Wellmer A, Misra VP, Sharief MK, Kopelman PG, Anand P. A double-blind placebo-controlled clinical trial of recombinant human brain-derived neurotrophic factor (rhBDNF) in diabetic polyneuropathy. J Peripher Nerv Syst 2001; 6(4): 204-10.
[http://dx.doi.org/10.1046/j.1529-8027.2001.01019.x] [PMID: 11800042]

[222] Sidorova YA, Saarma M. Small Molecules and Peptides Targeting Glial Cell Line-Derived Neurotrophic Factor Receptors for the Treatment of Neurodegeneration. Int J Mol Sci 2020; 21(18): 6575.
[http://dx.doi.org/10.3390/ijms21186575] [PMID: 32911810]

[223] Piltonen M, Bespalov MM, Ervasti D, *et al.* Heparin-binding determinants of GDNF reduce its tissue distribution but are beneficial for the protection of nigral dopaminergic neurons. Neurodegeneration 2009; 219(2): 499-506.
[http://dx.doi.org/10.1016/j.expneurol.2009.07.002] [PMID: 19615368]

[224] Runeberg-Roos P, Piccinini E, Penttinen A-M, *et al.* Developing therapeutically more efficient Neurturin variants for treatment of Parkinson's disease. Neurobiol Dis 2016; 96: 335-45.
[http://dx.doi.org/10.1016/j.nbd.2016.07.008] [PMID: 27425888]

[225] Larsen KE, Benn SC, Ay I, *et al.* A glial cell line-derived neurotrophic factor (GDNF):tetanus toxin fragment C protein conjugate improves delivery of GDNF to spinal cord motor neurons in mice. Brain Res 2006; 1120(1): 1-12.
[http://dx.doi.org/10.1016/j.brainres.2006.08.079] [PMID: 17020749]

[226] Ciriza J, Moreno-Igoa M, Calvo AC, *et al.* A genetic fusion GDNF-C fragment of tetanus toxin prolongs survival in a symptomatic mouse ALS model. Restor Neurol Neurosci 2008; 26(6): 459-65.
[PMID: 19096133]

[227] Zhou Q-H, Boado RJ, Lu JZ, Hui EK-W, Pardridge WM. Monoclonal antibody-glial-derived neurotrophic factor fusion protein penetrates the blood-brain barrier in the mouse. Drug Metab Dispos 2010; 38(4): 566-72.
[http://dx.doi.org/10.1124/dmd.109.031534] [PMID: 20075191]

[228] Zhou Q-H, Boado RJ, Hui EK-W, Lu JZ, Pardridge WM. Chronic dosing of mice with a transferrin receptor monoclonal antibody-glial-derived neurotrophic factor fusion protein. Drug Metab Dispos 2011; 39(7): 1149-54.
[http://dx.doi.org/10.1124/dmd.111.038349] [PMID: 21502195]

[229] Dietz GPH, Valbuena PC, Dietz B, *et al.* Application of a blood-brain-barrier-penetrating form of GDNF in a mouse model for Parkinson's disease. Brain Res 2006; 1082(1): 61-6.
[http://dx.doi.org/10.1016/j.brainres.2006.01.083] [PMID: 16703672]

[230] Fu A, Zhou Q-H, Hui EK-W, Lu JZ, Boado RJ, Pardridge WM. Intravenous treatment of experimental Parkinson's disease in the mouse with an IgG-GDNF fusion protein that penetrates the blood-brain barrier. Brain Res 2010; 1352: 208-13.
[http://dx.doi.org/10.1016/j.brainres.2010.06.059] [PMID: 20599807]

[231] Stenslik MJ, Potts LF, Sonne JWH, *et al.* Methodology and effects of repeated intranasal delivery of DNSP-11 in a rat model of Parkinson's disease. J Neurosci Methods 2015; 251: 120-9.
[http://dx.doi.org/10.1016/j.jneumeth.2015.05.006] [PMID: 25999268]

[232] Immonen T, Alakuijala A, Hytönen M, *et al.* A proGDNF-related peptide BEP increases synaptic excitation in rat hippocampus. Exp Neurol 2008; 210(2): 793-6.
[http://dx.doi.org/10.1016/j.expneurol.2007.12.018] [PMID: 18280470]

[233] Nielsen J, Gotfryd K, Li S, *et al*. Role of glial cell line-derived neurotrophic factor (GDNF)-neural cell adhesion molecule (NCAM) interactions in induction of neurite outgrowth and identification of a binding site for NCAM in the heel region of GDNF. J Neurosci 2009; 29(36): 11360-76.
[http://dx.doi.org/10.1523/JNEUROSCI.3239-09.2009] [PMID: 19741142]

[234] Ilieva M, Nielsen J, Korshunova I, *et al*. Artemin and an Artemin-Derived Peptide, Artefin, Induce Neuronal Survival, and Differentiation Through Ret and NCAM. Front Mol Neurosci 2019; 12: 47.
[http://dx.doi.org/10.3389/fnmol.2019.00047] [PMID: 30853893]

[235] Jmaeff S, Sidorova Y, Nedev H, Saarma M, Saragovi HU. Small-molecule agonists of the RET receptor tyrosine kinase activate biased trophic signals that are influenced by the presence of GFRa1 co-receptors. J Biol Chem 2020; 295(19): 6532-42.
[http://dx.doi.org/10.1074/jbc.RA119.011802] [PMID: 32245892]

[236] Jmaeff S, Sidorova Y, Lippiatt H, *et al*. Small-Molecule Ligands that Bind the RET Receptor Activate Neuroprotective Signals Independent of but Modulated by Coreceptor GFRα1. Mol Pharmacol 2020; 98(1): 1-12.
[http://dx.doi.org/10.1124/mol.119.118950] [PMID: 32362584]

[237] Sidorova YA, Bespalov MM, Wong AW, *et al*. A Novel Small Molecule GDNF Receptor RET Agonist, BT13, Promotes Neurite Growth from Sensory Neurons *in Vitro* and Attenuates Experimental Neuropathy in the Rat. Front Pharmacol 2017; 8: 365.
[http://dx.doi.org/10.3389/fphar.2017.00365] [PMID: 28680400]

[238] Viisanen H, Nuotio U, Kambur O, *et al*. Novel RET agonist for the treatment of experimental neuropathies. Mol Pain 2020; 16: 1744806920950866.
[http://dx.doi.org/10.1177/1744806920950866] [PMID: 32811276]

[239] Renko J-M, Voutilainen MH, Visnapuu T, Sidorova YA, Saarma M, Tuominen RK. GDNF Receptor Agonist Alleviates Motor Imbalance in Unilateral 6-Hydroxydopamine Model of Parkinson's Disease. Frontiers in Neurology and Neuroscience Research 2020; 1: 100004.
[PMID: 33479704]

[240] Mahato AK, Kopra J, Renko J-M, *et al*. Glial cell line-derived neurotrophic factor receptor Rearranged during transfection agonist supports dopamine neurons *in Vitro* and enhances dopamine release In Vivo. Mov Disord 2020; 35(2): 245-55.
[http://dx.doi.org/10.1002/mds.27943] [PMID: 31840869]

[241] Bespalov MM, Sidorova YA, Suleymanova I, Thompson J, Kambur O, Jokinen V, *et al*. Novel agonist of GDNF family ligand receptor RET for the treatment of experimental neuropathy bioRxiv 2016.
[http://dx.doi.org/10.1101/061820]

[242] Tokugawa K, Yamamoto K, Nishiguchi M, *et al*. XIB4035, a novel nonpeptidyl small molecule agonist for GFRalpha-1. Neurochem Int 2003; 42(1): 81-6.
[http://dx.doi.org/10.1016/S0197-0186(02)00053-0] [PMID: 12441171]

[243] Hedstrom KL, Murtie JC, Albers K, Calcutt NA, Corfas G. Treating small fiber neuropathy by topical application of a small molecule modulator of ligand-induced GFRα/RET receptor signaling. Proc Natl Acad Sci USA 2014; 111(6): 2325-30.
[http://dx.doi.org/10.1073/pnas.1308889111] [PMID: 24449858]

[244] Ivanova L, Tammiku-Taul J, Sidorova Y, Saarma M, Karelson M. Small-Molecule Ligands as Potential GDNF Family Receptor Agonists. ACS Omega 2018; 3(1): 1022-30.
[http://dx.doi.org/10.1021/acsomega.7b01932] [PMID: 30023796]

[245] Xie Y, Tisi MA, Yeo TT, Longo FM. Nerve growth factor (NGF) loop 4 dimeric mimetics activate ERK and AKT and promote NGF-like neurotrophic effects. J Biol Chem 2000; 275(38): 29868-74.
[http://dx.doi.org/10.1074/jbc.M005071200] [PMID: 10896671]

[246] https://onlinelibrary.wiley.com/doi/abs/10.1002/(SICI)1097-4547(19970401)48:1%3C1:AI--JNR1%3E3.0.CO;2-K

[247] Colangelo AM, Bianco MR, Vitagliano L, *et al.* A new nerve growth factor-mimetic peptide active on neuropathic pain in rats. J Neurosci 2008; 28(11): 2698-709.
[http://dx.doi.org/10.1523/JNEUROSCI.5201-07.2008] [PMID: 18337399]

[248] Triaca V, Fico E, Sposato V, *et al.* hNGF Peptides Elicit the NGF-TrkA Signalling Pathway in Cholinergic Neurons and Retain Full Neurotrophic Activity in the DRG Assay. Biomolecules 2020; 10(2): 216.
[http://dx.doi.org/10.3390/biom10020216] [PMID: 32024191]

[249] Seredenin SB, Gudasheva TA. The development of a pharmacologically active low-molecular mimetic of the nerve growth factor Z nevrol psikhiatr im SS Korsakova 2015; 115(6): 63.

[250] Scarpi D, Cirelli D, Matrone C, *et al.* Low molecular weight, non-peptidic agonists of TrkA receptor with NGF-mimetic activity. Cell Death Dis 2012; 3(7): e339-9.
[http://dx.doi.org/10.1038/cddis.2012.80] [PMID: 22764098]

[251] Maliartchouk S, Feng Y, Ivanisevic L, *et al.* A designed peptidomimetic agonistic ligand of TrkA nerve growth factor receptors. Mol Pharmacol 2000; 57(2): 385-91.
[PMID: 10648649]

[252] Meerovitch K, Torkildsen G, Lonsdale J, *et al.* Safety and efficacy of MIM-D3 ophthalmic solutions in a randomized, placebo-controlled Phase 2 clinical trial in patients with dry eye. Clin Ophthalmol 2013; 7: 1275-85.
[http://dx.doi.org/10.2147/OPTH.S44688] [PMID: 23836957]

[253] http://www.mimetogen.com/news-publications/press-releases/3910-mimetogen-pharmaceuticals-comp letes-enrollment-of-mim-728-phase-3-trial-for-tavilermide-for-dry-eye-disease.html

[254] Massa SM, Xie Y, Yang T, *et al.* Small, nonpeptide p75NTR ligands induce survival signaling and inhibit proNGF-induced death. J Neurosci 2006; 26(20): 5288-300.
[http://dx.doi.org/10.1523/JNEUROSCI.3547-05.2006] [PMID: 16707781]

[255] Simmons DA, Belichenko NP, Ford EC, *et al.* A small molecule p75NTR ligand normalizes signalling and reduces Huntington's disease phenotypes in R6/2 and BACHD mice. Hum Mol Genet 2016; 25(22): 4920-38.
[PMID: 28171570]

[256] Knowles JK, Simmons DA, Nguyen T-VV, *et al.* Small molecule p75NTR ligand prevents cognitive deficits and neurite degeneration in an Alzheimer's mouse model. Neurobiol Aging 2013; 34(8): 2052-63.
[http://dx.doi.org/10.1016/j.neurobiolaging.2013.02.015] [PMID: 23545424]

[257] Simmons DA, Knowles JK, Belichenko NP, *et al.* A small molecule p75NTR ligand, LM11A-31, reverses cholinergic neurite dystrophy in Alzheimer's disease mouse models with mid- to late-stage disease progression. PLoS One 2014; 9(8): e102136.
[http://dx.doi.org/10.1371/journal.pone.0102136] [PMID: 25153701]

[258] James ML, Belichenko NP, Shuhendler AJ, *et al.* [^{18}F]GE-180 PET Detects Reduced Microglia Activation After LM11A-31 Therapy in a Mouse Model of Alzheimer's Disease. Theranostics 2017; 7(6): 1422-36.
[http://dx.doi.org/10.7150/thno.17666] [PMID: 28529627]

[259] Nguyen T-VV, Shen L, Vander Griend L, *et al.* Small molecule p75NTR ligands reduce pathological phosphorylation and misfolding of tau, inflammatory changes, cholinergic degeneration, and cognitive deficits in AβPP(L/S) transgenic mice. J Alzheimers Dis 2014; 42(2): 459-83.
[http://dx.doi.org/10.3233/JAD-140036] [PMID: 24898660]

[260] Xie Y, Meeker RB, Massa SM, Longo FM. Modulation of the p75 neurotrophin receptor suppresses age-related basal forebrain cholinergic neuron degeneration. Sci Rep 2019; 9(1): 5273.
[http://dx.doi.org/10.1038/s41598-019-41654-8] [PMID: 30918278]

[261] PharmatrophiX Inc. A 6-months Prospective, Multi-center, Double-blind, Placebo-controlled,

Randomized, Adaptive-trial-design Study to Evaluate Safety, Tolerability and Exploratory Endpoints of Either Placebo or Two Different Oral Doses of LM11A-31-BHS in Patients With Mild to Moderate Probable Alzheimer's Disease clinicaltrialsgov 2020.

[262]   Jang S-W, Liu X, Yepes M, *et al.* A selective TrkB agonist with potent neurotrophic activities by 7,8-dihydroxyflavone. Proc Natl Acad Sci USA 2010; 107(6): 2687-92.
[http://dx.doi.org/10.1073/pnas.0913572107] [PMID: 20133810]

[263]   Liu X, Obianyo O, Chan CB, *et al.* Biochemical and biophysical investigation of the brain-derived neurotrophic factor mimetic 7,8-dihydroxyflavone in the binding and activation of the TrkB receptor. J Biol Chem 2014; 289(40): 27571-84.
[http://dx.doi.org/10.1074/jbc.M114.562561] [PMID: 25143381]

[264]   Li X-H, Dai C-F, Chen L, Zhou W-T, Han H-L, Dong Z-F. 7,8-dihydroxyflavone Ameliorates Motor Deficits *Via* Suppressing α-synuclein Expression and Oxidative Stress in the MPTP-induced Mouse Model of Parkinson's Disease. CNS Neurosci Ther 2016; 22(7): 617-24.
[http://dx.doi.org/10.1111/cns.12555] [PMID: 27079181]

[265]   Luo D, Shi Y, Wang J, *et al.* 7,8-dihydroxyflavone protects 6-OHDA and MPTP induced dopaminergic neurons degeneration through activation of TrkB in rodents. Neurosci Lett 2016; 620: 43-9.
[http://dx.doi.org/10.1016/j.neulet.2016.03.042] [PMID: 27019033]

[266]   Jiang M, Peng Q, Liu X, *et al.* Small-molecule TrkB receptor agonists improve motor function and extend survival in a mouse model of Huntington's disease. Hum Mol Genet 2013; 22(12): 2462-70.
[http://dx.doi.org/10.1093/hmg/ddt098] [PMID: 23446639]

[267]   Zhang Z, Liu X, Schroeder JP, *et al.* 7,8-dihydroxyflavone prevents synaptic loss and memory deficits in a mouse model of Alzheimer's disease. Neuropsychopharmacology 2014; 39(3): 638-50.
[http://dx.doi.org/10.1038/npp.2013.243] [PMID: 24022672]

[268]   Gao L, Tian M, Zhao H-Y, *et al.* TrkB activation by 7, 8-dihydroxyflavone increases synapse AMPA subunits and ameliorates spatial memory deficits in a mouse model of Alzheimer's disease. J Neurochem 2016; 136(3): 620-36.
[http://dx.doi.org/10.1111/jnc.13432] [PMID: 26577931]

[269]   Devi L, Ohno M. 7,8-dihydroxyflavone, a small-molecule TrkB agonist, reverses memory deficits and BACE1 elevation in a mouse model of Alzheimer's disease. Neuropsychopharmacology 2012; 37(2): 434-44.
[http://dx.doi.org/10.1038/npp.2011.191] [PMID: 21900882]

[270]   Aytan N, Choi J-K, Carreras I, *et al.* Protective effects of 7,8-dihydroxyflavone on neuropathological and neurochemical changes in a mouse model of Alzheimer's disease. Eur J Pharmacol 2018; 828: 9-17.
[http://dx.doi.org/10.1016/j.ejphar.2018.02.045] [PMID: 29510124]

[271]   English AW, Liu K, Nicolini JM, Mulligan AM, Ye K. Small-molecule trkB agonists promote axon regeneration in cut peripheral nerves. Proc Natl Acad Sci USA 2013; 110(40): 16217-22.
[http://dx.doi.org/10.1073/pnas.1303646110] [PMID: 24043773]

[272]   Chen C, Wang Z, Zhang Z, *et al.* The prodrug of 7,8-dihydroxyflavone development and therapeutic efficacy for treating Alzheimer's disease. Proc Natl Acad Sci USA 2018; 115(3): 578-83.
[http://dx.doi.org/10.1073/pnas.1718683115] [PMID: 29295929]

[273]   Massa SM, Yang T, Xie Y, *et al.* Small molecule BDNF mimetics activate TrkB signaling and prevent neuronal degeneration in rodents. J Clin Invest 2010; 120(5): 1774-85.
[http://dx.doi.org/10.1172/JCI41356] [PMID: 20407211]

[274]   Simmons DA, Belichenko NP, Yang T, *et al.* A small molecule TrkB ligand reduces motor impairment and neuropathology in R6/2 and BACHD mouse models of Huntington's disease. J Neurosci 2013; 33(48): 18712-27.
[http://dx.doi.org/10.1523/JNEUROSCI.1310-13.2013] [PMID: 24285878]

[275] Fan C-H, Lin C-W, Huang H-J, *et al.* LMDS-1, a potential TrkB receptor agonist provides a safe and neurotrophic effect for early-phase Alzheimer's disease. Psychopharmacology (Berl) 2020; 237(10): 3173-90.
[http://dx.doi.org/10.1007/s00213-020-05602-z] [PMID: 32748031]

[276] Jang S-W, Liu X, Chan CB, *et al.* Deoxygedunin, a natural product with potent neurotrophic activity in mice. PLoS One 2010; 5(7): e11528.
[http://dx.doi.org/10.1371/journal.pone.0011528] [PMID: 20644624]

[277] Jang S-W, Liu X, Chan C-B, *et al.* Amitriptyline is a TrkA and TrkB receptor agonist that promotes TrkA/TrkB heterodimerization and has potent neurotrophic activity. Chem Biol 2009; 16(6): 644-56.
[http://dx.doi.org/10.1016/j.chembiol.2009.05.010] [PMID: 19549602]

[278] Sudarshan K, Boda AK, Dogra S, Bose I, Yadav PN, Aidhen IS. Discovery of an isocoumarin analogue that modulates neuronal functions via neurotrophin receptor TrkB. Bioorg Med Chem Lett 2019; 29(4): 585-90.
[http://dx.doi.org/10.1016/j.bmcl.2018.12.057] [PMID: 30600206]

[279] Boltaev U, Meyer Y, Tolibzoda F, Jacques T, Gassaway M, Xu Q, *et al.* Multiplex quantitative assays indicate a need for reevaluating reported small-molecule TrkB agonists. Sci Signal 10(493)2017;
[http://dx.doi.org/10.1126/scisignal.aal1670]

[280] Todd D, Gowers I, Dowler SJ, *et al.* A monoclonal antibody TrkB receptor agonist as a potential therapeutic for Huntington's disease. PLoS One 2014; 9(2): e87923.
[http://dx.doi.org/10.1371/journal.pone.0087923] [PMID: 24503862]

[281] Cardenas-Aguayo M del C, Kazim SF, Grundke-Iqbal I, Iqbal K. Neurogenic and neurotrophic effects of BDNF peptides in mouse hippocampal primary neuronal cell cultures. PLoS One 2013; 8(1): e53596.
[http://dx.doi.org/10.1371/journal.pone.0053596] [PMID: 23320097]

[282] O'Leary PD, Hughes RA. Design of Potent Peptide Mimetics of Brain-derived Neurotrophic Factor http://www.jbc.org
[http://dx.doi.org/10.1074/jbc.M303209200]

[283] Wong AW, Giuffrida L, Wood R, *et al.* TDP6, a brain-derived neurotrophic factor-based trkB peptide mimetic, promotes oligodendrocyte myelination. Mol Cell Neurosci 2014; 63: 132-40.
[http://dx.doi.org/10.1016/j.mcn.2014.10.002] [PMID: 25461619]

[284] Ohnishi T, Sakamoto K, Asami-Odaka A, *et al.* Generation of a novel artificial TrkB agonist, BM17d99, using T7 phage-displayed random peptide libraries. Biochem Biophys Res Commun 2017; 483(1): 101-6.
[http://dx.doi.org/10.1016/j.bbrc.2016.12.186] [PMID: 28043792]

[285] Fletcher JL, Wood RJ, Nguyen J, *et al.* Targeting TrkB with a Brain-Derived Neurotrophic Factor Mimetic Promotes Myelin Repair in the Brain. J Neurosci 2018; 38(32): 7088-99.
[http://dx.doi.org/10.1523/JNEUROSCI.0487-18.2018] [PMID: 29976621]

[286] Huang YZ, Hernandez FJ, Gu B, *et al.* RNA aptamer-based functional ligands of the neurotrophin receptor, TrkB. Mol Pharmacol 2012; 82(4): 623-35.
[http://dx.doi.org/10.1124/mol.112.078220] [PMID: 22752556]

[287] Perreault M, Feng G, Will S, *et al.* Activation of TrkB with TAM-163 results in opposite effects on body weight in rodents and non-human primates. PLoS One 2013; 8(5): e62616.
[http://dx.doi.org/10.1371/journal.pone.0062616] [PMID: 23700410]

[288] Cazorla M, Arrang JM, Prémont J. Pharmacological characterization of six trkB antibodies reveals a novel class of functional agents for the study of the BDNF receptor. Br J Pharmacol 2011; 162(4): 947-60.
[http://dx.doi.org/10.1111/j.1476-5381.2010.01094.x] [PMID: 21039416]

[289] Rathje M, Pankratova S, Nielsen J, Gotfryd K, Bock E, Berezin V. A peptide derived from the CD

loop-D helix region of ciliary neurotrophic factor (CNTF) induces neuronal differentiation and survival by binding to the leukemia inhibitory factor (LIF) receptor and common cytokine receptor chain gp130. Eur J Cell Biol 2011; 90(12): 990-9.
[http://dx.doi.org/10.1016/j.ejcb.2011.08.001] [PMID: 22000729]

[290] Russmann V, Seeger N, Zellinger C, *et al.* The CNTF-derived peptide mimetic Cintrofin attenuates spatial-learning deficits in a rat post-status epilepticus model. Neurosci Lett 2013; 556: 170-5.
[http://dx.doi.org/10.1016/j.neulet.2013.10.003] [PMID: 24120433]

[291] Leyton-Jaimes MF, Ivert P, Hoeber J, *et al.* Empty mesoporous silica particles significantly delay disease progression and extend survival in a mouse model of ALS. Sci Rep 2020; 10(1): 20675.
[http://dx.doi.org/10.1038/s41598-020-77578-x] [PMID: 33244084]

[292] Chohan MO, Li B, Blanchard J, *et al.* Enhancement of dentate gyrus neurogenesis, dendritic and synaptic plasticity and memory by a neurotrophic peptide. Neurobiol Aging 2011; 32(8): 1420-34.
[http://dx.doi.org/10.1016/j.neurobiolaging.2009.08.008] [PMID: 19767127]

[293] Blanchard J, Chohan MO, Li B, Liu F, Iqbal K, Grundke-Iqbal I. Beneficial effect of a CNTF tetrapeptide on adult hippocampal neurogenesis, neuronal plasticity, and spatial memory in mice. J Alzheimers Dis 2010; 21(4): 1185-95.
[http://dx.doi.org/10.3233/JAD-2010-1000069] [PMID: 20952820]

[294] Bolognin S, Buffelli M, Puolivӓli J, Iqbal K. Rescue of cognitive-aging by administration of a neurogenic and/or neurotrophic compound. Neurobiol Aging 2014; 35(9): 2134-46.
[http://dx.doi.org/10.1016/j.neurobiolaging.2014.02.017] [PMID: 24702821]

[295] Kazim SF, Blanchard J, Dai C-L, *et al.* Disease modifying effect of chronic oral treatment with a neurotrophic peptidergic compound in a triple transgenic mouse model of Alzheimer's disease. Neurobiol Dis 2014; 71: 110-30.
[http://dx.doi.org/10.1016/j.nbd.2014.07.001] [PMID: 25046994]

[296] Wei W, Wang Y, Liu Y, *et al.* Prenatal to early postnatal neurotrophic treatment prevents Alzheimer-like behavior and pathology in mice. Alzheimers Res Ther 2020; 12(1): 102.
[http://dx.doi.org/10.1186/s13195-020-00666-7] [PMID: 32854771]

[297] Taupin JL, Legembre P, Bitard J, *et al.* Identification of agonistic and antagonistic antibodies against gp190, the leukemia inhibitory factor receptor, reveals distinct roles for its two cytokine-binding domains. J Biol Chem 2001; 276(51): 47975-81.
[http://dx.doi.org/10.1074/jbc.M105476200] [PMID: 11606572]

[298] Gu ZJ, De Vos J, Rebouissou C, *et al.* Agonist anti-gp130 transducer monoclonal antibodies are human myeloma cell survival and growth factors. Leukemia 2000; 14(1): 188-97.
[http://dx.doi.org/10.1038/sj.leu.2401632] [PMID: 10637495]

[299] Kim JW, Marquez CP, Sperberg RAP, *et al.* Engineering a potent receptor superagonist or antagonist from a novel IL-6 family cytokine ligand. Proc Natl Acad Sci USA 2020; 117(25): 14110-8.
[http://dx.doi.org/10.1073/pnas.1922729117] [PMID: 32522868]

[300] Paul G, Zachrisson O, Varrone A, *et al.* Safety and tolerability of intracerebroventricular PDGF-BB in Parkinson's disease patients. J Clin Invest 2015; 125(3): 1339-46.
[http://dx.doi.org/10.1172/JCI79635] [PMID: 25689258]

[301] Hellman M, Arumäe U, Yu L, Lindholm P, Peränen J, Saarma M, *et al.* Mesencephalic Astrocyte-derived Neurotrophic Factor (MANF) Has a Unique Mechanism to Rescue Apoptotic Neurons. J Biol Chem 2011; 286(4): 2675-80.

CHAPTER 4

Neural Bases of Executive Function in ADHD Children as Assessed Using fNIRS

Takahiro Ikeda[1], Akari Inoue[2], Masako Nagashima-Kawada[1], Tatsuya Tokuda[2], Takanori Yamagata[1], Ippeita Dan[2,3] and Yukifumi Monden[1,*]

[1] *Department of Pediatrics, Jichi Medical University, Tochigi, Japan*

[2] *Applied Cognitive Neuroscience Laboratory, Chuo University, Tokyo, Japan*

[3] *Center for Development of Advanced Medical Technology, Jichi Medical University, Tochigi, Japan*

Abstract: Attention deficit hyperactivity disorder (ADHD) is a neurodevelopmental disorder characterized by hyperactivity, impulsivity, and inattention and affects between 2 and 9% of school-aged children.

Accumulating evidence indicates that ADHD is caused due to imbalances of the dopamine (DA) and noradrenaline (NA) systems and provides support for the recommendation of medications such as methylphenidate (MPH) and atomoxetine (ATX), which work on each system, respectively. MPH inhibits the reuptake of catecholamines, especially DA, and ATX inhibits the reuptake of NA predominantly in the prefrontal cortices to improve ADHD symptoms.

There are no objective methods for evaluating the effects of medications. Clinicians usually refer to the severity levels of symptoms listed on rating questionnaires for subjective measures of ADHD symptoms conducted by children's parents or teachers. Therefore, more objective approaches are urgently required.

The use of non-invasive functional neuroimaging modalities has been reported as a method for visualizing neural function, and such modalities include functional magnetic resonance imaging (fMRI), positron emission tomography (PET), magnetoencephalography (MEG), and, lately, functional near-infrared spectroscopy (fNIRS). Since the 2000s, fMRI research has revealed pharmacological effects of ADHD therapeutic agents in the PFC. However, elimination rates for fMRI measurements have been high, especially in pediatric cases. In this regard, fNIRS, which measures cortical oxyhemoglobin concentration changes associated with neuronal activation, has several advantages, such as its usefulness in accessibility and tolerance of body motion.

* **Corresponding Author Yukifumi Monden:** Department of Pediatrics, Jichi Medical University, Tochigi, Japan; Tel: +81-285-58-7366; E-mail: mon4441977319@jichi.ac.jp

Atta-ur-Rahman & Zareen Amtul (Eds.)

Our fNIRS-based method with young ADHD children showed the neural dysfunction and neurofunctional modulation induced by medication. Hemodynamic responses in the right PFC during an inhibition task, a go/no-go task, were robustly lower in ADHD children than in controls. These responses could differentiate ADHD from controls individually, with a sensitivity of 90% and a specificity of 70%. Regarding pharmacological effects, randomized, double-blind, placebo-controlled, crossover studies have revealed normalized hemodynamic responses in the right PFC caused by MPH and ATX in ADHD children during go/no-go tasks. On the other hand, during an attentional task, an oddball task, ATX led to activation in the right inferior parietal cortex (IPC) in addition to the PFC, while MPH induced hemodynamic responses only in the right PFC. These responses reflect neuropharmacological effects on each of these two neural networks. MPH upregulates the dopamine system, and ATX affects the noradrenergic system. Furthermore, MPH-induced neuropharmacological effects in ADHD children with or without comorbid autism spectrum disorder (ASD) provided evidence for a differing neurofunctional pathology between the two groups.

This evidence provides a promising possibility to enable diagnosing and evaluating treatments for ADHD patients at the clinical level.

Keywords: Attention Deficit Hyperactivity Disorder, Autism Spectrum Disorder, Cortical Hemodynamics, Developmental Disorder, Dorsolateral Prefrontal Cortex, Discrimination Analysis, Dopamine, Optical Topography, Response Inhibition, Noradrenaline.

INTRODUCTION

Attention deficit hyperactivity disorder (ADHD), according to the fifth edition of the Diagnostic and Statistical Manual of Mental Disorders (DSM-5) [1], is a pediatric neurodevelopmental disorder that is characterized by inattention, hyperactivity, and impulsivity. It is the most prevalent behavioral disorder of childhood, affecting between 2 and 9% of school-aged children [2, 3]. ADHD symptoms are usually identified during the early elementary school years [4]. ADHD children tend to suffer from emotional problems, leading to academic difficulties and mental health problems. Furthermore, such ADHD symptoms often persist into adulthood and result in difficulty in educational and vocational performance and increased risk of depression, suicide, difficulties with employment and relationships, and criminality [5 - 8]. Indeed, its prevalence in adults has recently been reported to be 4-5% [9]. This evidence suggests a need for early diagnosis and intervention for a more positive long-term prognosis. Behavioral therapy and pharmacotherapy are both recommended in all ADHD clinical guidelines for ADHD children [10, 11]. In particular, pharmacotherapy has been reported to have an impact on not only the improvement of ADHD symptoms but also on a decreased risk of negative outcomes [12].

In Japan, methylphenidate (MPH) and atomoxetine (ATX) were approved for the treatment of ADHD in 2007 and 2009, respectively. Subsequently, Guanfacine (GXR) was approved in 2017, and Lisdexamfetamine Mesylate (LDX) was approved in 2019. Among these, we have focused on the neuropharmacological effects of MPH and ATX. In this chapter, we will introduce the results of these studies.

Based on several guidelines for medication treatment, the stimulant drug MPH and the non-stimulant drug ATX have been recommended as first-line therapy for the improvement of ADHD symptoms [13 - 15]. MPH inhibits the reuptake of dopamine (DA) and stimulates the DA system mainly in the prefrontal cortex (PFC) [16 - 19]. ATX inhibits the reuptake of noradrenaline (NA) and acts on the NA system, mainly located in the locus coeruleus with axonal projections to various cortical regions, including the prefrontal and parietal cortices [18 - 20]. Since imbalances of the DA and NA systems are the cause of deficits in executive function with ADHD, MPH and ATX work on each of these systems and improve ADHD symptoms [21].

In order to confirm the diagnosis and assess the effectiveness of medication treatment, the identification of objective biological markers [22] has been eagerly awaited [23 - 25]. ADHD is mainly diagnosed by interview-based evaluation of the degrees of the symptoms listed in the diagnostic criteria of the DSM-5, and the evaluation is based on observation by the patient's parents or teachers [23, 25]. However, the interview-based assessment often depends on subjective evaluation, which has a risk of underdiagnosis or overdiagnosis of ADHD symptoms [26, 27]. Due to the technical limitations of relying on clinical interviews and clinical observation, a more objective approach using biological markers for early diagnosis and assessing treatment efficacy is desirable to complement the current interview-based evaluation method [22, 23, 25].

Non-invasive functional neuroimaging in combination with neuropsychological testing is a promising approach to detect biomarkers of ADHD symptoms. Since the 2000s, a number of functional magnetic resonance imaging (fMRI) studies have revealed less prefrontal brain activation in ADHD participants compared to that of control groups during motor response inhibition tasks [28 - 30]. fMRI-based neuropharmacological studies using MPH have demonstrated acute functional upregulation and normalization of the right middle and inferior frontal gyri [31 - 33]. However, because of the high elimination rate of fMRI measurements, amounting to the rejection of 50% of ADHD children due to excessive motion artifacts [30], fMRI may prove limited in its usefulness for clinical applications. In addition, an fMRI scan and the ensuing off-line analyses take substantial time and effort. Conventional imaging modalities such as single-photon emission computed tomography, positron emission tomography, and magnetoencephalography, also have similar problems.

On the other hand, functional near-infrared spectroscopy (fNIRS), which is a non-invasive neuroimaging tool for continuously measuring hemodynamics in the cerebral cortex, could be an ideal neuroimaging tool for children, especially pediatric ADHD cases who have difficulty in performing active cognitive tasks in the enclosed environments required for other imaging modalities. It is relatively forgiving of body motion and has been successfully implemented in tasks involving body movement [34 - 40]. It utilizes the tight coupling between neural activity and regional cerebral hemodynamic changes with a high affinity for studying developing brains [41 - 45]. Moreover, fNIRS entails other advantages, including its usefulness in an ordinary examination room, compact size (useful in confined experimental settings), affordable price, unrestrictiveness, and accessibility. Indeed, fNIRS has been implemented in various clinical domains, including pathological gate monitoring in neurologic rehabilitation [46], monitoring of ischemia [47], assessment of language dominance before neurosurgery [48], detection of epileptic focus [49, 50], and diagnosis of various psychiatric diseases [40, 51]. Among these, an increasing number of fNIRS studies have started to investigate the cortical hemodynamics of ADHD patients, including children [52 - 55]. In addition, research using fNIRS to monitor the effects of neuro stimulants on cortical hemodynamic changes in ADHD children has been undertaken [56 - 65]. These studies report that neuro stimulants, such as MPH and ATX, induce hemodynamic responses mainly in the PFC during cognitive tasks. All of the studies described above included children with ADHD aged 6–16 years as subjects.

In line with these studies, we expect fNIRS to occupy a unique position among neuroimaging modalities because its convenience and robustness merits would be highly appreciated in the functional monitoring of ADHD, especially in younger children, who are difficult to assess with other modalities, including fMRI. Thus, we performed a series of fNIRS studies which assessed the acute neuropharmacological effects of MPH and ATX on the inhibitory and attentional functions of ADHD children, the difference of neural responses between ADHD and control subjects, and that between ADHD with and without comorbid autism spectrum disorder (ASD).

MONITORING NEUROPHARMACOLOGICAL EFFECTS OF MPH AND ATX USING FNIRS MEASUREMENT DURING A RESPONSE INHIBITION TASK (GO/NO-GO TASK) AND AN ATTENTION TASK (ODDBALL TASK)

Introduction

Symptoms of ADHD are reported to arise from primary deficits in executive functions (EFs). EFs are defined as functions regulating various cognitive processes necessary for goal-directed behavior, including planning, working memory, flexibility, response inhibition, and attention [66, 67]. In particular, impairment of response inhibition and attention are related to the major symptoms of ADHD, which is to say hyperactivity, impulsivity, and inattention [68 - 72]. A number of neuroimaging studies have been undertaken to reveal the neural basis of these brain functions.

Regarding the neural substrates underlying response inhibition, several neuroimaging studies have reported that the right PFC is a candidate brain region [73, 74]. A review of functional neuroimaging studies with healthy controls reveals that the PFC, especially the right inferior frontal gyrus (IFG), is associated with response inhibition [16]. fMRI studies have shown neural responses in the dorsolateral prefrontal cortex (DLPFC), ventrolateral prefrontal cortex (VLPFC), premotor cortex, inferior parietal lobe, lingual gyrus, caudate, and right anterior cingulate during go/no-go tasks [75]. The go/no-go task is one of the principal paradigms for response inhibition function. Former neuroimaging studies have used go/no-go, stop signal, and Stroop tasks as response inhibition tasks [22, 33, 56, 76, 77]. Performance of stop signal and Stroop tasks matures at the age of 13 and over [78, 79], while that of the go/no-go task matures at approximately 12 years of age [80]. Thus, we selected a go/no-go task as the experimental task for our current study of school-aged children.

Neuroimaging studies of ADHD patients, including children, have revealed right middle and inferior frontal hypoactivation during response inhibition tasks, including the go/no-go task [33, 53, 56, 74, 81 - 86]. In addition, fMRI-based neuropharmacological studies on ADHD children have shown acute functional upregulation of the frontal cortices and striatum after MPH administration during go/no-go tasks [33]. In line with these fMRI studies, we measured the cortical hemodynamics of ADHD children (6 to 14 years old) during a go/no-go task before and 1.5 h after MPH administration. A significantly increased oxy-Hb signal in the right lateral PFC was observed after MPH administration [59]. We demonstrated that fNIRS combined with a go/no-go task could monitor the cortical hemodynamics of ADHD children.

On the other hand, several neuroimaging studies have suggested that dysfunction of the fronto-parietal network exists in ADHD patients who mainly suffer from inattention. The fronto-parietal, network has been implicated as one of the main neuro-networks for attention [87, 88]. Two fMRI studies using attention tasks demonstrated the neuropharmacological effect of MPH in ADHD subjects, both utilizing a double-blind, placebo-controlled design [30, 89]. Shafritz *et al.* examined the neural responses for the effects of MPH during attention tasks for adolescents with ADHD and control subjects [89]. They reported significantly less activation of the left ventral basal ganglia and middle temporal gyrus induced by attentional tasks in unmedicated ADHD subjects than in control subjects. Neural activation was normalized after MPH administration in the left ventral basal ganglia but not in the middle temporal gyrus. Rubia *et al.* also investigated the effects of MPH on adolescents with ADHD and revealed that neural responses were reduced under the placebo condition while they became normalized after MPH medication [30].

Although these neuroimaging studies have revealed the neural basis of inhibition and attention in ADHD, these studies have several limitations, including the age of their participants and the kinds of medications examined. Participants in fMRI studies were often over 10 years old. In addition, the effects of non-stimulant drugs such as atomoxetine (ATX) have not been fully verified, despite which they have been recommended as first-line therapy for ADHD, as is MPH [13, 15]. Thus, to explore the neural basis of the effects of MPH and ATX on inhibitory and attentional control in school-aged ADHD children, we conducted a randomized, double-blind, placebo-controlled study employing a go/no-go task and an oddball task and assessed specific neural responses associated with MPH and ATX in children using fNIRS analysis.

Method

Experimental Design

The experimental procedures are summarized in Fig. (**1**) . The effects of MPH and ATX were evaluated in a randomized, double-blind, placebo-controlled, crossover study during a go/no-go task Fig. (**1a**) or an oddball task Fig. (**1b**) . All ADHD subjects were pre-medicated with MPH (go/no-go task *n* = 16, oddball task *n* = 22) or ATX (go/no-go task *n* = 16, oddball task *n* = 15) as part of their regular medication regimen and were examined twice. Both days of fNIRS measurements were scheduled after a washout period of 2 days and to be as close as possible, from 2 to 30 days apart. On each test day, participants received a total of two sessions, one before MPH/ATX or placebo administration and the other 1.5 h after administration. To compare the neural responses for each task, control subjects participated and underwent a single, non-medicated session.

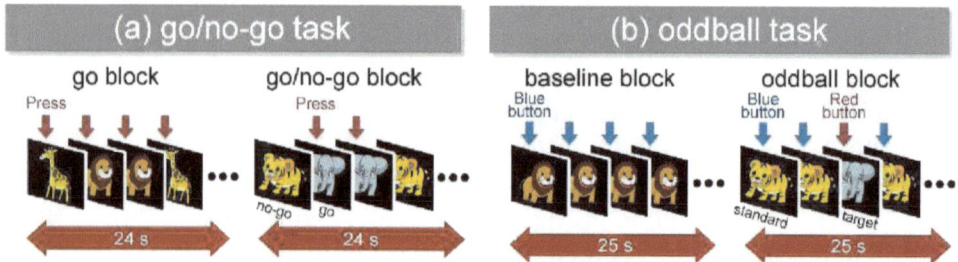

Fig. (1). Summary of experimental procedure for the go/no-go task (a) and the oddball task (b). The neural activity of ADHD and control subjects was measured during each task.

Task Design

Go/No-Go Task

We adopted the go/no-go task in a block design [22, 33, 35, 75, 90]. Each session consisted of 6 block sets, each containing alternating go (baseline) blocks, where participants were randomly presented with two different pictures of animals on the computer screen using E-prime 2.0.® (Psychology Software Tools, Pittsburgh, Pennsylvania) and instructed to respond to both pictures, and go/no-go (target) blocks, in which participants were presented with no-go pictures at a ratio of 50% randomly and asked not to respond to the no-go pictures, but to respond to the rest of the pictures. Each block lasted for 24 seconds and was preceded by instructions displayed for 3 seconds. The overall block-set time was 54 seconds, and the total session time was 6 minutes. After the first session, ADHD subjects were administered MPH, ATX, or placebo orally. Stimuli were displayed on a 17" desktop computer screen using E-prime 2.0 (Psychology Software Tools). The screen was located at about 50 cm from the subject's eyes.

Oddball Task

The experimental method for the oddball task was similar to the go/no-go task depicted in Fig. (**1a**) (see Fig. (**1b**)).

Each session contained 6 block sets, including 3 alternating baseline and 3 oddball blocks. Each block was preceded by 3 seconds of instructions to explain the subject of the following block and lasted 25 seconds. The overall block-set time was 56 seconds, and the session duration was 6 minutes. In the baseline block, participants were instructed to view a series of two kinds of animal pictures once every second and to press a red button for each picture. Following the baseline block, two new pictures of different animals were presented.

Pictures of a tiger (standard stimulus, 80% of trials) or an elephant (target stimulus, 20% of trials) were displayed sequentially for 200 ms with an 800 ms interstimulus interval. The ratio of target vs. standard was determined as 2:8 to maintain consistency with previous studies [34 - 36, 51, 53, 91, 92]. Participants were instructed to press a red button when the standard stimuli (tiger) appeared and to press a blue button when the target stimuli (elephant) appeared on the computer screen. The red and blue buttons were located side by side on the response box. A practice block was performed before each standard and target block to ensure that the participants understood the instructions.

fNIRS Measurement

The multichannel fNIRS system ETG-4000 (Hitachi Corporation, Kashiwa, Japan) was used for the measurement of neural responses. Near-infrared lights at two wavelengths (695 and 830 nm) were irradiated, and signals reflecting changes of oxygenated hemoglobin (oxy-Hb) and deoxygenated hemoglobin (deoxy-Hb) concentrations in the brain tissue at 25~30 mm deep from the surface were measured based on the modified Beer-Lambert Law [93] as previously described [94]. Accordingly, we calculated signals reflecting oxy-Hb and deoxy-Hb concentration changes, obtained in units of millimolar·millimeter (mM·mm) [94, 95].

Two sets of 3×5 multichannel probe holders were placed on the heads of the participants to cover the lateral prefrontal cortices and inferior parietal lobe as described in a previous study [59]. The holders consisted of 8 illuminating and 7 detecting probes arranged alternately at an inter-probe distance of 3 cm, and the midpoint of a pair of illuminating and detecting probes was defined as a channel location. Twenty-two channels (CH) were set on each holder (Fig. **2**) . The bilateral probe holders were attached as follows: (1) the upper anterior corners of the left and right probe holders connected by a belt were placed symmetrically on the sagittal midline. (2) the lower anterior corners of the probe holder were placed on the supraorbital ridge, and (3) the lower ends of the probe holders were placed on the upper part of the auricles.

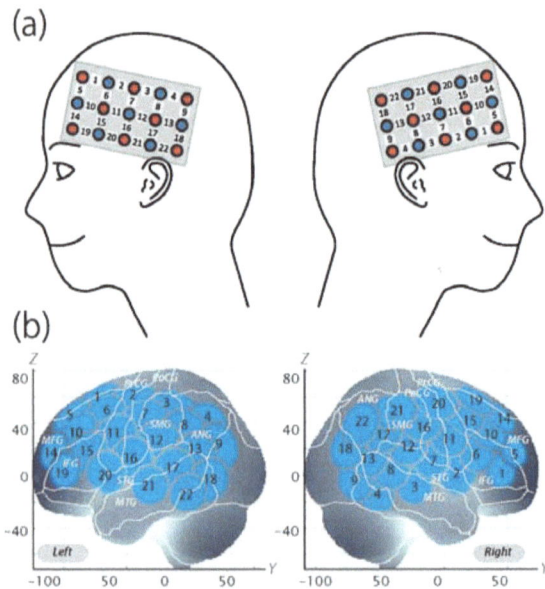

Fig. (2). Spatial profiles of fNIRS channels. **(a)** Side views of the probe and fNIRS channel locations on the scalp. Blue circles indicate detectors, red circles indicate illuminators, and white squares indicate channels. Black numbers indicate channel numbers. **(b)** Channel locations on the brain exhibited for both left and right side views. Statistically estimated fNIRS channel locations (centers of blue circles) for all participants and their spatial variability (*SD* is the radii of the blue circles) associated with the estimation are depicted in Montral Neurological Institute space.

For all participants, spatial profiling of fNIRS data was measured by registering fNIRS data to the Montreal Neurological Institute (MNI) standard brain space using a probabilistic registration method [52 - 54]. Specifically, we used a 3D digitizer (Patriot Digitizer®, Polhemus) to measure the position of each channel and reference point (Nz (nasion), Cz (midline central), and right and left anterior auricles) in real-world (RW) space. Each reference point in RW was affine-transformed to the corresponding MRI database reference point and replaced with MNI space. By employing the same transformation parameters, we were able to obtain the MNI coordinate values for the fNIRS channels and the most likely estimates of the channel locations given for a group of participants, as well as the spatial variability associated with those estimates. Finally, we anatomically labeled the estimated locations using a MATLAB® function that reads anatomical labeling information encoded in the microanatomical brain atlas, LBPA40 [96], and Brodmann's atlas [97].

Analysis of fNIRS Data

Individual timeline raw data for the oxy-Hb of each channel were band-pass filters from 0.8 Hz to 0.01 Hz to remove baseline drift and heartbeat pulsations. In the current study, Hb signals do not represent cortical Hb concentration changes directly because they contain an unknown optical path length at each channel, and they cannot be measured. Therefore, a direct comparison of Hb signals among different channels and regions is difficult [92]. Hence, statistical analyses in a channel-wise manner were performed. From the preprocessed time series data, we computed channel-wise and subject-wise contrasts by calculating the inter-trial mean of differences between the peak Hb signals (4 - 24 seconds after go/no-go block onset) and baseline (14 - 24 seconds after going block onset) periods. For the six go/no-go blocks, blocks with marked motion-related artifacts were removed. We subjected the resulting contrasts to second-level, random-effects group analyses. Analysis of fNIRS data during the oddball task was performed using the same paradigm as was used for the go/no-go task.

Statistical Analyses

Oxy-Hb signals were statistically analyzed in a channel-wise manner. Control subjects were examined only once, and a target versus baseline contrast for the session was generated. ADHD subjects were examined using the following procedures: (1) pre-medication contrasts: target versus baseline contrasts for pre-medication conditions for the first day exclusively; (2) post-medication contrasts: respective target versus baseline contrasts for post-placebo and post-MPH/ATX conditions; and (3) inter-medication contrasts: differences between MPH/ATX $^{post-pre}$ and placebo $^{post-pre}$ contrasts.

Target versus baseline contrasts was subjected to paired t-tests (two-tailed) to screen for the channels involved in each task for control and ADHD contrasts (pre-/post-placebo and pre-/post-MPH/ATX conditions). The statistical threshold was set at 0.05, with the Bonferroni method for family-wise error correction. For the channels screened in this way, comparisons between control and ADHD groups were made for the following two ADHD contrasts: (1) post-placebo and (2) post-MPH/ATX. These were subjected to independent two-sample t-tests (two-tailed) were performed, and the statistical threshold was set at $p < .05$. In addition, to examine the effect of medication on ADHD subjects, paired t-tests (two-tailed) were performed to compare the intra-MPH/ATX and intra-placebo administration contrasts (i.e., inter-medication contrast) of the channels with a statistical threshold of $p < .05$. Statistical analyses of all channels were performed using PASW Statistics (version 18 for Windows) (SPSS Inc., Chicago) software.

Results

Go/No-Go Task

At first, we screened for all fNIRS channels involved in the go/no-go task for the normal controls. After family-wise error correction, a significant oxy-Hb increase was found in the right CH 10 (mean 0.075, *SD* 0.074, $p < .05$, Bonferroni-corrected, Cohen's $d = 1.009$), which was located in the border region between the right MFG and IFG (MNI coordinates *x, y, z* (*SD*): 46,43,30 (14), MFG 78%, IFG 22%) with reference to macro anatomical brain atlases [96, 97]. Therefore, we set the CH 10 on the right as a region-of-interest (ROI) for the rest of the study.

Among ADHD subjects, the right CH 10 exhibited significant oxy-Hb increase in the post-MPH ($t = 4.41$, $p = .0006$, Cohen's $d = 1.138$) and post-ATX ($t = 4.66$, $p = .0003$, Cohen's $d = 1.166$) conditions. On the other hand, no channels were significantly activated in the premedication and post-placebo conditions. Finally, we examined the effects of the medications investigated in the inter-medication contrast and found a significant difference in the comparison between the placebo condition and the MPH (paired *t*-test, $p < .05$, Cohen's $d = 0.952$) and ATX (paired *t*-test, $p < .05$, Cohen's $d = 0.663$) conditions at the right CH 10. These results indicate that MPH and ATX, but not the placebo, induced robust neural responses in the right MFG/IFG during the go/no-go task (Fig. **3**) .

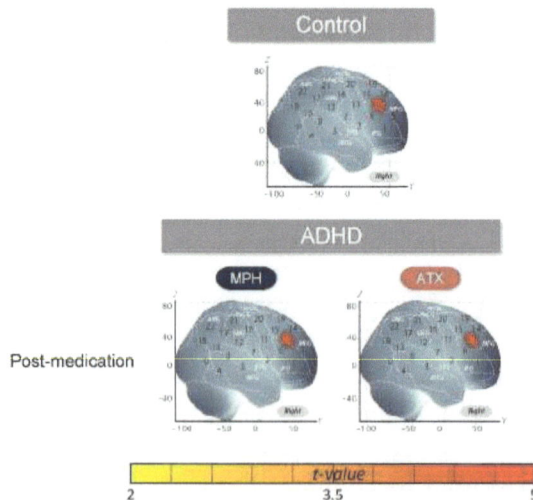

Fig. (3). Hemodynamic changes during go/no-go task. Control subjects and ADHD (post-MPH and post-ATX conditions) subjects are depicted: activation is shown as *t*-maps of oxy-Hb signal with significant *t*-values (one-sample *t*-test, $p < .05$ corrected) being shown according to the color bar.

First, all fNIRS channels were screened in control subjects and significant oxy-Hb increase in the right CH 10 ($t = 4.59$, $p = .0003$, Cohen's $d = 0.987$) and the right CH 22 ($t = 4.79$, $p = .0001$, Cohen's $d = 1.013$) were found. The right CH 10 was located in the border region between the right MFG and IFG (MNI coordinates x, y, z (SD): 48, 41, 39 (15), MFG 70%, IFG 31%) and the right CH 22 was located in the border region between the right angular gyrus (AG) and the right supramarginal gyrus (SMG) (MNI coordinates x, y, z (SD): 57, -60, 46 (18), AG 99%, SMG 1% with reference to macro anatomical brain atlases [96, 97]). Therefore, we set these two channels as ROIs for the rest of the study (Fig. **4**) .

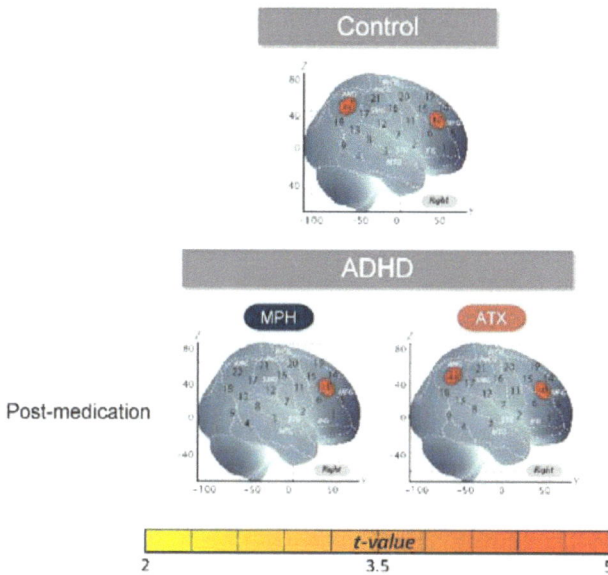

Fig. (4). Hemodynamic changes during oddball task. Control subjects and ADHD (post-MPH and post-ATX conditions) subjects are depicted; activations are shown as *t*-maps of oxy-Hb signal, with significant *t*-values (one-sample *t*-test, $p < .05$ corrected) according to the color bar.

In ADHD conditions, significant oxy-Hb increase was found in the right CH 10 ($t = 4.16$, $p = .0004$, Cohen's $d = 0.887$) in the post-MPH condition, while it was found in the right CH 10 ($t = 3.55$, $p = .0032$, Cohen's $d = 0.915$) and the right CH22 ($t = 2.55$, $p = .0231$, Cohen's $d = 0.658$) in the post-ATX condition. In the pre-medication or post-placebo conditions, no channels were significantly activated.

In the inter-medication contrast, we investigated the effects of medications. At the right CH 10, significant differences in the comparison between the placebo condition and the MPH (paired *t*-test, $p < .05$, Cohen's $d = 1.019$) and ATX (paired *t*-test, $p < .05$, Cohen's $d = 1.032$) conditions were found. Conversely, at the right CH 22, a significant difference was found only in the ATX (paired *t*-test, $p < .05$, Cohen's $d = 0.633$) condition, but not in the MPH condition.

Discussion

This study revealed that both MPH and ATX increased right IFG/MFG activation during a go/no-go task, as observed using fNIRS. On the other hand, during an oddball task, MPH normalized right IFG/MFG activation, but ATX increased right IFG/MFG and AG/SM activation.

In previous neuroimaging studies, the neural responses during go/no-go tasks have been reported at the bilateral IFG, the MFG and superior frontal gyrus (SFG), the inferior parietal lobule (IPL), the anterior cingulate gyrus, the caudate nucleus, the supplementary motor area, and the cerebellum [98, 99]. Although the MFG, SFG, and IPL have been reported to mediate more general meta-motor executive control functions such as attention, conflict monitoring, and response selection, it has been indicated that the IFG may be related to motor response inhibition specifically [100].

Animal studies have shown that both MPH and ATX increase dopamine (DA) and noradrenalin (NA) in the PFC [19, 101]. These results suggest that an increase of DA and NA concentration in the PFC induced by administration of either MPH or ATX leads to normalizing inhibitory function in ADHD children. However, this does not necessarily mean that these two medications affect functions of PFC through the same neuropharmacological mechanism. It should be noted that MPH and ATX have an almost opposite affinity for DA and NA transporters. MPH has a 10-fold higher affinity to DA than to NA transporters, while ATX has a 300-fold higher affinity to NA than to DA transporters [102]. Based on this evidence, it is speculated that MPH has a much greater impact on the DA system between the PFC and striatal region, while ATX has far larger effects on the NA system between the PFC and locus coeruleus region. Thus, similar activation patterns induced by MPH and ATX observed in the PFC may be due to different neural substrates.

On the other hand, previous neuroimaging studies using oddball tasks have elucidated neural correlates in several brain regions, including the bilateral inferior, dorsolateral and superior PFC, the anterior cingulate gyrus, the supplementary motor area, the parietal and temporal lobes, the amygdala, and the caudate nucleus [103 - 108]. The IFG, MFG, and AG are components of the attentional system and connect reciprocally [109]. The network of these regions is considered important in the executive control required to guide goal-directed and stimulus-driven attention. In addition, recent fMRI and event-related potential

(ERP) studies using oddball tasks with healthy adults have demonstrated the involvement of the networks between the prefrontal and parietal cortex [103 - 108]. In our current fNIRS-based study, also co-activation of the attention network between the PFC and IPL was detected in the control subjects as well.

In ADHD subjects, there is a difference in normalization induced by MPH and ATX, which appears to be relevant given the pharmacological effects of both substances. As mentioned above, although MPH is known to act on affect both the DA and NA systems [110, 111], it has a ten-fold higher affinity for DA receptors than for NA receptors [103], resulting in an overwhelmingly greater effect on the DA system. Given the predominant distribution of DA neurons in the right IFG/MFG and not in the right IPL reflecting the front-striatal dopamine system [112], it is reasonable to speculate that normalization of cortical activation in ADHD children induced by MPH occurs only in the right IFG/MFG. On the other hand, ATX has a predominantly higher affinity to NA than to DA transporters [102]. Considering the distribution of NA neurons in innervation from the locus coeruleus to the right IFG/MFG and IPL [64, 113], normalization of cortical activation in ADHD children seems to occur in the right IPL with ATX administration.

Through a series of studies, we confirmed the absence of right PFC activation during an inhibition task and of front-parietal activations during an attention task in pre-medicated ADHD children. In post-medicated ADHD children, the right IFG/MFG was activated during inhibition control with both medications, MPH and ATX, and in attention control with MPH; however, the right IFG/MFG and AG/SM were activated during attention control. These findings led us to conclude that we have successfully revealed the difference induced by MPH and ATX in the DA and NA systems in ADHD children.

INDIVIDUAL CLASSIFICATION OF ADHD CHILDREN USING FNIRS WITH A RESPONSE INHIBITION TASK (GO/NO-GO TASK)

Introduction

Our fNIRS studies discussed above have successfully visualized the neural substrates of inhibitory and attentional controls in school-aged ADHD children. We have investigated the neural substrate in a series of studies making the most of the advantages offered by fNIRS. Importantly, an overall total of 81 right-handed ADHD children and 69 control subjects, including 6-year-olds, have participated in our previous fNIRS studies, and the cumulative post-scan exclusion rate was less than 5% [59 - 64]. The rejection rate is far higher in fMRI studies, for which the rate is typically 50% in ADHD subjects older than 6 years and 30% in the corresponding normal controls. Motion and lack of compliance are the main reasons. Therefore, a series of studies have established neural response measurements in pediatric ADHD using fNIRS.

However, differences in neural substrates between children with ADHD and typically developing controls were evaluated in group analyses. Individual neural responses could not be examined.

At present, ADHD diagnosis relies primarily on the diagnostic criteria of the DSM-IV and DSM-5, which are interview-based assessments of the degrees of the symptoms observed by a patient's parents or teachers [23, 25]. However, the interview-based evaluations are often subjective and carry a risk of underdiagnosis or overdiagnosis of ADHD symptoms [26, 27]. Because of the limitations of relying solely on clinical interviews, it is necessary to identify a biological marker for early and objective diagnosis [23 - 25].

Results of our series of studies suggest that right IFG and MFG activations may be a neurofunctional biomarker to diagnose school-aged ADHD children. In group analyses, fNIRS has successfully visualized different neural responses in inhibitory and attentional functions between ADHD children and corresponding control subjects. Comparing activation in the right IFG/MFG during go/no-go and oddball tasks, the effect size was larger for the former (Cohen's d: 1.16, go/no-go task; 0.98, oddball task) [59, 61 - 63]. These results led us to postulate that neural responses in the right IFG/MFG measured by fNIRS, especially during a go/no-go task, might be used on the individual level as an objective biomarker to diagnose ADHD children. Thus, our next effort is to quantify the dysfunction associated with inhibition at an individual level in children. To find a highly accurate classification between the two groups at an individual level, we must first determine which regions of interest (ROIs) would best represent areas of inhibition-related cortical dysfunction in ADHD children. While patterns of activation during go/no-go tasks vary across individuals, in group analyses, activation in control subjects is significantly greater in the single-channel between the right MFG and IFG. The spatial variability between individuals necessitates the determination of which areas of the PFC should be assessed. Second, in order to distinguish between ADHD and non-ADHD children, the cut-off amplitude of cortical activation at each of the ROI mentioned above must be determined. Group analyses of activation in the PFC during go/no-go tasks in the control subjects reveal a large effect size, but the amplitude of activation varies across individuals. Therefore, we took full advantage of multichannel measurements and adopted well-formed formulae to analyze the optimized ROIs' constitutive CHs and evaluated whether a particular logic could improve the classification efficacy. Hence, in the current study, we explored a method of individual classification between ADHD children and typically developing children using fNIRS; in particular, we examined how the spatial distribution and amplitude of hemodynamic response in the PFC during a go/no-go task can be utilized. In doing so, we explored the possibility of developing clinically applicable objective diagnostic tools for school-aged ADHD children.

Method

Experimental Design and fNIRS Measurement

Pre-medicated ADHD and control subjects were examined once without medication. A washout period of more than 2 days for MPH was set as in our previous studies. Thirty ADHD children (mean age 9.1 (*SD* 2.6, range 6–15 years), gender 25 males and 5 females, IQ $\geq$ 70, right-handed) and 30 matched typically developing control subjects (mean age 9.7 (*SD* 2.3, range 6–14 years), gender 20 males and 10 females, IQ $\geq$ 70, right-handed) participated in the study. We measured inhibition-related hemodynamic cortical activation during a go/no-go task, using fNIRS following the same procedures as for our previous studies (see sections 1.2.2-1.2.3).

Statistical Analyses

Individual fNIRS-Based Classification

We explored a method for quantitative analysis to further improve disease classification between ADHD and typically developing children, based on the individual and channel-wise oxy-Hb signal contrast measured using fNIRS. Since cortical activation of ADHD children was statistically decreased in the right PFC in our previous studies [59, 60, 63], these areas were set as ROIs to investigate robust classification parameters. We adopted the following four anatomical ROIs: Region 1 was CHs 1 to 22, which covers all the channels in the right hemisphere. Region 2 was CHs 1, 2, 5, 6, 7, 10, 11, 14, 15, 16, 19, and 20, which includes all the channels probabilistically located in the right PFC. Region 3 was CHs which were significantly activated during the go/no-go task in the group analysis of control subjects. To select the experimental CHs, paired *t*-tests (two-tailed) for target versus baseline contrasts were performed with a statistical threshold of 0.05, and Bonferroni correction for family-wise errors. Region 4 was CHs which showed significantly greater go/no-go-task-related activation in control subjects than in ADHD children. Independent two-sample *t*-tests (two-tailed) with the statistical threshold at 0.05 for the integral value of oxy-Hb signal at each CH were performed. These values represent the size of the oxy-Hb signals during the inter-trial mean of differences between target (4–24 seconds after onset of go/no-go block) and baseline (14–24 seconds after onset of go block) periods in each CH. Each CH's integral value for the four ROIs presented above were averaged. Next, a cut-off value to more accurately distinguish between ADHD children and control children was explored. We performed a conservative receiver operating characteristic (ROC) analysis and generated a simple index of fNIRS-based oxy-Hb signal patterns for individual ADHD diagnosis. We classified the subjects and verified true positive

(TP), false positive (FP), false negative (FN), and true negative (TN). Sensitivity was calculated as TP / (TP + FN), and specificity was calculated as TN / (TN + FP). We also calculated the area under the resultant ROC curve to the optimal cut-off point for sensitivity and specificity.

Modified Individual fNIRS-Based Classification Using Well-Formed Formulae

We verified the formulae for the most effective ROI constituent CHs, for further optimization. AND logic was applied, and the system was classified as normal if the oxy-Hb signals of all CHs in the ROI exceeded a given threshold. OR logic was applied, and the system was classified as normal if the oxy-Hb signal of anyone CH in the ROI exceeded a given threshold. ROC analysis was performed as described above for each classification using well-formed formulae. All statistical analyses were performed with the PASW statistics (version 18 for Windows) (SPSS Inc., Chicago, USA) software package.

Results

ROC Analysis of the Integral Value of Oxy-Hb Signals

In region 1 (CHs 1 to 22), the area under the ROC curve (AUC) value was 75.1%. ADHD and control subjects could be distinguished with a sensitivity of 70.0% and a specificity of 86.7%, with the optimal cutoff value of 0.0120 mM·mm (Fig. **5a**). For a detailed description of the data, see our previous report [61].

In region 2 (CHs 1, 2, 5, 6, 7, 10, 11, 14, 15, 16, 19, and 20), the AUC value was 71.0%. ADHD and control subjects could be distinguished with a sensitivity of 63.3% and a specificity of 73.3% (Fig. **5b**), with the optimal cut-off value of 0.0120 mM·mm.

For region 3, all fNIRS channels were screened and oxy-Hb was significantly increased in three CHs: CH 5 (mean 0.057, *SD* 0.077, *p* < .05, Bonferroni-corrected, Cohen's *d* = 0.741), CH 6 (mean 0.046, *SD* 0.060, *p* < .05, Bonferroni-corrected, Cohen's *d* = 0.755), and CH 10 (mean 0.068, *SD* 0.065, *p* < .05, Bonferroni-corrected, Cohen's *d* = 1.046) in the control group but not in the ADHD group. Therefore, we set these channels as statistically specific ROIs. The AUC value was 79.2%. ADHD and control subjects could be distinguished with a sensitivity of 76.7% and a specificity of 76.7%, with the optimal cutoff value of 0.0240 mM·mm (Fig. **5c**).

For region 4, we set the right CHs 6 and 10 as ROI, because a significant group difference in oxy-Hb signals was found there (independent two-sample *t*-test; R CHs 6, *p* < .05 Bonferroni corrected, Cohen's *d* = 0.964; R CHs 10, *p* < .05 Bonferroni-corrected, Cohen's *d* = 0.699). They are located on the border of the right MFG and IFG. The AUC value was 84.7%. ADHD and control subjects could be distinguished with a sensitivity of 83.3% and a specificity of 73.3%, with the optimal cut-off value of 0.0374 mM·mm (Fig. **5d**).

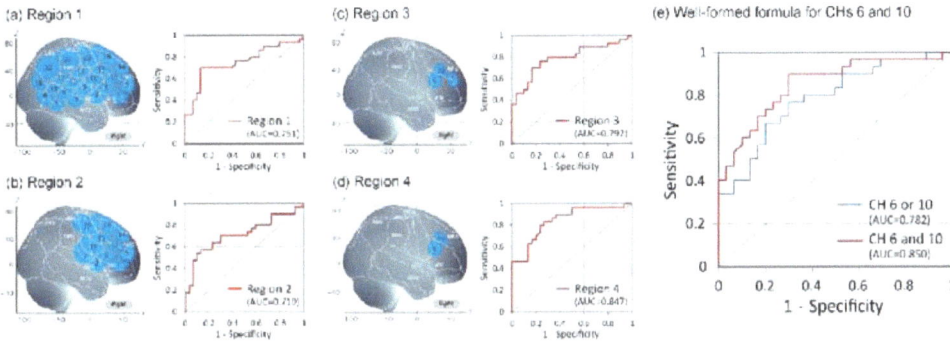

Fig. (5). The analysis of receiver operating characteristic (ROC) curves of the oxy-Hb signal contrast in each macro anatomical region of interest (ROIs). **(a)** Region 1 covers all channels in the right hemisphere. **(b)** Region 2 represents 12 channels covering the right prefrontal cortex (PFC).**(c)** Region 3 represents 3 channels in the right PFC activated in typically developing control subjects. **(d)** Region 4 represents 2 channels in the right PFC showing greater activation in typically developing control subjects than in ADHD children **(e)** The ROC curve adopting OR and logics using well-formed formulae. In OR logic, a patient was identified as normal when either CH 6 or 10 was activated. In AND logic, a patient was identified as normal when both CHs 6 and 10 were activated.

Modified Individual fNIRS-Based Classification Using Well-Formed Formulae

In addition, well-formed formulae were used for region 4 (CHs 6 and 10) to differentiate ADHD from normal control subjects at a higher rate. Adapting OR logic, the area under the AUC was 78.2%. ADHD and control subjects could be distinguished with a sensitivity of 76.7% and a specificity of 70.0%, with the optimal cut-off value of 0.0650 mM·mm (Fig. **6**). Next, adapting AND logic, the AUC value achieved 85.0%. This was the highest percentage among all classifications. ADHD and control subjects could be distinguished with a sensitivity of 90.0% and a specificity of 70.0%, with the optimal cut-off value of 0.0111 mM·mm (Fig. **5e**).

Discussion

The current study using fNIRS has successfully determined optimal ROI in the right IFG and MFG for differentiating ADHD children from typically developing children by individually assessing channel -wise oxy-Hb signal changes.

First, an analysis of the entire right hemisphere revealed a moderately efficient classification method for distinguishing ADHD children from typically developing children, with an AUC value of 0.751 and a sensitivity of 70%. Next, ROIs were narrowed sequentially to two CHs (CHs 6 and 10) in the right IFG/MFG, and an AUC value of 0.847 and 83.3% sensitivity was achieved.

In order to further utilize the oxy-Hb signal data obtained in the right IFG/ MFG, we applied well-formed formulae to CHs 6 and 10, which were anatomical neighbors. Specifically, when an individual's oxy-Hb signals for all the CHs in the ROI exceeded a predetermined threshold (AND logic), or their oxy-Hb signals of any CH in the ROI reached a predetermined threshold, the subject was classified as normal. Appling AND logic, we improved the AUC value to 0.850 and produced a sensitivity of 90%. Therefore, well-formed formulae on macroanatomically labeled fNIRS channels may be a practical option for enhancing the diagnostic prediction in ADHD children at an individual level. These two CHs are probabilistically located on the border of the right IFG and MFG, where fMRI studies have consistently reported abnormal function in ADHD patients during response inhibition tasks [95, 99, 100]. In addition, conclusive evidence for the involvement of the right-lateralized network, including the right IFG and MFG, which are associated with response inhibition, was revealed by an ALE meta-analysis of go/no-go tasks [114]. Therefore, the use of CHs in the right IFG and MFG to identify ADHD children can be considered reasonable from a neurofunctional point of view.

In addition, several other neuroimaging studies using fMRI have employed ROC analysis and attempted to identify biomarkers to classify ADHD children individually. One fMRI study evaluated the diagnosis of ADHD: Hart *et al.* measured neural responses during a motor-related inhibition task (stop task) in 30 adolescents (ages 10-17 years) with ADHD who were not receiving medication and 30 healthy comparison subjects (ages 10-17 years) [115]. They used ROC analysis to identify brain activation patterns that correctly classified the subjects with a sensitivity of 90%, specificity of 63%, and AUC value of 81%. In their study, the area of the identification network that optimally predicted controls included the right PFC, as in our current study. Our current study using fNIRS has, as described above, an advantage in investigating younger ADHD and control subjects, including 6-year-olds, and this is especially important because early identification and treatment are important in ADHD children.

DIFFERENTIATION OF NEUROPHARMACOLOGICAL EFFECT BETWEEN ADHD WITH AND WITHOUT ASD

Introduction

We have explored a stable biological marker using fNIRS as a clinical test method for the early diagnosis and treatment of children with ADHD. Specifically, we applied a clinically relevant assessment method using fNIRS young children with ADHD. These studies produced accumulative evidence for MPH- and ATX-induced neural function modulation of the right IFG/MFG during inhibitory function tasks [60, 63]. Based on these results, we concluded that the activation in the IFG/MFG could serve as an objective neurofunctional biomarker associated with inhibition function, which is one of the core neuropathological mechanisms of ADHD children. However, to establish our fNIRS measurement method as feasible for clinical application for the objective diagnosis and treatment of ADHD children, we consider the complication of autism spectrum disorder (ASD), which is an important issue.

ASD is classified as a neurodevelopmental disorder, as is ADHD, and is characterized by impairments in social interaction, communication, and non-social features, such as restricted and stereotyped behaviors [1]. In the DSM-5, a comorbid diagnosis of ADHD with ASD was allowed for the first time in 2013 [116]. Since then, there has been a growing number of clinical [117] and psychological studies [118] on the co-occurrence of ADHD and ASD.

Pharmacological studies have shown that the response rate to MPH differs in patients with ADHD and ASD compared to patients with ADHD alone [119]. In addition, side effects of MPH, such as irritability, may worsen in patients with ADHD and ASD than in patients with ADHD without ASD [119]. These studies may suggest that there are important differences in the core neurofunctional pathology between ADHD with and without ASD. However, no neurophysiological or neuroimaging evidence has yet been presented, with the exception of two pioneering studies using EEG [120] and fMRI [121].

In the EEG study, an analysis of ERPs during a continuous performance task revealed different neural activity among children aged 8-13 years with ADHD with ASD, ADHD without ASD, and ASD without ADHD [120]. In the fMRI study, brain discounting associations in several brain regions, including the PFC, in ADHD with ASD participants aged 11–17 years was more severe than that in control and ADHD without ASD groups [121]. Thus, elucidating the neuropharmacological mechanism of the comorbidity of ADHD with ASD is important because it could provide objective evidence for the presence of a clearly specific phenotype [118].

In the current study, we hypothesized that modulation of neural responses with MPH would be different between children with ADHD with ASD and those with ADHD without ASD. We applied our previous fNIRS study method with a randomized, double-blind, placebo-controlled, crossover design and investigated the neuropharmacological effect of MPH on the cortical blood flow of ADHD during a go/no-go task in medication-naïve ADHD children with and without comorbid ASD.

Experimental Design

We examined 32 clinically referred medication-naïve children with ADHD requiring MPH administration. Based on the DSM-5, participants were diagnosed with ADHD with or without ASD. As a result, 11 ADHD children with ASD (mean age 8.2, *SD* 2.1, range 7-14 years; 11 males; IQ$\geq$70, right-handed) and 21 ADHD children without ASD (mean age 7.8, *SD* 1.7, range 6-13 years; 17 males and 4 females; IQ $\geq$ 70, right-handed) were identified. In this study, medication-naïve ADHD children were selected. Although our previous studies have included both medicated and medication-naïve patients, animal studies [122] and human anatomical studies [123 - 125] indicate that brain function and structure may be altered by long-term administration of MPH. We measured inhibition-related hemodynamic cortical activation during a go/no-go task, using fNIRS with the same procedures as those of previous studies (see sections 1.2.2-1.2.3).

The effects of MPH were assessed using MPH (18 mg) in a randomized, double-blind, placebo-controlled, crossover design, as in our previous studies.

In addition, a Japanese version of the ADHD Rating Scale-IV (ADHDRS-IV-J) [126] was used to assess clinical pharmacological effects before the start of oral MPH administration and one month after participation in this study. ADHD-R--IV-J is an 18-item scale questionnaire, and each item corresponds to one of the 18 symptoms that reflect the diagnostic criteria of ADHD in DSM-IV (American

Psychiatric Association 1994 [127];). Parents of each patient answered the questionnaire.

fNIRS Measurement and Analysis of fNIRS Data

We adopted the go/no-go task and measured oxy-Hb using the multichannel fNIRS system, as in our previous studies [59 - 61]. These previous studies showed acute normalization of neural activation in the right IFG/MFG in ADHD children after MPH administration during a go/no-go task [59, 60, 64]. Therefore, we set the right CH 10 as an ROI for the rest of the study, located at the right IFG/MFG.

Statistical Analyses

ADHD-RS-IV-J scores (inattention, hyperactivity/impulsivity, and total) were compared between before and one month after oral MPH administration to investigate the effect of MPH. Paired *t*-tests (two-tailed) with an alpha level set at .05 were performed.

Statistical analyses were performed on oxy-Hb signals in a channel-wise manner. We generated the following contrasts for both groups, (a) first-day pre-medication target vs. baseline contrast (either placebo or MPH administration); (b) post-medication target vs. baseline contrasts (post-placebo and post-MPH contrasts); (c) intra-medication contrasts: difference between pre- and post-medication contrasts for each medication (both placebo and MPH); and (d) inter-medication contrast: difference between intra-MPH and intra-placebo contrasts. For the ROI, which was the right CH 10, paired *t*-tests (two-tailed) on target vs. baseline contrasts with an alpha level set at 0.05 were performed.

To examine the difference in the neuropharmacological effects of MPH between ADHD with and without ASD, we compared the following three contrasts for the right CH 10: (a) post-MPH, (b) intra-MPH, and (c) inter-medication. Two-sample *t*-tests (two-tailed) were performed on these contrasts, with an alpha level set at 0.05.

Results

Questionnaire Results

Table **1** summarizes the scores of the ADHD-RS-IV-J. In both ADHD with and without ASD, the ADHD-RS-IV-J scores before the start of oral MPH administration were significantly higher than the score one month after medication.

For region 3, all fNIRS channels were screened and oxy-Hb was significantly increased in three CHs: CH 5 (mean 0.057, *SD* 0.077, $p < .05$, Bonferroni-corrected, Cohen's $d = 0.741$), CH 6 (mean 0.046, *SD* 0.060, $p < .05$, Bonferroni-corrected, Cohen's $d = 0.755$), and CH 10 (mean 0.068, *SD* 0.065, $p < .05$, Bonferroni-corrected, Cohen's $d = 1.046$) in the control group but not in the ADHD group. Therefore, we set these channels as statistically specific ROIs. The AUC value was 79.2%. ADHD and control subjects could be distinguished with a sensitivity of 76.7% and a specificity of 76.7%, with the optimal cutoff value of 0.0240 mM·mm (Fig. **5c**).

For region 4, we set the right CHs 6 and 10 as ROI, because a significant group difference in oxy-Hb signals was found there (independent two-sample *t*-test; R CHs 6, $p < .05$ Bonferroni corrected, Cohen's $d = 0.964$; R CHs 10, $p < .05$ Bonferroni-corrected, Cohen's $d = 0.699$). They are located on the border of the right MFG and IFG. The AUC value was 84.7%. ADHD and control subjects could be distinguished with a sensitivity of 83.3% and a specificity of 73.3%, with the optimal cut-off value of 0.0374 mM·mm (Fig. **5d**).

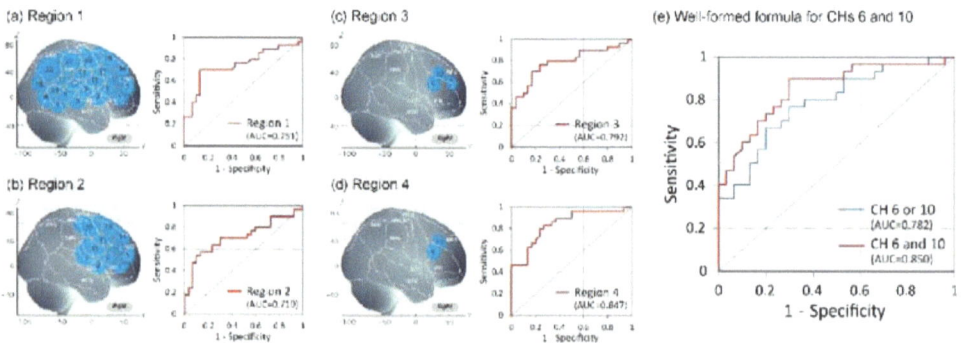

Fig. (5). The analysis of receiver operating characteristic (ROC) curves of the oxy-Hb signal contrast in each macro anatomical region of interest (ROIs). **(a)** Region 1 covers all channels in the right hemisphere. **(b)** Region 2 represents 12 channels covering the right prefrontal cortex (PFC).**(c)** Region 3 represents 3 channels in the right PFC activated in typically developing control subjects. **(d)** Region 4 represents 2 channels in the right PFC showing greater activation in typically developing control subjects than in ADHD children. **(e)** The ROC curve adopting OR and logics using well-formed formulae. In OR logic, a patient was identified as normal when either CH 6 or 10 was activated. In AND logic, a patient was identified as normal when both CHs 6 and 10 were activated.

Modified Individual fNIRS-Based Classification Using Well-Formed Formulae

In addition, well-formed formulae were used for region 4 (CHs 6 and 10) to differentiate ADHD from normal control subjects at a higher rate. Adapting OR logic, the area under the AUC was 78.2%. ADHD and control subjects could be

$p < .05$, Cohen's $d = 1.16$) and intra-MPH conditions (independent two-sample t-test, $p < .05$, Cohen's $d = 0.86$). However, no statistical difference was found in the inter-medication contrast (independent two-sample t-test, $p > .05$, Cohen's $d = 0.64$).

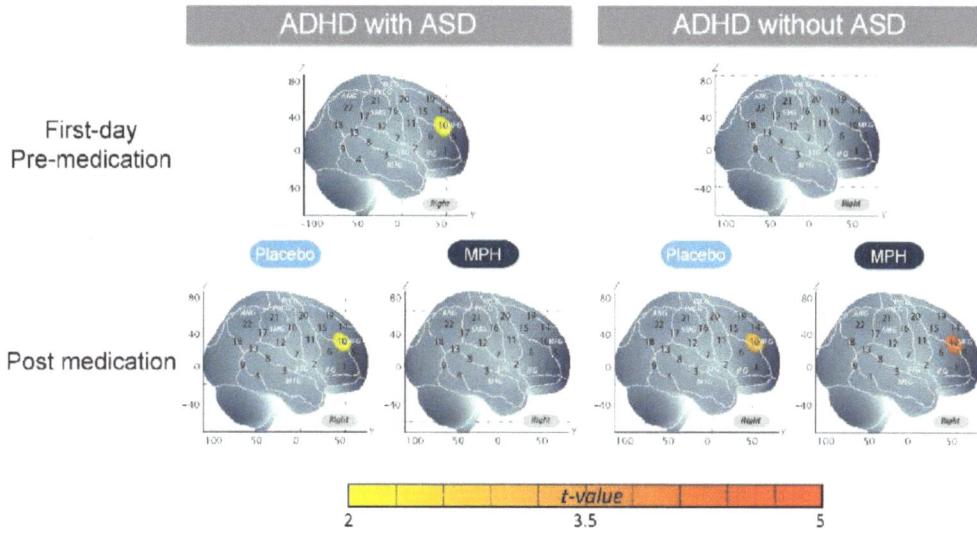

Fig. (6). Right PFC activation patterns for both ADHD with ASD and ADHD without ASD groups. t-maps of oxy-Hb signals are shown. Significant t-values are displayed according to the color bar (paired t-test). All coordinates are exhibited in MNI space.

Discussion

Our current fNIRS study has provided the first fNIRS -based neuro pharmacological evidence for a difference in the neural substrate of MPH effects in the right IFG/MFG during inhibition control tasks in medication-naïve school-aged children with ADHD (without ASD ($n = 21$) and with ASD ($n = 11$)).

The ADHD without ASD group exhibited MPH-induced activation in the right IFG/MFG (CH 10). In the pre-medicated condition, no significant activation in the right IFG/MFG was found, but significant activation was detected in the post-MPH conditions. These results suggest that the right IFG/MFG activation during a go/no-go task, as observed using fNIRS, may be impaired in medication-naïve ADHD children without ASD.

With the exception of the placebo-related conditions, the current results are consistent with our previous study [59]. In addition, these subject- and condition-dependent patterns of activation are consistent with the results of previous neuroimaging studies, using fMRI, of children with ADHD in general and of children with ADHD without ASD [76, 99, 114, 128 - 131]. Taken together, this suggests that the recovered right IFG/MFG activation induced by MPH during the go/no-go task robustly applies to medication-naïve ADHD children without ASD.

Conversely, the neural response in the right IFG/MFG decreased in the ADHD with ASD group after MPH administration. There was marginal activation of the right IFG/MFG in the pre-medicated condition, but the activation was reduced in the post-medicated conditions.

In addition, comparing the hemodynamic responses of post-/pre-MPH in ADHD patients without ASD and of post-/pre-MPH in ADHD patients with ASD revealed significantly different changes; however, due to the small number of subjects, the current fNIRS study was too limited to provide a rigorous statistical inference. However, we have succeeded in visualizing clear patterns of neuropharmacological activation between medication-naïve ADHD children with and without ASD.

Finally, a placebo-induced neuropharmacological effect in the IFG/MFG was found in both groups. In our previous studies, which employed the same protocol as the current study, placebo-induced activation was not observed [59, 60]. In the previous study, nine out of sixteen participants were not medication-naïve, while all participants were medication-naïve in the present study. In general, more than 50% of clinical trials of new psychiatric drugs have demonstrated the statistical superiority of placebo over actual medication [132, 133]. Heightened expectations of patients and researchers for drug efficacy have suggested an increased placebo response [134, 135]. Thus, the placebo effect found in the current study may reflect the increased expectations for the drug among medication-naïve ADHD patients.

Although the method of assessing the effect of MPH, as demonstrated by the current fNIRS study, could be applicable to medication-naïve ADHD children with and without ASD, several problems need to be resolved in order to establish its clinical utility. First, the DSM-5 criteria were used to diagnose ADHD and ASD [121]. To objectively assess ASD characteristics, a semi-structured assessment, such as the Autism Diagnostic Observation Schedule, Module 4 [136] and Autism Diagnostic Interview-Revised [137], which can more appropriately confirm the diagnosis of ASD, is required in future studies. In addition, the sample size of ADHD participants should be increased in order to strengthen the conclusions drawn. Nevertheless, we believe that the present study is meaningful because it reflects the actual distribution of ADHD children with and without ASD [138, 139].

GENERAL DISCUSSION AND CONCLUSION

ADHD is characterized by hyperactivity, impulsivity, and inattention from early childhood [140, 141]. To investigate the specific mechanisms of ADHD in children, this chapter provides evidence about the functional aspects of executive function, especially inhibition and attention. We assessed neural responses induced by MPH and ATX and visualized each monoamine network in the right PFC and IPC using fNIRS (section 1). Next, we compared hemodynamic responses between ADHD children and controls and revealed that the right prefrontal hypoactivation assessed by fNIRS would serve as a potentially effective biomarker for classifying ADHD at the individual level (section 2). Finally, we assessed the neuropharmacological effect of MPH in medication-naïve children with ADHD with or without comorbid ASD and evaluated the differences in neurofunctional pathology between the two groups (section 3).

We will discuss the main results and present an interpretation, with emphasis on the clinical implications, methodological strengths, limitations, and prospects for the future, and, finally, formulate the overall conclusions of this chapter.

fNIRS-Based Differentiation Between DA and NA Systems in ADHD Children

Our fNIRS studies using a go/no-go task and an oddball task showed the specific neuropharmacological mechanisms of MPH and ATX underlying the functional normalization of the inhibition and attention network components in the prefrontal and parietal brain regions in children with ADHD [59, 60, 63, 64]. Furthermore, the current results extend our previous conclusions, and our combined fNIRS studies show that we can distinguish between DA and NA neural substrates involved in the control of inhibition and attention (Fig. **7**) .

fNIRS-Based Differentiation Between ADHD Children and Control

By using fNIRS and individually evaluating channel-wise oxy-Hb signal changes, we found a simple and sufficiently robust tool which can classify ADHD and typically developing children individually with 90% sensitivity and 70% specificity. By monitoring hypoactivation of the two channels in the right IFG/MFG during an inhibitory control task, the two groups were most efficiently classified, and this was further enhanced by the adaptation of well-formulae. These results suggest that hypoactivation in the right IFFG/MFG may be a potential biomarker for individual diagnosis of ADHD patients, typically developing children. Conversely, ADHD children with ASD had alternative patterns of neural response. These differences suggest that ADHD with ASD is not just a mixture of two pure disorders but has its own neuropathological features.

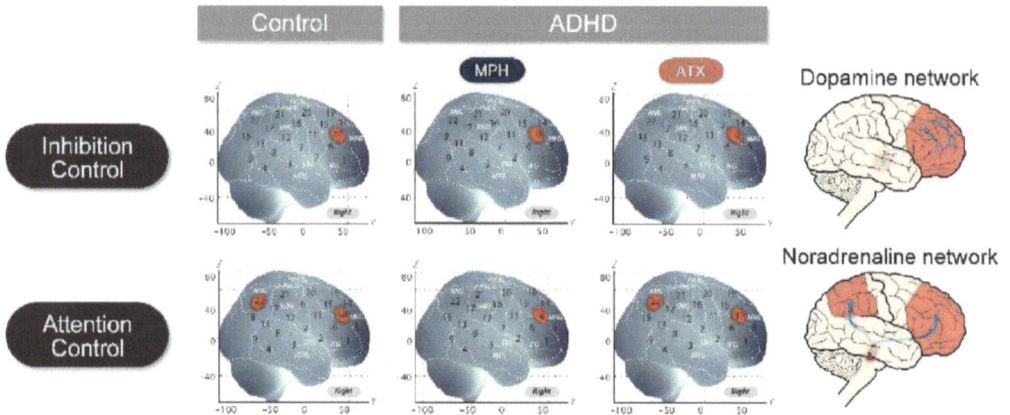

Fig. (7). Altered activation patterns of frontal and parietal brain regions in control and ADHD children during inhibition and attentional control in the post-medication condition.

Clinical Implications

It is clinically important to note that our series of fNIRS studies enrolled ADHD children as young as 6 years old with IQs over 70 with a quite low data exclusion rate. In our studies, no subjects were rejected due to problems with fNIRS measurements, while in an fMRI study, compliance and motion were issues, and the data exclusion rate was 50% with a similar patient population [56]. The shortness of the tasks is also believed to have contributed to the success of the measurements. The go/no-go task and the oddball task each took six minutes, and the total measurement time was less than fifteen minutes. The fNIRS-based examination was applicable to young ADHD children and thus would contribute to the early clinical diagnosis and treatment of ADHD children.

CONCLUSION

In our series of fNIRS studies on ADHD children, we were able to prove the neuropharmacological functional differences of ADHD medications and reveal the differences in the neurofunctional pathology between ADHD patients with and without ASD at a group level. Furthermore, the individual-level analysis allowed discrimination between ADHD and control subjects. These results suggest that neural activation could serve as an objective neurofunctional biomarker for ADHD children. The current method enables discrimination between responders and non-responders at the youngest possible age, thereby serving as a reasonable clinical choice for effective treatment.

In future studies, we plan to investigate the neuropharmacological effect of other ADHD medications, including guanfacine and lisdexamfetamine mesylate. Furthermore, we will explore the neural function of patients who discontinue oral ADHD treatment and the relationship between monoamine genes and neural activation.

CONSENT FOR PUBLICATION

Not applicable.

CONFLICT OF INTEREST

The authors declare no conflict of interest, financial or otherwise.

ACKNOWLEDGEMENTS

We would like to thank ELCS (English Language Consultation Services) for the English proofreading. We appreciate Illpop (http://illpop.com/animal_top01.htm) for providing source pictures for the experimental materials. These works were supported in part by the Grant-in-Aid for Scientific Research from the Japan Society for Promotion of Science (80382951 to YM, 70438662 to MN, and 90570383 to TI).

REFERENCES

[1]   DSM-5 Task Force Diagnostic and statistical manual of mental disorders: DSM-5. 5th ed., Washington, D.C.: American Psychiatric Publishing 2013.

[2]   Sayal K, Prasad V, Daley D, Ford T, Coghill D. ADHD in children and young people: prevalence, care pathways, and service provision. Lancet Psychiatry 2018; 5(2): 175-86.
[http://dx.doi.org/10.1016/S2215-0366(17)30167-0] [PMID: 29033005]

[3]   Danielson ML, Bitsko RH, Ghandour RM, Holbrook JR, Kogan MD, Blumberg SJ. Prevalence of Parent-Reported ADHD Diagnosis and Associated Treatment Among U.S. Children and Adolescents, 2016. J Clin Child Adolesc Psychol 2018; 47(2): 199-212.
[http://dx.doi.org/10.1080/15374416.2017.1417860] [PMID: 29363986]

[4]   Drechsler R, Brandeis D, Földényi M, Imhof K, Steinhausen HC. The course of neuropsychological functions in children with attention deficit hyperactivity disorder from late childhood to early adolescence. J Child Psychol Psychiatry 2005; 46(8): 824-36.
[http://dx.doi.org/10.1111/j.1469-7610.2004.00384.x] [PMID: 16033631]

[5]   Taylor E, Chadwick O, Heptinstall E, Danckaerts M. Hyperactivity and conduct problems as risk factors for adolescent development. J Am Acad Child Adolesc Psychiatry 1996; 35(9): 1213-26.
        [http://dx.doi.org/10.1097/00004583-199609000-00019] [PMID: 8824065]

[6]   Willoughby MT. Developmental course of ADHD symptomatology during the transition from childhood to adolescence: a review with recommendations. J Child Psychol Psychiatry 2003; 44(1): 88-106.
        [http://dx.doi.org/10.1111/1469-7610.t01-1-00104] [PMID: 12553414]

[7]   Shaw M, Hodgkins P, Caci H, *et al.* A systematic review and analysis of long-term outcomes in attention deficit hyperactivity disorder: effects of treatment and non-treatment. BMC Med 2012; 10: 99.
        [http://dx.doi.org/10.1186/1741-7015-10-99] [PMID: 22947230]

[8]   Classi P, Milton D, Ward S, Sarsour K, Johnston J. Social and emotional difficulties in children with ADHD and the impact on school attendance and healthcare utilization. Child Adolesc Psychiatry Ment Health 2012; 6(1): 33.
        [http://dx.doi.org/10.1186/1753-2000-6-33] [PMID: 23035861]

[9]   Safren SA, Sprich S, Mimiaga MJ, *et al.* Cognitive behavioral therapy vs relaxation with educational support for medication-treated adults with ADHD and persistent symptoms: a randomized controlled trial. JAMA 2010; 304(8): 875-80.
        [http://dx.doi.org/10.1001/jama.2010.1192] [PMID: 20736471]

[10]   Hodgkins P, Shaw M, Coghill D, Hechtman L. Amphetamine and methylphenidate medications for attention-deficit/hyperactivity disorder: complementary treatment options. Eur Child Adolesc Psychiatry 2012; 21(9): 477-92.
        [http://dx.doi.org/10.1007/s00787-012-0286-5] [PMID: 22763750]

[11]   Pliszka S. Practice parameter for the assessment and treatment of children and adolescents with attention-deficit/hyperactivity disorder. J Am Acad Child Adolesc Psychiatry 2007; 46(7): 894-921.
        [http://dx.doi.org/10.1097/chi.0b013e318054e724] [PMID: 17581453]

[12]   Boland H, DiSalvo M, Fried R, *et al.* A literature review and meta-analysis on the effects of ADHD medications on functional outcomes. J Psychiatr Res 2020; 123: 21-30.
        [http://dx.doi.org/10.1016/j.jpsychires.2020.01.006] [PMID: 32014701]

[13]   Cubillo A, Smith AB, Barrett N, *et al.* Shared and drug-specific effects of atomoxetine and methylphenidate on inhibitory brain dysfunction in medication-naive ADHD boys. Cereb Cortex 2014; 24(1): 174-85.
        [http://dx.doi.org/10.1093/cercor/bhs296] [PMID: 23048018]

[14]   Faraone SV, Buitelaar J. Comparing the efficacy of stimulants for ADHD in children and adolescents using meta-analysis. Eur Child Adolesc Psychiatry 2010; 19(4): 353-64.
        [http://dx.doi.org/10.1007/s00787-009-0054-3] [PMID: 19763664]

[15]   Newcorn JH, Kratochvil CJ, Allen AJ, *et al.* Atomoxetine and osmotically released methylphenidate for the treatment of attention deficit hyperactivity disorder: acute comparison and differential response. Am J Psychiatry 2008; 165(6): 721-30.
        [http://dx.doi.org/10.1176/appi.ajp.2007.05091676] [PMID: 18281409]

[16]   Aron AR, Poldrack RA. The cognitive neuroscience of response inhibition: relevance for genetic research in attention-deficit/hyperactivity disorder. Biol Psychiatry 2005; 57(11): 1285-92.
        [http://dx.doi.org/10.1016/j.biopsych.2004.10.026] [PMID: 15950000]

[17]   Gatley SJ, Pan D, Chen R, Chaturvedi G, Ding Y-S. Affinities of methylphenidate derivatives for dopamine, norepinephrine and serotonin transporters. Life Sci 1996; 58(12): 231-9.
        [http://dx.doi.org/10.1016/0024-3205(96)00052-5] [PMID: 8786705]

[18]   Easton N, Steward C, Marshall F, Fone K, Marsden C. Effects of amphetamine isomers, methylphenidate and atomoxetine on synaptosomal and synaptic vesicle accumulation and release of

dopamine and noradrenaline in vitro in the rat brain. Neuropharmacology 2007; 52(2): 405-14.
[http://dx.doi.org/10.1016/j.neuropharm.2006.07.035] [PMID: 17020775]

[19]   Koda K, Ago Y, Cong Y, Kita Y, Takuma K, Matsuda T. Effects of acute and chronic administration of atomoxetine and methylphenidate on extracellular levels of noradrenaline, dopamine and serotonin in the prefrontal cortex and striatum of mice. J Neurochem 2010; 114(1): 259-70.
[http://dx.doi.org/10.1111/j.1471-4159.2010.06750.x] [PMID: 20403082]

[20]   Del Campo N, Chamberlain SR, Sahakian BJ, Robbins TW. The roles of dopamine and noradrenaline in the pathophysiology and treatment of attention-deficit/hyperactivity disorder. Biol Psychiatry 2011; 69(12): e145-57.
[http://dx.doi.org/10.1016/j.biopsych.2011.02.036] [PMID: 21550021]

[21]   Arnsten AF. The Emerging Neurobiology of Attention Deficit Hyperactivity Disorder: The Key Role of the Prefrontal Association Cortex. J Pediatr 2009; 154(5): I-S43.
[PMID: 20596295]

[22]   Dillo W, Göke A, Prox-Vagedes V, *et al.* Neuronal correlates of ADHD in adults with evidence for compensation strategies–a functional MRI study with a Go/No-Go paradigm. GMS Ger Medical Sci 2010; p. 8.

[23]   Wehmeier PM, Schacht A, Barkley RA. Social and emotional impairment in children and adolescents with ADHD and the impact on quality of life. J Adolesc Health 2010; 46(3): 209-17.
[http://dx.doi.org/10.1016/j.jadohealth.2009.09.009] [PMID: 20159496]

[24]   Wolraich ML, Bard DE, Neas B, Doffing M, Beck L. The psychometric properties of the Vanderbilt attention-deficit hyperactivity disorder diagnostic teacher rating scale in a community population. J Dev Behav Pediatr 2013; 34(2): 83-93.
[http://dx.doi.org/10.1097/DBP.0b013e31827d55c3] [PMID: 23363973]

[25]   Zhu C-Z, Zang Y-F, Cao Q-J, *et al.* Fisher discriminative analysis of resting-state brain function for attention-deficit/hyperactivity disorder. Neuroimage 2008; 40(1): 110-20.
[http://dx.doi.org/10.1016/j.neuroimage.2007.11.029] [PMID: 18191584]

[26]   Batstra L, Hadders-Algra M, Nieweg E, Van Tol D, Pijl SJ, Frances A. Childhood emotional and behavioral problems: reducing overdiagnosis without risking undertreatment. Dev Med Child Neurol 2012; 54(6): 492-4.
[http://dx.doi.org/10.1111/j.1469-8749.2011.04176.x] [PMID: 22571729]

[27]   Bruchmüller K, Margraf J, Schneider S. Is ADHD diagnosed in accord with diagnostic criteria? Overdiagnosis and influence of client gender on diagnosis. J Consult Clin Psychol 2012; 80(1): 128-38.
[http://dx.doi.org/10.1037/a0026582] [PMID: 22201328]

[28]   Rubia K, Halari R, Cubillo A, *et al.* Methylphenidate normalizes fronto-striatal underactivation during interference inhibition in medication-naïve boys with attention-deficit hyperactivity disorder. Neuropsychopharmacology 2011; 36(8): 1575-86.
[http://dx.doi.org/10.1038/npp.2011.30] [PMID: 21451498]

[29]   Rubia K, Halari R, Mohammad A-M, Taylor E, Brammer M. Methylphenidate normalizes frontocingulate underactivation during error processing in attention-deficit/hyperactivity disorder. Biol Psychiatry 2011; 70(3): 255-62.
[http://dx.doi.org/10.1016/j.biopsych.2011.04.018] [PMID: 21664605]

[30]   Rubia K, Halari R, Cubillo A, Mohammad AM, Brammer M, Taylor E. Methylphenidate normalises activation and functional connectivity deficits in attention and motivation networks in medication-naïve children with ADHD during a rewarded continuous performance task. Neuropharmacology 2009; 57(7-8): 640-52.
[http://dx.doi.org/10.1016/j.neuropharm.2009.08.013] [PMID: 19715709]

[31]   Epstein JN, Casey BJ, Tonev ST, *et al.* ADHD- and medication-related brain activation effects in concordantly affected parent-child dyads with ADHD. J Child Psychol Psychiatry 2007; 48(9): 899-

913.
[http://dx.doi.org/10.1111/j.1469-7610.2007.01761.x] [PMID: 17714375]

[32] Marquand AF, O'Daly OG, De Simoni S, *et al.* Dissociable effects of methylphenidate, atomoxetine and placebo on regional cerebral blood flow in healthy volunteers at rest: a multi-class pattern recognition approach. Neuroimage 2012; 60(2): 1015-24.
[http://dx.doi.org/10.1016/j.neuroimage.2012.01.058] [PMID: 22266414]

[33] Vaidya CJ, Austin G, Kirkorian G, *et al.* Selective effects of methylphenidate in attention deficit hyperactivity disorder: a functional magnetic resonance study. Proc Natl Acad Sci USA 1998; 95(24): 14494-9.
[http://dx.doi.org/10.1073/pnas.95.24.14494] [PMID: 9826728]

[34] Herrmann MJ, Ehlis A-C, Fallgatter AJ. Bilaterally reduced frontal activation during a verbal fluency task in depressed patients as measured by near-infrared spectroscopy. J Neuropsychiatry Clin Neurosci 2004; 16(2): 170-5.
[http://dx.doi.org/10.1176/jnp.16.2.170] [PMID: 15260368]

[35] Herrmann MJ, Plichta MM, Ehlis A-C, Fallgatter AJ. Optical topography during a Go-NoGo task assessed with multi-channel near-infrared spectroscopy. Behav Brain Res 2005; 160(1): 135-40.
[http://dx.doi.org/10.1016/j.bbr.2004.11.032] [PMID: 15836908]

[36] Hock C, Villringer K, Müller-Spahn F, *et al.* Decrease in parietal cerebral hemoglobin oxygenation during performance of a verbal fluency task in patients with Alzheimer's disease monitored by means of near-infrared spectroscopy (NIRS)--correlation with simultaneous rCBF-PET measurements. Brain Res 1997; 755(2): 293-303.
[http://dx.doi.org/10.1016/S0006-8993(97)00122-4] [PMID: 9175896]

[37] Matsuo K, Kato T, Fukuda M, Kato N. Alteration of hemoglobin oxygenation in the frontal region in elderly depressed patients as measured by near-infrared spectroscopy. J Neuropsychiatry Clin Neurosci 2000; 12(4): 465-71.
[http://dx.doi.org/10.1176/jnp.12.4.465] [PMID: 11083163]

[38] Matsuo K, Taneichi K, Matsumoto A, *et al.* Hypoactivation of the prefrontal cortex during verbal fluency test in PTSD: a near-infrared spectroscopy study. Psychiatry Res 2003; 124(1): 1-10.
[http://dx.doi.org/10.1016/S0925-4927(03)00093-3] [PMID: 14511791]

[39] Shinba T, Nagano M, Kariya N, *et al.* Near-infrared spectroscopy analysis of frontal lobe dysfunction in schizophrenia. Biol Psychiatry 2004; 55(2): 154-64.
[http://dx.doi.org/10.1016/S0006-3223(03)00547-X] [PMID: 14732595]

[40] Suto T, Fukuda M, Ito M, Uehara T, Mikuni M. Multichannel near-infrared spectroscopy in depression and schizophrenia: cognitive brain activation study. Biol Psychiatry 2004; 55(5): 501-11.
[http://dx.doi.org/10.1016/j.biopsych.2003.09.008] [PMID: 15023578]

[41] Ferrari M, Quaresima V. A brief review on the history of human functional near-infrared spectroscopy (fNIRS) development and fields of application. Neuroimage 2012; 63(2): 921-35.
[http://dx.doi.org/10.1016/j.neuroimage.2012.03.049] [PMID: 22510258]

[42] Minagawa-Kawai Y, Mori K, Hebden JC, Dupoux E. Optical imaging of infants' neurocognitive development: recent advances and perspectives. Dev Neurobiol 2008; 68(6): 712-28.
[http://dx.doi.org/10.1002/dneu.20618] [PMID: 18383545]

[43] Lloyd-Fox S, Blasi A, Elwell CE. Illuminating the developing brain: the past, present and future of functional near infrared spectroscopy. Neurosci Biobehav Rev 2010; 34(3): 269-84.
[http://dx.doi.org/10.1016/j.neubiorev.2009.07.008] [PMID: 19632270]

[44] Obrig H, Villringer A. Beyond the visible--imaging the human brain with light. J Cereb Blood Flow Metab 2003; 23(1): 1-18.
[http://dx.doi.org/10.1097/01.WCB.0000043472.45775.29] [PMID: 12500086]

[45] Strangman G, Boas DA, Sutton JP. Non-invasive neuroimaging using near-infrared light. Biol

Psychiatry 2002; 52(7): 679-93.
[http://dx.doi.org/10.1016/S0006-3223(02)01550-0] [PMID: 12372658]

[46]   Miyai I, Suzuki M, Hatakenaka M, Kubota K. Effect of body weight support on cortical activation during gait in patients with stroke. Exp Brain Res 2006; 169(1): 85-91.
[http://dx.doi.org/10.1007/s00221-005-0123-x] [PMID: 16237521]

[47]   Murata Y, Sakatani K, Katayama Y, Fukaya C. Increase in focal concentration of deoxyhaemoglobin during neuronal activity in cerebral ischaemic patients. J Neurol Neurosurg Psychiatry 2002; 73(2): 182-4.
[http://dx.doi.org/10.1136/jnnp.73.2.182] [PMID: 12122179]

[48]   Watanabe E, Maki A, Kawaguchi F, *et al.* Non-invasive assessment of language dominance with near-infrared spectroscopic mapping. Neurosci Lett 1998; 256(1): 49-52.
[http://dx.doi.org/10.1016/S0304-3940(98)00754-X] [PMID: 9832214]

[49]   Watanabe E, Nagahori Y, Mayanagi Y. Focus diagnosis of epilepsy using near-infrared spectroscopy. Epilepsia 2002; 43 (Suppl. 9): 50-5.
[http://dx.doi.org/10.1046/j.1528-1157.43.s.9.12.x] [PMID: 12383281]

[50]   Nguyen DK, Tremblay J, Pouliot P, *et al.* Non-invasive continuous EEG-fNIRS recording of temporal lobe seizures. Epilepsy Res 2012; 99(1-2): 112-26.
[http://dx.doi.org/10.1016/j.eplepsyres.2011.10.035] [PMID: 22100148]

[51]   Hahn T, Marquand AF, Plichta MM, *et al.* A novel approach to probabilistic biomarker-based classification using functional near-infrared spectroscopy. Hum Brain Mapp 2013; 34(5): 1102-14.
[http://dx.doi.org/10.1002/hbm.21497] [PMID: 22965654]

[52]   Ehlis AC, Bähne CG, Jacob CP, Herrmann MJ, Fallgatter AJ. Reduced lateral prefrontal activation in adult patients with attention-deficit/hyperactivity disorder (ADHD) during a working memory task: a functional near-infrared spectroscopy (fNIRS) study. J Psychiatr Res 2008; 42(13): 1060-7.
[http://dx.doi.org/10.1016/j.jpsychires.2007.11.011] [PMID: 18226818]

[53]   Inoue Y, Sakihara K, Gunji A, *et al.* Reduced prefrontal hemodynamic response in children with ADHD during the Go/NoGo task: a NIRS study. Neuroreport 2012; 23(2): 55-60.
[http://dx.doi.org/10.1097/WNR.0b013e32834e664c] [PMID: 22146580]

[54]   Negoro H, Sawada M, Iida J, Ota T, Tanaka S, Kishimoto T. Prefrontal dysfunction in attention-deficit/hyperactivity disorder as measured by near-infrared spectroscopy. Child Psychiatry Hum Dev 2010; 41(2): 193-203.
[http://dx.doi.org/10.1007/s10578-009-0160-y] [PMID: 19826946]

[55]   Schecklmann M, Schaldecker M, Aucktor S, *et al.* Effects of methylphenidate on olfaction and frontal and temporal brain oxygenation in children with ADHD. J Psychiatr Res 2011; 45(11): 1463-70.
[http://dx.doi.org/10.1016/j.jpsychires.2011.05.011] [PMID: 21689828]

[56]   Durston S, Tottenham NT, Thomas KM, *et al.* Differential patterns of striatal activation in young children with and without ADHD. Biol Psychiatry 2003; 53(10): 871-8.
[http://dx.doi.org/10.1016/S0006-3223(02)01904-2] [PMID: 12742674]

[57]   Araki A, Ikegami M, Okayama A, *et al.* Improved prefrontal activity in AD/HD children treated with atomoxetine: a NIRS study. Brain Dev 2015; 37(1): 76-87.
[http://dx.doi.org/10.1016/j.braindev.2014.03.011] [PMID: 24767548]

[58]   Ishii-Takahashi A, Takizawa R, Nishimura Y, *et al.* Neuroimaging-aided prediction of the effect of methylphenidate in children with attention-deficit hyperactivity disorder: a randomized controlled trial. Neuropsychopharmacology 2015; 40(12): 2676-85.
[http://dx.doi.org/10.1038/npp.2015.128] [PMID: 25936640]

[59]   Monden Y, Dan H, Nagashima M, *et al.* Clinically-oriented monitoring of acute effects of methylphenidate on cerebral hemodynamics in ADHD children using fNIRS. Clin Neurophysiol 2012; 123(6): 1147-57.

[http://dx.doi.org/10.1016/j.clinph.2011.10.006] [PMID: 22088661]

[60] Monden Y, Dan H, Nagashima M, *et al*. Right prefrontal activation as a neuro-functional biomarker for monitoring acute effects of methylphenidate in ADHD children: An fNIRS study. Neuroimage Clin 2012; 1(1): 131-40.
[http://dx.doi.org/10.1016/j.nicl.2012.10.001] [PMID: 24179746]

[61] Monden Y, Dan I, Nagashima M, *et al*. Individual classification of ADHD children by right prefrontal hemodynamic responses during a go/no-go task as assessed by fNIRS. Neuroimage Clin 2015; 9: 1-12.
[http://dx.doi.org/10.1016/j.nicl.2015.06.011] [PMID: 26266096]

[62] Nagashima M, Monden Y, Dan I, *et al*. Neuropharmacological effect of atomoxetine on attention network in children with attention deficit hyperactivity disorder during oddball paradigms as assessed using functional near-infrared spectroscopy. Neurophotonics 2014; 1(2): 025007.
[http://dx.doi.org/10.1117/1.NPh.1.2.025007] [PMID: 26157979]

[63] Nagashima M, Monden Y, Dan I, *et al*. Acute neuropharmacological effects of atomoxetine on inhibitory control in ADHD children: a fNIRS study. Neuroimage Clin 2014; 6: 192-201.
[http://dx.doi.org/10.1016/j.nicl.2014.09.001] [PMID: 25379431]

[64] Nagashima M, Monden Y, Dan I, *et al*. Neuropharmacological effect of methylphenidate on attention network in children with attention deficit hyperactivity disorder during oddball paradigms as assessed using functional near-infrared spectroscopy. Neurophotonics 2014; 1(1): 015001.
[http://dx.doi.org/10.1117/1.NPh.1.1.015001] [PMID: 26157971]

[65] Matsuura N, Ishitobi M, Arai S, *et al*. Effects of methylphenidate in children with attention deficit hyperactivity disorder: a near-infrared spectroscopy study with CANTAB®. Child Adolesc Psychiatry Ment Health 2014; 8(1): 273.
[http://dx.doi.org/10.1186/s13034-014-0032-5] [PMID: 25606052]

[66] Stuss DT, Alexander MP. Executive functions and the frontal lobes: a conceptual view. Psychol Res 2000; 63(3-4): 289-98.
[http://dx.doi.org/10.1007/s004269900007] [PMID: 11004882]

[67] Hill EL. Executive dysfunction in autism. Trends Cogn Sci 2004; 8(1): 26-32.
[http://dx.doi.org/10.1016/j.tics.2003.11.003] [PMID: 14697400]

[68] Castellanos FX, Tannock R. Neuroscience of attention-deficit/hyperactivity disorder: the search for endophenotypes. Nat Rev Neurosci 2002; 3(8): 617-28.
[http://dx.doi.org/10.1038/nrn896] [PMID: 12154363]

[69] Pennington BF, Ozonoff S. Executive functions and developmental psychopathology. J Child Psychol Psychiatry 1996; 37(1): 51-87.
[http://dx.doi.org/10.1111/j.1469-7610.1996.tb01380.x] [PMID: 8655658]

[70] Sinzig J, Morsch D, Bruning N, Schmidt MH, Lehmkuhl G. Inhibition, flexibility, working memory and planning in autism spectrum disorders with and without comorbid ADHD-symptoms. Child Adolesc Psychiatry Ment Health 2008; 2(1): 4.
[http://dx.doi.org/10.1186/1753-2000-2-4] [PMID: 18237439]

[71] Schachar R, Tannock R, Marriott M, Logan G. Deficient inhibitory control in attention deficit hyperactivity disorder. J Abnorm Child Psychol 1995; 23(4): 411-37.
[http://dx.doi.org/10.1007/BF01447206] [PMID: 7560554]

[72] Barkley RA. Behavioral inhibition, sustained attention, and executive functions: constructing a unifying theory of ADHD. Psychol Bull 1997; 121(1): 65-94.
[http://dx.doi.org/10.1037/0033-2909.121.1.65] [PMID: 9000892]

[73] Aron AR, Fletcher PC, Bullmore ET, Sahakian BJ, Robbins TW. Stop-signal inhibition disrupted by damage to right inferior frontal gyrus in humans. Nat Neurosci 2003; 6(2): 115-6.
[http://dx.doi.org/10.1038/nn1003] [PMID: 12536210]

[74] Aron AR, Robbins TW, Poldrack RA. Inhibition and the right inferior frontal cortex. Trends Cogn Sci

2004; 8(4): 170-7.
[http://dx.doi.org/10.1016/j.tics.2004.02.010] [PMID: 15050513]

[75]   Menon V, Adleman NE, White CD, Glover GH, Reiss AL. Error-related brain activation during a Go/NoGo response inhibition task. Hum Brain Mapp 2001; 12(3): 131-43.
[http://dx.doi.org/10.1002/1097-0193(200103)12:3<131::AID-HBM1010>3.0.CO;2-C]   [PMID: 11170305]

[76]   Bush G, Frazier JA, Rauch SL, *et al.* Anterior cingulate cortex dysfunction in attention-deficit/hyperactivity disorder revealed by fMRI and the Counting Stroop. Biol Psychiatry 1999; 45(12): 1542-52.
[http://dx.doi.org/10.1016/S0006-3223(99)00083-9] [PMID: 10376114]

[77]   Rubia K, Overmeyer S, Taylor E, *et al.* Hypofrontality in attention deficit hyperactivity disorder during higher-order motor control: a study with functional MRI. Am J Psychiatry 1999; 156(6): 891-6.
[http://dx.doi.org/10.1176/ajp.156.6.891] [PMID: 10360128]

[78]   Williams BR, Ponesse JS, Schachar RJ, Logan GD, Tannock R. Development of inhibitory control across the life span. Dev Psychol 1999; 35(1): 205-13.
[http://dx.doi.org/10.1037/0012-1649.35.1.205] [PMID: 9923475]

[79]   Comalli PE Jr, Wapner S, Werner H. Interference effects of Stroop color-word test in childhood, adulthood, and aging. J Genet Psychol 1962; 100: 47-53.
[http://dx.doi.org/10.1080/00221325.1962.10533572] [PMID: 13880724]

[80]   Levin HS, Culhane KA, Hartmann J, *et al.* Developmental changes in performance on tests of purported frontal lobe functioning. Dev Neuropsychol 1991; 7: 377-95.
[http://dx.doi.org/10.1080/87565649109540499]

[81]   Beauregard M, Lévesque J. Functional magnetic resonance imaging investigation of the effects of neurofeedback training on the neural bases of selective attention and response inhibition in children with attention-deficit/hyperactivity disorder. Appl Psychophysiol Biofeedback 2006; 31(1): 3-20.
[http://dx.doi.org/10.1007/s10484-006-9001-y] [PMID: 16552626]

[82]   Derefinko KJ, Adams ZW, Milich R, Fillmore MT, Lorch EP, Lynam DR. Response style differences in the inattentive and combined subtypes of attention-deficit/hyperactivity disorder. J Abnorm Child Psychol 2008; 36(5): 745-58.
[http://dx.doi.org/10.1007/s10802-007-9207-3] [PMID: 18175214]

[83]   Ma J, Lei D, Jin X, *et al.* Compensatory brain activation in children with attention deficit/hyperactivity disorder during a simplified Go/No-go task. J Neural Transm (Vienna) 2012; 119(5): 613-9.
[http://dx.doi.org/10.1007/s00702-011-0744-0] [PMID: 22139325]

[84]   Smith AB, Taylor E, Brammer M, Toone B, Rubia K. Task-specific hypoactivation in prefrontal and temporoparietal brain regions during motor inhibition and task switching in medication-naive children and adolescents with attention deficit hyperactivity disorder. Am J Psychiatry 2006; 163(6): 1044-51.
[http://dx.doi.org/10.1176/ajp.2006.163.6.1044] [PMID: 16741205]

[85]   Solanto MV, Schulz KP, Fan J, Tang CY, Newcorn JH. Event-related FMRI of inhibitory control in the predominantly inattentive and combined subtypes of ADHD. J Neuroimaging 2009; 19(3): 205-12.
[http://dx.doi.org/10.1111/j.1552-6569.2008.00289.x] [PMID: 19594667]

[86]   Rubia K, Smith AB, Brammer MJ, Toone B, Taylor E. Abnormal brain activation during inhibition and error detection in medication-naive adolescents with ADHD. Am J Psychiatry 2005; 162(6): 1067-75.
[http://dx.doi.org/10.1176/appi.ajp.162.6.1067] [PMID: 15930054]

[87]   Peers PV, Ludwig CJ, Rorden C, *et al.* Attentional functions of parietal and frontal cortex. Cereb Cortex 2005; 15(10): 1469-84.
[http://dx.doi.org/10.1093/cercor/bhi029] [PMID: 15689522]

[88]   Rivera SM, Reiss AL, Eckert MA, Menon V. Developmental changes in mental arithmetic: evidence

for increased functional specialization in the left inferior parietal cortex. Cereb Cortex 2005; 15(11): 1779-90.
[http://dx.doi.org/10.1093/cercor/bhi055] [PMID: 15716474]

[89] Shafritz KM, Marchione KE, Gore JC, Shaywitz SE, Shaywitz BA. The effects of methylphenidate on neural systems of attention in attention deficit hyperactivity disorder. Am J Psychiatry 2004; 161(11): 1990-7.
[http://dx.doi.org/10.1176/appi.ajp.161.11.1990] [PMID: 15514398]

[90] Liddle PF, Kiehl KA, Smith AM. Event-related fMRI study of response inhibition. Hum Brain Mapp 2001; 12(2): 100-9.
[http://dx.doi.org/10.1002/1097-0193(200102)12:2<100::AID-HBM1007>3.0.CO;2-6] [PMID: 11169874]

[91] Hale TS, Hariri AR, McCracken JT. Attention-deficit/hyperactivity disorder: perspectives from neuroimaging. Ment Retard Dev Disabil Res Rev 2000; 6(3): 214-9.
[http://dx.doi.org/10.1002/1098-2779(2000)6:3<214::AID-MRDD9>3.0.CO;2-M] [PMID: 10982499]

[92] Katagiri A, Dan I, Tuzuki D, et al. Mapping of optical pathlength of human adult head at multi-wavelengths in near infrared spectroscopy. Adv Exp Med Biol 2010; 662: 205-12.
[http://dx.doi.org/10.1007/978-1-4419-1241-1_29] [PMID: 20204793]

[93] Cope M, Delpy DT, Reynolds EO, Wray S, Wyatt J, van der Zee P. Methods of quantitating cerebral near infrared spectroscopy data. Adv Exp Med Biol 1988; 222: 183-9.
[http://dx.doi.org/10.1007/978-1-4615-9510-6_21] [PMID: 3129910]

[94] Maki A, Yamashita Y, Ito Y, Watanabe E, Mayanagi Y, Koizumi H. Spatial and temporal analysis of human motor activity using noninvasive NIR topography. Med Phys 1995; 22(12): 1997-2005.
[http://dx.doi.org/10.1118/1.597496] [PMID: 8746704]

[95] Garavan H, Ross TJ, Stein EA. Right hemispheric dominance of inhibitory control: an event-related functional MRI study. Proc Natl Acad Sci USA 1999; 96(14): 8301-6.
[http://dx.doi.org/10.1073/pnas.96.14.8301] [PMID: 10393989]

[96] Shattuck DW, Mirza M, Adisetiyo V, et al. Construction of a 3D probabilistic atlas of human cortical structures. Neuroimage 2008; 39(3): 1064-80.
[http://dx.doi.org/10.1016/j.neuroimage.2007.09.031] [PMID: 18037310]

[97] Rorden C, Brett M. Stereotaxic display of brain lesions. Behav Neurol 2000; 12(4): 191-200.
[http://dx.doi.org/10.1155/2000/421719] [PMID: 11568431]

[98] Rubia K, Smith AB, Brammer MJ, Taylor E. Right inferior prefrontal cortex mediates response inhibition while mesial prefrontal cortex is responsible for error detection. Neuroimage 2003; 20(1): 351-8.
[http://dx.doi.org/10.1016/S1053-8119(03)00275-1] [PMID: 14527595]

[99] Simmonds DJ, Pekar JJ, Mostofsky SH. Meta-analysis of Go/No-go tasks demonstrating that fMRI activation associated with response inhibition is task-dependent. Neuropsychologia 2008; 46(1): 224-32.
[http://dx.doi.org/10.1016/j.neuropsychologia.2007.07.015] [PMID: 17850833]

[100] Rubia K, Russell T, Overmeyer S, et al. Mapping motor inhibition: conjunctive brain activations across different versions of go/no-go and stop tasks. Neuroimage 2001; 13(2): 250-61.
[http://dx.doi.org/10.1006/nimg.2000.0685] [PMID: 11162266]

[101] Ago Y, Umehara M, Higashino K, et al. Atomoxetine-induced increases in monoamine release in the prefrontal cortex are similar in spontaneously hypertensive rats and Wistar-Kyoto rats. Neurochem Res 2014; 39(5): 825-32.
[http://dx.doi.org/10.1007/s11064-014-1275-5] [PMID: 24634253]

[102] Bymaster FP, Katner JS, Nelson DL, et al. Atomoxetine increases extracellular levels of norepinephrine and dopamine in prefrontal cortex of rat: a potential mechanism for efficacy in

attention deficit/hyperactivity disorder. Neuropsychopharmacology 2002; 27(5): 699-711.
[http://dx.doi.org/10.1016/S0893-133X(02)00346-9] [PMID: 12431845]

[103] Adler CM, Sax KW, Holland SK, Schmithorst V, Rosenberg L, Strakowski SM. Changes in neuronal activation with increasing attention demand in healthy volunteers: an fMRI study. Synapse 2001; 42(4): 266-72.
[http://dx.doi.org/10.1002/syn.1112] [PMID: 11746725]

[104] Ardekani BA, Choi SJ, Hossein-Zadeh G-A, *et al.* Functional magnetic resonance imaging of brain activity in the visual oddball task. Brain Res Cogn Brain Res 2002; 14(3): 347-56.
[http://dx.doi.org/10.1016/S0926-6410(02)00137-4] [PMID: 12421658]

[105] Bledowski C, Prvulovic D, Goebel R, Zanella FE, Linden DE. Attentional systems in target and distractor processing: a combined ERP and fMRI study. Neuroimage 2004; 22(2): 530-40.
[http://dx.doi.org/10.1016/j.neuroimage.2003.12.034] [PMID: 15193581]

[106] Clark VP, Fannon S, Lai S, Benson R, Bauer L. Responses to rare visual target and distractor stimuli using event-related fMRI. J Neurophysiol 2000; 83(5): 3133-9.
[http://dx.doi.org/10.1152/jn.2000.83.5.3133] [PMID: 10805707]

[107] Gur RC, Turetsky BI, Loughead J, *et al.* Hemodynamic responses in neural circuitries for detection of visual target and novelty: An event-related fMRI study. Hum Brain Mapp 2007; 28(4): 263-74.
[http://dx.doi.org/10.1002/hbm.20319] [PMID: 17133387]

[108] Stevens AA, Skudlarski P, Gatenby JC, Gore JC. Event-related fMRI of auditory and visual oddball tasks. Magn Reson Imaging 2000; 18(5): 495-502.
[http://dx.doi.org/10.1016/S0730-725X(00)00128-4] [PMID: 10913710]

[109] Petrides M, Pandya DN. Projections to the frontal cortex from the posterior parietal region in the rhesus monkey. J Comp Neurol 1984; 228(1): 105-16.
[http://dx.doi.org/10.1002/cne.902280110] [PMID: 6480903]

[110] Volkow ND, Wang G-J, Ma Y, *et al.* Activation of orbital and medial prefrontal cortex by methylphenidate in cocaine-addicted subjects but not in controls: relevance to addiction. J Neurosci 2005; 25(15): 3932-9.
[http://dx.doi.org/10.1523/JNEUROSCI.0433-05.2005] [PMID: 15829645]

[111] Arnsten AF, Dudley AG. Methylphenidate improves prefrontal cortical cognitive function through $\alpha2$ adrenoceptor and dopamine D1 receptor actions: Relevance to therapeutic effects in Attention Deficit Hyperactivity Disorder. Behav Brain Funct 2005; 1(1): 2.
[http://dx.doi.org/10.1186/1744-9081-1-2] [PMID: 15916700]

[112] Arnsten AF, Pliszka SR. Catecholamine influences on prefrontal cortical function: relevance to treatment of attention deficit/hyperactivity disorder and related disorders. Pharmacol Biochem Behav 2011; 99(2): 211-6.
[http://dx.doi.org/10.1016/j.pbb.2011.01.020] [PMID: 21295057]

[113] Singh-Curry V, Husain M. The functional role of the inferior parietal lobe in the dorsal and ventral stream dichotomy. Neuropsychologia 2009, 47(6): 1434-48.
[http://dx.doi.org/10.1016/j.neuropsychologia.2008.11.033] [PMID: 19138694]

[114] Buchsbaum BR, Greer S, Chang WL, Berman KF. Meta-analysis of neuroimaging studies of the Wisconsin card-sorting task and component processes. Hum Brain Mapp 2005; 25(1): 35-45.
[http://dx.doi.org/10.1002/hbm.20128] [PMID: 15846821]

[115] Hart H, Chantiluke K, Cubillo AI, *et al.* Pattern classification of response inhibition in ADHD: toward the development of neurobiological markers for ADHD. Hum Brain Mapp 2014; 35(7): 3083-94.
[http://dx.doi.org/10.1002/hbm.22386] [PMID: 24123508]

[116] Leitner Y. The co-occurrence of autism and attention deficit hyperactivity disorder in children - what do we know? Front Hum Neurosci 2014; 8: 268.
[http://dx.doi.org/10.3389/fnhum.2014.00268] [PMID: 24808851]

[117] Craig F, Lamanna AL, Margari F, Matera E, Simone M, Margari L. Overlap Between Autism Spectrum Disorders and Attention Deficit Hyperactivity Disorder: Searching for Distinctive/Common Clinical Features. Autism Res 2015; 8(3): 328-37.
[http://dx.doi.org/10.1002/aur.1449] [PMID: 25604000]

[118] Craig F, Margari F, Legrottaglie AR, Palumbi R, de Giambattista C, Margari L. A review of executive function deficits in autism spectrum disorder and attention-deficit/hyperactivity disorder. Neuropsychiatr Dis Treat 2016; 12: 1191-202.
[PMID: 27274255]

[119] Arnold LE, Aman MG, Li X, *et al.* Research Units of Pediatric Psychopharmacology (RUPP) autism network randomized clinical trial of parent training and medication: one-year follow-up. J Am Acad Child Adolesc Psychiatry 2012; 51(11): 1173-84.
[http://dx.doi.org/10.1016/j.jaac.2012.08.028] [PMID: 23101743]

[120] Tye C, Asherson P, Ashwood KL, Azadi B, Bolton P, McLoughlin G. Attention and inhibition in children with ASD, ADHD and co-morbid ASD + ADHD: an event-related potential study. Psychol Med 2014; 44(5): 1101-16.
[http://dx.doi.org/10.1017/S0033291713001049] [PMID: 23673307]

[121] Chantiluke K, Christakou A, Murphy CM, *et al.* Disorder-specific functional abnormalities during temporal discounting in youth with Attention Deficit Hyperactivity Disorder (ADHD), Autism and comorbid ADHD and Autism. Psychiatry Res 2014; 223(2): 113-20.
[http://dx.doi.org/10.1016/j.pscychresns.2014.04.006] [PMID: 24929553]

[122] Moll GH, Hause S, Rüther E, Rothenberger A, Huether G. Early methylphenidate administration to young rats causes a persistent reduction in the density of striatal dopamine transporters. J Child Adolesc Psychopharmacol 2001; 11(1): 15-24.
[http://dx.doi.org/10.1089/104454601750143366] [PMID: 11322741]

[123] Nakao T, Radua J, Rubia K, Mataix-Cols D. Gray matter volume abnormalities in ADHD: voxel-based meta-analysis exploring the effects of age and stimulant medication. Am J Psychiatry 2011; 168(11): 1154-63.
[http://dx.doi.org/10.1176/appi.ajp.2011.11020281] [PMID: 21865529]

[124] Bledsoe J, Semrud-Clikeman M, Pliszka SR. A magnetic resonance imaging study of the cerebellar vermis in chronically treated and treatment-naïve children with attention-deficit/hyperactivity disorder combined type. Biol Psychiatry 2009; 65(7): 620-4.
[http://dx.doi.org/10.1016/j.biopsych.2008.11.030] [PMID: 19150052]

[125] Shaw P, Sharp WS, Morrison M, *et al.* Psychostimulant treatment and the developing cortex in attention deficit hyperactivity disorder. Am J Psychiatry 2009; 166(1): 58-63.
[http://dx.doi.org/10.1176/appi.ajp.2008.08050781] [PMID: 18794206]

[126] Yamazaki K. ADHD-RS-IV Japanese version.Japanese guideline for the diagnosis and treatment of attention deficit hyperactivity disorder (ADHD). Tokyo: Jiho 2013; pp. 48-54. (in Japanese)

[127] DuPaul GJ, Anastopoulos AD, Power TJ, Reid R, Ikeda MJ, McGoey KE. Parent ratings of attention-deficit/hyperactivity disorder symptoms: Factor structure and normative data. J Psychopathol Behav Assess 1998; 20: 83-102.
[http://dx.doi.org/10.1023/A:1023087410712]

[128] Asahi S, Okamoto Y, Okada G, Yamawaki S, Yokota N. Negative correlation between right prefrontal activity during response inhibition and impulsiveness: a fMRI study. Eur Arch Psychiatry Clin Neurosci 2004; 254(4): 245-51.
[http://dx.doi.org/10.1007/s00406-004-0488-z] [PMID: 15309395]

[129] Tamm L, Menon V, Ringel J, Reiss AL. Event-related FMRI evidence of frontotemporal involvement in aberrant response inhibition and task switching in attention-deficit/hyperactivity disorder. J Am Acad Child Adolesc Psychiatry 2004; 43(11): 1430-40.
[http://dx.doi.org/10.1097/01.chi.0000140452.51205.8d] [PMID: 15502603]

[130] Chikazoe J. Localizing performance of go/no-go tasks to prefrontal cortical subregions. Curr Opin Psychiatry 2010; 23(3): 267-72.
[http://dx.doi.org/10.1097/YCO.0b013e3283387a9f] [PMID: 20308899]

[131] Niendam TA, Laird AR, Ray KL, Dean YM, Glahn DC, Carter CS. Meta-analytic evidence for a superordinate cognitive control network subserving diverse executive functions. Cogn Affect Behav Neurosci 2012; 12(2): 241-68.
[http://dx.doi.org/10.3758/s13415-011-0083-5] [PMID: 22282036]

[132] Khan A, Khan S, Brown WA. Are placebo controls necessary to test new antidepressants and anxiolytics? Int J Neuropsychopharmacol 2002; 5(3): 193-7.
[http://dx.doi.org/10.1017/S1461145702002912] [PMID: 12366872]

[133] Khan A, Warner HA, Brown WA. Symptom reduction and suicide risk in patients treated with placebo in antidepressant clinical trials: an analysis of the Food and Drug Administration database. Arch Gen Psychiatry 2000; 57(4): 311-7.
[http://dx.doi.org/10.1001/archpsyc.57.4.311] [PMID: 10768687]

[134] Rutherford BR, Roose SP. A model of placebo response in antidepressant clinical trials. Am J Psychiatry 2013; 170(7): 723-33.
[http://dx.doi.org/10.1176/appi.ajp.2012.12040474] [PMID: 23318413]

[135] Sinyor M, Levitt AJ, Cheung AH, *et al.* Does inclusion of a placebo arm influence response to active antidepressant treatment in randomized controlled trials? Results from pooled and meta-analyses. J Clin Psychiatry 2010; 71(3): 270-9.
[http://dx.doi.org/10.4088/JCP.08r04516blu] [PMID: 20122371]

[136] Lord C, Risi S, Lambrecht L, *et al.* The autism diagnostic observation schedule-generic: a standard measure of social and communication deficits associated with the spectrum of autism. J Autism Dev Disord 2000; 30(3): 205-23.
[http://dx.doi.org/10.1023/A:1005592401947] [PMID: 11055457]

[137] Lord C, Rutter M, Le Couteur A. Autism Diagnostic Interview-Revised: a revised version of a diagnostic interview for caregivers of individuals with possible pervasive developmental disorders. J Autism Dev Disord 1994; 24(5): 659-85.
[http://dx.doi.org/10.1007/BF02172145] [PMID: 7814313]

[138] Rommelse NN, Franke B, Geurts HM, Hartman CA, Buitelaar JK. Shared heritability of attention-deficit/hyperactivity disorder and autism spectrum disorder. Eur Child Adolesc Psychiatry 2010; 19(3): 281-95.
[http://dx.doi.org/10.1007/s00787-010-0092-x] [PMID: 20148275]

[139] Jang J, Matson JL, Williams LW, Tureck K, Goldin RL, Cervantes PE. Rates of comorbid symptoms in children with ASD, ADHD, and comorbid ASD and ADHD. Res Dev Disabil 2013; 34(8): 2369-78.
[http://dx.doi.org/10.1016/j.ridd.2013.04.021] [PMID: 23708709]

[140] Dittmann RW, Wehmeier PM, Schacht A, *et al.* Atomoxetine treatment and ADHD-related difficulties as assessed by adolescent patients, their parents and physicians. Child Adolesc Psychiatry Ment Health 2009; 3(1): 21.
[http://dx.doi.org/10.1186/1753-2000-3-21] [PMID: 19703299]

[141] Thomas R, Sanders S, Doust J, Beller E, Glasziou P. Prevalence of attention-deficit/hyperactivity disorder: a systematic review and meta-analysis. Pediatrics 2015; 135(4): e994-e1001.
[http://dx.doi.org/10.1542/peds.2014-3482] [PMID: 25733754]

CHAPTER 5

Modulation of Mesenchymal Stem Cells, Glial Cells and the Immune System by Oligodeoxynucleotides as a Novel Multi-target Therapeutic Approach Against Chronic Pain

Pablo R. Brumovsky[1,*], Mailín Casadei[1], Candelaria Leiguarda[1], María Florencia Coronel[1], Julia Rubione[1], Alejandro Montaner[2] and Marcelo J. Villar[1]

[1] *Instituto de Investigaciones en Medicina Traslacional (IIMT), CONICET-Universidad Austral, Buenos Aires, Argentina*

[2] *Instituto de Ciencia y Tecnología "Dr. César Milstein", CONICET- Fundación Pablo Cassará, Buenos Aires, Argentina*

Abstract: Despite our growing understanding of chronic pain mechanisms, an alarming proportion of patients worldwide remains refractory to treatment. Chronic pain is complex, involving the interaction of both neuronal and non-neuronal systems. Several studies focused on immune, glial and mesenchymal stem cells (MSCs) have recently revealed key roles of these non-neuronal players in the initiation and perpetuation of chronic pain. The complexity of chronic pain is reflected by the difficulty of its therapeutic control, in particular when using mono-target drugs. A good proportion of these drugs target neuronal pathways, and serious concerns arise when it comes to the use of opioids and abuse liability. In contrast, novel pain drugs targeting non-neuronal components of chronic pain are scarce. Exceptions include classical non-steroidal anti-inflammatory drugs, or those modulating trophic factors, although their use remains restricted to the presence of appropriate targets. Synthetic oligodeoxynucleotides have been used as immune system modulators for the last 15 years. One of them, IMT504, a non-CpG oligodeoxynucleotide, exhibits remarkable, long-lasting anti-allodynic and anti-inflammatory properties upon single-dose systemic administration in rodent models of inflammatory or neuropathic pain. Mounting evidence suggests that the beneficial effects of IMT504 relate to actions on the immune system, glial cells and MSCs. In this state-of-the-art chapter, we address the current knowledge of the role of IMT504 over non-neuronal cells, its impact on chronic pain, and its translational potential. We also propose that further analysis on its mechanisms of action will be key to the identification of novel and effective multi-target pain drugs without abuse liability.

* Corresponding Author Pablo R. Brumovsky: Instituto de Investigaciones en Medicina Traslacional, Universidad Austral Av. Juan D. Perón 1500 B1629AHJ, Pilar, Buenos Aires, Argentina; Tel: +54-023-448-7424; email: pbrumovs@austral.edu.ar

Atta-ur-Rahman & Zareen Amtul (Eds.)
All rights reserved-© 2022 Bentham Science Publishers

Keywords: Astrocytes, B Lymphocytes, Chronic Pain, Dorsal Root Ganglia, Drugs, Glial Cells, Inflammation, Inflammatory Cells, Macrophages, Mesenchymal Stem Cells, Microglia, Neuroimmune Interaction, Nociception, Oligodeoxynucleotides, Pain, Spinal Cord, T Lymphocytes.

THE CHRONIC PAIN PROBLEM

The International Association for the Study of Pain (IASP) defines pain as "an unpleasant sensory and emotional experience associated with, or resembling that associated with, actual or potential tissue damage" [1, 2]. That is to say that pain is a personal experience that includes the conscious perception of a stimulus capable of generating tissue damage and may also depend on cognitive and emotional components [3 - 6].

Pain can be divided into two main categories: acute and chronic. Acute pain has a protective function to avoid potential damage and is concerned with the sensing of noxious stimuli by primary sensory neurons involved in pain perception, also known as nociceptors. It is therefore known as nociceptive pain, referring to pain evoked by intense stimuli [7]. The protective role of acute pain, associated with the experience of unpleasantness and emotional anguish, demands immediate attention and action, ultimately resulting in withdrawal reflexes.

Chronic pain instead is a maladaptive and pathological sensation that represents one of the most prevalent and disabling health conditions; it has no beneficial biologic significance and is characterized by spontaneous pain as well as evoked pain in response to noxious (hyperalgesia) or non-noxious (allodynia) stimuli [2, 8, 9].

Over 20% of the world population is estimated to suffer from chronic pain [10, 11], and only in the United States, medical care costs plus the lower economic productivity derived from the condition were estimated to represent between 560-635 billion $USD [12]. Moreover, besides the pain they experience [13 - 18], patients suffering from chronic pain undergo progressive deterioration of their general health and quality of life, sleep, eating and memory disturbances, functional disability and even severe depression [19 - 21]. Altogether, while chronic undertreated pain does not kill the patient, at least not directly [22], it remains a catastrophic burden in society [23, 24].

The occurrence of chronic pain has been related to neuronal plasticity which consists of peripheral sensitization in primary sensory neurons of dorsal root ganglia, somatosensory neurons in cranial nerve ganglia [7, 25, 26] and central sensitization of pain-processing neurons in the spinal cord and brain [27 - 30]. Among the several types of presentation of chronic pain, neuropathic pain, characterized by injury or disease that directly affects the somatosensory nervous system, including peripheral fibers and central neurons, stands out as particularly challenging [2]. The mechanisms involved in the development and maintenance of neuropathic pain were initially characterized as series of neuronal dysfunctions. Indeed, after nerve injury, various modifications occur across the pain pathway that include alterations in ion channel expression and function, and upregulation and downregulation of a series of neuropeptides and neurotransmitters and their receptors, leading to neuronal hyperexcitability [7, 31 - 42].

There is, however, a growing body of evidence indicating that the cause of neuropathic pain is not only restricted to changes in neuronal activity but, in addition, involves interactions with glial and immune cells [3, 43 - 49]. These cells interact with neuronal cell bodies and their fibers distributed throughout the peripheral and central nervous system (PNS and CNS, respectively), and upon nerve or spinal cord injury are activated for the release of inflammatory mediators that strongly modify neuronal function, culminating in alterations of painful perception [44 - 46] (Fig. **1**). Macrophages [47 - 49] and MSCs [50 - 54] are increasingly described as key cell subpopulations involved in neuroimmune interactions associated with neuropathic pain. However, other immune/glial cells are emerging as highly relevant, as will be discussed in the present chapter.

Pain perception is also typically associated with inflammation, a complex biological response of the somatosensory, immune, neuronal, autonomic, and vascular/circulatory system to tissue damage, pathogens, or irritants [4]. Conversely, nociceptors are also capable of modulating the immune response during inflammation [55, 56]. Just like immune cells, nociceptors express cytokine receptors, chemokines, and Toll-like receptors (TLRs) that are essential for immune modulation [57 - 60].

Finally, from a therapeutic standpoint, acute pain due to surgery or trauma can typically be treated with timely administered conventional analgesics, including opioids and non-steroidal anti-inflammatory drugs (NSAIDs), because these drugs reduce inflammation and/or nociceptive input. Notoriously, the majority of new analgesics developed over the past five decades fall into these two categories [61]. However, with the exclusion of opioids, conventional analgesics are often less effective or totally useless in alleviating chronic pain. Therefore, the need for effective treatment options for chronic pain remains imperative, without falling into the use of opioids and their well-known collateral effects.

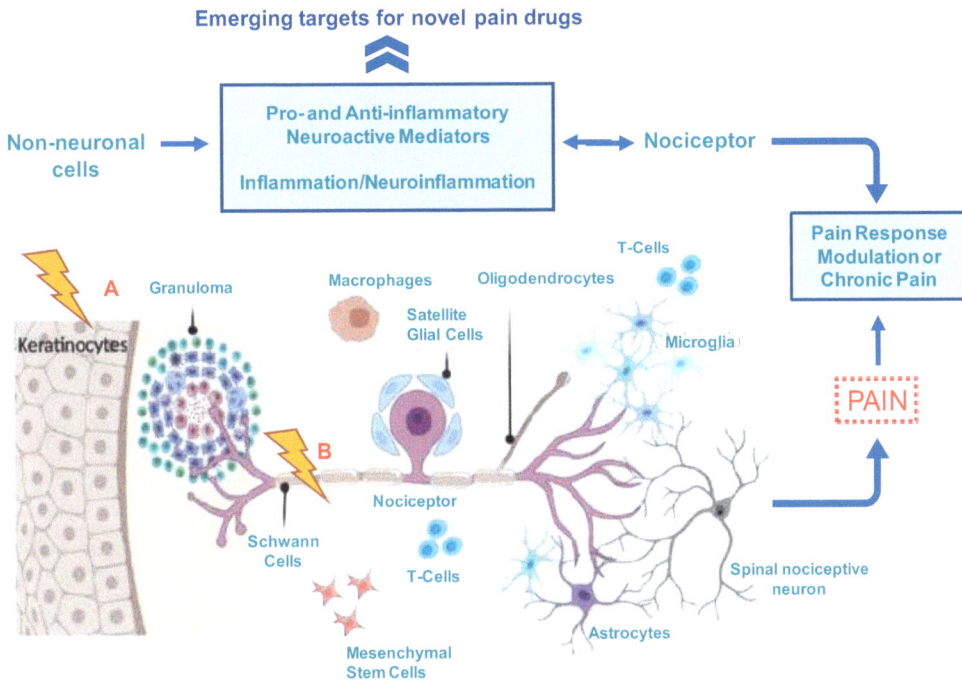

Fig. (1). Diagram highlighting the several sites of possible interaction between non-neuronal cells and primary and spinal nociceptive neurons. **A** and **B** represent the two main causes of chronic pain analyzed in the present chapter, namely inflammation (A) and neuropathy (B). The combined action of these different cellular "actors" results in the production of signals that are processed at the spinal cord level, and later transmitted to upper levels of the CNS, ultimately evoking painful conscious sensations. Non-neuronal cells participate in the origin and propagation of pain signals by producing pro- and anti-inflammatory neuroactive mediators, which then act upon receptors expressed in nociceptive neurons, modulating their excitability. This type of interaction is not only central to the mechanisms leading to pain during tissue inflammation (A), it is also relevant during direct nerve injury (B), in a process that can also be described as neuroinflammation, impacting the activity of nociceptors and being conducive of pain response modulation (in cases when self-limiting mechanisms are in place), or to chronic pain if the response becomes maladaptive. Importantly, a better knowledge of the neuroimmune interactions taking place during painful conditions emerges as an interesting source of novel targets against chronic pain (Created with Biorender and based in part on [62]).

In the following sections, we will analyze in more detail the emerging roles of non-neuronal cells (mainly immune cells, glial cells and MSCs) as key players in the mechanisms of chronic inflammatory and neuropathic pain. This will help us to understand the multifactorial nature of chronic pain. And, in this context, we will introduce IMT504, a very promising molecule with translational potential for the treatment of chronic pain involving the regulation of neuroimmune interactions (Fig. **1**).

THE "NEW KIDS ON THE BLOCK"

The Immune System: An Introduction

Immunology, or the study of the so called "defense system", has grown exponentially over the last several decades. Although originally viewed as being only responsible for preventing infections, it is now known that the immune system contributes to maintaining homeostasis in various tissues. Interestingly, its dysfunction has been implicated in neurological (*e.g.* multiple sclerosis and Alzheimer disease), cardiovascular (*e.g.* atherosclerosis and vasculitis) and gastrointestinal disorders (*e.g.* Crohn´s disease and ulcerative colitis), among others that are independent of infections [63].

First notions of an immune system appeared in the late 18th century when Edward Jenner demonstrated that inoculation with cowpox or vaccinia virus offered protection against the often-fatal disease of smallpox [64]. He called this experimental procedure "vaccination" which would become the cornerstone of one of the greatest achievements of modern medicine when it allowed for the eradication of smallpox almost two centuries later. Nevertheless, Jenner's observations were ignorant of the causative agent of the disease (the smallpox virus) and it wasn't until late in the 19th century when Robert Koch introduced the concept of microorganisms responsible for infectious diseases [65]. Alongside came Emil von Behring and Shibasaburo Kitasato who discovered a special short termed protection conferred by serum of diphtheria or tetanus infected animals to people who came in contact with either [66]. This was due to what we now know as antibodies.

Nowadays, the knowledge about the immune system is growing at a great pace as new components and functions are discovered. Schematically, two major compartments are recognized: the innate and the adaptive immune system, that act in an orchestrated manner [67]. The innate component is known as the "first line of defense" and, as such, is activated promptly and quickly. Initially, any pathogen encounters the physical and anatomical barriers of the skin and mucosal epithelia (*e.g.* respiratory, gastrointestinal and genitourinary) that have both passive (mainly the integrity of the epithelia) and active counteracting mechanisms. The latter is characterized by secretion of soluble mediators with antimicrobial activity, pro-inflammatory cytokines (Interleukin (IL)-1β, IL-6 and tumor necrosis-factor (TNF)-α), and a variety of immune cell chemical attractants termed chemokines. This induced response is possible thanks to the ability of the epithelia to recognize **P**athogen-**A**ssociated **M**olecular **P**atterns (also known as PAMPs) through **P**attern **R**ecognition **R**eceptors (PRR), that are expressed by virtually all cell types, and that can also recognize **D**anger-**A**ssociated **M**olecular

Patterns (DAMPs; usually intracellular components that are released or expressed upon tissue damage) [68, 69]. This strategy of non-self and altered self-identification is key to maintaining a state of surveillance that can very quickly recruit specific immune cells. These cells are monocytes, neutrophils, mastocytes, eosinophils, basophils and natural killer (NK) cells, which are normally found in circulation. At the site of infection two other cell types are found, that are tissue resident and play a fundamental role in the activation and functionality of the innate response, and the coordination and activation of the adaptive immune response: macrophages and dendritic cells, respectively. Both exhibit the ability to function in different profiles (*i.e.*, pro- and anti-inflammatory) and help modulate the inflammatory *milieu* [70, 71].

The adaptive immune response is governed by lymphocytes: T cells (CD4$^+$ and CD8$^+$) and B cells. Unlike the innate response, their activation takes between 7 and 15 days. However, an important feature of the adaptive immune response is the possibility of developing memory, leading to faster and more efficient future responses [72]. This is enabled by their non-self-recognition strategy that is antigen specific and allows for a pathogen-specific response. Both T and B cells are activated in lymph nodes adjacent to the site of infection and will migrate to peripheral tissues or bone marrow to initiate their response. Dendritic cells carrying processed antigens from the periphery migrate to lymph nodes where they encounter naïve CD4$^+$ and CD8$^+$ T cells and can activate them by presenting the antigen and expressing co-stimulatory molecules and cytokines [73]. After a phase of clonal expansion and differentiation to effector cells, they exit the lymph nodes and are attracted to the site of infection. From there, they can exacerbate the inflammatory response by secreting soluble mediators and elicit direct cytotoxic activity. Additionally, a subset of cells will differentiate to memory cells and return to circulation [74]. On the other hand, B cells recognize antigens without any intermediates at the lymph nodes, going through a phase of clonal expansion. Ultimately, they will differentiate to long-lived plasmocytes that secrete immunoglobulins (especially IgG) from niches in the bone marrow. However, a subset of B cells can differentiate to short-lived plasmocytes and secrete IgM early on [75].

Finally, regulatory mechanisms will take over to limit and contain the inflammatory response until homeostasis is restored, the initial insult resolved, and any tissue damage repaired. In case of infection, removal of antigens will promote an anti-inflammatory environment in which macrophages and dendritic cells can acquire tolerogenic activity, characterized by secretion of anti-inflammatory cytokines like IL-10 and transforming growth factor (TGF)-β [76]. This will result in the expansion of regulatory T cells and a decreased extravasation of leucocytes from circulation. Failure to correctly activate these regulatory mechanisms can give rise to chronic inflammation and autoimmune diseases, like rheumatoid arthritis or inflammatory bowel disease, among many others [77].

The Immune System and Pain

Besides its role against pathogens, and as already advanced, the immune system is intimately related to the pathophysiology of pain. This is particularly noticeable when considering pain due to peripheral tissue inflammation. Here, tissue aggression triggers innate and adaptive immune responses, activating resident immune cells, facilitating the recruitment of circulating cells towards the injured site, and resulting in the secretion of a variety of molecular mediators, including bradykinin, prostaglandins, nerve growth factor, and pro-inflammatory cytokines [4, 78, 79] among many others, with the capacity to sensitize the peripheral nerve terminals of nociceptive neurons [80, 81] (Fig. **1**); Table **1**. Sensitization is possible because nociceptive neurons and their nerve terminals express membrane receptors reactive to cytokines, lipids, and growth factors [7, 28] that, upon activation, lead to phosphorylation of intracellular pathways and/or gating of ion channels, resulting in increased action potential generation and pain sensitivity [82 - 84]. As stated above, if regulatory mechanisms are in place, and inflammation is resolved, tissue swelling, cell infiltration and pain decline and disappear. But if inflammation persists, the constant release of neurotransmitters and neuromodulators by primary afferent neurons and other cellular types within the spinal cord (see below) causes hyperactivity of second order neurons, central sensitization and chronic nociceptive pain [7].

Significantly, the immune system has been identified as a key factor also in the occurrence of neuropathic pain [80]. In fact, immediately after peripheral nerve injury, resident immune cells and injured tissue cells release neuroactive mediators such as alarmins (a subset of endogenous DAMPS) [85] and a variety of cytokines and chemokines with the capacity to activate injured axons and surrounding Schwann cells, ultimately inducing the recruitment of neutrophils. In hours to days, monocyte-derived macrophages will also infiltrate the damaged nerve; T cells usually arrive days to weeks post-injury, first infiltrating the site of injury and distal part of the nerve, but later also DRGs and the dorsal horn of the spinal cord (see [86] for further details). Altogether, these observations highlight the importance of the inflammatory process unfolding after tissue or nerve injury, and the relevance of the immune system as potential target of new drugs against a large variety of chronic pain conditions (see also [87]) (Fig. **1**; Table **1**).

Table 1. A number of secreted products by immune cells in peripheral tissues, the dorsal root ganglion and the spinal cord during inflammatory or neuropathic pain processes are listed (in red, pro-inflammatory/nociceptive secreted products; in blue, anti-inflammatory/nociceptive secreted products). The interaction of these different proteins with nociceptors is the basis for neuroimmune interactions conditioning the occurrence of nociceptor sensitization and pain (Based on [82 - 84, 88]).

	Source	Secreted products	Examples
Periphery	Immune cells at large	ROS, H+, NO, cytokines, prostaglandins, growth factors, proteases, pain mediators, chemokines, SPM	IL-1β, IL-6, NGF, TNF-α, PGE2, ATP, NO, ROS, histamine, bradykinin, prostanoids, IL-10, LE, β-END, SPM
	Endothelial cells	Kinins	Bradykinin
	Schwann cells	Cytokines	TNF-α, IL-1β, IL-10
	Keratinocytes	Prostaglandins, endothelins	PGE2, ET, ATP, β-END
Dorsal Root Ganglion	Macrophages	Cytokines, SPM	TNF-α, IL-1β, IL-10, SPM
	Satellite glial cells	Cytokines	TNFα, IL-1β, ATP
	T-cells	Cytokines, LE	LE, IL-17, IL-10, IL-4
Spinal cord	Nociceptor central nerve endings	Cytokines, neuropeptides	Casp6, ATP, CCL2, TNF-α, CSF-1, CGRP, SP
	Astrocytes	Cytokines, chemokines, glutamate	IL-1β, IL-6, TNF-α, PGE2, NO, CCL2, CXCL1, MMP2, TSP4, Glu, IFN-α, CX3CL1
	Microglia	Cytokines, NO, ROS, neuropeptides, chemokines, prostaglandins	TNF-α, IFN-γ, NO, ROS, IL-18, IL-23, IL-1β, IL-6, PGE2, IGF-1, VEGF, PDGF, BDNF, IL-10, IL-4, TGF-β
	Oligodendrocytes	Cytokines	IL-33
	T-cells	Cytokines	IFN-γ, IL-10, IL-4

ATP, Adenosine triphosphate; BDNF, Brain-derived neurotrophic factor; Casp6, Caspase 6; CCL, C-C Motif chemokine ligand; H+, Protons; CGRP, Calcitonin-gene related peptide; IFN, Interferon; CSF-1, Colony stimulating factor 1; IL, Interleukin; CXCL, C-X-C Motif chemokine ligand; NGF, Nerve growth factor; Glu, Glutamate; NO, Nitric Oxide; LE, Leukocyte elastase; PGE2, Prostaglandin E2; MMP2, Matrix metalloproteinase-2; ROS, Reactive oxygen species; PDGF, Platelet-derived growth factor; SPM, Specialized pro-resolving lipid mediators; SP, Substance P; TNF, Tumor necrosis factor; TSP4, Thrombospondin-4; β-END, beta-endorphin; VEGF, Vascular endothelial growth factor.

Virtually all types of immune cells have been ascribed a pro-nociceptive role in inflammatory and neuropathic pain, including granulocytes, macrophages, T cells and even mast cells (see [80, 84, 89] for extensive revision). In their activated state, these cells acquire the capacity to release a number of key pro-inflammatory mediators: TNF-α, IL-1β, IL-6, IL-17, IFN-γ, histamine, serotonin, reactive oxygen species and prostaglandins [80, 84, 89] Table **1**. These different mediators are known to sensitize primary afferent neurons, through activation of ion channels directly involved in pain transmission, such as the transient receptor potential vanilloid one (TRPV1), the transient receptor potential ankyrin 1 (TRPA1), and the sodium channels Nav1.7-1.9 [7, 90].

As already advanced, immune cells go through phenotypical changes that, depending on the stage of inflammation and the type of environment that surrounds them, may acquire an pro-inflammatory or an anti-inflammatory phenotype, as is the case for macrophages [91]. In fact, monocytes and macrophages not only are able to typically acquire an M1 (pro-inflammatory) phenotype, they can also promote analgesia and insult resolution when they turn into an M2 (anti-inflammatory) phenotype [49], exhibiting low levels of immune activators such as TNF-α, IL-1β, IL-6, IL-12, IL-23, CD86 and major histocompatibility complex-II, and releasing high levels of anti-inflammatory mediators such as IL-10, IL-4, TGF-β and specialized pro-resolving lipid mediators [26, 90, 92] Table **1**. This explains, contrary to expectations, why transgenic [49] or chemical [49] depletion of monocytes and macrophages results in significant increases of mechanical and thermal hypersensitivity, and delays resolution of pain in models of inflammatory pain.

CD4$^+$ T cells (also called T-helper cells) have also been shown to exhibit pro- and anti-inflammatory phenotypes influencing neuropathic pain. This was shown in pioneer studies using T cell-deficient animals (athymic rats) showing reduced thermal and mechanical pain hypersensitivity compared to WT mice [93]. Importantly, reconstitution with Th1 cells (pro-inflammatory T-helper) has been shown to restore pain behavior, while reconstitution with Th2 cells (anti-inflammatory) further reduces thermal pain sensitivity [93]. CD8$^+$ T cells can also be differentiated into cytotoxic (CTL) or suppressor/regulatory T cells (T-reg) [91], respectively exhibiting cytotoxic capacity, as well as protective and analgesic roles. Interestingly, their protective actions include the attenuation of the clonal expansion of activated CD4$^+$ T cells [94], the secretion of IL-10 [95], and the promotion of reductions in the number of pro-inflammatory Th17$^+$ T cells [96], ultimately supporting anti-inflammatory and anti-nociceptive roles. Likewise, CD8$^+$ T cell KO mice with hind paw inflammation have shown an impaired recovery from mechanical allodynia and thermal hyperalgesia, together with a significant reduction in the hind paw content of the opioid peptide Met-enkephalin [97].

Finally, it has been suggested that, in response to injury, T cells migrate into the spinal cord to reach the cerebral spinal fluid (CSF) as they infiltrate the dorsal root leptomeninges [98, 99]. Importantly, T cell infiltration into the spinal cord has been linked to hypersensitivity, as it has been demonstrated in recombination activating protein (Rag)-1 T-cells deficient mice with spared nerve injury [100]. Accordingly, the intrathecal injection of CTL cells has been shown to worsen pain hypersensitivity, while the intrathecal injection of T-reg cells in mice with chemotherapy-induced neuropathic pain results in reduced mechanical allodynia [59].

In conclusion, great progress has been made to demonstrating the critical role of immune cells in the pathogenesis and resolution of chronic pain. These non-neuronal cells can interact with nociceptive neurons, modulating their excitability by releasing neuroactive pro- and anti-inflammatory mediators. It should be noted, however, that the cells addressed here are not the only ones with the capacity to interact with nociceptors, and thus keratinocytes, Schwann cells, satellite glial cells (SCGs) and oligodendrocytes should also be taken into consideration (see [62, 101] for further review).

Glial Cells

Activation of glial cells and neuro-glial interactions have emerged as key mechanisms underlying pathological and chronic pain [62, 102, 103]. Today it is widely recognized that neurons are not the only cells involved in the generation of persistent pain. Glial cells have well documented roles in pain facilitation: they can modulate neuronal synaptic function and neuronal excitability by various mechanisms [62, 88, 104]. Activated glial cells are capable of producing and releasing proalgesic factors that can modulate the activity of both neurons and other glial cells thus impacting on the generation and / or maintenance of pain [88, 104, 105].

A vast amount of pre-clinical evidence has implicated 3 types of glial cells in the development and maintenance of chronic pain: microglia and astrocytes in the CNS, and SGCs in DRGs and cranial ganglia [62, 102]. Although these cells have unique roles in the modulation of neuronal function, they also participate in some overlapping functions such as the neuroimmune response. Once activated, these cells respond to and release a number of signaling molecules, including various pro-inflammatory cytokines, chemokines and several other mediators which have protective and/or pathological functions [88, 103].

In normal conditions, astrocytes, the most abundant non-neuronal cells in the CNS, fulfill various relevant functions for the maintenance of local homeostasis, since they are part of the blood brain barrier (BBB), regulate blood flow, maintain ionic and fluid balance and influence synaptic plasticity [106, 107]. Astrocytes also express different glutamate transporters such as the glutamate transporter 1 and the glutamate aspartate transporter, thus mediating glutamate uptake and regulating excitatory synaptic transmission [106, 107].

Astrocytes communicate with each other, with other glial cells and with neighboring neurons through gap junctions. Gap junction communication in astrocyte-coupled networks is mediated by homo- and heteromeric associations of hemichannels, such as connexin-43 (Cx43), the predominant connexin expressed in astrocytes [106, 107]. Among other functions, gap junctions facilitate intercellular transmission of Ca2+ signaling and exchange of cytosolic contents.

Furthermore, astrocytes enwrap neuronal somata and synapses, supporting and nourishing neurons, and regulating the external chemical environment during synaptic transmission [106]. Since astrocyte branches can make close contact with many different neurons and/or microglial cells, enabling astrocyte-neuron and astrocyte-microglia communication, their activation is key to initiating multicellular responses [102, 108]. Astrocytes express various functional neurotransmitter receptors such as ionotropic and metabotropic glutamate, purinergic and substance P receptors and therefore, the close astrocyte–neuron contact allows for astrocyte activation by neurotransmission [106].

After persistent noxious stimulation or different types of nervous system injuries, astrocytes become activated, triggering the process of reactive gliosis, which is characterized by a series of significant morphological and functional adaptations, including increased expression of glial fibrillary acidic protein (GFAP) and production of cytokines and chemokines [106, 107]. Such activation can be beneficial or harmful according to its duration and characteristics [107]. Reactive astrogliosis can increase neuroprotection and nutritional support for damaged neurons. In fact, one of the consequences of the reactive astrogliosis process is the formation of the glial scar, which allows the inflammatory mediators to be isolated at the site of the injury and prevents them from spreading to intact tissue, minimizing the extent of secondary damage [107]. Furthermore, activated astrocytes can reconstruct the damaged BBB. Thus, astrogliosis is an initial defense mechanism for repairing damage. However, astrogliosis can also be deleterious, *i.e.* activated astrocytes may encourage the development and maintenance of chronic pain by releasing proalgesic signaling molecules [106, 107].

Similar to astrocytes, SGCs, one prominent glial cell-type in the PNS, also express GFAP and form gap junctions [109]. Unlike astrocytes, each SGC contacts only one neuron, with a close interaction that enables effective intercellular signaling [109]. SGCs are activated after peripheral nerve injuries and/or peripheral inflammatory states and play an active role in the development of persistent pain [102].

Upon astrocyte/SGCs activation, the extracellular signal-regulated kinase (ERK) and c-Jun N-terminal kinase (JNK) signaling pathways are activated leading to an increase in the synthesis of pro-inflammatory factors (*e.g.* for example, IL-1 β, IL-6, TNF α, prostaglandin E2 (PGE2) and nitric oxide (NO)), which ultimately alter glial glutamate transporter function and gap-junction proteins, and also modulate neuronal activity [103, 107] (Table **1**. Although similar pathways are activated in microglia after injury, the temporal patterns of enzyme activation and pro-inflammatory cytokine release are distinct for astrocytes and microglia [106].

Microglial cells represent between 10 and 20% of the total number of cells in the CNS, where they are responsible for orchestrating and carrying out the innate immune response [110, 111]. They derive from the mononuclear phagocytic lineage, originating from monocytes derived from the bone marrow during perinatal development. Before the closure of the BBB, these cells migrate and distribute throughout the CNS [110].

Under physiological conditions, microglial cells are in a permanent "state of vigilance" and present a morphology characterized by a small soma and numerous thin processes that facilitate their interaction with neurons and astrocytes and allow them to census the microenvironment to maintain homeostasis [111, 112].

Following an injury affecting the nervous system or during a persistent peripheral inflammatory state, microglial cells respond by changing their morphology (a retraction of their processes and an increase in the size of the soma are observed), changing their gene expression profile and releasing pro-inflammatory molecules such as cytokines and chemokines [112, 113] Table **1**. Dysregulation of microglial activity has been associated with the development of chronic pain following injury to the peripheral nerve [112], the spinal cord [114], or peripheral inflammation [115].

In recent years, numerous studies have proposed that following activation, microglia responds by polarizing within an activation spectrum compatible with the M1 and M2 phenotypes already described for macrophages (see above) [116, 117]. Stimulation by DAMPs, free radicals or pro-inflammatory cytokines induces the "classical" activation phenotype or M1, associated with the expression of cell surface markers such as CD86, the secretion of pro-inflammatory cytokines such as IL-1β, the expression of the inducible form of nitric oxide synthase (iNOS), the production of ROS and the decrease in the secretion of neurotrophic factors [116, 117]. Although M1 pro-inflammatory microglial cells appear to be harmful, a regulated or controlled M1 response could be neuroprotective after a nervous system injury. Conversely, an exacerbated or prolonged M1 response can lead to secondary damage [116, 117].

The activated "alternative" phenotype or M2 is characterized by the expression of the mannose receptor CD206 and the Arginase 1 enzyme, and the secretion of anti-inflammatory cytokines and mediators such as IL-4 and TGF-β1 [116]. There are 3 variants within this phenotype: an M2a phenotype is generated in response to IL-14 and IL-13, is associated with the production of anti-inflammatory cytokines and increased phagocytic activity; and M2c phenotype is acquired upon exposure to IL-10, glucocorticoids, or apoptotic cells, and regulates tissue repair and modeling; an M2b phenotype is considered intermediate, is stimulated by exposure to immune complexes or TLR ligands and has both pro- and anti-inflammatory effects (secretion of IL-1β, IL-6, TNFα and IL-10) [117].

The existence of these divergent responses determines whether microglial activation leads to tissue clearance and resolution of the inflammatory response on the one hand, or chronic neuroinflammation on the other. The clearance mechanisms seek to eliminate the cellular debris generated, as a consequence of the injury or the subsequent inflammatory response, leaving the neural network intact, while chronic neuroinflammation can cause neuronal dysfunction [103, 106, 117] and lead to persistent pain.

Interestingly, reactive astrocytes have been also recently classified into A1 astrocytes and A2 astrocytes according to their functions [107]. After nervous system injury, A1 astrocytes can secrete neurotoxins that induce rapid death of neurons and oligodendrocytes, whereas A2 astrocytes promote neuronal survival and tissue repair [107]. These findings can well explain the previously mentioned dual effects of reactive astrocytes in central nervous system injury and diseases.

As previously referenced, SGCs, astrocytes and microglial cells are activated after an injury and trigger the process of neuroinflammation. Glial cells undergo a broad spectrum of changes in function and morphology regulated by transcriptional, translational and post-translational mechanisms, leading to proliferation, morphological changes (eg, hypertrophy), upregulation of glial markers (CD11b, IBA1, GFAP) and production of pro- and anti-inflammatory mediators [62, 102, 103]. Thus, upon activation, these cells upregulate the expression of receptors and ion channels such as TLRs and purinergic P2X and P2Y receptors, and release neurotransmitters, cytokines, and ROS, mediators that contribute to neuronal hyperexcitability and neuropathic behavior, though a neuron-glia-neuron bidirectional communication [102, 118, 119].

Among the glial mediators released within the CNS, a special emphasis has been placed on the pro-inflammatory cytokines IL-1β and TNFα, possibly for their involvement in a plethora of CNS diseases and neurotoxic conditions [108, 120]. These cytokines also facilitate pain *via* neural-glia interactions [62, 102, 121]. In particular, IL-1β plays a pivotal role in pain mechanisms. Since IL-1β functional receptor (IL-1RI) is not only found in glial cells but is also located in neurons [122], IL-1β may act directly on neurons to modulate their activity. In fact, IL-1β has been shown to enhance synaptic transmission and neuronal activity in the superficial dorsal horn [121, 123] and IL-1RI acts as a coordinating factor for the functional interaction between IL-1β and NMDA receptor [124, 125], a key player in pain transmission. It has been shown that IL-1β, *via* IL-1RI signaling, modulates glutamatergic response and facilitates NMDA receptor activation by increasing NMDA-subunit phosphorylation [119], enhancing the NMDA receptor-induced Ca2+ influx and modulating NMDA receptor expression and membrane distribution in neurons [122, 126, 127]. In this way, cytokines such as IL-1β behave as key signaling molecules between glia and neurons, capable of modulating neuronal function [120, 121] and enhancing pain transmission [62, 102].

In conclusion, accumulating evidence suggests that non-neuronal cells in the nervous system such as SGCs, astrocytes and microglial cells play active roles in the pathogenesis and resolution of pain [62, 107], and this would happen through secretion of neuroactive signaling molecules that are able to modulate pain [62, 102].

Mesenchymal Stem Cells

Mesenchymal stem cells are described as a heterogeneous population of multipotent cells, which are widely distributed in various tissues, such as bone marrow, adipose tissue and umbilical cord. MSCs can differentiate *in vitro* into various cells of the mesodermal lineage which include osteoblasts, adipocytes and chondrocytes and express specific surface antigens, such as CD105, CD73 and CD90. Additionally, these cells lack expression (<2% positive) of CD45, CD34, CD11b, CD19 and HLA class II which set them apart from hematopoietic stem cells (HSC) [128, 129].

Because of their high expansion potential, genetic stability, stable phenotype, and strong immunosuppressive properties, MSCs can be exploited for successful autologous and heterologous transplantation, without requiring immune suppressants. Initially, MSCs were mostly considered with respect to in their potential to reconstruct damaged or diseased tissues, due to their differentiation competence [130]. However, recent studies have shown that MSCs can affect a variety of physiological and pathological processes, including immune and inflammatory responses, by releasing cytokines, chemokines and trophic factors [50]. Moreover, the ability of MSCs to alter the inflammatory *milieu* has made them an attractive therapeutic strategy for treatment of various painful states such as inflammatory pain [131 - 133], neuropathic pain [51, 134 - 136] and cancer pain [137].

One important mechanism that is essential for MSCs therapy is their homing capacity, which is proving to be relevant in the modulation of neuropathic pain. Thus, we showed, using the single ligature nerve constriction (SLNC) model [138], that the intraganglionic injection of MSCs into the ipsilateral L4 DRG results in their selective migration and engraftment exclusively into injured DRGs (ipsilateral L4-6), where they acquire a striking perineuronal localization, resembling glial/satellite cells [53]. Moreover, when MSCs were injected into the non-affected (contralateral) L4 DRG, MSCs migrated specifically to the ipsilateral L4-6 DRGs, showing a selective migratory tropism for the injured ganglia [53]. Importantly, intraganglionic administration of MSCs also resulted in the prevention of both mechanical and cold allodynia in rats with SLNC [51], and an attenuation of the injury-induced changes in galanin, neuropeptide tyrosine (NPY) and NPY Y1 receptor expression [52].

Research over the last few decades has revealed the involvement of a number of mediators and receptors in the homing behavior of MSCs. Peripheral nerve injury is known to lead to the release of stromal cell derived factor-1 (SDF-1), also known as CXCL12, the monocyte chemoattractant protein-1a (MCP-1a) and the C-X-C motif chemokine ligand 13 (CXCL13). Acting through the C-X-C chemokine receptor type 4 (CXCR4), the C-C chemokine receptor type 2 (CCR2) and the CXCR5, respectively, these ligands induce engrafting of circulating MSCs. Particularly the CXCL12/CXCR4 axis has become the focus of attention as an emerging neuromodulator in pathological pain [139, 140]. The CXCL12/CXCR4 axis is widely distributed on nociceptive structures in the PNS and CNS. In uninjured DRGs, CXCL12 and CXCR4 share similar expression patterns, are predominantly localized in small- and medium-diameter neurons of either peptidergic (calcitonin gene-related peptide (CGRP)-containing) or non-peptidergic phenotype and can also be found in their peripheral axonal projections [141]. CXCR4 expression has also been demonstrated in Schwann cells in the sciatic nerve [142]. Importantly, chronic constriction injury (CCI) of the sciatic nerve causes a marked increase in CXCL12 levels in L4–L6 DRGs, where it attracts CXCR4-expressing MSCs [143]. Moreover, intrathecal administration of CXCR4-deficient MSCs was found not only to block the migration of MSCs to DRGs, but also to reduce the duration of MSC-induced analgesia [143].

The potential therapeutic benefit of exogenous MSCs is currently under study in a number of clinical trials addressing diverse clinical conditions, such as graft *versus* host disease [144], acute myocardial infarct [145], cut and contusion of the spinal cord [146], burns and wound healing [147], rheumatoid arthritis and lupus

[148], amyotrophic lateral sclerosis [149] and sepsis [150]. Interestingly, almost all preclinical and clinical studies suggest that rather than differentiating into phenotypic lineages related to the type of tissue sought to be restored, MSCs appear to improve the situation of the damaged tissues in two manners that involve paracrine actions: trophism and inmmunomodulation [151].

In fact, MSCs have an important role in sensing and switching immune responses. They can stimulate inflammation when the immune system is underactivated or restrain inflammation while the immune system is overactivated to avoid self-over attack [152]. The polarization process of MSCs toward either an anti-inflammatory or a pro-inflammatory phenotype is mediated by TLRs. MSCs express TLR2, TLR3, TLR4, TLR7 and TLR9. The activation of TLR3 induces an anti-inflammatory phenotype of MSCs, through the induction of anti-inflammatory cytokines expression, such as IL4, indoleamine 2,3-dioxygenase (IDO) or PGE2, while the activation of TLR4 induces a pro-inflammatory phenotype through upregulation of pro-inflammatory factors such as IL-6 or IL-8.

The MSC-dependent immunomodulation has been shown to influence a variety of immune cell types. Thus, MSCs suppress T cell proliferation and also significantly reduce the expression of activation markers including CD25, CD38, and CD69 on Phytohemagglutinin (PHA)-stimulated lymphocytes. They also suppress proliferation of both $CD4^+$ and $CD8^+$ T cells and are able to abrogate the response of memory T cells to their activating antigen [153]. Also, it has been demonstrated that MSCs inhibit B cell differentiation, proliferation, activation and IgG secretion [154, 155]. Interestingly, the modulation of monocytes by MSCs appears to be a critical step in their immunomodulatory mechanisms. Thus, MSCs have been shown to induce M2 macrophage polarization, through TGF-β and IL-10 pathways. In addition, MSCs have also been shown to inhibit differentiation of monocytes into dendritic cells and alter the cytokine secretion profile of dendritic cells toward upregulation of IL-10 and downregulation of IFN-γ, IL-12, and TNF-α, inducing a more tolerant dendritic cell phenotype [156]. Finally, and highlighting the relevance of the MSCs-monocyte interaction and its immunosuppressive effects, monocyte (but not B cell) depletion from human peripheral blood mononuclear cells reduced the immunosuppressive effects of MSCs over T cell proliferation [157].

Several are the immune modulators expressed and secreted by MSCs, including IDO, PGE2, inducible iNOS, TGF-β, IL-10, hepatocyte growth factor (HGF), histocompatibility locus antigen-G (HLA-G), CD39 and CD73, galectins, C-C motif chemokine ligand 2 (CCL2), programmed cell death ligands 1 and 2 (PD-L1 and PD-L2), heme oxygenase 1 (HO-1), tumor necrosis factor-stimulated gene 6 (TSG6), interleukin-1 receptor antagonist (IL1RA) and complement system–related factors. However, TGF-β1 and IL-10 are currently the center of attention as main immune-regulatory cytokines produced by quiescent MSCs and further upregulated by inflammatory factors.

On the one hand, TGF-β1 controls the proliferation of neurons and regulates neuronal survival, blocks microglial proliferation and has immunosuppressive effects on astroglia. Several groups focused on its spinal action have demonstrated that intrathecal infusion of recombinant TGF-β1 not only prevents the development of neuropathic pain following nerve injury, but also reverses previously established neuropathic pain conditions in rats [158, 159]. TGF-β1 inhibits peripheral nerve injury-induced spinal microglial and astrocytic activation. Also, it prevents the induction of ATF3 in neurons following nerve ligation and suppresses nerve injury-induced inflammatory response in the spinal cord [159]. Importantly, intrathecal administration of MSCs reduced CCI-induced neuropathic pain and axonal injury of DRG neurons and inhibited neuroinflammation in DRGs and spinal cord *via* TGF-β1 secretion [143].

On the other hand, IL-10 actions include the decrease in nuclear factor kappa B (NFκB) activity, resulting in the attenuation of the synthesis of pro-inflammatory cytokines such as TNF-α and IL-1β [160]. Accordingly, the endoneural injection of IL-10 into the sciatic nerve significantly attenuates (but does not prevent) thermal hyperalgesia following CCI injury in rats [161]. Moreover, intrathecal IL-10 gene therapy prevented and progressively reversed paclitaxel-induced allodynia, upregulating IL-10 mRNA levels in lumbar DRGs and meninges, and downregulating the expression of IL-1β, TNF-α and CD11b mRNAs in lumbar DRGs [162]. Interestingly, in mice with partial sciatic nerve ligation (PSL), MSCs transplantation or treatment using conditioned medium obtained from their own MSCs induced a robust and long-lasting anti-nociceptive effect, along with downregulated expression of IL-1β, TNF-α and IL-6, and increased levels of IL-10 in the sciatic nerve and the L4-5 spinal cord segments [163]. Finally, supporting the role of TGF-β1 and IL-10 in the mechanisms of action of MSCs, a study in rats with L5 spinal nerve ligation revealed that intrathecal administration of IL-1β pre-treated MSCs abolishes microglial activation and neuropathic pain not only by releasing TGF-β1, but also by decreasing CCL7 level in spinal cord, in a process likely mediated through the release of IL-10 [164].

In summary, MSC transplantation induces a long-lasting antinociceptive effect in models of peripheral neuropathic pain in rats and mice, in association with modulation of the immune and neurochemical reactions to injury. These MSC dependent actions appear to relate to their capacity to migrate to injured tissue, where they function as biological "pumps", releasing antinociceptive molecules or trophic factors that act at the site of injury or at pain processing centers, mediating therapeutic effects.

IMMUNOMODULATORY SYNTHETIC ODNS

As it may clearly appear so far, chronic pain is a complex biological entity, involving a wide variety of mechanisms that include a diverse array of non-neuronal cells that provide a rich range of interactions with neurons, important alterations during injury that ultimately translate into sensitization and pain. Such complexity calls for new and innovative analgesic drugs, and in view of the present discussion, with the capacity to multitarget a variety of non-neuronal cells involved in neuroimmune interactions critically relevant in pathological pain transmission.

Chemically, oligonucleotide drugs fall into two categories: double-stranded RNA (dsRNA)-based drugs and single-stranded DNA (ssDNA)-based drugs. Double-stranded RNA, are employed to modulate RNA function in cells with different therapeutic strategies, which include small interfering RNAs (siRNAs), gene therapy, genome editing, delivery of exogenously expressed mRNAs, and synthetic antisense oligonucleotides (ASOs). On the other hand, ssDNA have a completely different mechanism of action, by modulating the immune system. In general, they bind to TLRs or form aptamers [165].

A major breakthrough in the development of oligonucleotide drugs came with the introduction of the phosphorothioate (PS) backbone modification [166], which replaces one oxygen in the phosphate residue with sulfur. PS modification increases nuclease resistance and facilitates association with carrier proteins in the blood, leading to increased absorption, slower excretion through the kidneys and longer half-life [167].

Another key discovery in the field was the identification of polyinosinic: polycytidylic acid (poly I:C) as an inducer of interferon (IFN) [168]. The biological basis for this observation was later understood when TLR3 was shown to be the receptor for dsRNA [169]. Related to these findings was the notion that bacterial DNA has an antitumoral effect [170]. Then, Krieg *et al.* [171] demonstrated that CpG motifs in bacterial DNA trigger the activation of B-cells. That finding was followed by the identification of its receptor, named TLR9, and the corresponding intracellular signaling pathways [172, 173].

So far, two kinds of synthetic immunomodulatory oligodeoxynucleotides (ODNs) have been identified according to their CpG content: CpG and non-CpG ODNs. CpG ODNs are characterized by the presence of at least one active site that bears an unmethylated cytosine-guanine (CpG) dinucleotide in a given context [174]. The hallmark of CpG ODNs is their ability to induce secretion of IFN-α by plasmacytoid dendritic cells (pDCs) interacting with the endosome associated TLR9 [175, 176]. Nonetheless, both CpG and non-CpG ODNs can also directly co-stimulate mouse and human cells through TLR9-independent mechanisms [177].

Significantly, we previously identified a new family of non-CpG ODNs named PyNTTTTGT after its active motif, in which Py is pyrimidine [cytosine (C) or thymine (T)] and N is adenine (A), T, C, or guanine (G) [178, 179]. CpG and PyNTTTTGT ODNs share some common features but also have remarkable differences. *In vitro*, both kinds of ODNs act on B cells and pDCs, causing activation, proliferation, immunoglobulin secretion, and expression of costimulatory molecules, respectively. However, CpG ODNs induce the secretion of IFNα [174], while PS PyNTTTTGT ODNs do not [178]. On the other hand, PyNTTTTGT ODNs, but not CpG ODNs, increase the number of adult bone marrow (BM) MSC precursors *in vivo* and *in vitro* [180]. Furthermore, in the presence of IL-2, PyNTTTTGT ODNs induce the secretion of granulocyte macrophage-colony stimulating factor (GM-CSF) by NK and NK-T cells [181]. Interestingly, IFN-α inhibits the PyNTTTTGT ODNs dependent secretion of GM-CSF, and conversely, PyNTTTTGT ODNs inhibit the TLR9-dependent secretion of IFN-α by CpG-stimulated PDCs. The mutual interference between these major classes of ODNs suggests that they may stimulate different and incompatible immune response pathways [181].

During the characterization of the PyNTTTTGT ODNs, we formerly reported that PS ODN IMT504 (5'-TCATCATTTGTCATTTGTCATT-3'), the prototype of this family, has an immune stimulating profile by acting on peripheral blood mononuclear cells, PDCs, B lymphocytes [178, 179] and NK-T cells [181], with potential to be used as adjuvant for vaccines [179, 182 - 185] and cancer [186]. However, it has also been shown that IMT504 acts as an immune modulator in chronic toxic and auto-immune models of Type 1 Diabetes (T1D) by reducing the islet infiltration and restoring beta cell function [187 - 189]. Also, ODN IMT504 prevents deaths in neutropenic rats undergoing fatal acute bacteremia and sepsis, in association with reductions in the plasmatic levels of the pro-inflammatory IL-6 [190].

As mentioned above, we also reported that the subcutaneous injection of IMT504 induces the expansion of adult MSC both in BM and peripheral blood in rats [180]. It is worth noting that MSCs are known to interrelate with constituents of the immune system, exhibiting anti-inflammatory or pro-inflammatory properties depending on the *milieu* composition [191 - 193]. In general, MSCs first adopt a pro-inflammatory phenotype (MSC1) during early microbial invasion or trauma, when the concentration of pro-inflammatory cytokines in the *milieu* is relatively low. As inflammation proceeds, pro-inflammatory cytokines accumulate up to a critical level that switches differentiation of MSCs to an anti-inflammatory phenotype (MSC2). Based on these facts, we hypothesized that IMT504 induced a switch towards anti-inflammatory states involving MSCs and potentially also immune cells, mimicking a DAMP potentially recognized by a cytosolic DNA sensor [183], and thus facilitating reconstructive responses [194]. We also hypothesized that the IMT504-dependent expansion of MSC observed in peripheral blood could play a role in the modulation of chronic pain.

This latter hypothesis, and several other novel findings on the mechanisms of IMT504, will be addressed in the following section.

IMT504 – THE EMERGENCE OF A NON-CPG ODN WITH ANTI-NOCICEPTIVE PROPERTIES THROUGH MODULATION OF NON-NEURONAL CELLS

Modulation of Mesenchymal Stem Cells

The first evidence of the anti-nociceptive role of IMT504 came from a study where we employed a daily subcutaneous administration of the ODN during 5 days in rats with unilateral sciatic nerve crush [195]. We followed two approaches: an early treatment protocol, where administration begun on the same day of injury, and a delayed treatment protocol, where rats were injected starting on day 4 after injury. In addition, we tested two different IMT504 doses in the delayed treatment protocol, namely 5 or 20 mg/kg *per* dose (in a constant volume of 250-300 µl *per* dose, depending on individual weight at the time of injection). In rats with sciatic nerve crush alone or receiving saline or phosphate-buffered saline (PBS) as vehicle (used as controls), a progressive increase in mechanical and cold allodynia was evident, with a peak within the first week after injury, a progressive recovery starting on day 14 after injury, reaching basal levels 21 days after the insult. In contrast, the early administration of IMT504 virtually prevented mechanical allodynia and strongly reduced cold allodynia. Moreover, rats on the delayed treatment protocol, which by day 4 after injury had developed clear signs of mechanical and cold allodynia, showed a quick return to basal withdrawal thresholds as soon as the treatment with IMT504 ended (day 8 after injury), and reaching basal values by day 11 after the insult; these effects were observed both when using 20 or 5 mg/kg doses. Remarkably, a single intravenous administration of MSCs on the day of sciatic nerve crushing in a parallel group of rats resulted in virtually identical anti-allodynic effects as observed in injured rats receiving IMT504 on an early treatment protocol. This observation suggested two main hypotheses: 1) a potential role of MSCs in the mechanisms of anti-allodynic action of IMT504, including the promotion of their proliferation and migration towards injured nerves, and 2) the possibility of IMT504-dependent influences on the immune-chemical environment at the site of nerve damage [195].

Current research in our laboratory is providing strong support to both hypotheses (manuscript in preparation). And in the following sections, we will describe recently published results suggesting that, using a rat model of chronic non-resolving granulomatous inflammatory insult caused by the intraplantar injection of complete Freund′s adjuvant (CFA), IMT504 in fact modulates pain by acting upon non-neuronal cells that include immune and glial cells, and by influencing the inflammatory *milieu* at the site of injury [196 - 198].

Modulation of Immune Cells

The anti-allodynic properties upon systemic administration of IMT504 observed in rats with sciatic nerve crush [195] were put to the test in rats undergoing CFA-induced chronic inflammatory pain [196, 197]. We found a number of novel and interesting features.

First, in a series of behavioral tests characterizing the administration of IMT504 at different concentrations and number of doses (in a constant volume of 250-300 µl *per* dose, depending on individual weight at the time of injection), we determined that a single dose of 6 mg/kg IMT504, 7 days after injury [197], is as effective as 5 consecutive administrations of the ODN at a dose of 20 mg/kg [196] to reduce and eliminate mechanical and cold allodynia. Moreover, the effect observed with only one dose of IMT504 was long-lasting, with rats exhibiting mechanical allodynia-free behavior for up to 6 weeks after treatment; vehicle-treated rats remained allodynic during that time-period. Six weeks after treatment, IMT504 rats showed a return to withdrawal thresholds comparable to vehicle-treated animals. Importantly, we also observed that treatment with IMT504 did not impair nocifensive responses to heat [196, 197] nor intraplantar formalin [197], nor locomotion [197], with no evidence suggesting a sedative effect, and therefore supporting its safety if potentially used as an analgesic drug.

Second, the long-lasting anti-nociceptive effects of multiple [196] or single [197] doses of IMT504 were associated with long-term local resolution of cellular infiltrate and edema at the site of injury. This was confirmed by the observation of decreases in hind paw thickness (at the expense of reductions in the thickness of the dermis), and full resolution of the abundant cellular infiltrate that otherwise remains in rats receiving vehicle [196, 197]. A closer look at the cellular infiltrate revealed that IMT504 modulates both infiltrating myeloid and lymphoid cells, particularly macrophages, B cells and CD8$^+$ T cells. Thus, IMT504 was shown to reduce the number of infiltrating macrophages, as determined 21 days after treatment initiation (this is, 28 days after injury), and to induce important reductions in B cell counts, both 1 and 21 days after treatment. In contrast, IMT504 induced an increased number of CD8$^+$ T cells, also observed at early and late times after treatment, as compared to vehicle-treated animals [197]. It is relevant to mention that while B cells, PDCs [178, 179] and NK-T cells [181] have previously been found to be IMT504 targets, the finding of effects upon macrophages and CD8+ T cells establishes newly found potential cell targets of the ODN.

Third, IMT504 not only influenced cellular infiltration and edema, it also profoundly modulated the inflammatory *milieu* typically participating in the pathophysiology of inflammatory pain [84, 89]. Thus, already 1 day after treatment, leukocyte adhesion molecules such as the soluble intercellular adhesion molecule 1 (sICAM1) and CD62L, leukocyte chemoattractants including CXCL5, CXCL7 and CCL5, and TIMP-1, a matrix metalloproteinase inhibitor involved in tissue protection, were all found to be downregulated when compared to values detected in vehicle-treated rats. These downregulating effect of IMT504 was maintained by 21 days after treatment, extending also to other chemoattractants such as CCL3 and CCL20, the ciliary neurotrophic factor (CNTF), produced by Schwann cells and participating in the modulation of neuroinflammatory processes [199], the pro-inflammatory interleukins IL-1α and –β, and IL1-RA, the natural inhibitor of IL-1β. Conversely, a mild upregulation of the anti-inflammatory interleukin IL-10 was detected 21 days after IMT504 treatment in injured rats. Finally, adding to the molecular actions of IMT504, it was observed that the ODN modulates the protein expression levels of β-endorphin at the inflamed hind paw, in a process that is speculated to derive from the also observed decreased numbers of β-endorphin-secreting immune cells such as macrophages and B cells (see [197]).

Altogether, an immunomodulating effect for IMT504 begins to emerge, influencing the mechanisms of action of immune cells in the innate and adaptive systems during an inflammatory insult, and seemingly favoring a switch away from pro-inflammatory conditions.

Modulation of Glial Cells

The observation of clear and long-lasting effects derived from the systemic administration of IMT504 in rats undergoing sciatic nerve crush [195] and persistent inflammatory pain [196, 197], including profound effects both in pain behavior and the activity of the immune system, led us to two questions: 1) is the activity of IMT504 restricted to actions in the periphery, or can it also be effective upon intrathecal administration; and 2) are glial cells a target of IMT504, particularly microglía, classically described as a modified immune cell in the CNS? In our most recent publication, we provide evidence positively supporting both questions [198].

Answering the first question, intrathecally applied IMT504 in rats with CFA-induced injury revealed that while exhibiting a much shorter-lasting anti-allodynic and anti-hyperalgesic effect (about 48 h), the direct administration of the ODN to the spinal cord is able to block inflammatory pain in a dose-dependent fashion. Accordingly, IMT504 induced a slow but steady downregulation in wind-up responses normally observed in rats with hind paw inflammation. These results: a) proved that IMT504 can exert an anti-nociceptive effect by acting directly on the spinal cord, suggesting the existence of specific targets within the SNC, b) revealed that the peripheral (and not the central) actions of IMT504 are key to inducing the long-lasting effects observed after systemic administration of the ODN, and c) provided further support to the non-neuronal basis of the mechanisms of action of IMT504, since the downregulation in mechanical allodynia and hyperalgesia, and of wind-up responses was considerably slow (full and statistically significant wind-up depression took 90 min to develop) [195 - 197].

Answering the second question, we identified microglia and astrocytes as spinal targets of IMT504 [198]. Thus, injured rats receiving intrathecal IMT504 exhibited reduced reactive spinal astrogliosis and microgliosis, as evidenced by considerable reductions in the immunohistochemical signal of GFAP and CD11b/c, respectively. Further morphometric analysis of glial cells showed that the IMT504-induced reductions in GFAP and CD11b/c in injured rats were at the expense of decreases in the number of resting astroglía, a virtual depletion of hypertrophied astroglia and a pronounced decrease in the number of hypertrophied/amoeboid microglia. These effects on glial cells were also accompanied by important reductions in the spinal protein expression of TLR4 and the phosphorylated p65 subunit of NFκB. Some of these observations were also replicated *in vitro*, in mixed glial cell cultures, where the upregulation of GFAP and TLR4 induced by incubation with LPS was blocked by the co-incubation with IMT504. Interestingly, in these analyses it was also demonstrated that IMT504 is actively incorporated by astroglia and microglia, supporting a previously unknown direct interaction and action on spinal glial cells.

Taken together, this evidence suggests that IMT504 does indeed modulate spinal mechanisms of pain transduction, most likely through modulation of spinal glia and by limiting their conversion to a pro-inflammatory phenotype.

A "NEW HOPE" AGAINST CHRONIC PAIN

It is extremely worrisome to realize that many patients suffering chronic pain remain refractory to treatment, or do not receive proper care. Currently, a rather limited battery of pharmaceutical drugs is available and broadly used to treat a variety of chronic pain conditions. This includes opioids (targeting μ-opioid receptors), NSAIDs and ciclo-oxygenase type 2 (COX2) inhibitors, antiepileptics such as gabapentin or carbamazepine (acting upon voltage-gated sodium channels), anti-depressants such as duloxetine (acting as serotonin-norepinephrin reuptake inhibitors), triptans (acting as serotonin receptor agonists), and more recently, Erenumab, an CGRP receptor antagonist [200]. Topical applications of capsaicin or botulinum toxin are other options currently used, for instance, to treat neuropathic pain [15, 201]. However, even when resorting to more aggressive methods such as surgical corrections or the implantation of peristaltic pumps or neuromodulators, for many patients, this is only a palliative solution to their problem [15, 201]. Moreover, and sadly, results of these different therapeutic approaches will be variable, and when present, serious adverse effects (gastrointestinal, hepatic, renal or neurological alterations) will be cause enough to reject the medication [15, 201].

One aspect that is common for many of the therapeutic agents mentioned above is that they aim at single targets. Naturally, in such scenario, the more the knowledge of the mechanisms of action involved in a particular painful condition, the more the chances to be therapeutically successful; as a counterpart, lacking the specific target leads to therapeutic failure [200]. Also relevant, many of the pain drugs devised so far typically target neurons, when there is a growing understanding of the neuroimmune nature of chronic pain [80, 89]. This concept has not only led to modifying the normally "neuronocentric" vision of pain, it begins to be addressed as an opportunity to develop the next generation of drugs targeting chronic pain.

An additional factor to be considered in this chapter is the escalating opioid crisis affecting the global community. This situation has exposed three critical needs [202]: 1) reversing opioid overdose, 2) treating opioid addiction, and 3) finding safe, new approaches for effective pain relief without the side effects of opioids. Unfortunately, efficacious analgesics offering opioid-like pain relief without associated adverse effects are not available, although efforts are currently underway to develop new opioid receptor agonist-based analgesics with limited tolerance, opioid-induced hyperalgesia, physical dependence, and tolerance [203]. Current examples under experimentation include the use of antibodies against NGF [204, 205], CGRP [206, 207] or pharmacological blockade of sodium channels [208, 209], all of which are presumably devoid of abuse liability.

Finally, in recent years research has focused on female-male differences in the mechanisms of chronic pain. A growing number of differences in genetic, neural and neuroimmune mediations, as well as in cognitive, social and environmental factors influenced by sex are regularly being identified [210] (however, see also [211], where a stronger sex-related impact is placed onto "the peripheral immune system or in more complex central nervous system processes" than on basic nociceptive neuron biology, since no sex-related differences were found in the transcriptional expression of more than 12.000 genes in naïve or injured mouse nociceptors, including select nociception-related genes). Regarding neuroimmune interactions, initially, this was reported for spinal microglía, where their role in neuropathic pain was found to be sex-dependent [212]. However, such differences are increasingly found also in relation to the immune system [210, 211].

Altogether, it is evident that chronic pain is a multifactorial problem, where both neuronal and non-neuronal cells have critical roles and are subject to gender influence in manners that have just begun to be exposed. It is then not unexpected that targeting such a complex problem with single-target approaches is prone to fail. This may not only be important from a pathophysiological point of view, it could also certainly influence differences in the way male and female subjects respond to any given drug.

Certainly, the complexity and wide variety of mechanisms involved in the generation of chronic pain are reasons to explain the difficulty in generating effective analgesic drugs. The understanding of those mechanisms at cellular and molecular levels should provide opportunities for therapeutic interventions, although the pharmaceutical industry has not yet taken advantage of the tremendous amount of new knowledge currently being developed.

In this chapter, we discussed the current evidence regarding the cellular and molecular interactions between primary sensory and dorsal horn neurons, and a variety of non-neuronal cells, including MSCs that might play a crucial role in the development of neuropathic pain. Also, we addressed IMT504 as a strong candidate with multiple targets that efficiently modulates both, inflammatory and neuropathic pain. As of now, MSCs, macrophages, B- and T-cells, as well as on glial cells, both astrocytes and microglía have been identified as IMT504 targets. The list of molecules modulated by IMT504 begins to grow, with impact in intracellular pathways (TLR4 and NFκB) or affecting the inflammatory milieu, altogether inducing a switch from pro-to-anti-inflammatory states. Such a broad (although still emerging) array of actions may be one reason for the high efficacy of this ODN for the control of pathological inflammatory and neuropathic pain, supporting the need for multi-target drugs to efficiently tackle these problems. Also important, IMT504 treatment did not seem to alter regular nocifensive responses, nor induce sedative effects. The latter feature is particularly relevant, as it would support the use of IMT504 to provide long-lasting analgesic effects in chronic pain patients, while avoiding the chances of abuse liability (Fig. **2**).

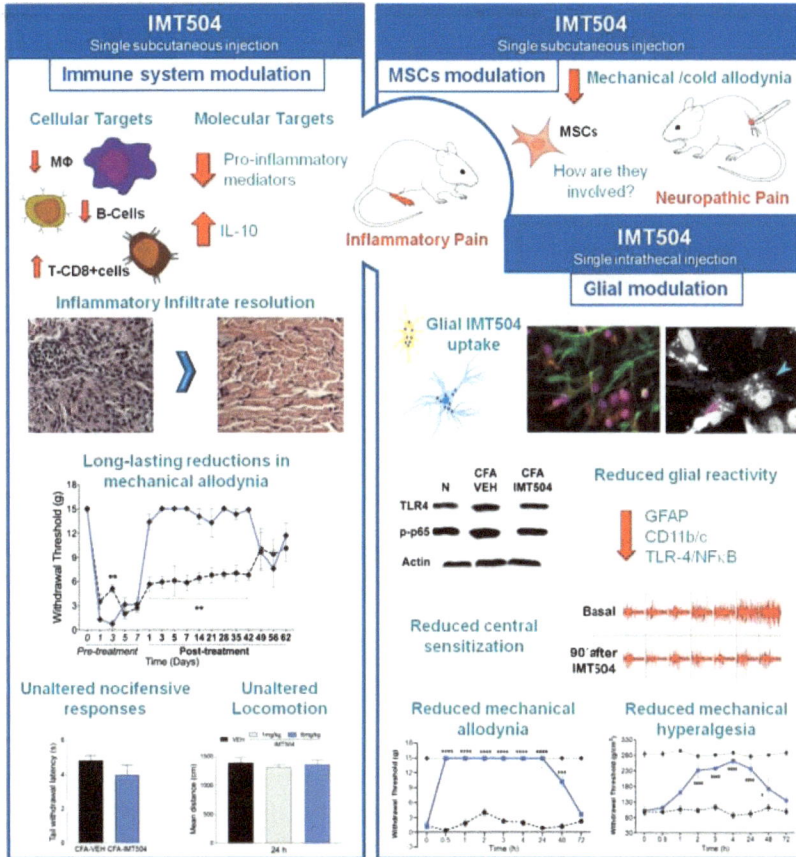

Fig. (2). Summary of known mechanisms of action and anti-nociceptive roles of IMT504 in models of inflammatory and neuropathic pain. In models of inflammatory pain, systemic IMT504 modulates a variety of immune cells, with macrophages and T-CD8+ cells emerging as novel cellular targets. In addition, IMT504 treatment induces modulation of a number of molecular targets, including downregulation of several pro-inflammatory mediators and the upregulation of the anti-inflammatory IL-10. These actions combined lead to tissue inflammation resolution and long-lasting mechanical and cold (not shown) allodynia (blue and black lines denote IMT504- or vehicle-treated rats, respectively, undergoing persistent hindpaw inflammation). Importantly, these effects seem not to affect regular nocifensive responses, neither locomotor activity. Intrathecal administration of IMT504 confirmed the effect upon non-neuronal cells of this ODN (which in addition exhibits intraglial internalization), as shown by reductions in glial reactivity accompanied by the blockade of the upregulation of key components of the TLR4-NFκB pro-inflammatory pathway. These effects also associate with reduced central sensitization and lead to profound, although short-lasting (around 48 h) reductions in mechanical allodynia and hyperalgesia. Finally, systemic administration of IMT504 in rats with neuropathic pain also result in decreased mechanical allodynia, in a manner that implies the action of mesenchymal stem cells (Mφ: macrophages; p-p65: phosphorylated p65 subunit of NFκB) (Images and information were modified and organized with permission from [195, 197, 198]).

However, and despite these recent advances, the evidence on the mechanisms of action of IMT504 is still incipient and many questions remain. What is the sub-cellular target of IMT504? Is the ODN acting with the same efficacy in female rats, as observed in male rats? How do other types of pain conditions, such as antineoplastic-derived neuropathy, diabetes, or visceral pain, react to IMT504? Is IMT504 modulating the phenotype of MSCs and cells in the immune system, and how? Is there a "one cell rules them all" being modulated IMT504 (perhaps the MSCs)? Or is there a combination of IMT504-dependent actions upon different cellular types that determines the anti-nociceptive actions of IMT504? Is the action of IMT504 only based on the modulation of neuroimmune interactions, or is there also a direct neuronal action? Will it be possible to replicate these findings in humans? Could a better understanding of IMT504 lead to the identification of new cellular and molecular targets that may modulate chronic pain, leading to the development of novel and more efficient, abuse liability-free, pain drugs?

Here we argue that IMT504 is a "golden vein" to be dug in search for better pain drugs, with a high translational potential as evidenced by correlations between data obtained both from rat and human cells, and that continued efforts to better characterize its mechanisms of action will help to establish this ODN, and any new target derived from its analysis, as the next generation of pharmaceutical compounds against a variety of pain conditions.

CONCLUSION

In this chapter, we present IMT504, a non-CpG, non-coding ODN with immunomodulating capabilities and remarkable anti-nociceptive and anti-inflammatory actions, as revealed in models of neuropathic and inflammatory pain. Current knowledge suggests that the ability of IMT504 to reduce pain rests on its modulation of a number of non-neuronal cells, including cells in the innate and adaptive immune system, as well as MSCs. Such an emerging mechanism of action is particularly important, when contemplating the complexity and wide variety of mechanisms involved in the generation of chronic pain. With chronic pain remaining an ever-present unmet therapeutic challenge, we conclude that IMT504 may be providing clues for the effective design of drugs that tackle a multifactorial problem such as this.

LIST OF ABBREVIATIONS

A, Adenine; **A1**, Astrocyte phenotype 1; **A2**, Astrocyte phenotype 2; **ASO**, Antisense synthetic oligonucleotides; **ATF3**, Activating transcription factor 3; **ATP**, Adenosine triphosphate; **BBB**, Blood brain barrier; **β-END**, beta-endorphin; **BDNF**; Brain-derived nerve growth factor; **BM**, Bone marrow; **C**, Cytosine; **Ca2+**, Calcium ion; **Casp6**, Caspase 6; **CCI**, Chronic constriction injury;

CCL(x), C-C motif chemokine ligand (x number); **CCR2**, C-C chemokine receptor type 2; **CD(x)**, Cluster of differentiation (x number); **CFA**, Complete Freund´s adjuvant; **CGRP**, Calcitonin gene–related peptide; **CNS**, Central nervous system; **CNTF**, Ciliary neurotrophic factor; **COX2**, Cyclo-oxygenase type 2; **CpG**, Cytosine phosphate guanine; **CSF**, Cerebral spinal fluid; **CTL**, Cytotoxic T lymphocyte; **Cx43**, Connexin-43; **CXC**, Chemokine; **CXCR(x)**, Chemokine receptor (x number); **DAMP**, Damage associated molecular pattern; **DNA**, Deoxyribonucleic acid; **DRG**, Dorsal root ganglion; **dsRNA**, Double-stranded ribonucleic acid; *e.g.*, Exempli gratia; **ERK**, Extracellular signal-regulated kinase; **G**, Guanine; **GFAP**, Glial fibrillary acidic protein, **Glu**, glutamate; **GM-CSF**, Granulocyte macrophage-colony stimulating factor; **HGF**, Hepatocyte growth factor; **HLA**, Histocompatibility locus antigen; **HLA-G**, Histocompatibility locus antigen-G; **HO-1**, Heme oxygenase-1; **HSC**, Hematopoietic stem cells; **H+**, Protons; *i.e.*, that is; **IASP**, International Association for the Study of Pain; **IBA1**, Ionized calcium-binding adaptor molecule 1; **IDO**, Indoleamine 2,3-dioxygenase; **IFN**, Interferon; **IgG**, Immunoglobulin G; **IgM**, Immunoglobulin M; **IL-1**, Interleukin; **IL1RA**, Interleukin-1 receptor antagonist; **IL-1RI**, Interleukin 1 receptor type I; **iNOS**, Inducible form of nitric oxide synthase; **JNK**, Jun N-terminal kinase; **KO**, Knock-out, **LE**, Leucocyte elastase; **L4-6**, Lumbar 4-6; **M1**, Macrophage phenotype 1; **M2**, Macrophage phenotype 2; **M2b**, Macrophage phenotype 2b; **M2c**, Macrophage phenotype 2c; **MCP-1a**, Monocyte chemoattractant protein-1a; **MMP2**, matrix metalloproteinase-2; **mRNA**, Messenger ribonucleic acid; **MSC**, Mesenchymal Stem Cell; **MSC1**, Mesenchymal stem cell phenotype 1; **MSC2**, Mesenchymal stem cell phenotype 2; **Nav1.7**, Voltage-gated sodium channel 1.7; **Nav1.9**, Voltage-gated sodium channel 1.9; **NFkB**, Nuclear factor kappa-ligh--chain-enhancer of activated B cells; **NGF**, Nerve growth factor; **NK**, Natural killer cell; **NK-T**, Natural killer-T cell; **NMDA**, N-methyl-D-aspartate; **NO**, Nitric oxide; **NPY**, Neuropeptide tyrosine; **NSAID**, Non-steroidal anti-inflammatory drug; **P2X**, Purinergic 2X receptor; **P2Y**, Purinergic 2Y receptor; **PAMP**, Pathogen-associated molecular pattern; **PBS**, Phosphate buffered saline; **pDC**, Plasmacytoid dendritic cell; **PDGF**, Platelet-derived growth factor; **PD-L1**, Programmed cell death ligands 1; **PD-L2** Programmed cell death ligands 2; **PGE2**, Prostaglandin E2; **PHA**, Phytohemagglutinin; **PNS**, Peripheral nervous system; **poly I:C**, Polyinosinic: polycytidylic acid; **PRR**, Pathogen recognition receptor; **PS**, Phosphorothioate; **PSL**, Partial sciatic nerve ligation; **Py**, Pyrimidine; **ROS**, Reactive oxygen species; **RNA**, Ribonucleic acid; **SCG**, Satellite glial cell; **SDF-1**, Stromal cell derived factor-1; **SP**, Substance P; **SPM**, Specialized pro-resolving lipid mediators; **sICAM1**, Soluble intercellular adhesion molecule 1; **siRNA** Small interfering ribonucleic acid; **SLNC**, Single ligature nerve constriction; **ssDNA**, Single-stranded deoxyribonucleic acid; **T**,

Thymine; **T1D**, Type 1 Diabetes; **TGF-β**, Transforming growth factor-β; **Th1**, T helper type 1 cell; **Th2**, T helper type 2 cell; **Th17**, T helper type 17 cell; **TIMP-1**, Tissue inhibitor of metalloproteinase-1; **TLR**, Toll-like receptor; **TNF-α**, Tumour necrosis factor-α; **T-reg**, Regulatory T cells; **TRPA1**, Transient receptor potential ankyrin 1; **TRPV1**, Transient receptor potential cation channel subfamily V member 1; **TSG6**, Tumor necrosis factor stimulated gene 6; **TSP4**, Thrombospondin-4; **USD**, United States dollar; **VEGF**, Vascular endothelial growth factor; **WT**, Wild type.

CONSENT FOR PUBLICATION

Not Applicable.

CONFLICT OF INTEREST

The authors declare no conflict of interest, financial or otherwise.

ACKNOWLEDGEMENTS

Research summarized in this book chapter was carried out with the support from the Argentinean National Agency for the Promotion of Science and Technology (PICTO-Startup 2016-0091 and PICT 2017 – 0969), Austral University (UA---80020160200010UA-01) and the International Brain Research Organization (IBRO).

REFERENCES

[1]   Raja SN, Carr DB, Cohen M, *et al.* The revised International Association for the Study of Pain definition of pain: concepts, challenges, and compromises. Pain 2020; 161(9): 1976-82.
[http://dx.doi.org/10.1097/j.pain.0000000000001939] [PMID: 32694387]

[2]   Treede RD, Rief W, Barke A, *et al.* Chronic pain as a symptom or a disease: the IASP Classification of Chronic Pain for the International Classification of Diseases (ICD-11). Pain 2019; 160(1): 19-27.
[http://dx.doi.org/10.1097/j.pain.0000000000001384] [PMID: 30586067]

[3]   Colloca L, Ludman T, Bouhassira D, *et al.* Neuropathic pain. Nat Rev Dis Primers 2017; 3: 17002.
[http://dx.doi.org/10.1038/nrdp.2017.2] [PMID: 28205574]

[4]   Julius D, Basbaum AI. Molecular mechanisms of nociception. Nature 2001; 413(6852): 203-10.
[http://dx.doi.org/10.1038/35093019] [PMID: 11557989]

[5]   Von Korff M, Scher AI, Helmick C, *et al.* United States National Pain Strategy for Population Research: Concepts, Definitions, and Pilot Data. J Pain 2016; 17(10): 1068-80.
[http://dx.doi.org/10.1016/j.jpain.2016.06.009] [PMID: 27377620]

[6]   Wieseler-Frank J, Maier SF, Watkins LR. Glial activation and pathological pain. Neurochem Int 2004; 45(2-3): 389-95.
[http://dx.doi.org/10.1016/j.neuint.2003.09.009] [PMID: 15145553]

[7]   Basbaum AI, Bautista DM, Scherrer G, Julius D. Cellular and molecular mechanisms of pain. Cell 2009; 139(2): 267-84.
[http://dx.doi.org/10.1016/j.cell.2009.09.028] [PMID: 19837031]

[8]   Loeser JD, Treede RD. The Kyoto protocol of IASP Basic Pain Terminology. Pain 2008; 137(3): 473-7.
[http://dx.doi.org/10.1016/j.pain.2008.04.025] [PMID: 18583048]

[9]   Woolf CJ. What is this thing called pain? J Clin Invest 2010; 120(11): 3742-4.
[http://dx.doi.org/10.1172/JCI45178] [PMID: 21041955]

[10]   Elzahaf RA, Tashani OA, Unsworth BA, Johnson MI. The prevalence of chronic pain with an analysis of countries with a Human Development Index less than 0.9: a systematic review without meta-analysis. Curr Med Res Opin 2012; 28(7): 1221-9.
[http://dx.doi.org/10.1185/03007995.2012.703132] [PMID: 22697274]

[11]   Jackson T, Thomas S, Stabile V, Shotwell M, Han X, McQueen K. A Systematic Review and Meta-Analysis of the Global Burden of Chronic Pain Without Clear Etiology in Low- and Middle-Income Countries: Trends in Heterogeneous Data and a Proposal for New Assessment Methods. Anesth Analg 2016; 123(3): 739-48.
[http://dx.doi.org/10.1213/ANE.0000000000001389] [PMID: 27537761]

[12]   Clauw DJ, Essex MN, Pitman V, Jones KD. Reframing chronic pain as a disease, not a symptom: rationale and implications for pain management. Postgrad Med 2019; 131(3): 185-98.
[http://dx.doi.org/10.1080/00325481.2019.1574403] [PMID: 30700198]

[13]   Jensen TS, Finnerup NB. Allodynia and hyperalgesia in neuropathic pain: clinical manifestations and mechanisms. Lancet Neurol 2014; 13(9): 924-35.
[http://dx.doi.org/10.1016/S1474-4422(14)70102-4] [PMID: 25142459]

[14]   Finnerup NB, Haroutounian S, Kamerman P, *et al.* Neuropathic pain: an updated grading system for research and clinical practice. Pain 2016; 157(8): 1599-606.
[http://dx.doi.org/10.1097/j.pain.0000000000000492] [PMID: 27115670]

[15]   Gilron I, Baron R, Jensen T. Neuropathic pain: principles of diagnosis and treatment. Mayo Clin Proc 2015; 90(4): 532-45.
[http://dx.doi.org/10.1016/j.mayocp.2015.01.018] [PMID: 25841257]

[16]   Lawrence RC, Felson DT, Helmick CG, *et al.* Estimates of the prevalence of arthritis and other rheumatic conditions in the United States. Part II. Arthritis Rheum 2008; 58(1): 26-35.
[http://dx.doi.org/10.1002/art.23176] [PMID: 18163497]

[17]   Helmick CG, Felson DT, Lawrence RC, *et al.* Estimates of the prevalence of arthritis and other rheumatic conditions in the United States. Part I. Arthritis Rheum 2008; 58(1): 15-25.
[http://dx.doi.org/10.1002/art.23177] [PMID: 18163481]

[18]   Litwic A, Edwards MH, Dennison EM, Cooper C. Epidemiology and burden of osteoarthritis. Br Med Bull 2013; 105: 185-99.
[http://dx.doi.org/10.1093/bmb/lds038] [PMID: 23337796]

[19]   Nicholson B, Verma S. Comorbidities in chronic neuropathic pain. Pain Med 2004; 5 (Suppl. 1): S9-S27.
[http://dx.doi.org/10.1111/j.1526-4637.2004.04019.x] [PMID: 14996227]

[20]   Vos T, Flaxman AD, Naghavi M, *et al.* Years lived with disability (YLDs) for 1160 sequelae of 289 diseases and injuries 1990-2010: a systematic analysis for the Global Burden of Disease Study 2010. Lancet 2012; 380(9859): 2163-96.
[http://dx.doi.org/10.1016/S0140-6736(12)61729-2] [PMID: 23245607]

[21]   Neogi T. The epidemiology and impact of pain in osteoarthritis. Osteoarthritis Cartilage 2013; 21(9): 1145-53.
[http://dx.doi.org/10.1016/j.joca.2013.03.018] [PMID: 23973124]

[22]   Johnson MI. The Landscape of Chronic Pain: Broader Perspectives. Medicina (Kaunas) 2019; 55(5): E182.
[http://dx.doi.org/10.3390/medicina55050182] [PMID: 31117297]

[23]   Gaskin DJ, Richard P. The economic costs of pain in the United States. J Pain 2012; 13(8): 715-24.
        [http://dx.doi.org/10.1016/j.jpain.2012.03.009] [PMID: 22607834]

[24]   Breivik H, Eisenberg E, O'Brien T. The individual and societal burden of chronic pain in Europe: the
        case for strategic prioritisation and action to improve knowledge and availability of appropriate care.
        BMC Public Health 2013; 13: 1229.
        [http://dx.doi.org/10.1186/1471-2458-13-1229] [PMID: 24365383]

[25]   Hucho T, Levine JD. Signaling pathways in sensitization: toward a nociceptor cell biology. Neuron
        2007; 55(3): 365-76.
        [http://dx.doi.org/10.1016/j.neuron.2007.07.008] [PMID: 17678851]

[26]   Gold MS, Gebhart GF. Nociceptor sensitization in pain pathogenesis. Nat Med 2010; 16(11): 1248-57.
        [http://dx.doi.org/10.1038/nm.2235] [PMID: 20948530]

[27]   Ji RR, Kohno T, Moore KA, Woolf CJ. Central sensitization and LTP: do pain and memory share
        similar mechanisms? Trends Neurosci 2003; 26(12): 696-705.
        [http://dx.doi.org/10.1016/j.tins.2003.09.017] [PMID: 14624855]

[28]   Woolf CJ. Central sensitization: implications for the diagnosis and treatment of pain. Pain 2011;
        152(3) (Suppl.): S2-S15.
        [http://dx.doi.org/10.1016/j.pain.2010.09.030] [PMID: 20961685]

[29]   Ossipov MH, Dussor GO, Porreca F. Central modulation of pain. J Clin Invest 2010; 120(11): 3779-
        87.
        [http://dx.doi.org/10.1172/JCI43766] [PMID: 21041960]

[30]   Luo C, Kuner T, Kuner R. Synaptic plasticity in pathological pain. Trends Neurosci 2014; 37(6): 343-
        55.
        [http://dx.doi.org/10.1016/j.tins.2014.04.002] [PMID: 24833289]

[31]   Dubin AE, Patapoutian A. Nociceptors: the sensors of the pain pathway. J Clin Invest 2010; 120(11):
        3760-72.
        [http://dx.doi.org/10.1172/JCI42843] [PMID: 21041958]

[32]   England JD, Happel LT, Kline DG, *et al.* Sodium channel accumulation in humans with painful
        neuromas. Neurology 1996; 47(1): 272-6.
        [http://dx.doi.org/10.1212/WNL.47.1.272] [PMID: 8710095]

[33]   Fitzgerald DC, Zhang GX, El-Behi M, *et al.* Suppression of autoimmune inflammation of the central
        nervous system by interleukin 10 secreted by interleukin 27-stimulated T cells. Nat Immunol 2007;
        8(12): 1372-9.
        [http://dx.doi.org/10.1038/ni1540] [PMID: 17994023]

[34]   Le Bars D, Gozariu M, Cadden SW. Animal models of nociception. Pharmacol Rev 2001; 53(4): 597-
        652.
        [PMID: 11734620]

[35]   Milligan ED, Penzkover KR, Soderquist RG, Mahoney MJ. Spinal interleukin-10 therapy to treat
        peripheral neuropathic pain. Neuromodulation 2012; 15(6): 520-6.
        [http://dx.doi.org/10.1111/j.1525-1403.2012.00462.x] [PMID: 22672183]

[36]   Tsujino H, Kondo E, Fukuoka T, *et al.* Activating transcription factor 3 (ATF3) induction by axotomy
        in sensory and motoneurons: A novel neuronal marker of nerve injury. Mol Cell Neurosci 2000; 15(2):
        170-82.
        [http://dx.doi.org/10.1006/mcne.1999.0814] [PMID: 10673325]

[37]   Hökfelt T. Phenotypic changes in peripheral and central neurons induced by nerve injury: focus on
        neuropeptidesChallenges for Neurosciences in the 21st Century. The Twenty-second International
        Symposium 1999; 63-87. Karger.

[38]   Xu Z, Cortés R, Villar M, Morino P, Castel MN, Hökfelt T. Evidence for upregulation of galanin

synthesis in rat glial cells in vivo after colchicine treatment. Neurosci Lett 1992; 145(2): 185-8.
[http://dx.doi.org/10.1016/0304-3940(92)90018-3] [PMID: 1281534]

[39]   Hökfelt T, Zhang X, Verge V, *et al.* Coexistence and interaction of neuropeptides with substance P in primary sensory neurons, with special reference to galanin. Regul Pept 1993; 46(1-2): 76-80.
[http://dx.doi.org/10.1016/0167-0115(93)90015-Z] [PMID: 7692570]

[40]   Hökfelt T, *et al.* Messenger plasticity in primary sensory neurons. International Symposium on Neuropeptides, Nociception and Pain. Mainz, Germany. 1992.

[41]   Hökfelt T, *et al.* Plasticity in expression of neuropeptides, in 6th ECNP. Budapest, Hungary 1993; pp. 162-3.

[42]   Zhang X, Ji RR, Nilsson S, *et al.* Neuropeptide Y and galanin binding sites in rat and monkey lumbar dorsal root ganglia and spinal cord and effect of peripheral axotomy. Eur J Neurosci 1995; 7(3): 367-80.
[http://dx.doi.org/10.1111/j.1460-9568.1995.tb00332.x] [PMID: 7539691]

[43]   Zigmond RE, Echevarria FD. Macrophage biology in the peripheral nervous system after injury. Prog Neurobiol 2019; 173: 102-21.
[http://dx.doi.org/10.1016/j.pneurobio.2018.12.001] [PMID: 30579784]

[44]   Peng J, Gu N, Zhou L, *et al.* Microglia and monocytes synergistically promote the transition from acute to chronic pain after nerve injury. Nat Commun 2016; 7: 12029.
[http://dx.doi.org/10.1038/ncomms12029] [PMID: 27349690]

[45]   White FA, Sun J, Waters SM, *et al.* Excitatory monocyte chemoattractant protein-1 signaling is up-regulated in sensory neurons after chronic compression of the dorsal root ganglion. Proc Natl Acad Sci USA 2005; 102(39): 14092-7.
[http://dx.doi.org/10.1073/pnas.0503496102] [PMID: 16174730]

[46]   Zelenka M, Schäfers M, Sommer C. Intraneural injection of interleukin-1beta and tumor necrosis factor-alpha into rat sciatic nerve at physiological doses induces signs of neuropathic pain. Pain 2005; 116(3): 257-63.
[http://dx.doi.org/10.1016/j.pain.2005.04.018] [PMID: 15964142]

[47]   Ghasemlou N, Chiu IM, Julien JP, Woolf CJ. CD11b+Ly6G- myeloid cells mediate mechanical inflammatory pain hypersensitivity. Proc Natl Acad Sci USA 2015; 112(49): E6808-17.
[http://dx.doi.org/10.1073/pnas.1501372112] [PMID: 26598697]

[48]   Old EA, Nadkarni S, Grist J, *et al.* Monocytes expressing CX3CR1 orchestrate the development of vincristine-induced pain. J Clin Invest 2014; 124(5): 2023-36.
[http://dx.doi.org/10.1172/JCI71389] [PMID: 24743146]

[49]   Willemen HL, Eijkelkamp N, Garza Carbajal A, *et al.* Monocytes/Macrophages control resolution of transient inflammatory pain. J Pain 2014; 15(5): 496-506.
[http://dx.doi.org/10.1016/j.jpain.2014.01.491] [PMID: 24793056]

[50]   Caplan AI, Correa D. The MSC: an injury drugstore. Cell Stem Cell 2011; 9(1): 11-5.
[http://dx.doi.org/10.1016/j.stem.2011.06.008] [PMID: 21726829]

[51]   Musolino PL, Coronel MF, Hökfelt T, Villar MJ. Bone marrow stromal cells induce changes in pain behavior after sciatic nerve constriction. Neurosci Lett 2007; 418(1): 97-101.
[http://dx.doi.org/10.1016/j.neulet.2007.03.001] [PMID: 17379405]

[52]   Coronel MF, Musolino PL, Brumovsky PR, Hökfelt T, Villar MJ. Bone marrow stromal cells attenuate injury-induced changes in galanin, NPY and NPY Y1-receptor expression after a sciatic nerve constriction. Neuropeptides 2009; 43(2): 125-32.
[http://dx.doi.org/10.1016/j.npep.2008.12.003] [PMID: 19168218]

[53]   Coronel MF, Musolino PL, Villar MJ. Selective migration and engraftment of bone marrow mesenchymal stem cells in rat lumbar dorsal root ganglia after sciatic nerve constriction. Neurosci Lett 2006; 405(1-2): 5-9.

[http://dx.doi.org/10.1016/j.neulet.2006.06.018] [PMID: 16806704]

[54]   Usach V, Malet M, López M, *et al.* Systemic Transplantation of Bone Marrow Mononuclear Cells Promotes Axonal Regeneration and Analgesia in a Model of Wallerian Degeneration. Transplantation 2017; 101(7): 1573-86.
[http://dx.doi.org/10.1097/TP.0000000000001478] [PMID: 27607534]

[55]   Chiu IM, von Hehn CA, Woolf CJ. Neurogenic inflammation and the peripheral nervous system in host defense and immunopathology. Nat Neurosci 2012; 15(8): 1063-7.
[http://dx.doi.org/10.1038/nn.3144] [PMID: 22837035]

[56]   Liu T, Ji RR. New insights into the mechanisms of itch: are pain and itch controlled by distinct mechanisms? Pflugers Arch 2013; 465(12): 1671-85.
[http://dx.doi.org/10.1007/s00424-013-1284-2] [PMID: 23636773]

[57]   Liu XJ, Liu T, Chen G, *et al.* TLR signaling adaptor protein MyD88 in primary sensory neurons contributes to persistent inflammatory and neuropathic pain and neuroinflammation. Sci Rep 2016; 6: 28188.
[http://dx.doi.org/10.1038/srep28188] [PMID: 27312666]

[58]   Liu T, Berta T, Xu ZZ, *et al.* TLR3 deficiency impairs spinal cord synaptic transmission, central sensitization, and pruritus in mice. J Clin Invest 2012; 122(6): 2195-207.
[http://dx.doi.org/10.1172/JCI45414] [PMID: 22565312]

[59]   Liu XJ, Zhang Y, Liu T, *et al.* Nociceptive neurons regulate innate and adaptive immunity and neuropathic pain through MyD88 adapter. Cell Res 2014; 24(11): 1374-7.
[http://dx.doi.org/10.1038/cr.2014.106] [PMID: 25112711]

[60]   Talbot S, Foster SL, Woolf CJ. Neuroimmunity: Physiology and Pathology. Annu Rev Immunol 2016; 34: 421-47.
[http://dx.doi.org/10.1146/annurev-immunol-041015-055340] [PMID: 26907213]

[61]   Kissin I. The development of new analgesics over the past 50 years: a lack of real breakthrough drugs. Anesth Analg 2010; 110(3): 780-9.
[http://dx.doi.org/10.1213/ANE.0b013e3181cde882] [PMID: 20185657]

[62]   Ji RR, Chamessian A, Zhang YQ. Pain regulation by non-neuronal cells and inflammation. Science 2016; 354(6312): 572-7.
[http://dx.doi.org/10.1126/science.aaf8924] [PMID: 27811267]

[63]   Sattler S. The Role of the Immune System Beyond the Fight Against Infection. Adv Exp Med Biol 2017; 1003: 3-14.
[http://dx.doi.org/10.1007/978-3-319-57613-8_1] [PMID: 28667551]

[64]   Jenner E. History of the Inoculation of the Cow-Pox: Further Observations on the Variolæ Vaccinæ, or Cow-Pox. Med Phys J 1799; 1(4): 313-8.
[PMID: 30489938]

[65]   Koch R. An Address on Bacteriological Research. BMJ 1890; 2(1546): 380-3.
[http://dx.doi.org/10.1136/bmj.2.1546.380] [PMID: 20753110]

[66]   von Behring E, Kitasato S. The mechanism of diphtheria immunity and tetanus immunity in animals. Mol Immunol 1991; 28(12): 1317-1319-1320.
[PMID: 1749380]

[67]   Parkin J, Cohen B. An overview of the immune system. Lancet 2001; 357(9270): 1777-89.
[http://dx.doi.org/10.1016/S0140-6736(00)04904-7] [PMID: 11403834]

[68]   Gong T, Liu L, Jiang W, Zhou R. DAMP-sensing receptors in sterile inflammation and inflammatory diseases. Nat Rev Immunol 2020; 20(2): 95-112.
[http://dx.doi.org/10.1038/s41577-019-0215-7] [PMID: 31558839]

[69]   Zindel J, Kubes P. DAMPs, PAMPs, and LAMPs in Immunity and Sterile Inflammation. Annu Rev

Pathol 2020; 15: 493-518.
[http://dx.doi.org/10.1146/annurev-pathmechdis-012419-032847] [PMID: 31675482]

[70]   Shapouri-Moghaddam A, Mohammadian S, Vazini H, *et al.* Macrophage plasticity, polarization, and function in health and disease. J Cell Physiol 2018; 233(9): 6425-40.
[http://dx.doi.org/10.1002/jcp.26429] [PMID: 29319160]

[71]   Waisman A, Lukas D, Clausen BE, Yogev N. Dendritic cells as gatekeepers of tolerance. Semin Immunopathol 2017; 39(2): 153-63.
[http://dx.doi.org/10.1007/s00281-016-0583-z] [PMID: 27456849]

[72]   Natoli G, Ostuni R. Adaptation and memory in immune responses. Nat Immunol 2019; 20(7): 783-92.
[http://dx.doi.org/10.1038/s41590-019-0399-9] [PMID: 31213714]

[73]   Raeber ME, Zurbuchen Y, Impellizzieri D, Boyman O. The role of cytokines in T-cell memory in health and disease. Immunol Rev 2018; 283(1): 176-93.
[http://dx.doi.org/10.1111/imr.12644] [PMID: 29664568]

[74]   Omilusik KD, Goldrath AW. Remembering to remember: T cell memory maintenance and plasticity. Curr Opin Immunol 2019; 58: 89-97.
[http://dx.doi.org/10.1016/j.coi.2019.04.009] [PMID: 31170601]

[75]   Suan D, Sundling C, Brink R. Plasma cell and memory B cell differentiation from the germinal center. Curr Opin Immunol 2017; 45: 97-102.
[http://dx.doi.org/10.1016/j.coi.2017.03.006] [PMID: 28319733]

[76]   Iberg CA, Hawiger D. Natural and Induced Tolerogenic Dendritic Cells. J Immunol 2020; 204(4): 733-44.
[http://dx.doi.org/10.4049/jimmunol.1901121] [PMID: 32015076]

[77]   Headland SE, Norling LV. The resolution of inflammation: Principles and challenges. Semin Immunol 2015; 27(3): 149-60.
[http://dx.doi.org/10.1016/j.smim.2015.03.014] [PMID: 25911383]

[78]   Oh SB, Tran PB, Gillard SE, Hurley RW, Hammond DL, Miller RJ. Chemokines and glycoprotein120 produce pain hypersensitivity by directly exciting primary nociceptive neurons. J Neurosci 2001; 21(14): 5027-35.
[http://dx.doi.org/10.1523/JNEUROSCI.21-14-05027.2001] [PMID: 11438578]

[79]   Dawes JM, Calvo M, Perkins JR, *et al.* CXCL5 mediates UVB irradiation-induced pain. Sci Transl Med 2011; 3(90): 90ra60.
[http://dx.doi.org/10.1126/scitranslmed.3002193] [PMID: 21734176]

[80]   Pinho-Ribeiro FA, Verri WA Jr, Chiu IM. Nociceptor Sensory Neuron-Immune Interactions in Pain and Inflammation. Trends Immunol 2017; 38(1): 5-19.
[http://dx.doi.org/10.1016/j.it.2016.10.001] [PMID: 27793571]

[81]   Yang QQ, Zhou JW. Neuroinflammation in the central nervous system: Symphony of glial cells. Glia 2019; 67(6): 1017-35.
[http://dx.doi.org/10.1002/glia.23571] [PMID: 30548343]

[82]   Ji RR, Nackley A, Huh Y, Terrando N, Maixner W. Neuroinflammation and Central Sensitization in Chronic and Widespread Pain. Anesthesiology 2018; 129(2): 343-66.
[http://dx.doi.org/10.1097/ALN.0000000000002130] [PMID: 29462012]

[83]   Sugimoto MA, Vago JP, Perretti M, Teixeira MM. Mediators of the Resolution of the Inflammatory Response. Trends Immunol 2019; 40(3): 212-27.
[http://dx.doi.org/10.1016/j.it.2019.01.007] [PMID: 30772190]

[84]   Cook AD, Christensen AD, Tewari D, McMahon SB, Hamilton JA. Immune Cytokines and Their Receptors in Inflammatory Pain. Trends Immunol 2018; 39(3): 240-55.
[http://dx.doi.org/10.1016/j.it.2017.12.003] [PMID: 29338939]

[85]   Kapurniotu A, Gokce O, Bernhagen J. The Multitasking Potential of Alarmins and Atypical Chemokines. Front Med (Lausanne) 2019; 6(3): 3.
[http://dx.doi.org/10.3389/fmed.2019.00003] [PMID: 30729111]

[86]   Laumet G, Ma J, Robison AJ, Kumari S, Heijnen CJ, Kavelaars A. T Cells as an Emerging Target for Chronic Pain Therapy. Front Mol Neurosci 2019; 12: 216.
[http://dx.doi.org/10.3389/fnmol.2019.00216] [PMID: 31572125]

[87]   Austin PJM-T. Animal models of neuropathic pain due to nerve injury, in Stimulation and inhibition of neurons, P.M.F. Pilowsky. New York: Humana Press 2013; pp. 239-60.

[88]   Gao YJ, Ji RR. Chemokines, neuronal-glial interactions, and central processing of neuropathic pain. Pharmacol Ther 2010; 126(1): 56-68.
[http://dx.doi.org/10.1016/j.pharmthera.2010.01.002] [PMID: 20117131]

[89]   Gonçalves Dos Santos G, Delay L, Yaksh TL, Corr M. Neuraxial Cytokines in Pain States. Front Immunol 2020; 10: 3061.
[http://dx.doi.org/10.3389/fimmu.2019.03061] [PMID: 32047493]

[90]   Ji RR, Xu ZZ, Gao YJ. Emerging targets in neuroinflammation-driven chronic pain. Nat Rev Drug Discov 2014; 13(7): 533-48.
[http://dx.doi.org/10.1038/nrd4334] [PMID: 24948120]

[91]   Mousset CM, Hobo W, Woestenenk R, Preijers F, Dolstra H, van der Waart AB. Comprehensive Phenotyping of T Cells Using Flow Cytometry. Cytometry A 2019; 95(6): 647-54.
[http://dx.doi.org/10.1002/cyto.a.23724] [PMID: 30714682]

[92]   Abdi R, Fiorina P, Adra CN, Atkinson M, Sayegh MH. Immunomodulation by mesenchymal stem cells: a potential therapeutic strategy for type 1 diabetes. Diabetes 2008; 57(7): 1759-67.
[http://dx.doi.org/10.2337/db08-0180] [PMID: 18586907]

[93]   Moalem G, Xu K, Yu L. T lymphocytes play a role in neuropathic pain following peripheral nerve injury in rats. Neuroscience 2004; 129(3): 767-77.
[http://dx.doi.org/10.1016/j.neuroscience.2004.08.035] [PMID: 15541898]

[94]   Hu D, Ikizawa K, Lu L, Sanchirico ME, Shinohara ML, Cantor H. Analysis of regulatory CD8 T cells in Qa-1-deficient mice. Nat Immunol 2004; 5(5): 516-23.
[http://dx.doi.org/10.1038/ni1063] [PMID: 15098030]

[95]   Filaci G, Fravega M, Negrini S, *et al.* Nonantigen specific CD8+ T suppressor lymphocytes originate from CD8+CD28- T cells and inhibit both T-cell proliferation and CTL function. Hum Immunol 2004; 65(2): 142-56.
[http://dx.doi.org/10.1016/j.humimm.2003.12.001] [PMID: 14969769]

[96]   Harrington LE, Mangan PR, Weaver CT. Expanding the effector CD4 T-cell repertoire: the Th17 lineage. Curr Opin Immunol 2006; 18(3): 349-56.
[http://dx.doi.org/10.1016/j.coi.2006.03.017] [PMID: 16616472]

[97]   Baddack-Werncke U, Busch-Dienstfertig M, González-Rodríguez S, *et al.* Cytotoxic T cells modulate inflammation and endogenous opioid analgesia in chronic arthritis. J Neuroinflammation 2017; 14(1): 30.
[http://dx.doi.org/10.1186/s12974-017-0804-y] [PMID: 28166793]

[98]   Du B, Ding YQ, Xiao X, Ren HY, Su BY, Qi JG. CD4+ αβ T cell infiltration into the leptomeninges of lumbar dorsal roots contributes to the transition from acute to chronic mechanical allodynia after adult rat tibial nerve injuries. J Neuroinflammation 2018; 15(1): 81.
[http://dx.doi.org/10.1186/s12974-018-1115-7] [PMID: 29544518]

[99]   Schläger C, Körner H, Krueger M, *et al.* Effector T-cell trafficking between the leptomeninges and the cerebrospinal fluid. Nature 2016; 530(7590): 349-53.
[http://dx.doi.org/10.1038/nature16939] [PMID: 26863192]

[100] Costigan M, Scholz J, Woolf CJ. Neuropathic pain: a maladaptive response of the nervous system to damage. Annu Rev Neurosci 2009; 32: 1-32.
[http://dx.doi.org/10.1146/annurev.neuro.051508.135531] [PMID: 19400724]

[101] Kocot-Kępska M, Zajączkowska R, Mika J, Wordliczek J, Dobrogowski J, Przeklasa-Muszyńska A. Peripheral Mechanisms of Neuropathic Pain-the Role of Neuronal and Non-Neuronal Interactions and Their Implications for Topical Treatment of Neuropathic Pain. Pharmaceuticals (Basel) 2021; 14(2): 77.
[http://dx.doi.org/10.3390/ph14020077] [PMID: 33498496]

[102] Ji RR, Berta T, Nedergaard M. Glia and pain: is chronic pain a gliopathy? Pain 2013; 154 (Suppl. 1): S10-28.
[http://dx.doi.org/10.1016/j.pain.2013.06.022] [PMID: 23792284]

[103] Milligan ED, Watkins LR. Pathological and protective roles of glia in chronic pain. Nat Rev Neurosci 2009; 10(1): 23-36.
[http://dx.doi.org/10.1038/nrn2533] [PMID: 19096368]

[104] Mika J, Zychowska M, Popiolek-Barczyk K, Rojewska E, Przewlocka B. Importance of glial activation in neuropathic pain. Eur J Pharmacol 2013; 716(1-3): 106-19.
[http://dx.doi.org/10.1016/j.ejphar.2013.01.072] [PMID: 23500198]

[105] Rothman SM, Winkelstein BA. Cytokine antagonism reduces pain and modulates spinal astrocytic reactivity after cervical nerve root compression. Ann Biomed Eng 2010; 38(8): 2563-76.
[http://dx.doi.org/10.1007/s10439-010-0012-8] [PMID: 20309734]

[106] Ji RR, Donnelly CR, Nedergaard M. Astrocytes in chronic pain and itch. Nat Rev Neurosci 2019; 20(11): 667-85.
[http://dx.doi.org/10.1038/s41583-019-0218-1] [PMID: 31537912]

[107] Li T, Chen X, Zhang C, Zhang Y, Yao W. An update on reactive astrocytes in chronic pain. J Neuroinflammation 2019; 16(1): 140.
[http://dx.doi.org/10.1186/s12974-019-1524-2] [PMID: 31288837]

[108] Burda JE, Sofroniew MV. Reactive gliosis and the multicellular response to CNS damage and disease. Neuron 2014; 81(2): 229-48.
[http://dx.doi.org/10.1016/j.neuron.2013.12.034] [PMID: 24462092]

[109] Hanani M, Verkhratsky A. Satellite Glial Cells and Astrocytes, a Comparative Review. Neurochem Res 2021.
[http://dx.doi.org/10.1007/s11064-021-03255-8] [PMID: 33523395]

[110] Loane DJ, Byrnes KR. Role of microglia in neurotrauma. Neurotherapeutics 2010; 7(4): 366-77.
[http://dx.doi.org/10.1016/j.nurt.2010.07.002] [PMID: 20880501]

[111] Beggs S, *et al.* The known knowns of microglia-neuronal signalling in neuropathic pain. Neurosci Lett. \ 2013; pp. 37-42.

[112] Calvo M, Bennett DL. The mechanisms of microgliosis and pain following peripheral nerve injury. Exp Neurol 2012; 234(2): 271-82.
[http://dx.doi.org/10.1016/j.expneurol.2011.08.018] [PMID: 21893056]

[113] Tsuda M. Microglia in the spinal cord and neuropathic pain. J Diabetes Investig 2016; 7(1): 17-26.
[http://dx.doi.org/10.1111/jdi.12379] [PMID: 26813032]

[114] Gwak YS, Kang J, Unabia GC, Hulsebosch CE. Spatial and temporal activation of spinal glial cells: role of gliopathy in central neuropathic pain following spinal cord injury in rats. Exp Neurol 2012; 234(2): 362-72.
[http://dx.doi.org/10.1016/j.expneurol.2011.10.010] [PMID: 22036747]

[115] Gu N, *et al.* Spinal Microglia Contribute to Sustained Inflammatory Pain *via* Amplifying Neuronal Activity bioRxiv 2019.

[http://dx.doi.org/10.1101/2019.12.16.878728]

[116]   Kigerl KA, Gensel JC, Ankeny DP, Alexander JK, Donnelly DJ, Popovich PG. Identification of two distinct macrophage subsets with divergent effects causing either neurotoxicity or regeneration in the injured mouse spinal cord. J Neurosci 2009; 29(43): 13435-44.
[http://dx.doi.org/10.1523/JNEUROSCI.3257-09.2009] [PMID: 19864556]

[117]   Orihuela R, McPherson CA, Harry GJ. Microglial M1/M2 polarization and metabolic states. Br J Pharmacol 2016; 173(4): 649-65.
[http://dx.doi.org/10.1111/bph.13139] [PMID: 25800044]

[118]   Scholz J, Woolf CJ. The neuropathic pain triad: neurons, immune cells and glia. Nat Neurosci 2007; 10(11): 1361-8.
[http://dx.doi.org/10.1038/nn1992] [PMID: 17965656]

[119]   Guo W, Wang H, Watanabe M, *et al.* Glial-cytokine-neuronal interactions underlying the mechanisms of persistent pain. J Neurosci 2007; 27(22): 6006-18.
[http://dx.doi.org/10.1523/JNEUROSCI.0176-07.2007] [PMID: 17537972]

[120]   Vezzani A, *et al.* Neuromodulatory properties of inflammatory cytokines and their impact on neuronal excitability. Neuropharmacology, 2015. 96(Pt A). 2015; 70-82.

[121]   Kawasaki Y, Zhang L, Cheng JK, Ji RR. Cytokine mechanisms of central sensitization: distinct and overlapping role of interleukin-1beta, interleukin-6, and tumor necrosis factor-alpha in regulating synaptic and neuronal activity in the superficial spinal cord. J Neurosci 2008; 28(20): 5189-94.
[http://dx.doi.org/10.1523/JNEUROSCI.3338-07.2008] [PMID: 18480275]

[122]   Gardoni F, Boraso M, Zianni E, *et al.* Distribution of interleukin-1 receptor complex at the synaptic membrane driven by interleukin-1β and NMDA stimulation. J Neuroinflammation 2011; 8(1): 14.
[http://dx.doi.org/10.1186/1742-2094-8-14] [PMID: 21314939]

[123]   Pedersen LM, Jacobsen LM, Mollerup S, Gjerstad J. Spinal cord long-term potentiation (LTP) is associated with increased dorsal horn gene expression of IL-1beta, GDNF and iNOS. Eur J Pain 2010; 14(3): 255-60.
[http://dx.doi.org/10.1016/j.ejpain.2009.05.016] [PMID: 19596210]

[124]   Zhang RX, Li A, Liu B, *et al.* IL-1ra alleviates inflammatory hyperalgesia through preventing phosphorylation of NMDA receptor NR-1 subunit in rats. Pain 2008; 135(3): 232-9.
[http://dx.doi.org/10.1016/j.pain.2007.05.023] [PMID: 17689191]

[125]   Fogal B, Hewett SJ. Interleukin-1beta: a bridge between inflammation and excitotoxicity? J Neurochem 2008; 106(1): 1-23.
[http://dx.doi.org/10.1111/j.1471-4159.2008.05315.x] [PMID: 18315560]

[126]   Viviani B, Bartesaghi S, Gardoni F, *et al.* Interleukin-1beta enhances NMDA receptor-mediated intracellular calcium increase through activation of the Src family of kinases. J Neurosci 2003; 23(25): 8692-700.
[http://dx.doi.org/10.1523/JNEUROSCI.23-25-08692.2003] [PMID: 14507968]

[127]   Viviani B, Gardoni F, Bartesaghi S, *et al.* Interleukin-1 beta released by gp120 drives neural death through tyrosine phosphorylation and trafficking of NMDA receptors. J Biol Chem 2006; 281(40): 30212-22.
[http://dx.doi.org/10.1074/jbc.M602156200] [PMID: 16887807]

[128]   Dominici M, Le Blanc K, Mueller I, *et al.* Minimal criteria for defining multipotent mesenchymal stromal cells. The International Society for Cellular Therapy position statement. Cytotherapy 2006; 8(4): 315-7.
[http://dx.doi.org/10.1080/14653240600855905] [PMID: 16923606]

[129]   Lindner U, Kramer J, Rohwedel J, Schlenke P. Mesenchymal Stem or Stromal Cells: Toward a Better Understanding of Their Biology? Transfus Med Hemother 2010; 37(2): 75-83.
[http://dx.doi.org/10.1159/000290897] [PMID: 20737049]

[130] Pittenger MF, Mackay AM, Beck SC, *et al.* Multilineage potential of adult human mesenchymal stem cells. Science 1999; 284(5411): 143-7.
[http://dx.doi.org/10.1126/science.284.5411.143] [PMID: 10102814]

[131] Yang F, Leung VY, Luk KD, Chan D, Cheung KM. Mesenchymal stem cells arrest intervertebral disc degeneration through chondrocytic differentiation and stimulation of endogenous cells. Mol Ther 2009; 17(11): 1959-66.
[http://dx.doi.org/10.1038/mt.2009.146] [PMID: 19584814]

[132] Pettine K, Suzuki R, Sand T, Murphy M. Treatment of discogenic back pain with autologous bone marrow concentrate injection with minimum two year follow-up. Int Orthop 2016; 40(1): 135-40.
[http://dx.doi.org/10.1007/s00264-015-2886-4] [PMID: 26156727]

[133] Orozco L, Soler R, Morera C, Alberca M, Sánchez A, García-Sancho J. Intervertebral disc repair by autologous mesenchymal bone marrow cells: a pilot study. Transplantation 2011; 92(7): 822-8.
[http://dx.doi.org/10.1097/TP.0b013e3182298a15] [PMID: 21792091]

[134] Watanabe S, Uchida K, Nakajima H, *et al.* Early transplantation of mesenchymal stem cells after spinal cord injury relieves pain hypersensitivity through suppression of pain-related signaling cascades and reduced inflammatory cell recruitment. Stem Cells 2015; 33(6): 1902-14.
[http://dx.doi.org/10.1002/stem.2006] [PMID: 25809552]

[135] Klass M, Gavrikov V, Drury D, *et al.* Intravenous mononuclear marrow cells reverse neuropathic pain from experimental mononeuropathy. Anesth Analg 2007; 104(4): 944-8.
[http://dx.doi.org/10.1213/01.ane.0000258021.03211.d0] [PMID: 17377111]

[136] Siniscalco D, Giordano C, Galderisi U, *et al.* Long-lasting effects of human mesenchymal stem cell systemic administration on pain-like behaviors, cellular, and biomolecular modifications in neuropathic mice. Front Integr Nuerosci 2011; 5: 79.
[http://dx.doi.org/10.3389/fnint.2011.00079] [PMID: 22164136]

[137] Durand C, Pezet S, Eutamène H, *et al.* Persistent visceral allodynia in rats exposed to colorectal irradiation is reversed by mesenchymal stromal cell treatment. Pain 2015; 156(8): 1465-76.
[http://dx.doi.org/10.1097/j.pain.0000000000000190] [PMID: 25887464]

[138] Brumovsky PR, Bergman E, Liu HX, Hökfelt T, Villar MJ. Effect of a graded single constriction of the rat sciatic nerve on pain behavior and expression of immunoreactive NPY and NPY Y1 receptor in DRG neurons and spinal cord. Brain Res 2004; 1006(1): 87-99.
[http://dx.doi.org/10.1016/j.brainres.2003.09.085] [PMID: 15047027]

[139] Luo X, Tai WL, Sun L, *et al.* Central administration of C-X-C chemokine receptor type 4 antagonist alleviates the development and maintenance of peripheral neuropathic pain in mice. PLoS One 2014; 9(8): e104860.
[http://dx.doi.org/10.1371/journal.pone.0104860] [PMID: 25119456]

[140] Shen W, Hu XM, Liu YN, *et al.* CXCL12 in astrocytes contributes to bone cancer pain through CXCR4-mediated neuronal sensitization and glial activation in rat spinal cord. J Neuroinflammation 2014; 11: 75.
[http://dx.doi.org/10.1186/1742-2094-11-75] [PMID: 24735601]

[141] Reaux-Le Goazigo A, Rivat C, Kitabgi P, Pohl M, Melik Parsadaniantz S. Cellular and subcellular localization of CXCL12 and CXCR4 in rat nociceptive structures: physiological relevance. Eur J Neurosci 2012; 36(5): 2619-31.
[http://dx.doi.org/10.1111/j.1460-9568.2012.08179.x] [PMID: 22694179]

[142] Küry P, Greiner-Petter R, Cornely C, Jürgens T, Müller HW. Mammalian achaete scute homolog 2 is expressed in the adult sciatic nerve and regulates the expression of Krox24, Mob-1, CXCR4, and p57kip2 in Schwann cells. J Neurosci 2002; 22(17): 7586-95.
[http://dx.doi.org/10.1523/JNEUROSCI.22-17-07586.2002] [PMID: 12196582]

[143] Chen G, Park CK, Xie RG, Ji RR. Intrathecal bone marrow stromal cells inhibit neuropathic pain via

TGF-β secretion. J Clin Invest 2015; 125(8): 3226-40.
[http://dx.doi.org/10.1172/JCI80883] [PMID: 26168219]

[144]   Zhang Y, Zheng H, Ren J, *et al.* Mesenchymal stem cells enhance the impact of KIR receptor-ligand mismatching on acute graft-*versus*-host disease following allogeneic hematopoietic stem cell transplantation in patients with acute myeloid leukemia but not in those with acute lymphocytic leukemia. Hematol Oncol 2021; 39(3): 380-9.
[http://dx.doi.org/10.1002/hon.2867] [PMID: 33848027]

[145]   Bolli R, Mitrani RD, Hare JM, *et al.* A Phase II study of autologous mesenchymal stromal cells and c-kit positive cardiac cells, alone or in combination, in patients with ischaemic heart failure: the CCTRN CONCERT-HF trial. Eur J Heart Fail 2021; 23(4): 661-74.
[http://dx.doi.org/10.1002/ejhf.2178] [PMID: 33811444]

[146]   Johnson LDV, Pickard MR, Johnson WEB. The Comparative Effects of Mesenchymal Stem Cell Transplantation Therapy for Spinal Cord Injury in Humans and Animal Models: A Systematic Review and Meta-Analysis. Biology (Basel) 2021; 10(3): 230.
[http://dx.doi.org/10.3390/biology10030230] [PMID: 33809684]

[147]   Sierra-Sánchez Á, Montero-Vilchez T, Quiñones-Vico MI, Sanchez-Diaz M, Arias-Santiago S. Current Advanced Therapies Based on Human Mesenchymal Stem Cells for Skin Diseases. Front Cell Dev Biol 2021; 9: 643125.
[http://dx.doi.org/10.3389/fcell.2021.643125] [PMID: 33768095]

[148]   El-Jawhari JJ, El-Sherbiny Y, McGonagle D, Jones E. Multipotent Mesenchymal Stromal Cells in Rheumatoid Arthritis and Systemic Lupus Erythematosus; From a Leading Role in Pathogenesis to Potential Therapeutic Saviors? Front Immunol 2021; 12: 643170.
[http://dx.doi.org/10.3389/fimmu.2021.643170] [PMID: 33732263]

[149]   Morata-Tarifa C, Azkona G, Glass J, Mazzini L, Sanchez-Pernaute R. Looking backward to move forward: a meta-analysis of stem cell therapy in amyotrophic lateral sclerosis. NPJ Regen Med 2021; 6(1): 20.
[http://dx.doi.org/10.1038/s41536-021-00131-5] [PMID: 33795700]

[150]   Premer C, *et al.* The role of mesenchymal stem/stromal cells in the acute clinical setting. Am J Emerg Med 2020.
[PMID: 33279332]

[151]   Caplan AI, Dennis JE. Mesenchymal stem cells as trophic mediators. J Cell Biochem 2006; 98(5): 1076-84.
[http://dx.doi.org/10.1002/jcb.20886] [PMID: 16619257]

[152]   Wang Y, Chen X, Cao W, Shi Y. Plasticity of mesenchymal stem cells in immunomodulation: pathological and therapeutic implications. Nat Immunol 2014; 15(11): 1009-16.
[http://dx.doi.org/10.1038/ni.3002] [PMID: 25329189]

[153]   Krampera M, Glennie S, Dyson J, *et al.* Bone marrow mesenchymal stem cells inhibit the response of naive and memory antigen-specific T cells to their cognate peptide. Blood 2003; 101(9): 3722-9.
[http://dx.doi.org/10.1182/blood-2002-07-2104] [PMID: 12506037]

[154]   Che N, Li X, Zhou S, *et al.* Umbilical cord mesenchymal stem cells suppress B-cell proliferation and differentiation. Cell Immunol 2012; 274(1-2): 46-53.
[http://dx.doi.org/10.1016/j.cellimm.2012.02.004] [PMID: 22414555]

[155]   Rosado MM, Bernardo ME, Scarsella M, *et al.* Inhibition of B-cell proliferation and antibody production by mesenchymal stromal cells is mediated by T cells. Stem Cells Dev 2015; 24(1): 93-103.
[http://dx.doi.org/10.1089/scd.2014.0155] [PMID: 25036865]

[156]   Chen HW, Chen HY, Wang LT, *et al.* Mesenchymal stem cells tune the development of monocyte-derived dendritic cells toward a myeloid-derived suppressive phenotype through growth-regulated oncogene chemokines. J Immunol 2013; 190(10): 5065-77.
[http://dx.doi.org/10.4049/jimmunol.1202775] [PMID: 23589610]

[157]　Cutler AJ, Limbani V, Girdlestone J, Navarrete CV. Umbilical cord-derived mesenchymal stromal cells modulate monocyte function to suppress T cell proliferation. J Immunol 2010; 185(11): 6617-23.
[http://dx.doi.org/10.4049/jimmunol.1002239] [PMID: 20980628]

[158]　Chen NF, Huang SY, Chen WF, *et al.* TGF-β1 attenuates spinal neuroinflammation and the excitatory amino acid system in rats with neuropathic pain. J Pain 2013; 14(12): 1671-85.
[http://dx.doi.org/10.1016/j.jpain.2013.08.010] [PMID: 24290447]

[159]　Echeverry S, Shi XQ, Haw A, Liu H, Zhang ZW, Zhang J. Transforming growth factor-beta1 impairs neuropathic pain through pleiotropic effects. Mol Pain 2009; 5: 16.
[http://dx.doi.org/10.1186/1744-8069-5-16] [PMID: 19327151]

[160]　Ouyang W, O'Garra A. IL-10 Family Cytokines IL-10 and IL-22: from Basic Science to Clinical Translation. Immunity 2019; 50(4): 871-91.
[http://dx.doi.org/10.1016/j.immuni.2019.03.020] [PMID: 30995504]

[161]　Wagner R, Janjigian M, Myers RR. Anti-inflammatory interleukin-10 therapy in CCI neuropathy decreases thermal hyperalgesia, macrophage recruitment, and endoneurial TNF-alpha expression. Pain 1998; 74(1): 35-42.
[http://dx.doi.org/10.1016/S0304-3959(97)00148-6] [PMID: 9514558]

[162]　Ledeboer A, Jekich BM, Sloane EM, *et al.* Intrathecal interleukin-10 gene therapy attenuates paclitaxel-induced mechanical allodynia and proinflammatory cytokine expression in dorsal root ganglia in rats. Brain Behav Immun 2007; 21(5): 686-98.
[http://dx.doi.org/10.1016/j.bbi.2006.10.012] [PMID: 17174526]

[163]　Gama KB, Santos DS, Evangelista AF, *et al.* Conditioned Medium of Bone Marrow-Derived Mesenchymal Stromal Cells as a Therapeutic Approach to Neuropathic Pain: A Preclinical Evaluation. Stem Cells Int 2018; 2018: 8179013.
[http://dx.doi.org/10.1155/2018/8179013] [PMID: 29535781]

[164]　Li J, Deng G, Wang H, *et al.* Interleukin-1β pre-treated bone marrow stromal cells alleviate neuropathic pain through CCL7-mediated inhibition of microglial activation in the spinal cord. Sci Rep 2017; 7: 42260.
[http://dx.doi.org/10.1038/srep42260] [PMID: 28195183]

[165]　Lundin KE, Gissberg O, Smith CI. Oligonucleotide Therapies: The Past and the Present. Hum Gene Ther 2015; 26(8): 475-85.
[http://dx.doi.org/10.1089/hum.2015.070] [PMID: 26160334]

[166]　Eckstein F. Nucleoside phosphorothioates. J Am Chem Soc 1970; 92(15): 4718-23.
[http://dx.doi.org/10.1021/ja00718a039] [PMID: 4316997]

[167]　Khorkova O, Wahlestedt C. Oligonucleotide therapies for disorders of the nervous system. Nat Biotechnol 2017; 35(3): 249-63.
[http://dx.doi.org/10.1038/nbt.3784] [PMID: 28244991]

[168]　Field AK, Tytell AA, Lampson GP, Hilleman MR. Inducers of interferon and host resistance. II. Multistranded synthetic polynucleotide complexes. Proc Natl Acad Sci USA 1967; 58(3): 1004-10.
[http://dx.doi.org/10.1073/pnas.58.3.1004] [PMID: 5233831]

[169]　Alexopoulou L, Holt AC, Medzhitov R, Flavell RA. Recognition of double-stranded RNA and activation of NF-kappaB by Toll-like receptor 3. Nature 2001; 413(6857): 732-8.
[http://dx.doi.org/10.1038/35099560] [PMID: 11607032]

[170]　Tokunaga T, Yamamoto H, Shimada S, *et al.* Antitumor activity of deoxyribonucleic acid fraction from Mycobacterium bovis BCG. I. Isolation, physicochemical characterization, and antitumor activity. J Natl Cancer Inst 1984; 72(4): 955-62.
[PMID: 6200641]

[171]　Krieg AM, Yi AK, Matson S, *et al.* CpG motifs in bacterial DNA trigger direct B-cell activation. Nature 1995; 374(6522): 546-9.

[http://dx.doi.org/10.1038/374546a0] [PMID: 7700380]

[172] Hemmi H, Takeuchi O, Kawai T, *et al.* A Toll-like receptor recognizes bacterial DNA. Nature 2000; 408(6813): 740-5.
[http://dx.doi.org/10.1038/35047123] [PMID: 11130078]

[173] Takeda K, *et al.* Toll-like receptors. Curr Protoc Immunol. 2015.

[174] Krieg AM. CpG motifs in bacterial DNA and their immune effects. Annu Rev Immunol 2002; 20: 709-60.
[http://dx.doi.org/10.1146/annurev.immunol.20.100301.064842] [PMID: 11861616]

[175] Krug A, Rothenfusser S, Hornung V, *et al.* Identification of CpG oligonucleotide sequences with high induction of IFN-alpha/beta in plasmacytoid dendritic cells. Eur J Immunol 2001; 31(7): 2154-63.
[http://dx.doi.org/10.1002/1521-4141(200107)31:7<2154::AID-IMMU2154>3.0.CO;2-U] [PMID: 11449369]

[176] Bauer S, Kirschning CJ, Häcker H, *et al.* Human TLR9 confers responsiveness to bacterial DNA *via* species-specific CpG motif recognition. Proc Natl Acad Sci USA 2001; 98(16): 9237-42.
[http://dx.doi.org/10.1073/pnas.161293498] [PMID: 11470918]

[177] Landrigan A, Wong MT, Utz PJ. CpG and non-CpG oligodeoxynucleotides directly costimulate mouse and human CD4+ T cells through a TLR9- and MyD88-independent mechanism. J Immunol 2011; 187(6): 3033-43.
[http://dx.doi.org/10.4049/jimmunol.1003414] [PMID: 21844387]

[178] Elias F, Flo J, Lopez RA, Zorzopulos J, Montaner A, Rodriguez JM. Strong cytosine-guanosin--independent immunostimulation in humans and other primates by synthetic oligodeoxynucleotides with PyNTTTTGT motifs. J Immunol 2003; 171(7): 3697-704.
[http://dx.doi.org/10.4049/jimmunol.171.7.3697] [PMID: 14500668]

[179] Rodriguez JM, Elías F, Fló J, López RA, Zorzopulos J, Montaner AD. Immunostimulatory PyNTTTTGT oligodeoxynucleotides: structural properties and refinement of the active motif. Oligonucleotides 2006; 16(3): 275-85.
[http://dx.doi.org/10.1089/oli.2006.16.275] [PMID: 16978090]

[180] Hernando Insúa A, Montaner AD, Rodriguez JM, *et al.* IMT504, the prototype of the immunostimulatory oligonucleotides of the PyNTTTTGT class, increases the number of progenitors of mesenchymal stem cells both in vitro and in vivo: potential use in tissue repair therapy. Stem Cells 2007; 25(4): 1047-54.
[http://dx.doi.org/10.1634/stemcells.2006-0479] [PMID: 17420228]

[181] Rodriguez JM, Marchicio J, López M, *et al.* PyNTTTTGT and CpG immunostimulatory oligonucleotides: effect on granulocyte/monocyte colony-stimulating factor (GM-CSF) secretion by human CD56+ (NK and NKT) cells. PLoS One 2015; 10(2): e0117484.
[http://dx.doi.org/10.1371/journal.pone.0117484] [PMID: 25706946]

[182] Elias F, Flo J, Rodriguez JM, *et al.* PyNTTTTGT prototype oligonucleotide IMT504 is a potent adjuvant for the recombinant hepatitis B vaccine that enhances the Th1 response. Vaccine 2005; 23(27): 3597-603.
[http://dx.doi.org/10.1016/j.vaccine.2004.12.030] [PMID: 15855019]

[183] Zhao G, Jin H, Li J, *et al.* PyNTTTTGT prototype oligonucleotide IMT504, a novel effective adjuvant of the FMDV DNA vaccine. Viral Immunol 2009; 22(2): 131-8.
[http://dx.doi.org/10.1089/vim.2008.0073] [PMID: 19327000]

[184] Montaner AD, Denichilo A, Rodríguez JM, *et al.* Addition of the immunostimulatory oligonucleotide IMT504 to a seasonal flu vaccine increases hemagglutinin antibody titers in young adult and elder rats, and expands the anti-hemagglutinin antibody repertoire. Nucleic Acid Ther 2011; 21(4): 265-74.
[http://dx.doi.org/10.1089/nat.2011.0284] [PMID: 21793787]

[185] Montaner AD, *et al.* IMT504: A New and Potent Adjuvant for Rabies Vaccines Permitting Significant

Dose Sparing. World J Vaccines 2012; 2: 182-8.
[http://dx.doi.org/10.4236/wjv.2012.24025]

[186]   Rodriguez JM, Elias F, Montaner A, *et al.* Oligonucleotide IMT504 induces an immunogenic phenotype and apoptosis in chronic lymphocytic leukemia cells. Medicina (B Aires) 2006; 66(1): 9-16.
[PMID: 16555722]

[187]   Bianchi MS, Bianchi S, Hernado-Insúa A, *et al.* Proposed mechanisms for oligonucleotide IMT504 induced diabetes reversion in a mouse model of immunodependent diabetes. Am J Physiol Endocrinol Metab 2016; 311(2): E380-95.
[http://dx.doi.org/10.1152/ajpendo.00104.2016] [PMID: 27329801]

[188]   Bianchi MS, Hernando-Insúa A, Chasseing NA, *et al.* Oligodeoxynucleotide IMT504 induces a marked recovery in a streptozotocin-induced model of diabetes in rats: correlation with an early increase in the expression of nestin and neurogenin 3 progenitor cell markers. Diabetologia 2010; 53(6): 1184-9.
[http://dx.doi.org/10.1007/s00125-010-1694-z] [PMID: 20221823]

[189]   Bianchi S, *et al.* Oligonucleotide IMT504 Improves Glucose Metabolism and Controls Immune Cell Mediators in Female Diabetic NOD Mice IMT504, a Potential Therapy for Type 1 Diabetes. Nucleic Acid Ther 2020.

[190]   Chahin A, Opal SM, Zorzopulos J, *et al.* The novel immunotherapeutic oligodeoxynucleotide IMT504 protects neutropenic animals from fatal *Pseudomonas aeruginosa* bacteremia and sepsis. Antimicrob Agents Chemother 2015; 59(2): 1225-9.
[http://dx.doi.org/10.1128/AAC.03923-14] [PMID: 25512413]

[191]   Le Blanc K, Mougiakakos D. Multipotent mesenchymal stromal cells and the innate immune system. Nat Rev Immunol 2012; 12(5): 383-96.
[http://dx.doi.org/10.1038/nri3209] [PMID: 22531326]

[192]   Prockop DJ, Oh JY. Medical therapies with adult stem/progenitor cells (MSCs): a backward journey from dramatic results *in vivo* to the cellular and molecular explanations. J Cell Biochem 2012; 113(5): 1460-9.
[http://dx.doi.org/10.1002/jcb.24046] [PMID: 22213121]

[193]   Keating A. Mesenchymal stromal cells: new directions. Cell Stem Cell 2012; 10(6): 709-16.
[http://dx.doi.org/10.1016/j.stem.2012.05.015] [PMID: 22704511]

[194]   Zorzopulos J, Opal SM, Hernando-Insúa A, *et al.* Immunomodulatory oligonucleotide IMT504: Effects on mesenchymal stem cells as a first-in-class immunoprotective/immunoregenerative therapy. World J Stem Cells 2017; 9(3): 45-67.
[http://dx.doi.org/10.4252/wjsc.v9.i3.45] [PMID: 28396715]

[195]   Coronel MF, Hernando-Insúa A, Rodriguez JM, *et al.* Oligonucleotide IMT504 reduces neuropathic pain after peripheral nerve injury. Neurosci Lett 2008; 444(1): 69-73.
[http://dx.doi.org/10.1016/j.neulet.2008.07.045] [PMID: 18672022]

[196]   Leiguarda C, Coronel MF, Montaner AD, Villar MJ, Brumovsky PR. Long-lasting ameliorating effects of the oligodeoxynucleotide IMT504 on mechanical allodynia and hindpaw edema in rats with chronic hindpaw inflammation. Neurosci Lett 2018; 666: 17-23.
[http://dx.doi.org/10.1016/j.neulet.2017.12.032] [PMID: 29248616]

[197]   Leiguarda C, *et al.* IMT504 Provides Analgesia by Modulating Cell Infiltrate and Inflammatory Milieu in a Chronic Pain Model. J Neuroimmune Pharmacol 2020; 16(3): 651-66.
[PMID: 33221983]

[198]   Leiguarda C, Villarreal A, Potilinski C, *et al.* Intrathecal Administration of an Anti-nociceptive Non-CpG Oligodeoxynucleotide Reduces Glial Activation and Central Sensitization. J Neuroimmune Pharmacol In press [p.].
[PMID: 33502706]

[199]   Hu Z, Deng N, Liu K, Zhou N, Sun Y, Zeng W. CNTF-STAT3-IL-6 Axis Mediates Neuroinflammatory Cascade across Schwann Cell-Neuron-Microglia. Cell Rep 2020; 31(7): 107657.
[http://dx.doi.org/10.1016/j.celrep.2020.107657] [PMID: 32433966]

[200]   Woolf CJ. Capturing Novel Non-opioid Pain Targets. Biol Psychiatry 2020; 87(1): 74-81.
[http://dx.doi.org/10.1016/j.biopsych.2019.06.017] [PMID: 31399256]

[201]   Finnerup NB, Attal N, Haroutounian S, *et al.* Pharmacotherapy for neuropathic pain in adults: a systematic review and meta-analysis. Lancet Neurol 2015; 14(2): 162-73.
[http://dx.doi.org/10.1016/S1474-4422(14)70251-0] [PMID: 25575710]

[202]   Volkow ND, Collins FS. The Role of Science in Addressing the Opioid Crisis. N Engl J Med 2017; 377(4): 391-4.
[http://dx.doi.org/10.1056/NEJMsr1706626] [PMID: 28564549]

[203]   Ding H, Kiguchi N, Yasuda D, *et al.* A bifunctional nociceptin and mu opioid receptor agonist is analgesic without opioid side effects in nonhuman primates. Sci Transl Med 2018; 10(456): eaar3483.
[http://dx.doi.org/10.1126/scitranslmed.aar3483] [PMID: 30158150]

[204]   Miller RE, *et al.* Current status of nerve growth factor antibodies for the treatment of osteoarthritis pain. Clin Exp Rheumatol. 35 Suppl 2017; 107(5): 85-7.

[205]   Bannwarth B, Kostine M. Nerve Growth Factor Antagonists: Is the Future of Monoclonal Antibodies Becoming Clearer? Drugs 2017; 77(13): 1377-87.
[http://dx.doi.org/10.1007/s40265-017-0781-6] [PMID: 28660479]

[206]   Yuan H, Spare NM, Silberstein SD. Targeting CGRP for the Prevention of Migraine and Cluster Headache: A Narrative Review. Headache 2019; 59 (Suppl. 2): 20-32.
[http://dx.doi.org/10.1111/head.13583] [PMID: 31291020]

[207]   Agostoni EC, Barbanti P, Calabresi P, *et al.* Current and emerging evidence-based treatment options in chronic migraine: a narrative review. J Headache Pain 2019; 20(1): 92.
[http://dx.doi.org/10.1186/s10194-019-1038-4] [PMID: 31470791]

[208]   Cardoso FC, Lewis RJ. Structure-Function and Therapeutic Potential of Spider Venom-Derived Cysteine Knot Peptides Targeting Sodium Channels. Front Pharmacol 2019; 10: 366.
[http://dx.doi.org/10.3389/fphar.2019.00366] [PMID: 31031623]

[209]   Maatuf Y, Geron M, Priel A. The Role of Toxins in the Pursuit for Novel Analgesics. Toxins (Basel) 2019; 11(2): E131.
[http://dx.doi.org/10.3390/toxins11020131] [PMID: 30813430]

[210]   Mogil JS. Qualitative sex differences in pain processing: emerging evidence of a biased literature. Nat Rev Neurosci 2020; 21(7): 353-65.
[http://dx.doi.org/10.1038/s41583-020-0310-6] [PMID: 32440016]

[211]   Lopes DM, Malek N, Edye M, *et al.* Sex differences in peripheral not central immune responses to pain-inducing injury. Sci Rep 2017; 7(1): 16460.
[http://dx.doi.org/10.1038/s41598-017-16664-z] [PMID: 29184144]

[212]   Sorge RE, Mapplebeck JC, Rosen S, *et al.* Different immune cells mediate mechanical pain hypersensitivity in male and female mice. Nat Neurosci 2015; 18(8): 1081-3.
[http://dx.doi.org/10.1038/nn.4053] [PMID: 26120961]

CHAPTER 6

Chronic Pain: Focus on Anticonvulsants

Erika I. Araya[1,*]**, Eder Gambeta**[1]**, Vanessa B. P. Lejeune**[1]**, Joelle M. Turnes**[1]**, Carlos H. A. Jesus**[1] **and Juliana G. Chichorro**[1]

[1] *Department of Pharmacology, Biological Sciences Sector, Federal University of Parana, Curitiba, Brazil*

Abstract: Chronic pain is a serious health problem that affects millions globally by decreasing their quality of life, social activities, and hours of restful sleep. It usually presents psychological comorbidities such as anxiety and depression, which persist in many cases throughout life. Satisfactory treatment of chronic pain is unattainable for the vast majority of patients and generally involves antidepressants and anticonvulsants. Among the anticonvulsants, phenytoin has been used in pain management since 1942, followed by the introduction of carbamazepine in the 1960s. Gabapentinoids (*i.e.*, gabapentin and pregabalin) have been extensively employed since 2002 in the treatment of several pain conditions and their comorbidities. Although pregabalin seems to have a better pharmacokinetic profile, gabapentin shows a superior therapeutic effect in some types of chronic pain. The analgesic potential of both is still being evaluated in clinical trials due to the need to develop more effective and safer medications for the interruption of treatment. The most common side effects are loss of coordination and drowsiness. Furthermore, T-channel blocking anticonvulsants were recently developed, but their efficacy seems to be inferior to gabapentinoids for chronic pain management. In addition to new strategies, some studies have suggested that the use of phenytoin in the treatment of trigeminal neuralgia should be reevaluated, as its topical use has been shown to diminish polyneuropathy and sciatic pain. This chapter will review the current application and efficacy of anticonvulsants in chronic pain management and will address new applications and drugs in advanced stages of development.

Keywords: Calcium channels, Carbamazepine, Flupirtine, GABA, Gabapentin, Glutamate, Neuropathic pain, Oxcarbazepine, Potassium channels, Potassium channels, Pregabalin, Retigabine, Sodium channels, Topiramate, Valproatess.

INTRODUCTION

Chronic pain is a frequent and disabling condition that globally affects around 30% of the population [1]. Systematic classification of chronic pain was

---

* **Corresponding author Erika I. Araya:** Department of Pharmacology, Biological Sciences Sector, Federal University of Parana, Curitiba, Brazil; Tel:+55 41 3361 1720, Email:erikaiaraya@gmail.com

Atta-ur-Rahman & Zareen Amtul (Eds.)

in 2015 by the International Association for the Study of Pain (IASP) according to the content model of the World Health Organization (WHO) for International Classification of Diseases (ICD), ICD-11, including pain severity, its temporal progression, and evidence of psychological and social factors [2]. In this sense, proper categorization of chronic pain is essential for effective pain therapy. Its control is challenging, often requiring multidisciplinary efforts, as it is usually refractory to multiple treatments, in addition to having a high impact on the patient's quality of life, with development of affective disorders and increased risk of drug dependence [3]. It is known that up to 85% of chronic pain patients have depression, and half of the adults with anxiety or mood disorders reported chronic pain [4 - 6]. In addition, drug addiction was reported in 40% of patients with chronic non-cancer pain [7]. Pharmacologic treatment for chronic pain conditions includes anticonvulsants, antidepressants, nonsteroidal anti-inflammatory drugs, opioids, alone or in combination, among others. Chronic pain includes non-specific low back pain, fibromyalgia, osteoarthritis, rheumatoid arthritis, chronic posttraumatic and postsurgical pain, neuropathic pain, as well as other forms that manifest secondarily to an underlying disease. Anticonvulsants are the first line therapy for neuropathic pain, which is a pain arising as a direct consequence of a lesion or disease affecting the somatosensory system. It is often undertreated, thus causing suffering, disability, impaired quality of life, and increased costs [8].

Anticonvulsants in Chronic Pain Treatment

The mechanism of action of anticonvulsants consists, in general, the inhibition of neuronal hyperactivity by blocking ion channels, including voltage-gated sodium and calcium channels. In addition, they can also inhibit the release of excitatory amino acids (*i.e.*, glutamine and aspartate) or improve the inhibitory activity of γ-aminobutyric acid (GABA) [9]. Nowadays, anticonvulsant drugs are mainly used in the treatment of neuropathic pain [10] and increasingly in the treatment of chronic pain [11]. However, the adverse events of anticonvulsant therapy that have been reported [12] include drowsiness, dizziness, fatigue, ataxia, and changes in body weight, which are the most common [12], in addition to an increased suicide risk with the misuse of some anticonvulsant drugs [13].

The aim of this chapter is to review the pharmacokinetics, mechanisms of action, efficacy and adverse events of the anticonvulsant drugs used in chronic pain treatment, and to present the perspectives on the development and study of novel therapies.

HISTORY OF ANTICONVULSANT USE IN PAIN MANAGEMENT

The analgesic effect of anticonvulsants was first described in 1942 by Bergouignan from studies carried out in patients with trigeminal neuralgia, suggesting that phenytoin was able to reduce paroxysms [14]. Before the 1950s, several studies reported a potential analgesic role of phenytoin in trigeminal neuralgia, which were corroborated by several case series indicating its efficacy (for review, see [15]). However, the literature still lacks randomized controlled clinical trials evaluating the efficacy and safety of phenytoin in trigeminal neuralgia, including comparisons with other anticonvulsants. In 1962, studies conducted by Blom demonstrated that carbamazepine presented a satisfactory outcome in the treatment of trigeminal neuralgia [16], and subsequent clinical studies confirmed this analgesic effect by decreasing the conductance in the sodium channels and inhibiting ectopic discharges (as will be further discussed in detail). The results of high-quality clinical trials with other anticonvulsants were encouraging in the treatment of trigeminal neuralgia, painful diabetic neuropathy, and post-herpetic neuralgia, thus marking a new phase in the treatment of neuropathic pain [16, 17]. In 1969, oxcarbazepine, a structural analog of carbamazepine, was patented, but its medical use started only in 1990 [18]. It is classified as a first-line treatment, though its effectiveness in trigeminal neuralgia needs more evidence [19].

A milestone in the history of anticonvulsant use in pain treatment was the clinical introduction of gabapentinoids, which include gabapentin and pregabalin. Gabapentin was discovered in 1975 by Parke-Davis and patented in 1977, having its first use approved by the FDA in 1993 [18]. It is a first-line treatment drug [20] for diabetic neuropathic pain and post-herpetic pain; however, its use in other painful neuropathies requires additional studies to certify its effectiveness [21, 22]. Pregabalin was synthesized in 1990 as an anti-epileptic and analgesic drug. It was developed as a successor to gabapentin and was approved in the United States in 2004. It is recommended as a first-line treatment for chronic pain [20] in diabetic neuropathic pain, postherpetic pain, and central neuropathic pain [17], but its use remains controversial for sciatica, low back pain, cancer-related pain, migraine, and prevention of chronic postoperative pain [18]. Another important anticonvulsant used in the treatment of somatic neuropathic pain is lamotrigine, which was approved for epilepsy in 1994 in the United States and has been used in pain management ever since. In addition to its use for the prevention and management of seizures, it is also approved as a mood stabilizer in bipolar disorder. The World Health Organization (WHO) lists lamotrigine as essential medicine and one of the most effective and safe medicine in a healthcare system [18]. However, its use for the treatment of neuropathic pain, migraine, peripheral neuropathy, and trigeminal neuralgia is off-label. Lacosamide indicated as monotherapy or adjuvant therapy in focal seizures or diabetic neuropathic pain [18], was discovered in 1996 by Harold Kohn, Shridhar Andurkar, and colleagues at the University of Houston. Valproate, first produced in 1881 by Beverly S. Burton as an analogue of valeric acid, naturally found in Valeriana officinalis, is also part of the list of essential medicines proposed by the WHO and used in some cases of chronic pain [18].

Therefore, since the discovery of the pain-relieving properties of phenytoin, researchers have constantly been investigating the use of anticonvulsants in the treatment of various painful conditions. However, their application in pain management is mostly off label and lacks evidence for their widespread use in several pain conditions [22].

CLASSIFICATION

Anticonvulsants have been used to treat nonepileptic central nervous system disorders. Most anticonvulsants present multiple mechanisms of action; however, they are routinely categorized by their main mechanisms (Table 1), which include modulation of γ-aminobutyric acid (GABA) GABAergic neurotransmission, changes in the function of voltage-gated ion channels, and intracellular signaling pathways [23, 24]. These specific mechanisms explain the efficacy of anticonvulsants in the treatment of neuropathic pain and other pain syndrome therapies. Their diverse chemical structures present various activity spectra and different tolerability profiles [25].

Table 1. Classification of some anticonvulsants used in pain management according to their mechanisms of action.

ANTICONVULSANT	Ion channel modulator Sodium	Ion channel modulator Calcium	Other targets Potassium	Other targets GABAergic	Other targets Modulators of Presynaptic Machinery	Other targets Selective inhibitors of excitatory neurotransmission
Phenytoin	X					
Carbamazepine	X					
Oxcarbazepine	X					
Lamotrigine	X					
Valproate	X			X		
Gabapentin		X				
Pregabalin		X				
Retigabine			X			
Topiramate	X	X	X	X	X	
Tiagabine				X		
Perampanel						X
Levetiracetam					X	

Neuropathic pain represents a heterogeneous group of conditions that differ widely in etiology (*i.e.,* diabetes, medication, viral infections, vitamin deficiency, trauma, *etc.*) [24]. Therefore, the underlying mechanisms and resulting body distribution and pain phenotype also vary widely. Moreover, patients' comorbidities, age, and general health conditions may affect the efficacy and safety of anticonvulsant treatment. Thus, all these factors have been taken into consideration to define the most appropriate anticonvulsant for pain treatment.

It is often unclear which is the most important mechanism for a particular effect of each anticonvulsant [9]. However, the depression of neuronal activity is considered the basis of anticonvulsant use in pain management, which is achieved through different mechanisms [18]. Deep knowledge of the pharmacology of the anticonvulsants combined with a careful patient evaluation are crucial aspects to achieve better treatment outcomes.

ION CHANNEL MODULATORS

Sodium Channels

Voltage gated sodium channels (VGSC) are widely expressed in the membrane of excitable cells such as neurons and cardiac cells. These channels are essential to trigger and propagate the action potential. The structure of VGSC is composed of an alpha (α) subunit, which can be associated with one or two beta (β) subunits [26]. The α subunit is composed of four homologous domains (DI-DIV), and each domain has six transmembrane segments (S1-S6). The α subunit is responsible for forming the selective pore for sodium ions (Fig. **1**). The S4, positively charged, is responsible for the voltage dependent activation, and the S5 and S6 integrate the channel pore [27]. Nine different isoforms for the α subunit have been identified depending on the gene that codifies the structure, which are named $Na_v1.1$ to $Na_v1.9$ [28, 29]. The $Na_v1.1$, $Na_v1.2$, $Na_v1.3$, and $Na_v1.6$ are commonly expressed in the central nervous system, while the $Na_v1.7$, $Na_v1.8$, and $Na_v1.9$ are expressed in the periphery. Meanwhile, $Na_v1.4$ is expressed in skeletal muscle cells and $Na_v1.5$ in cardiomyocytes [29]. However, $Na_v1.1$ and $Na_v1.6$ are also expressed in sensory neurons, and together with $Na_v1.7$, $Na_v1.8$, and $Na_v1.9$, they have an important role in pain processing [30]. Interestingly, the $Na_v1.3$ is highly expressed during the developmental stage while its expression decreases in adulthood; however, this channel can be highly reexpressed in sensory neurons after a lesion [30]. It is important to highlight that these channels present different biophysical properties. Moreover, they can be found in three conformational states: closed, open, and inactive [31]. At resting membrane potential, the VGSCs are in closed or resting states, and after the depolarization of the membrane, the channels open and let sodium ions flow inside the cell. Within milliseconds the channel inactivates, and there are two types of inactivation for VGSC: a fast and a slow inactivation [31]. In fast inactivation, usually between 1-2 ms after channel activation, the inactivation gate formed by the DIII and DIV linker region folds over the pore and prevents the influx of sodium [31]. Meanwhile, in states of prolonged depolarization, the channel enters a conformation of slow inactivation, which depends on prolonged repolarization to recover from inactivation [31]. Furthermore, one of the main differences between the fast and slow inactivation is the recovery time of the channel.

Due to their expression in peripheral nerves, the VGSCs are highly important in pain processing. Each channel has a different role in the conduction of the action potential. $Na_V1.9$ and $Na_V1.7$ are responsible for amplifying the subthreshold stimuli while $Na_V1.7$ and $Na_V1.6$ contribute to the initial phase of the rise of the action potential, and $Na_V1.8$ plays the major role in the rising phase [30]. Several studies have already demonstrated that chronic pain conditions are associated with changes in the expression and/or function of VGSCs, suggesting a crucial role in chronic pain physiopathology [32]. With these in mind, drugs that block or modulate VGSCs present as potential analgesics, and the ones commonly used in the clinical setting will be described below.

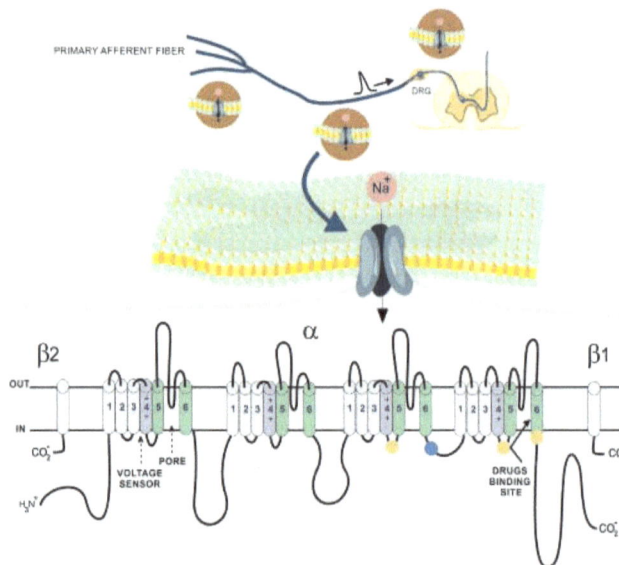

Fig. (1). Distribution and structure of voltage-gated sodium channels (VGSC). The upper panel shows that the VGSCs are expressed in the peripheral terminal, along the axon, and in the synaptic nerve terminals in the spinal dorsal horn. These channels are important to trigger action potentials and conduct the information to the central nervous system. The structure of the VGSCs is depicted in the lower panel. The VGSCs are composed of an alpha (α) subunit, which is composed of four domains with six transmembrane segments and one or two beta (β) subunits. The fourth transmembrane segment is responsible for voltage sensing, while the fifth and sixth form the pore region. The blue circle represents the inactivation gate, and the yellow circles the regions implicated in forming the inactivation gate receptor. The binding site for anticonvulsants is located in the external pore loop and the pore-lining of segment 6 in domain 4.

Pharmacokinetics Aspects

Carbamazepine (CBZ) has a bioavailability of approximately 80% after oral ingestion [33]. The volume of distribution is in the range of 0.7 to 1.4 L/Kg [34, 35]. CBZ is metabolized almost completely in the liver, and the major metabolite is carbamazepine-10,11-epoxide, which is the pharmacologically active component. This conversion is catalyzed mainly by CYP3A4 and CYP2C8 [36]. Other minor pathways can involve several CYP450 isoforms generating other metabolites, such as 3-hydroxycarbamazepine, 2-hydroxycarbamazepine, and *o*-quinone [37 - 39]. Moreover, other enzymes are involved in CBZ metabolism, including uridine diphosphate glucuronosyltransferase and myeloperoxidase [39 - 42]. After oral administration of CBZ, the time to reach the maximum concentration was 0.5 hours with a peak concentration of 37.80 µg/mL, and the elimination half-life was approximately 3.3 hours [43]. Furthermore, *in Vitro* application of CBZ at 40 µM resulted in around 70% plasma protein binding [43]. It is important to highlight that CBZ causes an increase in liver CYP3A4 expression, favouring its metabolism, which can reduce its half-life after repeated doses [44]. Thus, the initial dose of CBZ may become ineffective in 2-3 weeks after starting the treatment.

Oxcarbazepine (OXC) is completely absorbed after oral ingestion, and according to the FDA, the volume of distribution of OXC is 49 L/Kg. OXC is metabolized to 10,11-dihydro-10-hydroxy-carbazepine, a monohydroxy derivative (MHD), by aldo-keto reductase enzymes [44 - 46]. MHD is further metabolized to a glucuronide metabolite [47]. The peak concentration of OXC is reached within 1-3 hours, with a maximum concentration of 1.35 µg/mL with 2.45 hours as half-life elimination [45, 46]. As OXC does not interfere with the expression of CYP enzymes, its pharmacokinetic profile is considered more favourable.

After oral administration, phenytoin (PHT) is well absorbed and reaches a peak plasma concentration for up to 8 hours in its extended-released formulation [48 - 50]. With a volume of distribution of 0.8 L/kg, the drug is well distributed in the body. Around 90% of the drug is bound to plasma protein and can easily cross the blood brain barrier [50]. PHT is metabolized in the liver by the cytochrome P450 enzyme [50]. Moreover, its elimination kinetics present a non-linear elimination with a half life of 22 hours [51].

The peak concentration of lamotrigine (LTG) is reached within 1-3 hours after a single dose with a bioavailability of 98% [52, 53]. The volume of distribution of LTG ranged from 0.9 to 1.2 L/Kg [54, 55]. It is estimated that 55% binds to plasma protein [56]. LTG is metabolized in the liver by uridine-diphosphate glucuronosyltransferase (UGT) [57]. The elimination half-life of LTG ranges between 23-37 hours (single and multiple doses); however, enzyme-inducing drugs such as CBZ and PHT reduce the half-life to 15 hours [56].

Mechanisms of Action

The general mechanism of this class of anticonvulsants is to block VGSCs; however, some will modulate the fast inactivation by a hinge-lid mechanism, while others facilitate slow inactivation inducing a conformational change in the pore.

CBZ blocks VGSCs leading to inhibition of the action potential, which results in a decrease of the synaptic transmission and stabilization of the membrane potential in hyperexcited neurons [44, 58]. CBZ binds the channel in its inactive conformational state with high affinity and presents a use- and voltage-dependent effect [59, 60]. The interaction between the external pore loop and the pore-lining part of the S6 of the DIV, composed by the residues W1716 and F1764, respectively, forms the binding pocket for CBZ [60, 61]. The two phenyl groups in the CBZ structure are recognized by the narrow part of the channel pore and the junction of the widened external vestibule, which is the proposed binding site for CBZ [60, 62]. After binding in this region, the channel gating properties change, and the channel becomes more stable in its inactivated state, preventing sodium influx [60]. Using the heterologous expression of the human $Na_V1.7$ channel in HEK293 cell linage, it was demonstrated that external application of CBZ was able to block sodium current in whole-cell patch clamp recordings, while the internal administration did not have an effect [63]. However, when inside-out recordings were performed, CBZ was able to inhibit sodium current. This suggests that CBZ can reach its binding site when applied from either side of the membrane [63]. A similar effect was observed in isolated neurons from the DRG. The tetrodotoxin (TTX)-resistant sodium current from the $Na_V1.8$ channel was blocked by CBZ in a use-dependent manner and increased the time constant for recovery from inactivation, which suggests an interaction with the inactive state of the channel [64 - 67].

Since OXC is a variation from CBZ, it presents a similar mode of action that blocks the sodium current. The OXC effect differs from other sodium blockers anticonvulsants as it can achieve the current inhibition with a lower concentration [68]. Like CBZ, OXC is also able to inhibit $Na_V1.7$ channel currents, promoting a hyperpolarizing shift in the activation and inactivation curves of the channel [69]. One major difference between OXC and CBZ is in their actions on voltage-gated calcium channels (VGCCs). CBZ can block L-type channels, while the metabolite from OXC, MHD, blocks N-, P- and R-type channels [70 - 72].

PHT is a first-generation anticonvulsant drug with a mechanism of action similar to CBX and OXC. The mechanism of action for PHT has been studied for over 80 years, with the first proposed mechanism published in 1937 by Putnam and Merritt [73]. The authors developed an animal model for screening anticonvulsants by electroshock-induced convulsion, and they found that PHT was able to increase the threshold in this model [73]. It was only in 1949 that it was considered a blocker, somehow inhibiting and decreasing the influx of sodium ions [74]. Over 20 years later, in 1972, Lipicky and colleagues found that PHT decreased the sodium conductance in a dose dependent manner, suggesting a direct block of the sodium channels [75]. Several studies in the 1990s demonstrated that PHT was able to stabilize the channel in its inactive conformational state, blocking the sodium conductance [76 - 78]. One important feature of its mechanism of action is its slow onset and slow recovery from inactivation [79, 80]. A more detailed history of the discoveries regarding the mechanism of action of phenytoin was reviewed by Hasselink in 2017 [81].

LTG is a triazine with the main effect of reducing glutamate release by blocking VGSCs, VGCCs, and voltage-gated potassium channels [82]. It preferentially binds to the inactive conformation of VGSCs [79 - 83]. Moreover, it was demonstrated that LTG shares a common binding site with CBZ and PHT (77). It is also known that the binding sites for LTG in VGSCs are in the both S6 of the DIII and the DIV [84]. Additionally, LTG can also inhibit N- and P-type VGCCs, which can also contribute to its analgesic effect [57].

Indication and Efficacy

CBZ is the only drug approved by the FDA for the treatment of trigeminal neuralgia and, along with OXC, is recommended for long-term treatment of this condition [85, 86]. Moreover, CBZ is used off-label for the treatment of other neuropathic pain conditions, fibromyalgia, and central post-stroke pain [87, 88]. PHT and LTG may each be used as monotherapy or in combination with CBZ or OXC for trigeminal neuralgia [86, 89]. LTG is also indicated for postherpetic neuralgia [90].

A recent study assessed the effectiveness and tolerability of CBZ and OXC in trigeminal neuralgia [91]. The authors found that both drugs had high efficacy, and around 90% of patients responded to the treatment; however, CBZ presented more frequent side effects compared to OXC [91]. Additionally, clinical studies demonstrated that 12-week treatment with CBZ was able to reduce the pain in type 2 diabetic patients with neuropathic pain and in patients with erythromelalgia [92, 93]. Moreover, a comparative study evaluated the efficacy between LTG and CBZ in patients with trigeminal neuralgia. Around 90% of patients reported pain relief with CBZ, with only 21% reporting a "complete" pain relief using a visual analog scale. Furthermore, 57% of patients tolerated the side effects [94]. On the other hand, a systematic review found that in the studies analyzed, none provided first or second tier evidence for an efficacy outcome. Third tier evidence demonstrated that CBZ presented a better analgesic profile compared to placebo in three different conditions of chronic pain (trigeminal neuralgia, diabetic neuropathic pain, and post stroke pain) [87], with the authors estimating the NNT for CBZ as 1.7 (1.5 to 2.0). Besides being a first-line treatment for trigeminal neuralgia, a systematic review found little evidence to support the efficacy of OXC in other neuropathic pain conditions [95, 96].

Several studies have already demonstrated the efficacy of PHT for the treatment of trigeminal neuralgia, and it is recommended as a therapy for treatment-resistant trigeminal neuralgia [15]. A systematic review tried to assess the analgesic efficacy of PHT in neuropathic pain and fibromyalgia; however, the literature did not reveal any studies that met the inclusion criteria [97], while two studies found conflicting data regarding its efficacy on diabetic neuropathy [98, 99]. However, recent studies reported that a cream formulation of PHT reduced pain in patients with neuropathic pain [100 - 102]. On the other hand, no first-tier evidence for efficacy was provided for LTG in different neuropathic pain conditions [86, 103 - 105]. However, LTG is indicated as a first-line treatment in cases of central post-stroke pain [88].

Side Effects

The most common side effects of this class of anticonvulsants include dizziness, drowsiness, headache, nausea, and vomiting.

A systematic review that analyzes the efficacy and safety profile of CBZ concluded that around 50% of patients reported side effects such as cognitive impairment, drowsiness, somnolence, headaches, gastrointestinal symptoms, and mood change. Although rare, severe side effects, including upper gastrointestinal bleeding and cutaneous rashes, were described, the latter being very serious due to association with CBZ-induced Steven-Johnson syndrome [106]. A comparative study between CBZ and OCX found that the most common adverse effects were CNS depression leading to somnolence, loss of balance, and dizziness [91]. Moreover, these side effects recurred more frequently with CBZ treatment (17.3%) than with OXC treatment (5.7%), resulting in discontinuation of treatment or lowering the dosage to a substandard level [91]. It is important to highlight that patients carrying the gene variation *HLA-B*1502* have a higher risk of developing severe skin reactions (107). However, a meta-analysis demonstrated that patients carrying the variations *HLB-B*4001, HLB-B*4601,* and *HLB-*

B*5801 present a strong protective effect against CBZ-induced Steven-Johnson syndrome [108]. A frequent side effect from CBZ and OXC is hyponatremia, *i.e.,* a reduction of sodium levels, which can decrease the extracellular osmolarity and cell swelling, which could contribute to the central effects [109].

The most common side effects of PHT and LTG include drowsiness, fatigue, loss of body movement and coordination, restlessness, headache, nausea, vomiting, hyponatremia, and rashes [50, 110, 111]. Severe PHT side effects could lead to anaphylaxis or DRESS (drug reaction with eosinophilia and systemic symptoms) [50] associated with drug-induced hypersensitivity. Due to this, PHT could produce a severe cutaneous adverse reaction (SCAR) which can be life-threatening [50].

Perspectives

Eslicarbazepine acetate (ESL) is a carboxamide anticonvulsant that shares a similar structure with other carboxamides drugs such as CBZ and OXC [112]. ESL presents a more favorable metabolic profile with reduced enzymatic induction, and unlike CBZ, it is not metabolized into carbamazepine-10,1--epoxide, which produces unfavorable side effects [112]. The peak concentration of ESL is reached 2-3 hours after ingestion with a half-life up to 24 hours [113, 114]. With scarce clinical evidence regarding its efficacy, ESL treatment has shown good safety, tolerance, and effect profile in trigeminal patients and in diabetic neuropathy (NCT00980746) [115, 116]. In addition, a pre-clinical study found that in different models of pain (trigeminal nociception induced by formalin, streptozotocin-induced type 1 diabetic neuropathy, and visceral pain), treatment with ESL produced a dose-dependent antinociceptive effect [117]. Moreover, the authors suggest that ESL can also promote the antinociceptive effect by the serotonergic 5-HT$_{1B/1D}$ and CB$_1$/CB$_2$ cannabinoid receptors [117]. Furthermore, a study found that ESL was also able to block the voltage-gated calcium channel Ca$_v$3.2, which has been shown to be involved in chronic pain [118, 119]. A recent meta-analysis assessed the tolerability and safety of ESL, which revealed a higher occurrence of side effects and withdrawal with ESL compared to placebo [120]. The study included neuropathic pain, migraine, fibromyalgia, epilepsy, and bipolar I disorder, and similar to other VGSCs blockers, the most significant side effects were blurred vision, vertigo, nausea, vomiting, and dizziness [120].

Lacosamide (LCM) is a third-generation anticonvulsant with a better safety profile than first-generation anticonvulsants like carbamazepine and phenytoin [121]. LCM is rapidly absorbed with a peak concentration reached 4 hours after the intake and a maximum concentration of 7.4 μg/mL [121]. Several enzymes from the cytochrome P450 system are responsible for LCM metabolism [121]. With a very distinct mechanism of action compared to the classical sodium blocker anticonvulsants, LCM enhances the slow inactivation of the sodium channel [122 - 124]. However, a recent finding demonstrated that LCM could also bind to the fast inactivation state with slower kinetics (28119481) [125]. LCM is indicated for several types of chronic pain, including diabetic neuropathy, fibromyalgia, and neuropathic pain (NCT00546351, NCT00401830, NCT00237458, NCT03777956) [126, 127].

Calcium Channels

VGCCs are important regulators of brain, heart, and muscle functions, and their dysfunction can give rise to numerous pathophysiological conditions. Calcium channel blockers have been successfully developed for the treatment of absence seizures and are an emerging drug class for the treatment of neurological disorders and chronic pain [128, 129]. The gabapentinoid drugs gabapentin and pregabalin were originally developed as antiseizure drugs but currently represent the most frequently prescribed calcium channel modulators for neuropathic pain [129].

Gabapentinoids are GABA derivatives, but their mechanism of action has been considered unrelated to direct effects on the GABAergic system. The first FDA approval for gabapentin in 1993 as adjunctive therapy for partial seizures was followed by its indication for postherpetic neuralgia management in 2002. Likewise, the congener drug pregabalin was FDA approved for postherpetic neuralgia in 2004, followed by a more extensive list of approvals comparable to gabapentin, including generalized anxiety disorder, neuropathic pain associated with diabetes or spinal cord injuries, and fibromyalgia. It is noteworthy that pregabalin became the first FDA-approved drug for the specific treatment of this later condition [130, 131]. Moreover, both pregabalin and gabapentin represent first-line recommendations for neuropathic pain treatment, according to international guidelines [132, 133].

In spite of these limited indications, gabapentinoids are widely prescribed off-label for various pain syndromes, including trigeminal neuralgia, lower back pain, and pain from osteoarthritis and surgery. In fact, there is evidence that >95% of the prescriptions of this class are off label in nature. Additionally, a marked increase in their prescription has been reported in recent years [11, 134]. From 2012 to 2016, gabapentin prescription in the USA increased by 64%, and in 2017, it became the 10th most commonly prescribed medication [131]. Reasons proposed for such growth include the search for safer alternatives to opioids, extensive marketing strategies, and limited pharmacological options for pain treatment. However, this growth has recently become a matter of concern due to evidence of abuse liabilities, risk of respiratory depression when combined with opioids, and overdose fatalities with gabapentinoids, which will be further discussed in this chapter [11, 131, 134].

Pharmacokinetics Aspects

Some marked differences in the pharmacokinetic profile of gabapentin and pregabalin have implications for clinical practice. Both drugs are structurally similar to the amino acid leucine and undergo active uptake across cell membranes through l-amino acid transporters (LAT) to reach their intracellular site of action. Both are absorbed in the small intestine, but absorption of gabapentin depends exclusively on LAT, making the absorption process saturable and resulting in a nonlinear pharmacokinetic profile. Conversely, pregabalin may be transported by other carriers, in addition to LAT, resulting in non-saturable absorption and proportionality between dose and plasma concentration [135]. The clinical implication of this difference is faster absorption of pregabalin compared to gabapentin (1 *versus* 3 hours) and higher oral bioavailability (90% compared to 30-60% for gabapentin). Neither drug binds to plasma proteins and are highly water-soluble (volume of distribution estimated in 0.8 and 0.5L/kg for gabapentin and pregabalin, respectively) [136]. According to their characteristics and pre-clinical studies, their secretion into human breast milk is expected. Moreover, they are able to cross the blood-brain barrier, and at least for pregabalin, there is evidence that LAT1 is involved in its distribution to the spinal cord across the blood-spinal cord barrier and to the fetus across the placental barrier [136, 137]. The hepatic metabolism of gabapentinoids is similar and negligible. Additionally, they do not induce or inhibit the major cytochrome P450 (CYP) enzymes responsible for drug metabolism. Thus, neither metabolic drug-drug interactions nor pharmacokinetic variations in the function of genetic polymorphisms of metabolizing enzymes are expected [136]. There are slight differences in renal elimination, and for pregabalin, more than 90% of the drug is renally excreted, mostly without prior metabolization [136, 138]. The half-life parameter is similar and estimated in 5-7 hours for gabapentin and around 6 hours for pregabalin [136].

Mechanisms of Action

Gabapentin and pregabalin were originally designed as GABA mimetics, but in spite of their structural similarity, they do not affect or bind to GABA-A or GABA-B receptors [135]. Gabapentinoids selectively bind to the alpha-2-delta ($\alpha_2\delta$), which is an auxiliary subunit of VGCCs (Fig. **2**). The subtypes 1 and 2 (*i.e.,* $\alpha_2\delta$-1 and $\alpha_2\delta$-2), their main targets, are expressed in neurons, cardiac and skeletal muscle tissue, as well as presynaptic endings of neurons in the brain and spinal cord. The $\alpha_2\delta$-1 expression appears to co-localize mainly with excitatory neurons, whereas $\alpha_2\delta$-2 is found largely in inhibitory neurons. In line with this observation, it has been suggested that gabapentinoids binding to the $\alpha_2\delta$-1 subunit of VGCC contribute to the analgesic effects, whereas the $\alpha_2\delta$-2 subunit contributes to the central side effects, suggesting that ligand selectivity for $\alpha_2\delta$-1 and $\alpha_2\delta$-2 might result in different clinical outcomes [139 - 141].

The $\alpha_2\delta$-1 expression is enriched in many areas of the sensory and affective pain component, such as the spinal dorsal horn, periaqueductal gray (PAG), anterior cingulate cortex (ACC), insula, and amygdala with gabapentinoid analgesic effect due in part to its action on $\alpha_2\delta$-1 in these areas [141]. It is widely accepted that neuropathic pain conditions see increased expression of the $\alpha_2\delta$-1 subunit in DRG neurons and the spinal cord while gabapentinoids binding to the $\alpha_2\delta$-1subunit has been shown to inhibit trafficking of the subunit from DRG neurons to central terminals in the dorsal horn [141]. Moreover, gabapentinoids have been demonstrated to inhibit the release of several transmitters, including glutamate, but the precise mechanism remains unknown. Some studies suggest a direct inhibition of VGCC, while others propose that they restore the aberrant neurotransmission resultant of up-regulation of the $\alpha_2\delta$-1subunit associated with neuropathy [135, 141].

Another well documented consequence of increased expression of the $\alpha_2\delta$-1subunit is increased excitatory synaptogenesis, which results in augmented neuronal network excitability, an important contributor to central sensitization. This mechanism has been related to the ability of the $\alpha_2\delta$-1subunit to interact directly with extracellular thrombospondin adhesion molecules secreted by activated astrocytes. Thrombospondins are endogenous ligands of $\alpha_2\delta$-1 and have been demonstrated to up-regulate in neuropathic pain conditions. Thus, gabapentinoids have been suggested to prevent the formation of new excitatory synapses in response to astrocyte activation after neuronal damage or sustained neuronal activation [135, 139, 140].

It is also noteworthy that many studies have reported that central mechanisms contribute to the analgesia induced by gabapentinoids. Their ability to restore noradrenergic descending inhibition and inhibit descending serotonergic facilitation has been demonstrated in this regard.These mechanisms seem to involve the activity of gabapentinoids in the $\alpha_2\delta$-1subunit and may be of great importance in conditions associated with deficits in endogenous descending pain modulation, such as fibromyalgia [135, 140].

Another target increasingly implicated in gabapentinoid-induced analgesia is the N-methyl-D-aspartate (NMDA) receptor. Studies have shown that the $\alpha_2\delta$-1 may form a complex with NMDA receptors and regulate their synaptic trafficking and activity in the brain and dorsal root ganglia. Thus, the binding to the $\alpha_2\delta$-1subunit would result in reduced expression of $\alpha_2\delta$-1-NMDA receptor complexes. In line with this observation, in pre-clinical models of pain, the pre-treatment with an NMDA receptor antagonist potentiated the analgesic effect of pregabalin [139, 142, 143]. Likewise, recent findings indicate that the $\alpha_2\delta$-1 has the potential to interact with many other targets, including potassium channels, synaptic cell-adhesion molecules, GABA receptors, but the contribution of these mechanisms to gabapentinoid analgesic effects remains to be clarified (for review see [139]).

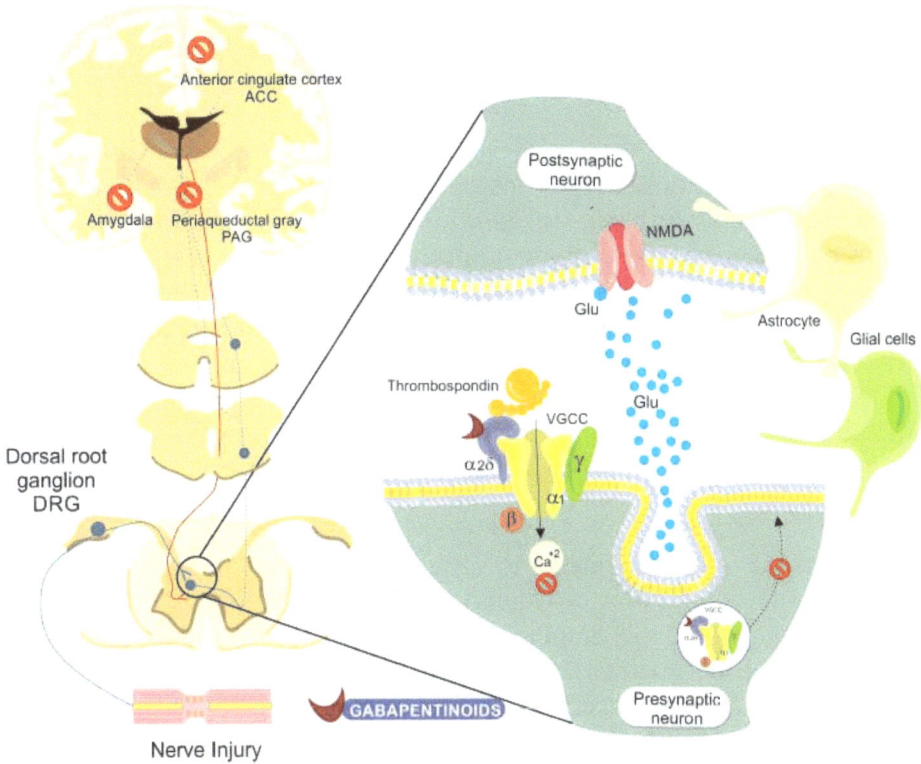

Fig. (2). Analgesic mechanisms of gabapentinoids. Gabapentinoids selectively bind to alpha-2-delta ($\alpha_2\delta$) subunit of voltage-gated calcium channels (VGCCs) in the presynaptic endings of neurons in the spinal cord. In this synapse (enlarged image), gabapentinoids binding to the $\alpha_2\delta$-1subunit results in inhibition of trafficking of the subunit and the release of several transmitters, including glutamate (Glu). Gabapentinoids may also prevent the formation of new excitatory synapses in response to astrocyte activation after neuronal damage or sustained neuronal activation. This mechanism results from the interaction of the $\alpha_2\delta$-1 subunit with thrombospondin, which may be disrupted in the presence of gabapentinoids. Central mechanisms also contribute to the analgesia induced by gabapentinoids. The $\alpha_2\delta$-1 expression is enriched in several brain areas such as periaqueductal gray (PAG), anterior cingulate cortex (ACC), and amygdala, which represent relevant sites for gabapentinoids analgesic effects.

Indication and Efficacy

Gabapentinoids represent first line medications for the treatment of several neuropathic pain conditions. According to a previously conducted study [133], the number needed to treat (NNT) for 50% pain relief in patients is 7.7 for pregabalin and 7.2 for gabapentin. It is important to mention that there are many different types of neuropathic pain conditions and high variability in pain phenotypes across patients, which may account for this high NNT value. In fact, systematic reviews have reported lower NNT values for pregabalin use in post-herpetic neuralgia and diabetic neuropathic pain (6 and below) and higher NNT values for fibromyalgia (7 and above) [144, 145]. Both pregabalin and gabapentin are considered equally effective in the treatment of patients with neuropathic pain associated with spinal cord injury while also improving sleep, anxiety, and general patient status [146].

In off label use of gabapentinoids, there is no evidence that pregabalin is effective in acute pain conditions such as postoperative pain or inflammatory chronic pain conditions, such as arthritis. In addition, it has been shown that these do not prevent chronic pain after surgery, but there is some modest evidence that they may reduce opioid consumption after surgery. As will be discussed in the next section, the risk of this association has to be taken into consideration [135, 145, 147, 148]. Pregabalin was not considered effective in pain related to HIV [133, 135]. Additionally, a systematic review of chronic low back pain studies concluded that gabapentinoids did not demonstrate benefits but was associated with an increased risk of adverse events in patients suffering from this condition [149].

Few studies have investigated the efficacy of gabapentinoids in other painful conditions such as migraine or trigeminal neuralgia, while the quality of evidence is low for many off label indications. Thus, considering the risks and adverse events associated with these drugs (see below), it is recommended that their prescription should be based on current scientific evidence combined with a complete history of the patient.

Side Effects

The clinical utility of pregabalin and gabapentin is limited by central nervous system (CNS) side effects, mainly dizziness, drowsiness, and somnolence. These effects are relatively common and may worsen when gabapentinoids are prescribed concomitantly with other drugs, which affect CNS normal functioning; however, they usually resolve with drug discontinuation. Other common side effects include nausea/vomiting, dry mouth, constipation, fatigue, nasopharyngitis, blurred vision, weight gain, confusion, memory loss, irritability, and insomnia [11, 135, 149].

Most adverse reactions involving the vestibulocerebellar/brainstem structures and higher cortical function have a clear dose-response relationship and are related to the mode of action of gabapentinoids. VGCCs (*i.e.,* P- Q- and N-type) are highly expressed in the cerebellum and hippocampus, and their dysfunction/inhibition has been linked to ataxia and cognitive impairment. Unlike central side effects, gastrointestinal and metabolic dysfunction related to gabapentinoid use are less common and do not share a clear dose-response relationship or well-defined pathogenesis [135, 150]. In this regard, the increased evidence of the association of pregabalin with peripheral edema and worsening of heart failure symptoms is noteworthy [151, 152]. The mechanisms underlying these effects are not totally clear, but the peripheral edema may be the result of antagonism of the L-type calcium channel in the vasculature causing vasodilation, similar to the mechanism of action of calcium channel blockers used to treat hypertension [153]. This greater incidence with pregabalin than gabapentin may be due to the greater potency of pregabalin for the inhibition of the $\alpha_2\delta$ subunit of VGCCs. Although less common than central nervous system toxicity, cardiovascular toxicity may lead to serious consequences and should be considered by clinicians, especially in patients already at risk.

The gabapentinoids safety profile has contributed to their wide use and extensive off-label prescribing. However, in the past ten years, increasing reports link gabapentin and pregabalin to abuse and dependence, life-threatening intoxications, and death [154]. The likelihood of gabapentinoids abuse is reportedly heightened among current or past users of opioids and benzodiazepines. A systematic review of gabapentin misuse reported a prevalence of 1% in the general population, 40–65% among individuals with prescriptions, and between 15–22% in populations who abuse opioids [155]. Reports of patients' misuse and abuse with gabapentinoids effects of dissociation, euphoria, sedation/relaxation/calmness, elevated mood, disinhibition, delirium, feeling "high" and feeling drunk [156]. This class of drugs is commonly abused using supratherapeutic doses in combination with alcohol, cannabis, Selective Serotonin Reuptake Inhibitors (SSRIs), amphetamines, and heroin, as well as to potentiate opioid-induced "highs". It has been suggested that "GABA-mimetic" properties of gabapentinoids cause these effects at doses well above those prescribed [157]. Though rarely addictive in the general population, it is recommended that their use should be avoided or closely monitored only in patients at risk, especially in those with a history of previous substance abuse or concomitant use of opioids and benzodiazepines [154].

Finally, clinicians should be aware that abrupt interruption of gabapentinoid treatment, especially at high doses, has been associated with withdrawal symptoms that resembled those of SSRIs or benzodiazepines. The most prevalent symptoms are agitation with confusion and disorientation, but there are reports of dysphoria, irritability, depersonalization, gait instability, vertigo, dizziness, fatigue, tremor, insomnia, myalgia, flu-like symptoms, abdominal discomfort, nausea, sleeplessness, palpitations, and tachycardia. Elderly patients and those with psychiatric disorders are at higher risk, and a slower tapering schedule is recommended to minimize it [135, 154, 156, 158].

Perspectives

Advances in the development of a ligand for the $\alpha_2\delta$ subunit of VGCCs include an extended-release formulation of gabapentin (*i.e.*, gabapentin ER), gabapentin enacarbil, and mirogabalin (*i.e.*, mirogabalin besylate).

A recent systematic review and meta-analysis of randomized controlled trials comparing gabapentin ER and gabapentin enacarbil concluded that gabapentin ER once daily treatment caused effective pain relief but was associated with a high incidence of adverse events. On the other hand, gabapentin enacarbil was more effective for pain relief and safer for the treatment of post-herpetic neuralgia [159]. Gabapentin enacarbil is an actively transported prodrug of gabapentin that provides sustained, dose-proportional exposure to gabapentin. This formulation is already recommended, together with gabapentin, as the first-line therapeutic option for neuropathic pain treatment [133].

Mirogabalin is an orally administered gabapentinoid developed for the treatment of peripheral diabetic neuropathic pain and post-herpetic neuralgia. First approved in Japan in 2019 for neuropathic pain, it is under clinical investigation in other Asian countries. However, clinical trials for fibromyalgia-related pain were discontinued in Europe and USA for not achieving the primary endpoint in the 3rd phase [160]. Studies suggest that mirogabalin binds selectively and with a high affinity to the $\alpha_2\delta$-1/-2 subunits of VGCCs, compared to pregabalin. In addition, it binds to the $\alpha_2\delta$-1 subunit for a much longer time resulting in a better analgesic and safety profile [161]. It has a favorable pharmacokinetic profile, being quickly absorbed after oral administration and reaching the maximal plasma concentration in 0.5–1.5 h after single or repeated doses [162]. Due to a low affinity and rapid dissociation from the $\alpha_2\delta$-2 subunits of VGCCs, some studies suggest it causes less adverse effects at the CNS level. The most frequent adverse reactions are similar to those observed with pregabalin and gabapentin, namely dizziness, somnolence, and headache, but their frequency seems to be lower [160]. Finally, it is noteworthy that mirogabalin seems to have a lower potential for abuse and is suggested to be better tolerated by persons with a history of multidrug use for recreational purposes compared to pregabalin [163]. Further studies are needed to explore the utility of mirogabalin in differing populations and pain conditions, but current evidence suggests it may represent an effective and safer therapeutic option, especially when abuse potential is a concern.

ANTICONVULSANTS WITH MULTIPLE MECHANISMS OF ACTION

Some anticonvulsants present multiple molecular targets (Fig. **3**), which may represent advantages over single target drugs in complex multifactorial   disorders such as chronic pain [164]. In this regard, topiramate (TOP) and valproate (VPA) deserve to be highlighted.

TOP is a sulfamate-substituted monosaccharide approved as a treatment for patients with epilepsy in the United Kingdom in 1995 and by the Food and Drug Administration (FDA) in 1996 [165, 166]. TOP is also indicated in cases of Lennox-Gastaut syndrome, obesity (under the name of QSYMIA®), and Alcohol Use Disorders [166]. TOP became one of the first line treatments for migraine prophylaxis for adults and adolescents older than 12 after the FDA approval in 2004 [165].

VPA is a simple branched-chained fatty acid with a structure different from other anticonvulsant drugs [167]. It was first discovered as a solvent for barbiturates and since 1970 as the most used antiepileptic and mood stabilizing drug worldwide [168, 169]. VPA is also indicated as a first line treatment for migraine prophylaxis [170 - 172].

TOP and VPA are prescribed off-label for several chronic pain disorders due to potential mechanisms of action in the nociceptive pathway. The current evidence on the effectiveness of these uses will be discussed below.

Pharmacokinetics Aspects

TOP is rapidly absorbed in the gastrointestinal tract, reaching peak plasma levels after approximately 2-3 hours [166]. The TOP oral dose has more than 80% of bioavailability. Protein binding is negligible (9-17%), and the volume of distribution ranges from 0.6-0.8 L/kg [166, 173]. TOP is not widely metabolized, and 75-80% is excreted unaltered in the urine [173]. TOP is metabolized through hydroxylation, hydrolysis, and glucuronidation and forms over six metabolites [166]. In drug-drug interactions, a synergic effect has been found with concomitant administration of topiramate or valproic acid with amitriptyline or venlafaxine for migraine prophylaxis. Another important interaction is with oral contraceptives. Topiramate caused the induction of hepatic CYP P450 and was responsible for the failure of contraceptive medications by reduction of estrogen levels [166]. TOP associated with antidepressants that cause weight gain (amitriptyline, mirtazapine, paroxetine) may decrease this effect [170]. This anticonvulsant is predominantly eliminated through the kidneys, and its half-life is estimated at 20-30 hours [166, 170].

Valproate formulations such as sodium valproate, valproic acid, and semi sodium valproate result in dissociation to the valproate ion in the gastrointestinal tract [174]. Absorption depends on the compound administered, its preparation, and the condition of the gastrointestinal tract [174]. The volume of distribution is 0.13 to 0.19 L/Kg [175]. Standard formulations of VPA have good bioavailability, ranging from 96 to 100% [167]. The drug passage through the blood-brain barrier is by passive diffusion and a bidirectional carrier-mediated transport [175]. VPA is widely metabolized by the liver, with <3% excreted unchanged in the urine. The main hepatic metabolic routes are mitochondrial β-oxidation and glucuronide conjugation, which account for 40 and 50% of the dose, respectively, and cytochrome P450-dependant oxidation (~10%) [176]. The CYP-mediated (CYP2C9, CYP2B6, CYP2A6) metabolism routes produce the 4-ene-VPA metabolite, which is possibly related to valproate-induced liver injury [176, 177]. The half-life of valproic acid is about 9 to 18 hours and shorter (5-12 hours) for patients using enzyme-inducing drugs [175].

Mechanisms of Action

Topiramate has been shown to block activity-dependent, voltage-gated Na^+ channels, enhances the action of GABA-A receptors, inhibits L-type voltage-gated Ca^{2+} channels, modulates voltage-gated K^+ channels, pre-synaptically reduces glutamate release, and post-synaptically blocks kainate/α-amino-3-hydroxy-5-methyl-4-isoxazolepropionic acid (AMPA) receptors, primarily the GluR5 (*i.e.*, GluK1) kainate receptor [10, 164, 178, 179]. The effects on GABA and Ca^{2+} channels could be related to the observed effects in migraine prophylaxis [164]. The Ca^{2+} channel blockage probably accounts for the inhibition of calcitonin gene-related peptide (CGRP) and glutamate release. Throughout several mechanisms described above, TOP inhibits nociceptive neuronal firing in the trigemino-cervical complex and dural vasodilation [164]. TOP also inhibited evoked cortical spreading depression in a non-clinical study which can be related to its effects in Na^+ channels and glutamate receptors [164].

Several VPA mechanisms of action common to TOP could be contributing to the antimigraine effects, although the specific mechanisms for its therapeutic effect are not fully elucidated [177, 178]. Valproate can increase the levels of GABA in the brain, enhancing the activity of glutamic acid decarboxylase and inhibiting

GABA degradation enzymes [180, 181]. VPA also suppresses voltage sensitive Na^+ channels, restraining neuronal repetitive firing [177].

Indication and Efficacy

Two clinical trials evaluating the TOP effectiveness in chronic migraine demonstrated reductions in migraine frequency in 22% of patients with $\geq$ 50% migraine frequency reduction and improvement in daily activities. These results were even better in the subset of patients with medication overuse headache [182 - 185].

Three multicenter, double-blind, placebo-controlled trials studied the efficacy of TOP in 1500 patients with episodic migraine. It was observed that the responder rate, *i.e.*, patients with $\geq$50% reduction in monthly migraine frequency ranged from 36 to 52% compared with 22-23% of placebo. Moreover, the efficacy was similar to other first-line treatments (amitriptyline, propranolol, and valproate) [164, 186, 187]. The efficacy of VPA was established in a long-term clinical study which demonstrated that the benefits for migraine prophylaxis were maintained for a period of three years [188]. Additionally, the TOP extended-release formulation (XR) seems to promote improvements in adherence, outcomes, and fewer cognitive effects [164].

TOP is also indicated as an alternative prophylactic treatment of cluster headache, although evidence of its effect is not clear [189]. Few studies were conducted with VPA to determine properly the efficacy in cluster headache treatment [189]. An evidence level 3 for TOP was found in phantom limb pain treatment, demonstrating 68-80% symptoms reduction [190]. A meta-analysis concluded that TOP has moderate to high effect in the treatment of chronic low back pain; nevertheless, it is not indicated due to the high risk of harm [191]. Studies evaluating TOP in trigeminal neuralgia and VPA in diabetic neuropathy are scarce, and its efficacy remains unknown [168, 192].

Side Effects

VPA may cause hair loss, weight gain, somnolence, encephalopathy, and hepatic failure.In most headache clinical trials, no difference was observed between adverse effects' incidence in patients taking placebo or VPA [169]. Importantly, due to fetal malformation, VPA is contraindicated in pregnant women and should be avoided during childbearing years [169, 172]. Co-administration of VPA and LTG may lead to Stevens-Johnson syndrome, which can be fatal. VPA interferes with LTG metabolism, resulting in decreased clearance, high serum concentrations of LTG, and augmented production of a highly toxic metabolite. Thus, this combination should be avoided [193].

The adverse effects associated with TOP are disorder dependent. A meta-analysis demonstrated that migraineurs are more susceptible to side effects of TOP than patients with epilepsy [194]. The adverse effects reported were paresthesia, fatigue, cognitive problems, weight loss, appetite loss, nausea, diarrhea, insomnia, hyperesthesia, somnolence, anxiety, and alteration of taste [194]. Paresthesia is a common side effect but is not associated with treatment discontinuation. However, cognitive problems, fatigue, insomnia, and anxiety occurred less frequently than paresthesia but were more likely to cause discontinuation [164].

Co-prescription of VPA or TOP with antidepressants for migraine prophylaxis or migraine comorbid with depression may occur [170]. An advantage of TOP in these conditions is that it may decrease the effect on the weight of antidepressants associated with weight gain (amitriptyline, mirtazapine, paroxetine). Moreover, TOP itself can promote weight loss, being a good candidate for obese patients [165]. Attention should be directed to the effects of TOP in carbonic anhydrase inhibition, which can favor metabolic acidosis and renal stones. Migraineurs are already more susceptible to the development of renal stones [164]. Symptoms of blurring, visual disturbances and/or ocular pain after beginning TOP treatment could be an indication of a rare sight-threatening idiosyncratic event [164].

In cluster headache, the most common side effects reported were paresthesia, depression, drowsiness, speech disturbances, dizziness, and altered taste [189].

Perspectives

Zonisamide is a synthetic 1,2-benzisoxazole-3-methanesulfonamide approved for partial-onset epilepsies by the FDA in 2000 [195, 196]. Various mechanisms seem to be related to its antinociceptive effects, including inhibition of nitric oxide synthase and repetitive firing of voltage-dependent sodium channels, changes in voltage-dependent T-type calcium channel currents, and scavenging of free radicals [195, 197]. Moreover, it has the potential for treating neuropathic pain through inhibition of microglia-mediated neuroinflammation, although the current evidence is limited [10, 195, 197]. A case history report found that 70% of the patients with cluster headache responded to the treatment, *i.e.*, pain improvement of more than 50% or remission [195]. These results indicate that zonisamide may be a good candidate for prophylactic treatment of chronic and episodic cluster headache disorder [195]. However, more studies are necessary to assess the efficacy of this drug in neuropathic and chronic pain.

OTHER TARGETS

Potassium Ion Channels

Voltage-gated potassium (K^+) channels are responsible for the repolarization process of the cell membrane after the firing of an action potential. They are a diverse and widely distributed family of ion channels in neurons, and studies have reported that changes in the expression of these channels may play an important role in the pathophysiology of neuropathic pain [198, 199]. Potassium channel activators such as retigabine (also known as ezogabine in the US) hold potential for the treatment of inflammatory and neuropathic pain [200].

Retigabine, under the trade name Potiga, was approved by the FDA and introduced in the US in 2011 to treat partial-onset seizures in epileptic patients [201]. Although studies in rodent models of inflammatory and neuropathic pain have shown an antinociceptive effect of retigabine [202, 203], in human trials, the results have been unsatisfying. In a report from 2009, GlaxoSmithKline/Valeant Pharmaceuticals announced inconclusive results from a phase IIa proof-o--concept clinical trial of retigabine for the treatment of neuropathic pain associated with postherpetic neuralgia [200]. The clinical use of retigabine is limited and has been discontinued for seizures due to its adverse effects that are mostly CNS-related, such as dizziness, headache, confusion, speech disorder, fatigue, somnolence, and others, as well as its short half-life of 6-8 hours, requiring constant dosing [204 - 206]. On the other hand, the structural analog of retigabine, flupirtine, has been used in several European countries for the treatment of cancer pain, menstrual pain, headache, neuralgias, and postsurgical pain for over 30 years, but it is not clinically available in the US [199, 204].

Retigabine and flupirtine exert an analgesic effect through activation of voltage-gated K^+ channels type 7 (K_V7 channels) [207]. They reduce the excitability of nociceptive fibers by enhancing non-inactivating currents, mostly *via* K_V7 channels activation [198]. Flupirtine also interacts with NMDA receptors as an indirect antagonist *via* activation of G-protein regulated inwardly rectifying K^+ channels (GIRK channels). These channels contribute to the maintenance of resting membrane potential and management of cell excitability, inhibiting nociceptive impulses [204]. Moreover, flupirtine has been shown to induce muscle relaxation and potentiate GABA-mediated analgesia [208].

Flupirtine is hydrophilic and completely absorbed in the gastrointestinal tract with a bioavailability of 90% with a 6.5-hour half-life by oral route. Flupirtine is around 80% bound to albumin and metabolized in the liver by the peroxidase enzyme. About 72% of the total dose is present in the urine as the parent drug and its metabolites 4-fluorohippuric acid and *N*-acetylated analogue D13223 [209].

Common long-term adverse effects of flupirtine may include dry mouth, gastric fullness, dizziness, nausea, pruritus, drowsiness, and muscular tremor [209, 210].

GABAergic Transmission

GABA is a neurotransmitter released at up to 40% of synapses in the brain. It produces inhibitory effects in different areas of the central nervous system, including the amygdala, hippocampus, cerebral cortex, and dorsal horn of the spinal cord. Furthermore, physiological and behavioral studies have shown that the inhibitory function of GABA plays an important role in the inhibition of nociceptive information throughout the pain transmission circuitry, representing an important target to treat painful conditions [198, 211]. Anticonvulsant drugs such as tiagabine are products initially developed to boost inhibitory neuro-transmission by GABA and have shown potential to treat some modalities of chronic pain [212].

Tiagabine is a selective, potent and competitive inhibitor of GABA transporter protein (GAT-1), leading to inhibition of GABA reuptake into nerve terminals and glial cells [198]. Functionally, tiagabine treatment leads to a postsynaptic increase in inhibitory transmission, and preclinical studies have demonstrated the analgesic potential of tiagabine in models of neuropathic pain [213]. Clinically, only a few studies have observed the beneficial effect of tiagabine in chronic pain (*e.g.,* fibromyalgia and peripheral neuropathy) [214, 215]. However, currently in the US, tiagabine may be prescribed off label not only for neuropathic pain patients but also to treat anxiety disorders [216].

The pharmacokinetics of tiagabine are known to be linear and predictable, with a half-life varying from 4 to 8 hours. It shows a rapid absorption (1h) orally with a bioavailability of 90%. Tiagabine is widely distributed in the organism and highly bound to plasma proteins. It is extensively metabolized, and only a small percentage of the drug is excreted unchanged in the urine or feces. Frequent adverse effects include dizziness, nausea, impaired concentration, asthenia, nervousness, and tremor [214, 217].

Modulators of Presynaptic Machinery

The synaptic vesicle glycoprotein SV2A is an integral member of synaptic vesicle membranes and has been shown to be involved in exocytosis, trafficking, vesicle release, and recycling, all crucial processes for neurotransmission. SV2A protein is highly expressed in presynaptic nerve terminals, and it is currently considered the main target of the anticonvulsant drug levetiracetam [218]. Levetiracetam is FDA-approved for adult patients with myoclonic and focal seizures. Although no specific functional mechanism has been described, levetiracetam seems to limit neurotransmitter release (*e.g.*, GABA and glutamate) by binding to the vesicle protein SV2A [198].

Studies have demonstrated a potential analgesic effect of levetiracetam in different animal models of neuropathic pain [219, 220], as well as in the open labeled and cohort studies in humans with lumbar radiculopathy, multiple sclerosis, postherpetic neuralgia, trigeminal neuralgia, and painful diabetic neuropathy [221 - 224]. However, no analgesic effect of levetiracetam in randomized placebo-controlled trials in patients with spinal cord injury pain and polyneuropathy has been reported [224 - 226]. Therefore, more controlled trials are required to elucidate the potential of levetiracetam as a treatment for chronic pain.

Pharmacokinetic studies of levetiracetam indicate that after oral ingestion, the drug is rapidly absorbed with a peak concentration at 1.3 hours. Levetiracetam has a bioavailability above 95%, and the half-life varies from 5 to 11 hours depending on the age of the patient. Levetiracetam is not metabolized in the liver but rather by hydrolysis in the bloodstream. Approximately 1/3 of the drug is metabolized, while 2/3 is excreted unchanged in the urine. Dosage adjustments are required for patients with renal impairment since clearance of the drug is dependent on creatinine clearance. Commonly reported adverse events may include nausea, dizziness, headache, diarrhea, and abdominal pain [227, 228].

Selective Inhibitors of Excitatory Neurotransmission

Ionotropic glutamate receptors such as AMPA are important members of excitatory synapses in areas of the central and peripheral nervous systems that transmit pain. AMPA receptors are mainly found in postsynaptic membranes of excitatory neurons. Glutamate, the primary excitatory neurotransmitter in the central and peripheral nervous systems, acts in AMPA receptors facilitating transmission of pain impulses as well as seizure discharges [198]. Anticonvulsants such as perampanel, tezampanel and selurampanel are antagonists of AMPA receptors. They bind to the receptor, limiting the translation of agonist binding, leading to reduced excitatory input and seizure and pain transmission.

Perampanel is the first selective AMPA receptor antagonist approved for epilepsy. Orally, perampanel is rapidly absorbed from the gastrointestinal tract and its bioavailability reaches 100%. Approximately 95% of the drug is bound to proteins. Oxidation is the main route of metabolization. Up to 70% of the drug is excreted *via* feces and 30% in the urine [228].

Perampanel has shown potential for the treatment of inflammatory and neuropathic pain in preclinical studies [229]. However, it has failed to demonstrate efficacy in clinical trials in patients with diabetic neuropathic pain and post-herpetic neuralgia [230, 231]. Adverse effects reported include dizziness, somnolence, and psychiatric disorders [228].

The other AMPA receptor antagonists, tezampanel and selurampanel, have shown promising results in clinical trials. Tezampanel, for instance, has been tested in phase II clinical trial of patients with acute migraine. In this trial, at a low dose of 40 mg, tezampanel induced a significant decrease in pain when compared to placebo (p=0.03). Higher doses (70 mg and 100 mg) did not induce significant differences [232]. Lastly, selurampanel, at a 250 mg dose, was also tested in a randomized, double-blind, proof-of-concept trial in patients with acute migraine pain. Although selurampanel effects favorably compared to patients in the same study receiving sumatriptan, a high rate of side effects was found (nervous system disorders, gastrointestinal disorders, cardiac disorders, and others) [233].

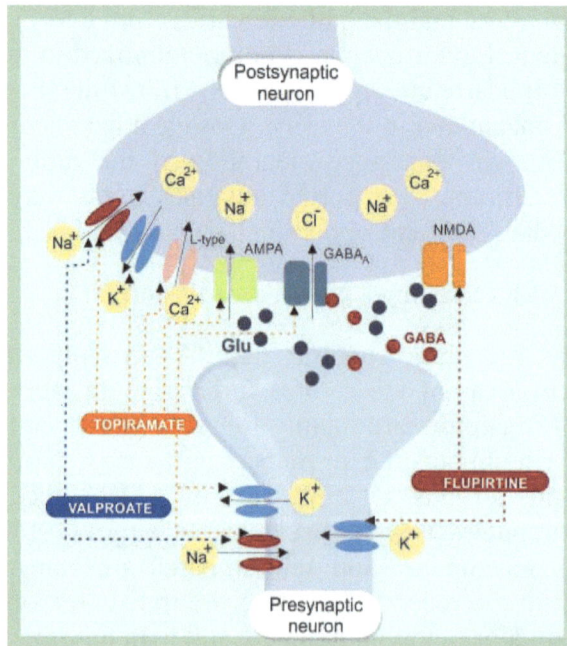

Fig. (3). Mechanisms that contribute to the analgesic effect of anticonvulsants that act on multiple targets. Topiramate actions are due to blockage of activity-dependent voltage-gated sodium (Na$^+$) channels, L-type voltage-gated calcium (Ca^{2+}) channels, post-synaptic kainite/α-amino-3-hydroxy-5-methyl-4-isoxazolepropionic acid (AMPA) receptors, increase the action of γ-aminobutyric acid (GABA)-A receptors and activation of potassium (K$^+$) conductance, resulting in cell hyperpolarization. Blockade of voltage-gated Na$^+$ channels accounts for the analgesic effect of valproate. Flupirtine acts on the presynaptic membrane by regulating inwardly rectifying K$^+$ channels and indirectly interacts with the N-methyl-D-aspartate (NMDA) receptor through activation of G protein in the postsynaptic membrane to reduce the glutamatergic activity and, consequently, decreases the excitability of nociceptive fibers.

ONGOING RESEARCH

Besides the classical anticonvulsants such as carbamazepine still being used and prescribed for chronic pain management, a new generation of anticonvulsants has been appeared for pain control. The third generation of anticonvulsants include mainly eslicarbazepine, lacosamide, vigabatrin and mirogabalin [234]. As previously described, these drugs are being evaluated in chronic pain conditions, and although they present a good analgesic profile with better pharmacokinetics compared to the classic anticonvulsants, the development of new anticonvulsants remains essential. Classic cases of molecule structure alterations, such as between carbamazepine and oxcarbazepine, show that it is possible to develop a new compound with a comparable effect with lower side effects. In sum, an effort in the development of new anticonvulsant compounds could present an improved outcome.

HSK16149 is a new ligand for the voltage gated calcium channel $\alpha_2\delta$ subunit. It was demonstrated that this compound presents a greater potency compared to pregabalin without affecting the other 105 targets, demonstrating an advantageous safety profile [235]. In addition to a good effect *in vitro*, it demonstrated a dose-dependent analgesic effect in several models of chronic pain, such as neuropathic pain by chronic constriction of the sciatic nerve, type 1 diabetic neuropathy, and fibromyalgia [235]. Moreover, the level of HSK16149 found in the brain was lower than pregabalin, and the authors suggested that in an equipotent dose between the drugs, HSK16149 could present fewer side effects [235]. This new $\alpha_2\delta$ subunit ligand presents a promising anticonvulsant and analgesic profile in non-clinical studies, and currently, a clinical trial is evaluating the efficacy and safety of HSK16149 in diabetic neuropathic pain patients (NCT04647773).

Several compounds based on succinimides (pyrrolidine-2,5-dione core) were identified as possible broad acting anticonvulsants. These compounds are hybrids of other anticonvulsant drugs, such as ethosuximide, levetiracetam, lacosamide, and tiagabine [236 - 239]. This process leads to a multitarget therapy, which could present the advantage of modulating several targets simultaneously, reducing the consumption of several drugs. Several of these compounds demonstrated analgesic effects in different models of pain, such as capsaicin-induced pain, formalin, and oxaliplatin-induced neuropathic pain [236, 240]. Among all the compounds developed and assessed, the compound KA-104 showed a favorable efficacy [241]. It was revealed that this compound was able to produce an antinociceptive effect in an acute model of pain induced by capsaicin and in a model of neuropathic pain induced by oxaliplatin [241]. The pharmacokinetic parameters were assessed, and it was observed that the maximum concentration found in plasma and brain was 823.7 and 565.7 ng/mL, respectively, 30- and 15-

minutes post systemic administration of 20 mg/kg [241]. The mechanism of action proposed for this compound was based on *in vitro* findings, where the compound was able to inhibit sodium and calcium currents (mainly L-type currents), as well as TRPV1 channels [242]. While showing promising effects in animal models and *in vitro* assays, this compound is still in clinical trials.

It is important to highlight that even with the current commercial anticonvulsants, we still have patients that are non-responsive or need combination therapy to reduce their pain. Keeping this in mind, the development and the research for new drugs to treat pain are crucial. Anticonvulsants are very promising due to their efficacy in several conditions of chronic pain.

CONCLUDING REMARKS

The analgesic properties of anticonvulsants were noticed in the 1960s, and subsequently, similarities between the pathophysiology and biochemical mechanisms were observed in epilepsy and neuropathic pain, mainly neuronal hyperexcitability. This resulted in the introduction of this class of drugs into the treatment of neuropathic pain and its progress to first-line therapy for most neuropathic pain conditions. Initially, carbamazepine showed an analgesic effect in the treatment of trigeminal neuralgia, a clinical condition that differs from other neuropathic pain and responds differently to pharmacological therapy. Carbamazepine is currently the first line treatment of trigeminal neuralgia. Later, gabapentinoids were increasingly employed in other neuropathic pain conditions due to their high efficacy and low adverse events compared with opioids. As such, they are currently recommended as frontline therapies in neuropathic pain and fibromyalgia pain syndromes. However, it is worth mentioning that the widespread use of gabapentinoids for a variety of chronic pain conditions lacks scientific evidence and has raised some concerns about its safety profile. Later, anticonvulsants with multiple mechanisms of action were implemented in clinical practice for chronic pain therapy. Topiramate and valproate have proven their efficacy and safety in the prophylactic treatment of migraine. Topiramate has minimal pharmacological interaction with other medications and is usually well tolerated by patients. Finally, other anticonvulsant molecules that modulate potassium channels, enhance GABAergic transmission, or inhibit excitatory neurotransmission have been used for chronic pain control; however, their efficacy in clinical trials is controversial.

Anticonvulsants are one of the most studied drugs among all other potential treatments for chronic pain, and it still remains unclear which pharmacological mechanism of action is responsible for their therapeutic effects on chronic pain. Certainly, in the future, further investigations of how effective anticonvulsants differ mechanistically at the molecular level in each neurological disorder may provide more accurate approaches regarding the development of new treatments for chronic pain syndromes.

CONSENT FOR PUBLICATION

Not Applicable.

CONFLICT OF INTEREST

The author declares no conflict of interest, financial or otherwise.

ACKNOWLEDGEMENTS

The authors are deeply grateful to Peg Davis for her assistance with English language editing and review.

REFERENCES

[1]   Elzahaf RA, Tashani OA, Unsworth BA, Johnson MI. The prevalence of chronic pain with an analysis of countries with a Human Development Index less than 0.9: a systematic review without meta-analysis. Curr Med Res Opin 2012; 28(7): 1221-9.
[http://dx.doi.org/10.1185/03007995.2012.703132] [PMID: 22697274]

[2]   Treede RD, Rief W, Barke A, *et al*. A classification of chronic pain for ICD-11. Pain 2015; 156(6): 1003-7.
[http://dx.doi.org/10.1097/j.pain.0000000000000160] [PMID: 25844555]

[3]   Yang S, Chang MC. Chronic Pain: Structural and Functional Changes in Brain Structures and Associated Negative Affective States. Int J Mol Sci 2019; 20(13): 3130.
[http://dx.doi.org/10.3390/ijms20133130] [PMID: 31248061]

[4]   Bair MJ, Robinson RL, Katon W, Kroenke K. Depression and pain comorbidity: a literature review. Arch Intern Med 2003; 163(20): 2433-45.
[http://dx.doi.org/10.1001/archinte.163.20.2433] [PMID: 14609780]

[5]   Dahan A, van Velzen M, Niesters M. Comorbidities and the complexities of chronic pain. Anesthesiology 2014; 121(4): 675-7.
[http://dx.doi.org/10.1097/ALN.0000000000000402] [PMID: 25099749]

[6]   Williams LS, Jones WJ, Shen J, Robinson RL, Weinberger M, Kroenke K. Prevalence and impact of depression and pain in neurology outpatients. J Neurol Neurosurg Psychiatry 2003; 74(11): 1587-9.
[http://dx.doi.org/10.1136/jnnp.74.11.1587] [PMID: 14617727]

[7]   Tetsunaga T, Tetsunaga T, Nishida K, *et al*. Drug dependence in patients with chronic pain: A retrospective study. Medicine (Baltimore) 2018; 97(40): e12748.
[http://dx.doi.org/10.1097/MD.0000000000012748] [PMID: 30290690]

[8]   Treede RD, Jensen TS, Campbell JN, *et al*. Neuropathic pain: redefinition and a grading system for clinical and research purposes. Neurology 2008; 70(18): 1630-5.
[http://dx.doi.org/10.1212/01.wnl.0000282763.29778.59] [PMID: 18003941]

[9]   Söderpalm B. Anticonvulsants: aspects of their mechanisms of action. Eur J Pain 2002; 6 Suppl A: 3-9.
[http://dx.doi.org/10.1053/eujp.2001.0315]

[10]   Wiffen PJ, Derry S, Moore RA, *et al.* Antiepileptic drugs for neuropathic pain and fibromyalgia - an overview of Cochrane reviews. Cochrane Database Syst Rev 2013; 2013(11): CD010567.
        [http://dx.doi.org/10.1002/14651858.CD010567.pub2] [PMID: 24217986]

[11]   Goodman CW, Brett AS. Gabapentin and Pregabalin for Pain - Is Increased Prescribing a Cause for Concern? N Engl J Med 2017; 377(5): 411-4.
        [http://dx.doi.org/10.1056/NEJMp1704633] [PMID: 28767350]

[12]   Perucca E, Meador KJ. Adverse effects of antiepileptic drugs. Acta Neurol Scand Suppl 2005; 181: 30-5.
        [http://dx.doi.org/10.1111/j.1600-0404.2005.00506.x] [PMID: 16238706]

[13]   Bailly F, Belaid H. Suicidal ideation and suicide attempt associated with antidepressant and antiepileptic drugs: Implications for treatment of chronic pain. Joint Bone Spine 2021; 88(1): 105005.
        [http://dx.doi.org/10.1016/j.jbspin.2020.04.016] [PMID: 32438065]

[14]   Vargas A, Thomas K. Intravenous fosphenytoin for acute exacerbation of trigeminal neuralgia: case report and literature review. Ther Adv Neurol Disord 2015; 8(4): 187-8.
        [http://dx.doi.org/10.1177/1756285615583202] [PMID: 26136846]

[15]   Keppel Hesselink JM, Schatman ME. Phenytoin and carbamazepine in trigeminal neuralgia: marketing-based *versus* evidence-based treatment. J Pain Res 2017; 10: 1663-6.
        [http://dx.doi.org/10.2147/JPR.S141896] [PMID: 28761370]

[16]   Tremont-Lukats IW, Megeff C, Backonja M-M. Anticonvulsants for neuropathic pain syndromes: mechanisms of action and place in therapy. Drugs 2000; 60(5): 1029-52.
        [http://dx.doi.org/10.2165/00003495-200060050-00005] [PMID: 11129121]

[17]   Binder A, Baron R. The pharmacological therapy of chronic neuropathic pain. Dtsch Arztebl Int 2016; 113(37): 616-25.
        [http://dx.doi.org/10.3238/arztebl.2016.0616] [PMID: 27697147]

[18]   Selvy M, Cuménal M, Kerckhove N, Courteix C, Busserolles J, Balayssac D. The safety of medications used to treat peripheral neuropathic pain, part 1 (antidepressants and antiepileptics): review of double-blind, placebo-controlled, randomized clinical trials. Expert Opin Drug Saf 2020; 19(6): 707-33.
        [http://dx.doi.org/10.1080/14740338.2020.1764934] [PMID: 32363948]

[19]   Gambeta E, Chichorro JG, Zamponi GW. Trigeminal neuralgia: An overview from pathophysiology to pharmacological treatments. Mol Pain 2020; 16: 1744806920901890.
        [http://dx.doi.org/10.1177/1744806920901890] [PMID: 31908187]

[20]   Gore M, Sadosky A, Tai K-S, Stacey B. A retrospective evaluation of the use of gabapentin and pregabalin in patients with postherpetic neuralgia in usual-care settings. Clin Ther 2007; 29(8): 1655-70.
        [http://dx.doi.org/10.1016/j.clinthera.2007.08.019] [PMID: 17919547]

[21]   Johannessen Landmark C. Antiepileptic drugs in non-epilepsy disorders: relations between mechanisms of action and clinical efficacy. CNS Drugs 2008; 22(1): 27-47.
        [http://dx.doi.org/10.2165/00023210-200822010-00003] [PMID: 18072813]

[22]   Perucca E. Antiepileptic drugs: evolution of our knowledge and changes in drug trials. Epileptic Disord 2019; 21(4): 319-29.
        [PMID: 31403463]

[23]   Bialer M. Why are antiepileptic drugs used for nonepileptic conditions? Epilepsia 2012; 53 (Suppl. 7): 26-33.
        [http://dx.doi.org/10.1111/j.1528-1167.2012.03712.x] [PMID: 23153207]

[24]   Jensen TS. Anticonvulsants in neuropathic pain: rationale and clinical evidence. Eur J Pain 2002; 6 (Suppl. A): 61-8.
        [http://dx.doi.org/10.1053/eujp.2001.0324] [PMID: 11888243]

[25]   Bialer M. Chemical properties of antiepileptic drugs (AEDs). Adv Drug Deliv Rev 2012; 64(10): 887-95.
[http://dx.doi.org/10.1016/j.addr.2011.11.006] [PMID: 22210279]

[26]   Isom LL. Sodium channel beta subunits: anything but auxiliary. Neuroscientist 2001; 7(1): 42-54.
[http://dx.doi.org/10.1177/107385840100700108] [PMID: 11486343]

[27]   Catterall WA. From ionic currents to molecular mechanisms: the structure and function of voltage-gated sodium channels. Neuron 2000; 26(1): 13-25.
[http://dx.doi.org/10.1016/S0896-6273(00)81133-2] [PMID: 10798388]

[28]   Catterall WA, Goldin AL, Waxman SG. International Union of Pharmacology. XLVII. Nomenclature and structure-function relationships of voltage-gated sodium channels. Pharmacol Rev 2005; 57(4): 397-409.
[http://dx.doi.org/10.1124/pr.57.4.4] [PMID: 16382098]

[29]   Catterall WA. Sodium channels, inherited epilepsy, and antiepileptic drugs. Annu Rev Pharmacol Toxicol 2014; 54: 317-38.
[http://dx.doi.org/10.1146/annurev-pharmtox-011112-140232] [PMID: 24392695]

[30]   Bennett DL, Clark AJ, Huang J, Waxman SG, Dib-Hajj SD. The Role of Voltage-Gated Sodium Channels in Pain Signaling. Physiol Rev 2019; 99(2): 1079-151.
[http://dx.doi.org/10.1152/physrev.00052.2017] [PMID: 30672368]

[31]   Catterall WA, Lenaeus MJ, Gamal El-Din TM. Structure and Pharmacology of Voltage-Gated Sodium and Calcium Channels. Annu Rev Pharmacol Toxicol 2020; 60: 133-54.
[http://dx.doi.org/10.1146/annurev-pharmtox-010818-021757] [PMID: 31537174]

[32]   Cardoso FC, Lewis RJ. Sodium channels and pain: from toxins to therapies. Br J Pharmacol 2018; 175(12): 2138-57.
[http://dx.doi.org/10.1111/bph.13962] [PMID: 28749537]

[33]   Tolou-Ghamari Z, Zare M, Habibabadi JM, Najafi MR. A quick review of carbamazepine pharmacokinetics in epilepsy from 1953 to 2012. J Res Med Sci 2013; 18 (Suppl. 1): S81-5.
[PMID: 23961295]

[34]   Marino SE, Birnbaum AK, Leppik IE, *et al.* Steady-state carbamazepine pharmacokinetics following oral and stable-labeled intravenous administration in epilepsy patients: effects of race and sex. Clin Pharmacol Ther 2012; 91(3): 483-8.
[http://dx.doi.org/10.1038/clpt.2011.251] [PMID: 22278332]

[35]   Rawlins MD, Collste P, Bertilsson L, Palmér L. Distribution and elimination kinetics of carbamazepine in man. Eur J Clin Pharmacol 1975; 8(2): 91-6.
[http://dx.doi.org/10.1007/BF00561556] [PMID: 1233212]

[36]   Kerr BM, Thummel KE, Wurden CJ, *et al.* Human liver carbamazepine metabolism. Role of CYP3A4 and CYP2C8 in 10,11-epoxide formation. Biochem Pharmacol 1994; 47(11): 1969-79.
[http://dx.doi.org/10.1016/0006-2952(94)90071-X] [PMID: 8010982]

[37]   Pearce RE, Vakkalagadda GR, Leeder JS. Pathways of carbamazepine bioactivation *in vitro* I. Characterization of human cytochromes P450 responsible for the formation of 2- and 3-hydroxylated metabolites. Drug Metab Dispos 2002; 30(11): 1170-9.
[http://dx.doi.org/10.1124/dmd.30.11.1170] [PMID: 12386121]

[38]   Pearce RE, Uetrecht JP, Leeder JS. Pathways of carbamazepine bioactivation *in vitro*: II. The role of human cytochrome P450 enzymes in the formation of 2-hydroxyiminostilbene. Drug Metab Dispos 2005; 33(12): 1819-26.
[http://dx.doi.org/10.1124/dmd.105.004861] [PMID: 16135660]

[39]   Pearce RE, Lu W, Wang Y, Uetrecht JP, Correia MA, Leeder JS. Pathways of carbamazepine bioactivation *in vitro*. III. The role of human cytochrome P450 enzymes in the formation of 2,3-dihydroxycarbamazepine. Drug Metab Dispos 2008; 36(8): 1637-49.

[http://dx.doi.org/10.1124/dmd.107.019562] [PMID: 18463198]

[40] Staines AG, Coughtrie MW, Burchell B. N-glucuronidation of carbamazepine in human tissues is mediated by UGT2B7. J Pharmacol Exp Ther 2004; 311(3): 1131-7.
[http://dx.doi.org/10.1124/jpet.104.073114] [PMID: 15292462]

[41] Lu W, Uetrecht JP. Peroxidase-mediated bioactivation of hydroxylated metabolites of carbamazepine and phenytoin. Drug Metab Dispos 2008; 36(8): 1624-36.
[http://dx.doi.org/10.1124/dmd.107.019554] [PMID: 18463199]

[42] Yip VLM, Pertinez H, Meng X, *et al.* Evaluation of clinical and genetic factors in the population pharmacokinetics of carbamazepine. Br J Clin Pharmacol 2021; 87(6): 2572-88.
[http://dx.doi.org/10.1111/bcp.14667] [PMID: 33217013]

[43] Fortuna A, Alves G, Soares-da-Silva P, Falcão A. Pharmacokinetics, brain distribution and plasma protein binding of carbamazepine and nine derivatives: new set of data for predictive *in silico* ADME models. Epilepsy Res 2013; 107(1-2): 37-50.
[http://dx.doi.org/10.1016/j.eplepsyres.2013.08.013] [PMID: 24050973]

[44] Maan JS, Duong Tv H, Saadabadi A. Carbamazepine. Treasure Island, FL: StatPearls Publishing 2021. [Internet]

[45] Antunes NJ, van Dijkman SC, Lanchote VL, *et al.* Population pharmacokinetics of oxcarbazepine and its metabolite 10-hydroxycarbazepine in healthy subjects. Eur J Pharm Sci 2017; 109S: S116-23.
[http://dx.doi.org/10.1016/j.ejps.2017.05.034] [PMID: 28528287]

[46] May TW, Korn-Merker E, Rambeck B. Clinical pharmacokinetics of oxcarbazepine. Clin Pharmacokinet 2003; 42(12): 1023-42.
[http://dx.doi.org/10.2165/00003088-200342120-00002] [PMID: 12959634]

[47] Flesch G, Czendlik C, Renard D, Lloyd P. Pharmacokinetics of the monohydroxy derivative of oxcarbazepine and its enantiomers after a single intravenous dose given as racemate compared with a single oral dose of oxcarbazepine. Drug Metab Dispos 2011; 39(6): 1103-10.
[http://dx.doi.org/10.1124/dmd.109.030593] [PMID: 21389120]

[48] Iorga A, Horowitz BZ. Phenytoin Toxicity. Treasure Island, FL: StatPearls Publishing 2021.

[49] Gupta M, Tripp J. Phenytoin. Treasure Island, FL: StatPearls 2021. [Internet]

[50] Patocka J, Wu Q, Nepovimova E, Kuca K. Phenytoin - An anti-seizure drug: Overview of its chemistry, pharmacology and toxicology. Food Chem Toxicol 2020; 142: 111393.
[http://dx.doi.org/10.1016/j.fct.2020.111393] [PMID: 32376339]

[51] Bergen DC. Pharmacokinetics of phenytoin: reminders and discoveries. Epilepsy Curr 2009; 9(4): 102-4.
[http://dx.doi.org/10.1111/j.1535-7511.2009.01307.x] [PMID: 19693326]

[52] Methaneethorn J, Leelakanok N. Sources of lamotrigine pharmacokinetic variability: A systematic review of population pharmacokinetic analyses. Seizure 2020; 82: 133-47.
[http://dx.doi.org/10.1016/j.seizure.2020.07.014] [PMID: 33060011]

[53] Goldsmith DR, Wagstaff AJ, Ibbotson T, Perry CM. Spotlight on lamotrigine in bipolar disorder. CNS Drugs 2004; 18(1): 63-7.
[http://dx.doi.org/10.2165/00023210-200418010-00007] [PMID: 14731061]

[54] Cohen AF, Land GS, Breimer DD, Yuen WC, Winton C, Peck AW. Lamotrigine, a new anticonvulsant: pharmacokinetics in normal humans. Clin Pharmacol Ther 1987; 42(5): 535-41.
[http://dx.doi.org/10.1038/clpt.1987.193] [PMID: 3677542]

[55] Goa KL, Ross SR, Chrisp P. Lamotrigine. A review of its pharmacological properties and clinical efficacy in epilepsy. Drugs 1993; 46(1): 152-76.
[http://dx.doi.org/10.2165/00003495-199346010-00009] [PMID: 7691504]

[56] Rambeck B, Wolf P. Lamotrigine clinical pharmacokinetics. Clin Pharmacokinet 1993; 25(6): 433-43.

[http://dx.doi.org/10.2165/00003088-199325060-00003] [PMID: 8119045]

[57]　Mitra-Ghosh T, Callisto SP, Lamba JK, *et al.* PharmGKB summary: lamotrigine pathway, pharmacokinetics and pharmacodynamics. Pharmacogenet Genomics 2020; 30(4): 81-90.
　　　[http://dx.doi.org/10.1097/FPC.0000000000000397] [PMID: 32187155]

[58]　Lin CH, Hsu SP, Cheng TC, *et al.* Effects of anti-epileptic drugs on spreading depolarization-induced epileptiform activity in mouse hippocampal slices. Sci Rep 2017; 7(1): 11884.
　　　[http://dx.doi.org/10.1038/s41598-017-12346-y] [PMID: 28928441]

[59]　Kuo CC, Chen RS, Lu L, Chen RC. Carbamazepine inhibition of neuronal Na+ currents: quantitative distinction from phenytoin and possible therapeutic implications. Mol Pharmacol 1997; 51(6): 1077-83.
　　　[http://dx.doi.org/10.1124/mol.51.6.1077] [PMID: 9187275]

[60]　Yang YC, Huang CS, Kuo CC. Lidocaine, carbamazepine, and imipramine have partially overlapping binding sites and additive inhibitory effect on neuronal Na+ channels. Anesthesiology 2010; 113(1): 160-74.
　　　[http://dx.doi.org/10.1097/ALN.0b013e3181dc1dd6] [PMID: 20526191]

[61]　Yang YC, Hsieh JY, Kuo CC. The external pore loop interacts with S6 and S3-S4 linker in domain 4 to assume an essential role in gating control and anticonvulsant action in the Na(+) channel. J Gen Physiol 2009; 134(2): 95-113.
　　　[http://dx.doi.org/10.1085/jgp.200810158] [PMID: 19635852]

[62]　Yang YC, Kuo CC. An inactivation stabilizer of the Na+ channel acts as an opportunistic pore blocker modulated by external Na+. J Gen Physiol 2005; 125(5): 465-81.
　　　[http://dx.doi.org/10.1085/jgp.200409156] [PMID: 15824190]

[63]　Jo S, Bean BP. Sidedness of carbamazepine accessibility to voltage-gated sodium channels. Mol Pharmacol 2014; 85(2): 381-7.
　　　[http://dx.doi.org/10.1124/mol.113.090472] [PMID: 24319110]

[64]　Cardenas CA, Cardenas CG, de Armendi AJ, Scroggs RS. Carbamazepine interacts with a slow inactivation state of NaV1.8-like sodium channels. Neurosci Lett 2006; 408(2): 129-34.
　　　[http://dx.doi.org/10.1016/j.neulet.2006.08.070] [PMID: 16978779]

[65]　Rush AM, Elliott JR. Phenytoin and carbamazepine: differential inhibition of sodium currents in small cells from adult rat dorsal root ganglia. Neurosci Lett 1997; 226(2): 95-8.
　　　[http://dx.doi.org/10.1016/S0304-3940(97)00258-9] [PMID: 9159498]

[66]　Bräu ME, Dreimann M, Olschewski A, Vogel W, Hempelmann G. Effect of drugs used for neuropathic pain management on tetrodotoxin-resistant Na(+) currents in rat sensory neurons. Anesthesiology 2001; 94(1): 137-44.
　　　[http://dx.doi.org/10.1097/00000542-200101000-00024] [PMID: 11135733]

[67]　Stummann TC, Salvati P, Fariello RG, Faravelli L. The anti-nociceptive agent ralfinamide inhibits tetrodotoxin-resistant and tetrodotoxin-sensitive Na+ currents in dorsal root ganglion neurons. Eur J Pharmacol 2005; 510(3): 197-208.
　　　[http://dx.doi.org/10.1016/j.ejphar.2005.01.030] [PMID: 15763243]

[68]　McLean MJ, Schmutz M, Wamil AW, Olpe HR, Portet C, Feldmann KF. Oxcarbazepine: mechanisms of action. Epilepsia 1994; 35 (Suppl. 3): S5-9.
　　　[http://dx.doi.org/10.1111/j.1528-1157.1994.tb05949.x] [PMID: 8156978]

[69]　Zhang S, Zhang Z, Shen Y, *et al.* SCN9A Epileptic Encephalopathy Mutations Display a Gain-o--function Phenotype and Distinct Sensitivity to Oxcarbazepine. Neurosci Bull 2020; 36(1): 11-24.
　　　[http://dx.doi.org/10.1007/s12264-019-00413-5] [PMID: 31372899]

[70]　Ambrósio AF, Silva AP, Malva JO, Soares-da-Silva P, Carvalho AP, Carvalho CM. Carbamazepine inhibits L-type Ca2+ channels in cultured rat hippocampal neurons stimulated with glutamate receptor agonists. Neuropharmacology 1999; 38(9): 1349-59.

[http://dx.doi.org/10.1016/S0028-3908(99)00058-1] [PMID: 10471089]

[71]   Stefani A, Pisani A, De Murtas M, Mercuri NB, Marciani MG, Calabresi P. Action of GP 47779, the active metabolite of oxcarbazepine, on the corticostriatal system. II. Modulation of high-voltag--activated calcium currents. Epilepsia 1995; 36(10): 997-1002.
[http://dx.doi.org/10.1111/j.1528-1157.1995.tb00958.x] [PMID: 7555964]

[72]   Schmidt D, Elger CE. What is the evidence that oxcarbazepine and carbamazepine are distinctly different antiepileptic drugs? Epilepsy Behav 2004; 5(5): 627-35.
[http://dx.doi.org/10.1016/j.yebeh.2004.07.004] [PMID: 15380112]

[73]   Putnam TJ, Merritt HH. Experimental Determination of the Anticonvulsant Properties of Some Phenyl Derivatives. Science 1937; 85(2213): 525-6.
[http://dx.doi.org/10.1126/science.85.2213.525] [PMID: 17750072]

[74]   Toman JE. The neuropharmacology of antiepileptics. Electroencephalogr Clin Neurophysiol 1949; 1(1): 33-44.
[http://dx.doi.org/10.1016/0013-4694(49)90161-3] [PMID: 18144159]

[75]   Lipicky RJ, Gilbert DL, Stillman IM. Diphenylhydantoin inhibition of sodium conductance in squid giant axon. Proc Natl Acad Sci USA 1972; 69(7): 1758-60.
[http://dx.doi.org/10.1073/pnas.69.7.1758] [PMID: 4505652]

[76]   Ragsdale DS, McPhee JC, Scheuer T, Catterall WA. Common molecular determinants of local anesthetic, antiarrhythmic, and anticonvulsant block of voltage-gated Na+ channels. Proc Natl Acad Sci USA 1996; 93(17): 9270-5.
[http://dx.doi.org/10.1073/pnas.93.17.9270] [PMID: 8799190]

[77]   Kuo CC. A common anticonvulsant binding site for phenytoin, carbamazepine, and lamotrigine in neuronal Na+ channels. Mol Pharmacol 1998; 54(4): 712-21.
[PMID: 9765515]

[78]   Kuo CC, Bean BP. Na+ channels must deactivate to recover from inactivation. Neuron 1994; 12(4): 819-29.
[http://dx.doi.org/10.1016/0896-6273(94)90335-2] [PMID: 8161454]

[79]   Kuo CC, Lu L. Characterization of lamotrigine inhibition of Na+ channels in rat hippocampal neurones. Br J Pharmacol 1997; 121(6): 1231-8.
[http://dx.doi.org/10.1038/sj.bjp.0701221] [PMID: 9249262]

[80]   Rogawski MA, Löscher W. The neurobiology of antiepileptic drugs. Nat Rev Neurosci 2004; 5(7): 553-64.
[http://dx.doi.org/10.1038/nrn1430] [PMID: 15208697]

[81]   Keppel Hesselink JM. Phenytoin: a step by step insight into its multiple mechanisms of action-80 years of mechanistic studies in neuropharmacology. J Neurol 2017; 264(9): 2043-7.
[http://dx.doi.org/10.1007/s00415-017-8465-4] [PMID: 28349209]

[82]   Grunze H, von Wegerer J, Greene RW, Walden J. Modulation of calcium and potassium currents by lamotrigine. Neuropsychobiology 1998; 38(3): 131-8.
[http://dx.doi.org/10.1159/000026528] [PMID: 9778600]

[83]   Xie X, Lancaster B, Peakman T, Garthwaite J. Interaction of the antiepileptic drug lamotrigine with recombinant rat brain type IIA Na+ channels and with native Na+ channels in rat hippocampal neurones. Pflugers Arch 1995; 430(3): 437-46.
[http://dx.doi.org/10.1007/BF00373920] [PMID: 7491269]

[84]   Yarov-Yarovoy V, Brown J, Sharp EM, Clare JJ, Scheuer T, Catterall WA. Molecular determinants of voltage-dependent gating and binding of pore-blocking drugs in transmembrane segment IIIS6 of the Na(+) channel alpha subunit. J Biol Chem 2001; 276(1): 20-7.
[http://dx.doi.org/10.1074/jbc.M006992200] [PMID: 11024055]

[85]   Maarbjerg S, Di Stefano G, Bendtsen L, Cruccu G. Trigeminal neuralgia - diagnosis and treatment.

Cephalalgia 2017; 37(7): 648-57.
[http://dx.doi.org/10.1177/0333102416687280] [PMID: 28076964]

[86]   Bendtsen L, Zakrzewska JM, Abbott J, *et al.* European Academy of Neurology guideline on trigeminal neuralgia. Eur J Neurol 2019; 26(6): 831-49.
[http://dx.doi.org/10.1111/ene.13950] [PMID: 30860637]

[87]   Wiffen PJ, Derry S, Moore RA, Kalso EA. Carbamazepine for chronic neuropathic pain and fibromyalgia in adults. Cochrane Database Syst Rev 2014; (4): CD005451.
[http://dx.doi.org/10.1002/14651858.CD005451.pub3] [PMID: 24719027]

[88]   Choi HR, Aktas A, Bottros MM. Pharmacotherapy to Manage Central Post-Stroke Pain. CNS Drugs 2021; 35(2): 151-60.
[http://dx.doi.org/10.1007/s40263-021-00791-3] [PMID: 33550430]

[89]   Bergouignan M. [Fifteen years of trial therapy of essential trigeminal neuralgia: the place of diphenylhydantoin and its derivatives]. Rev Neurol (Paris) 1958; 98(5): 414-6. [Fifteen years of trial therapy of essential trigeminal neuralgia: the place of diphenylhydantoin and its derivatives].
[PMID: 13579820]

[90]   Vinik A, Casellini C, Nevoret ML. Diabetic Neuropathies. 2000.
[http://dx.doi.org/10.1007/s001250051477]

[91]   Di Stefano G, De Stefano G, Leone C, *et al.* Real-world effectiveness and tolerability of carbamazepine and oxcarbazepine in 354 patients with trigeminal neuralgia. Eur J Pain 2021; 25(5): 1064-71.
[http://dx.doi.org/10.1002/ejp.1727] [PMID: 33428801]

[92]   Saeed T, Nasrullah M, Ghafoor A, *et al.* Efficacy and tolerability of carbamazepine for the treatment of painful diabetic neuropathy in adults: a 12-week, open-label, multicenter study. Int J Gen Med 2014; 7: 339-43.
[PMID: 25061334]

[93]   Geha P, Yang Y, Estacion M, *et al.* Pharmacotherapy for Pain in a Family With Inherited Erythromelalgia Guided by Genomic Analysis and Functional Profiling. JAMA Neurol 2016; 73(6): 659-67.
[http://dx.doi.org/10.1001/jamaneurol.2016.0389] [PMID: 27088781]

[94]   Shaikh S, Yaacob HB, Abd Rahman RB. Lamotrigine for trigeminal neuralgia: efficacy and safety in comparison with carbamazepine. J Chin Med Assoc 2011; 74(6): 243-9.
[http://dx.doi.org/10.1016/j.jcma.2011.04.002] [PMID: 21621166]

[95]   Bendtsen L, Zakrzewska JM, Heinskou TB, *et al.* Advances in diagnosis, classification, pathophysiology, and management of trigeminal neuralgia. Lancet Neurol 2020; 19(9): 784-96.
[http://dx.doi.org/10.1016/S1474-4422(20)30233-7] [PMID: 32822636]

[96]   Zhou M, Chen N, He L, Yang M, Zhu C, Wu F. Oxcarbazepine for neuropathic pain. Cochrane Database Syst Rev 2017; 12: CD007963.
[PMID: 29199767]

[97]   Birse F, Derry S, Moore RA. Phenytoin for neuropathic pain and fibromyalgia in adults. Cochrane Database Syst Rev 2012; (5): CD009485.
[http://dx.doi.org/10.1002/14651858.CD009485.pub2] [PMID: 22592741]

[98]   Saudek CD, Werns S, Reidenberg MM. Phenytoin in the treatment of diabetic symmetrical polyneuropathy. Clin Pharmacol Ther 1977; 22(2): 196-9.
[http://dx.doi.org/10.1002/cpt1977222196] [PMID: 328209]

[99]   Chadda VS, Mathur MS. Double blind study of the effects of diphenylhydantoin sodium on diabetic neuropathy. J Assoc Physicians India 1978; 26(5): 403-6.
[PMID: 365857]

[100]   Russell AL, Kopsky DJ, Hesselink JMK. Phenytoin Cream for the Treatment of Sciatic Pain: Clinical

Effects and Theoretical Considerations: Case Report. J Pain Palliat Care Pharmacother 2020; 34(2): 99-105.
[http://dx.doi.org/10.1080/15360288.2020.1733169] [PMID: 32118498]

[101] Kopsky DJ, Keppel Hesselink JM. Single-Blind Placebo-Controlled Response Test with Phenytoin 10% Cream in Neuropathic Pain Patients. Pharmaceuticals (Basel) 2018; 11(4): 122.
[http://dx.doi.org/10.3390/ph11040122] [PMID: 30424471]

[102] Kopsky DJ, Keppel Hesselink JM. Phenytoin Cream for the Treatment for Neuropathic Pain: Case Series. Pharmaceuticals (Basel) 2018; 11(2): E53.
[http://dx.doi.org/10.3390/ph11020053] [PMID: 29843362]

[103] Sidhu HS, Sadhotra A. Current Status of the New Antiepileptic Drugs in Chronic Pain. Front Pharmacol 2016; 7: 276.
[http://dx.doi.org/10.3389/fphar.2016.00276] [PMID: 27610084]

[104] Wiffen PJ, Derry S, Moore RA. Lamotrigine for chronic neuropathic pain and fibromyalgia in adults. Cochrane Database Syst Rev 2013; (12): CD006044.
[http://dx.doi.org/10.1002/14651858.CD006044.pub4] [PMID: 24297457]

[105] Zakrzewska JM, Linskey ME. Trigeminal neuralgia. BMJ Clin Evid 2014. 2014; 1207.

[106] Wiffen PJ, Derry S, Moore RA, McQuay HJ. Carbamazepine for acute and chronic pain in adults. Cochrane Database Syst Rev 2011; (1): CD005451.
[PMID: 21249671]

[107] Ferrell PB Jr, McLeod HL. Carbamazepine, HLA-B*1502 and risk of Stevens-Johnson syndrome and toxic epidermal necrolysis: US FDA recommendations. Pharmacogenomics 2008; 9(10): 1543-6.
[http://dx.doi.org/10.2217/14622416.9.10.1543] [PMID: 18855540]

[108] Wang Q, Sun S, Xie M, Zhao K, Li X, Zhao Z. Association between the HLA-B alleles and carbamazepine-induced SJS/TEN: A meta-analysis. Epilepsy Res 2017; 135: 19-28.
[http://dx.doi.org/10.1016/j.eplepsyres.2017.05.015] [PMID: 28618376]

[109] Lu X, Wang X. Hyponatremia induced by antiepileptic drugs in patients with epilepsy. Expert Opin Drug Saf 2017; 16(1): 77-87.
[http://dx.doi.org/10.1080/14740338.2017.1248399] [PMID: 27737595]

[110] Abou-Khalil BW. Update on Antiepileptic Drugs 2019. Continuum (Minneap Minn) 2019; 25(2): 508-36.
[http://dx.doi.org/10.1212/CON.0000000000000715] [PMID: 30921021]

[111] Li J, Sun M, Wang X. The adverse-effect profile of lacosamide. Expert Opin Drug Saf 2020; 19(2): 131-8.
[http://dx.doi.org/10.1080/14740338.2020.1713089] [PMID: 31914330]

[112] Galiana GL, Gauthier AC, Mattson RH. Eslicarbazepine Acetate: A New Improvement on a Classic Drug Family for the Treatment of Partial-Onset Seizures. Drugs R D 2017; 17(3): 329-39.
[http://dx.doi.org/10.1007/s40268-017-0197-5] [PMID: 28741150]

[113] Bialer M, Johannessen SI, Levy RH, Perucca E, Tomson T, White HS. Progress report on new antiepileptic drugs: a summary of the Ninth Eilat Conference (EILAT IX). Epilepsy Res 2009; 83(1): 1-43.
[http://dx.doi.org/10.1016/j.eplepsyres.2008.09.005] [PMID: 19008076]

[114] Verrotti A, Loiacono G, Rossi A, Zaccara G. Eslicarbazepine acetate: an update on efficacy and safety in epilepsy. Epilepsy Res 2014; 108(1): 1-10.
[http://dx.doi.org/10.1016/j.eplepsyres.2013.10.005] [PMID: 24225327]

[115] Sanchez-Larsen A, Sopelana D, Diaz-Maroto I, *et al.* Assessment of efficacy and safety of eslicarbazepine acetate for the treatment of trigeminal neuralgia. Eur J Pain 2018; 22(6): 1080-7.
[http://dx.doi.org/10.1002/ejp.1192] [PMID: 29369456]

[116] Garcia-Escrivà A, López-Herandez N, Lezcano M, Berenguer L. Experience with eslicarbazepine acetate as treatment for painful diabetic neuropathy. Neurologia 2016; 31(9): 639-40.
[http://dx.doi.org/10.1016/j.nrleng.2014.10.009] [PMID: 25681850]

[117] Tomić MA, Pecikoza UB, Micov AM, Stepanović-Petrović RM. The Efficacy of Eslicarbazepine Acetate in Models of Trigeminal, Neuropathic, and Visceral Pain: The Involvement of 5-HT1B/1D Serotonergic and CB1/CB2 Cannabinoid Receptors. Anesth Analg 2015; 121(6): 1632-9.
[http://dx.doi.org/10.1213/ANE.0000000000000953] [PMID: 26465930]

[118] Doeser A, Dickhof G, Reitze M, *et al.* Targeting pharmacoresistant epilepsy and epileptogenesis with a dual-purpose antiepileptic drug. Brain 2015; 138(Pt 2): 371-87.
[http://dx.doi.org/10.1093/brain/awu339] [PMID: 25472797]

[119] Cai S, Gomez K, Moutal A, Khanna R. Targeting T-type/CaV3.2 channels for chronic pain. Transl Res 2021; S1931-5244: 2-5.

[120] Zhu LN, Chen D, Tan G, Wang HJ, Chu S, Liu L. The tolerability and safety profile of eslicarbazepine acetate in neurological disorders. J Neurol Sci 2020; 413: 116772.
[http://dx.doi.org/10.1016/j.jns.2020.116772] [PMID: 32217376]

[121] Cawello W. Clinical pharmacokinetic and pharmacodynamic profile of lacosamide. Clin Pharmacokinet 2015; 54(9): 901-14.
[http://dx.doi.org/10.1007/s40262-015-0276-0] [PMID: 25957198]

[122] Beyreuther BK, Freitag J, Heers C, Krebsfänger N, Scharfenecker U, Stöhr T. Lacosamide: a review of preclinical properties. CNS Drug Rev 2007; 13(1): 21-42.
[http://dx.doi.org/10.1111/j.1527-3458.2007.00001.x] [PMID: 17461888]

[123] Errington AC, Stöhr T, Heers C, Lees G. The investigational anticonvulsant lacosamide selectively enhances slow inactivation of voltage-gated sodium channels. Mol Pharmacol 2008; 73(1): 157-69.
[http://dx.doi.org/10.1124/mol.107.039867] [PMID: 17940193]

[124] Rogawski MA, Tofighy A, White HS, Matagne A, Wolff C. Current understanding of the mechanism of action of the antiepileptic drug lacosamide. Epilepsy Res 2015; 110: 189-205.
[http://dx.doi.org/10.1016/j.eplepsyres.2014.11.021] [PMID: 25616473]

[125] Jo S, Bean BP. Lacosamide Inhibition of Nav1.7 Voltage-Gated Sodium Channels: Slow Binding to Fast-Inactivated States. Mol Pharmacol 2017; 91(4): 277-86.
[http://dx.doi.org/10.1124/mol.116.106401] [PMID: 28119481]

[126] Carona A, Bicker J, Silva R, Fonseca C, Falcão A, Fortuna A. Pharmacology of lacosamide: From its molecular mechanisms and pharmacokinetics to future therapeutic applications. Life Sci 2021; 275: 119342.
[http://dx.doi.org/10.1016/j.lfs.2021.119342] [PMID: 33713668]

[127] Kushnarev M, Pirvulescu IP, Candido KD, Knezevic NN. Neuropathic pain: preclinical and early clinical progress with voltage-gated sodium channel blockers. Expert Opin Investig Drugs 2020; 29(3): 259-71.
[http://dx.doi.org/10.1080/13543784.2020.1728254] [PMID: 32070160]

[128] Zamponi GW. Targeting voltage-gated calcium channels in neurological and psychiatric diseases. Nat Rev Drug Discov 2016; 15(1): 19-34.
[http://dx.doi.org/10.1038/nrd.2015.5] [PMID: 26542451]

[129] Patel R, Montagut-Bordas C, Dickenson AH. Calcium channel modulation as a target in chronic pain control. Br J Pharmacol 2018; 175(12): 2173-84.
[http://dx.doi.org/10.1111/bph.13789] [PMID: 28320042]

[130] Morrison EE, Sandilands EA, Webb DJ. Gabapentin and pregabalin: do the benefits outweigh the harms? J R Coll Physicians Edinb 2017; 47(4): 310-3.
[http://dx.doi.org/10.4997/JRCPE.2017.402] [PMID: 29537399]

[131] Peckham AM, Evoy KE, Ochs L, Covvey JR. Gabapentin for Off-Label Use: Evidence-Based or Cause for Concern? Subst Abuse 2018; 12: 1178221818801311.
[http://dx.doi.org/10.1177/1178221818801311] [PMID: 30262984]

[132] Attal N. Pharmacological treatments of neuropathic pain: The latest recommendations. Rev Neurol (Paris) 2019; 175(1-2): 46-50.
[http://dx.doi.org/10.1016/j.neurol.2018.08.005] [PMID: 30318260]

[133] Finnerup NB, Attal N, Haroutounian S, *et al.* Pharmacotherapy for neuropathic pain in adults: a systematic review and meta-analysis. Lancet Neurol 2015; 14(2): 162-73.
[http://dx.doi.org/10.1016/S1474-4422(14)70251-0] [PMID: 25575710]

[134] Anantharamu T, Govind MA. Managing chronic pain: are gabapentinoids being misused? Pain Manag (Lond) 2018; 8(5): 309-11.
[http://dx.doi.org/10.2217/pmt-2018-0018] [PMID: 30212257]

[135] Chincholkar M. Gabapentinoids: pharmacokinetics, pharmacodynamics and considerations for clinical practice. Br J Pain 2020; 14(2): 104-14.
[http://dx.doi.org/10.1177/2049463720912496] [PMID: 32537149]

[136] Bockbrader HN, Wesche D, Miller R, Chapel S, Janiczek N, Burger P. A comparison of the pharmacokinetics and pharmacodynamics of pregabalin and gabapentin. Clin Pharmacokinet 2010; 49(10): 661-9.
[http://dx.doi.org/10.2165/11536200-000000000-00000] [PMID: 20818832]

[137] Takahashi Y, Nishimura T, Higuchi K, *et al.* Transport of Pregabalin *Via* L-Type Amino Acid Transporter 1 (SLC7A5) in Human Brain Capillary Endothelial Cell Line. Pharm Res 2018; 35(12): 246.
[http://dx.doi.org/10.1007/s11095-018-2532-0] [PMID: 30374619]

[138] Schulze-Bonhage A. Pharmacokinetic and pharmacodynamic profile of pregabalin and its role in the treatment of epilepsy. Expert Opin Drug Metab Toxicol 2013; 9(1): 105-15.
[http://dx.doi.org/10.1517/17425255.2013.749239] [PMID: 23205518]

[139] Alles SRA, Cain SM, Snutch TP. Pregabalin as a Pain Therapeutic: Beyond Calcium Channels. Front Cell Neurosci 2020; 14: 83.
[http://dx.doi.org/10.3389/fncel.2020.00083] [PMID: 32351366]

[140] Stahl SM, Porreca F, Taylor CP, Cheung R, Thorpe AJ, Clair A. The diverse therapeutic actions of pregabalin: is a single mechanism responsible for several pharmacological activities? Trends Pharmacol Sci 2013; 34(6): 332-9.
[http://dx.doi.org/10.1016/j.tips.2013.04.001] [PMID: 23642658]

[141] Patel R, Dickenson AH. Mechanisms of the gabapentinoids and α 2 δ-1 calcium channel subunit in neuropathic pain. Pharmacol Res Perspect 2016; 4(2): e00205.
[http://dx.doi.org/10.1002/prp2.205] [PMID: 27069626]

[142] Chen J, Li L, Chen SR, *et al.* The $\alpha2\delta$-1-NMDA Receptor Complex Is Critically Involved in Neuropathic Pain Development and Gabapentin Therapeutic Actions. Cell Rep 2018; 22(9): 2307-21.
[http://dx.doi.org/10.1016/j.celrep.2018.02.021] [PMID: 29490268]

[143] Meymandi MS, Keyhanfar F, Sepehri GR, Heravi G, Yazdanpanah O. The Contribution of NMDA Receptors in Antinociceptive Effect of Pregabalin: Comparison of Two Models of Pain Assessment. Anesth Pain Med 2017; 7(3): e14602.
[http://dx.doi.org/10.5812/aapm.14602] [PMID: 28824867]

[144] Derry S, Cording M, Wiffen PJ, Law S, Phillips T, Moore RA. Pregabalin for pain in fibromyalgia in adults. Cochrane Database Syst Rev 2016; 9: CD011790.
[http://dx.doi.org/10.1002/14651858.CD011790.pub2] [PMID: 27684492]

[145] Moore RA, Straube S, Wiffen PJ, Derry S, McQuay HJ. Pregabalin for acute and chronic pain in adults. Cochrane Database Syst Rev 2009; 3(3): CD007076.

[http://dx.doi.org/10.1002/14651858.CD007076.pub2] [PMID: 19588419]

[146] Davari M, Amani B, Amani B, Khanijahani A, Akbarzadeh A, Shabestan R. Pregabalin and gabapentin in neuropathic pain management after spinal cord injury: a systematic review and meta-analysis. Korean J Pain 2020; 33(1): 3-12.
[http://dx.doi.org/10.3344/kjp.2020.33.1.3] [PMID: 31888312]

[147] Rai AS, Khan JS, Dhaliwal J, *et al.* Preoperative pregabalin or gabapentin for acute and chronic postoperative pain among patients undergoing breast cancer surgery: A systematic review and meta-analysis of randomized controlled trials. J Plast Reconstr Aesthet Surg 2017; 70(10): 1317-28.
[http://dx.doi.org/10.1016/j.bjps.2017.05.054] [PMID: 28751024]

[148] Chaparro LE, Smith SA, Moore RA, Wiffen PJ, Gilron I. Pharmacotherapy for the prevention of chronic pain after surgery in adults. Cochrane Database Syst Rev 2013; 2013;7 Cd008307.

[149] Shanthanna H, Gilron I, Rajarathinam M, *et al.* Benefits and safety of gabapentinoids in chronic low back pain: A systematic review and meta-analysis of randomized controlled trials. PLoS Med 2017; 14(8): e1002369.
[http://dx.doi.org/10.1371/journal.pmed.1002369] [PMID: 28809936]

[150] Zaccara G, Gangemi P, Perucca P, Specchio L. The adverse event profile of pregabalin: a systematic review and meta-analysis of randomized controlled trials. Epilepsia 2011; 52(4): 826-36.
[http://dx.doi.org/10.1111/j.1528-1167.2010.02966.x] [PMID: 21320112]

[151] Page RL II, Cantu M, Lindenfeld J, Hergott LJ, Lowes BD. Possible heart failure exacerbation associated with pregabalin: case discussion and literature review. J Cardiovasc Med (Hagerstown) 2008; 9(9): 922-5.
[http://dx.doi.org/10.2459/JCM.0b013e3282fb7629] [PMID: 18695430]

[152] Erdoğan G, Ceyhan D, Güleç S. Possible heart failure associated with pregabalin use: case report. Agri 2011; 23(2): 80-3.
[PMID: 21644108]

[153] Gallagher R, Apostle N. Peripheral edema with pregabalin. CMAJ 2013; 185(10): E506.
[http://dx.doi.org/10.1503/cmaj.121232] [PMID: 23128284]

[154] Bonnet U, Richter EL, Isbruch K, Scherbaum N. On the addictive power of gabapentinoids: a mini-review. Psychiatr Danub 2018; 30(2): 142-9.
[http://dx.doi.org/10.24869/psyd.2018.142] [PMID: 29930223]

[155] Smith RV, Havens JR, Walsh SL. Gabapentin misuse, abuse and diversion: a systematic review. Addiction 2016; 111(7): 1160-74.
[http://dx.doi.org/10.1111/add.13324] [PMID: 27265421]

[156] Vickers-Smith R, Sun J, Charnigo RJ, Lofwall MR, Walsh SL, Havens JR. Gabapentin drug misuse signals: A pharmacovigilance assessment using the FDA adverse event reporting system. Drug Alcohol Depend 2020; 206: 107709.
[http://dx.doi.org/10.1016/j.drugalcdep.2019.107709] [PMID: 31732295]

[157] Schifano F. Misuse and abuse of pregabalin and gabapentin: cause for concern? CNS Drugs 2014; 28(6): 491-6.
[http://dx.doi.org/10.1007/s40263-014-0164-4] [PMID: 24760436]

[158] McAnally H, Bonnet U, Kaye AD. Gabapentinoid Benefit and Risk Stratification: Mechanisms Over Myth. Pain Ther 2020; 9(2): 441-52.
[http://dx.doi.org/10.1007/s40122-020-00189-x] [PMID: 32737803]

[159] Wang J, Zhu Y. Different doses of gabapentin formulations for postherpetic neuralgia: A systematical review and meta-analysis of randomized controlled trials. J Dermatolog Treat 2017; 28(1): 65-77.
[http://dx.doi.org/10.3109/09546634.2016.1163315] [PMID: 27798973]

[160] Zajączkowska R, Mika J, Leppert W, Kocot-Kępska M, Malec-Milewska M, Wordliczek J. Mirogabalin-A Novel Selective Ligand for the α2δ Calcium Channel Subunit. Pharmaceuticals (Basel)

2021; 14(2): 14.
[http://dx.doi.org/10.3390/ph14020112] [PMID: 33572689]

[161] Domon Y, Arakawa N, Inoue T, *et al.* Binding Characteristics and Analgesic Effects of Mirogabalin, a Novel Ligand for the $\alpha_2\delta$ Subunit of Voltage-Gated Calcium Channels. J Pharmacol Exp Ther 2018; 365(3): 573-82.
[http://dx.doi.org/10.1124/jpet.117.247551] [PMID: 29563324]

[162] Jansen M, Warrington S, Dishy V, *et al.* A Randomized, Placebo-Controlled, Double-Blind Study of the Safety, Tolerability, Pharmacokinetics, and Pharmacodynamics of Single and Repeated Doses of Mirogabalin in Healthy Asian Volunteers. Clin Pharmacol Drug Dev 2018; 7(6): 661-9.
[http://dx.doi.org/10.1002/cpdd.448] [PMID: 29663714]

[163] Mendell J, Levy-Cooperman N, Sellers E, *et al.* Abuse potential of mirogabalin in recreational polydrug users. Ther Adv Drug Saf 2019; 10: 2042098619836032.
[http://dx.doi.org/10.1177/2042098619836032] [PMID: 31057786]

[164] Silberstein SD. Topiramate in Migraine Prevention: A 2016 Perspective. Headache 2017; 57(1): 165-78.
[http://dx.doi.org/10.1111/head.12997] [PMID: 27902848]

[165] Gupta A, Kulkarni A, Ramanujam V, Zheng L, Treacy E. Improvement in chronic low back pain in an obese patient with topiramate use. J Pain Palliat Care Pharmacother 2015; 29(2): 140-3.
[http://dx.doi.org/10.3109/15360288.2015.1035837] [PMID: 26095484]

[166] Khalil NY, AlRabiah HK, Al Rashoud SS, Bari A, Wani TA. Topiramate: Comprehensive profile. Profiles Drug Subst Excip Relat Methodol 2019; 44: 333-78.
[http://dx.doi.org/10.1016/bs.podrm.2018.11.005] [PMID: 31029222]

[167] Johannessen CU, Johannessen SI. Valproate: past, present, and future. CNS Drug Rev 2003; 9(2): 199-216.
[http://dx.doi.org/10.1111/j.1527-3458.2003.tb00249.x] [PMID: 12847559]

[168] Gill D, Derry S, Wiffen PJ, Moore RA. Valproic acid and sodium valproate for neuropathic pain and fibromyalgia in adults. Cochrane Database Syst Rev 2011; 2011; 10 Cd009183.

[169] Vajda FJ. Valproate under threat. Aust N Z J Psychiatry 2019; 53(9): 923.
[http://dx.doi.org/10.1177/0004867419834357] [PMID: 30848662]

[170] Italiano D, Spina E, de Leon J. Pharmacokinetic and pharmacodynamic interactions between antiepileptics and antidepressants. Expert Opin Drug Metab Toxicol 2014; 10(11): 1457-89.
[http://dx.doi.org/10.1517/17425255.2014.956081] [PMID: 25196459]

[171] Kouremenos E, Arvaniti C, Constantinidis TS, *et al.* Consensus of the Hellenic Headache Society on the diagnosis and treatment of migraine. J Headache Pain 2019; 20(1): 113.
[http://dx.doi.org/10.1186/s10194-019-1060-6] [PMID: 31835997]

[172] Silberstein SD, Holland S, Freitag F, Dodick DW, Argoff C, Ashman E. Evidence-based guideline update: pharmacologic treatment for episodic migraine prevention in adults: report of the Quality Standards Subcommittee of the American Academy of Neurology and the American Headache Society. Neurology 2012; 78(17): 1337-45.
[http://dx.doi.org/10.1212/WNL.0b013e3182535d20] [PMID: 22529202]

[173] Johannessen SI. Pharmacokinetics and interaction profile of topiramate: review and comparison with other newer antiepileptic drugs. Epilepsia 1997; 38 (Suppl. 1): S18-23.
[http://dx.doi.org/10.1111/j.1528-1157.1997.tb04512.x] [PMID: 9092953]

[174] Haddad PM, Das A, Ashfaq M, Wieck A. A review of valproate in psychiatric practice. Expert Opin Drug Metab Toxicol 2009; 5(5): 539-51.
[http://dx.doi.org/10.1517/17425250902911455] [PMID: 19409030]

[175] Perucca E. Pharmacological and therapeutic properties of valproate: a summary after 35 years of clinical experience. CNS Drugs 2002; 16(10): 695-714.

[http://dx.doi.org/10.2165/00023210-200216100-00004] [PMID: 12269862]

[176] Ghodke-Puranik Y, Thorn CF, Lamba JK, *et al*. Valproic acid pathway: pharmacokinetics and pharmacodynamics. Pharmacogenet Genomics 2013; 23(4): 236-41.
[http://dx.doi.org/10.1097/FPC.0b013e32835ea0b2] [PMID: 23407051]

[177] Zhu MM, Li HL, Shi LH, Chen XP, Luo J, Zhang ZL. The pharmacogenomics of valproic acid. J Hum Genet 2017; 62(12): 1009-14.
[http://dx.doi.org/10.1038/jhg.2017.91] [PMID: 28878340]

[178] Bagnato F, Good J. The Use of Antiepileptics in Migraine Prophylaxis. Headache 2016; 56(3): 603-15.
[http://dx.doi.org/10.1111/head.12781] [PMID: 26935348]

[179] Peres MF, Mercante JP, Tanuri FC, Nunes M, Zukerman E. Chronic migraine prevention with topiramate. J Headache Pain 2006; 7(4): 185-7.
[http://dx.doi.org/10.1007/s10194-006-0339-6] [PMID: 17016684]

[180] Löscher W, Hörstermann D. Differential effects of vigabatrin, gamma-acetylenic GABA, aminooxyacetic acid, and valproate on levels of various amino acids in rat brain regions and plasma. Naunyn Schmiedebergs Arch Pharmacol 1994; 349(3): 270-8.
[http://dx.doi.org/10.1007/BF00169293] [PMID: 8208305]

[181] Chong MS, Libretto SE. The rationale and use of topiramate for treating neuropathic pain. Clin J Pain 2003; 19(1): 59-68.
[http://dx.doi.org/10.1097/00002508-200301000-00008] [PMID: 12514458]

[182] Diener HC, Bussone G, Van Oene JC, Lahaye M, Schwalen S, Goadsby PJ. Topiramate reduces headache days in chronic migraine: a randomized, double-blind, placebo-controlled study. Cephalalgia 2007; 27(7): 814-23.
[http://dx.doi.org/10.1111/j.1468-2982.2007.01326.x] [PMID: 17441971]

[183] Diener HC, Dodick DW, Goadsby PJ, *et al*. Utility of topiramate for the treatment of patients with chronic migraine in the presence or absence of acute medication overuse. Cephalalgia 2009; 29(10): 1021-7.
[http://dx.doi.org/10.1111/j.1468-2982.2009.01859.x] [PMID: 19735529]

[184] Silberstein S, Lipton R, Dodick D, *et al*. Topiramate treatment of chronic migraine: a randomized, placebo-controlled trial of quality of life and other efficacy measures. Headache 2009; 49(8): 1153-62.
[http://dx.doi.org/10.1111/j.1526-4610.2009.01508.x] [PMID: 19719543]

[185] Dodick DW, Silberstein S, Saper J, *et al*. The impact of topiramate on health-related quality of life indicators in chronic migraine. Headache 2007; 47(10): 1398-408.
[http://dx.doi.org/10.1111/j.1526-4610.2007.00950.x] [PMID: 18052949]

[186] Brandes JL, Saper JR, Diamond M, *et al*. Topiramate for migraine prevention: a randomized controlled trial. JAMA 2004; 291(8): 965-73.
[http://dx.doi.org/10.1001/jama.291.8.965] [PMID: 14982912]

[187] Silberstein SD, Neto W, Schmitt J, Jacobs D. Topiramate in migraine prevention: results of a large controlled trial. Arch Neurol 2004; 61(4): 490-5.
[http://dx.doi.org/10.1001/archneur.61.4.490] [PMID: 15096395]

[188] Pini L-A, Lupo L. Anti-epileptic drugs in the preventive treatment of migraine headache: a brief review 2001; 2: 13-9.

[189] Brandt RB, Doesborg PGG, Haan J, Ferrari MD, Fronczek R. Pharmacotherapy for Cluster Headache. CNS Drugs 2020; 34(2): 171-84.
[http://dx.doi.org/10.1007/s40263-019-00696-2] [PMID: 31997136]

[190] McCormick Z, Chang-Chien G, Marshall B, Huang M, Harden RN. Phantom limb pain: a systematic neuroanatomical-based review of pharmacologic treatment. Pain Med 2014; 15(2): 292-305.
[http://dx.doi.org/10.1111/pme.12283] [PMID: 24224475]

[191] Enke O, New HA, New CH, *et al.* Anticonvulsants in the treatment of low back pain and lumbar radicular pain: a systematic review and meta-analysis. CMAJ 2018; 190(26): E786-93.
[http://dx.doi.org/10.1503/cmaj.171333] [PMID: 29970367]

[192] Araya EI, Claudino RF, Piovesan EJ, Chichorro JG. Trigeminal Neuralgia: Basic and Clinical Aspects. Curr Neuropharmacol 2020; 18(2): 109-19.
[http://dx.doi.org/10.2174/1570159X17666191010094350] [PMID: 31608834]

[193] Vázquez M, Maldonado C, Guevara N, *et al.* Lamotrigine-Valproic Acid Interaction Leading to Stevens-Johnson Syndrome. Case Rep Med 2018; 2018: 5371854.
[http://dx.doi.org/10.1155/2018/5371854] [PMID: 30228819]

[194] Luykx J, Mason M, Ferrari MD, Carpay J. Are migraineurs at increased risk of adverse drug responses? A meta-analytic comparison of topiramate-related adverse drug reactions in epilepsy and migraine. Clin Pharmacol Ther 2009; 85(3): 283-8.
[http://dx.doi.org/10.1038/clpt.2008.203] [PMID: 18987621]

[195] Limmer AL, Holland LC, Loftus BD. Zonisamide for Cluster Headache Prophylaxis: A Case Series. Headache 2019; 59(6): 924-9.
[http://dx.doi.org/10.1111/head.13546] [PMID: 31038740]

[196] Holder JL Jr, Wilfong AA. Zonisamide in the treatment of epilepsy. Expert Opin Pharmacother 2011; 12(16): 2573-81.
[http://dx.doi.org/10.1517/14656566.2011.622268] [PMID: 21967409]

[197] Koshimizu H, Ohkawara B, Nakashima H, *et al.* Zonisamide ameliorates neuropathic pain partly by suppressing microglial activation in the spinal cord in a mouse model. Life Sci 2020; 263: 118577.
[http://dx.doi.org/10.1016/j.lfs.2020.118577] [PMID: 33058918]

[198] Sills GJ, Rogawski MA. Mechanisms of action of currently used antiseizure drugs. 2020; 168 107966.
[http://dx.doi.org/10.1016/j.neuropharm.2020.107966]

[199] Abd-Elsayed A, Jackson M, Gu SL, Fiala K, Gu J. Neuropathic pain and Kv7 voltage-gated potassium channels: The potential role of Kv7 activators in the treatment of neuropathic pain 2019; 15: 1-8.

[200] Passmore G, Delmas P. Does cure for pain REST on Kv7 channels? Pain 2011; 152(4): 709-10.
[http://dx.doi.org/10.1016/j.pain.2011.02.040] [PMID: 21377798]

[201] Stafstrom CE, Grippon S, Kirkpatrick P. Ezogabine (retigabine). Nat Rev Drug Discov 2011; 10(10): 729-30.
[http://dx.doi.org/10.1038/nrd3561] [PMID: 21959281]

[202] Blackburn-Munro G, Jensen BS. The anticonvulsant retigabine attenuates nociceptive behaviours in rat models of persistent and neuropathic pain. Eur J Pharmacol 2003; 460(2-3): 109-16.
[http://dx.doi.org/10.1016/S0014-2999(02)02924-2] [PMID: 12559370]

[203] Abd-Elsayed AA, Ikeda R, Jia Z, *et al.* KCNQ channels in nociceptive cold-sensing trigeminal ganglion neurons as therapeutic targets for treating orofacial cold hyperalgesia. Mol Pain 2015; 11: 45.
[http://dx.doi.org/10.1186/s12990-015-0048-8] [PMID: 26227020]

[204] Tompson DJ, Crean CS. Clinical pharmacokinetics of retigabine/ezogabine. Curr Clin Pharmacol 2013; 8(4): 319-31.
[http://dx.doi.org/10.2174/15748847113089990053] [PMID: 23342983]

[205] Porter RJ, Partiot A, Sachdeo R, Nohria V, Alves WM. Randomized, multicenter, dose-ranging trial of retigabine for partial-onset seizures. Neurology 2007; 68(15): 1197-204.
[http://dx.doi.org/10.1212/01.wnl.0000259034.45049.00] [PMID: 17420403]

[206] Brodie MJ, Lerche H, Gil-Nagel A, *et al.* Efficacy and safety of adjunctive ezogabine (retigabine) in refractory partial epilepsy. Neurology 2010; 75(20): 1817-24.
[http://dx.doi.org/10.1212/WNL.0b013e3181fd6170] [PMID: 20944074]

[207] Tsantoulas C, McMahon SB. Opening paths to novel analgesics: The role of potassium channels in

chronic pain 2014; 37: 146-58.

[208] Klinger F, Geier P, Dorostkar MM, *et al.* Concomitant facilitation of GABAA receptors and KV7 channels by the non-opioid analgesic flupirtine. Br J Pharmacol 2012; 166(5): 1631-42.
[http://dx.doi.org/10.1111/j.1476-5381.2011.01821.x] [PMID: 22188423]

[209] Harish S, Bhuvana K, Bengalorkar GM, Kumar T. Flupirtine: Clinical pharmacology. J Anaesthesiol Clin Pharmacol 2012; 28(2): 172-7.
[http://dx.doi.org/10.4103/0970-9185.94833] [PMID: 22557738]

[210] Hummel T, Friedmann T, Pauli E, Niebch G, Borbe HO, Kobal G. Dose-related analgesic effects of flupirtine. Br J Clin Pharmacol 1991; 32(1): 69-76.
[http://dx.doi.org/10.1111/j.1365-2125.1991.tb05615.x] [PMID: 1888644]

[211] Li C, Lei Y, Tian Y, *et al.* The etiological contribution of GABAergic plasticity to the pathogenesis of neuropathic pain. 2019.
[http://dx.doi.org/10.1177/1744806919847366]

[212] Löscher W, Schmidt D. Strategies in antiepileptic drug development: is rational drug design superior to random screening and structural variation? : Elsevier; 1994; 17: 95-134.

[213] Giardina WJ, Decker MW, Porsolt RD, *et al.* An evaluation of the GABA uptake blocker tiagabine in animal models of neuropathic and nociceptive pain. Drug Dev Res 1998; 44: 106-13.
[http://dx.doi.org/10.1002/(SICI)1098-2299(199806/07)44:2/3<106::AID-DDR8>3.0.CO;2-Q]

[214] Novak V, Kanard R, Kissel JT, Mendell JR. Treatment of painful sensory neuropathy with tiagabine: a pilot study. Clin Auton Res 2001; 11(6): 357-61.
[http://dx.doi.org/10.1007/BF02292767] [PMID: 11794716]

[215] Todorov AA, Kolchev CB, Todorov AB. Tiagabine and gabapentin for the management of chronic pain. Clin J Pain 2005; 21(4): 358-61.
[http://dx.doi.org/10.1097/01.ajp.0000110637.14355.77] [PMID: 15951655]

[216] Jasmin L, Wu MV, Ohara PT. GABA puts a stop to pain 2004; 3: 487-505.
[http://dx.doi.org/10.2174/1568007043336716]

[217] Adkins JC, Noble S. A review of its pharmacodynamic and pharmacokinetic properties and therapeutic potential in the management of epilepsy. 1998; 55: 437-60.

[218] Löscher W, Gillard M, Sands ZA, Kaminski RM, Klitgaard H. Synaptic Vesicle Glycoprotein 2A Ligands in the Treatment of Epilepsy and Beyond. Springer International Publishing 2016; 30(11): 1055-77.
[http://dx.doi.org/10.1007/s40263-016-0384-x] [PMID: 27752944]

[219] Ozcan M, Ayar A, Canpolat S, Kutlu S. Antinociceptive efficacy of levetiracetam in a mice model for painful diabetic neuropathy. Acta Anaesthesiol Scand 2008; 52(7): 926-30.
[http://dx.doi.org/10.1111/j.1399-6576.2007.01578.x] [PMID: 18477089]

[220] Reda HM, Zaitone SA, Moustafa YM. Effect of levetiracetam *versus* gabapentin on peripheral neuropathy and sciatic degeneration in streptozotocin-diabetic mice: Influence on spinal microglia and astrocytes. Eur J Pharmacol 2016; 771: 162-72.
[http://dx.doi.org/10.1016/j.ejphar.2015.12.035] [PMID: 26712375]

[221] Hawker K, Frohman E, Racke M. Levetiracetam for phasic spasticity in multiple sclerosis. Arch Neurol 2003; 60(12): 1772-4.
[http://dx.doi.org/10.1001/archneur.60.12.1772] [PMID: 14676055]

[222] Hamza MS, Anderson DG, Snyder JW, Deschner S, Cifu DX. Effectiveness of levetiracetam in the treatment of lumbar radiculopathy: an open-label prospective cohort study. PM R 2009; 1(4): 335-9.
[http://dx.doi.org/10.1016/j.pmrj.2008.12.004] [PMID: 19627916]

[223] Rowbotham MC, Manville NS, Ren J. Pilot tolerability and effectiveness study of levetiracetam for postherpetic neuralgia. Neurology 2003; 61(6): 866-7.

[http://dx.doi.org/10.1212/01.WNL.0000079463.16377.07] [PMID: 14504347]

[224] Wiffen PJ, Derry S, Moore RA, Lunn MPT. Levetiracetam for neuropathic pain in adults. John Wiley and Sons Ltd 2014; 7 CD010943.
[http://dx.doi.org/10.1002/14651858.CD010943]

[225] Finnerup NB, Grydehøj J, Bing J, *et al.* Levetiracetam in spinal cord injury pain: a randomized controlled trial. Spinal Cord 2009; 47(12): 861-7.
[http://dx.doi.org/10.1038/sc.2009.55] [PMID: 19506571]

[226] Holbech JV, Otto M, Bach FW, Jensen TS, Sindrup SH. The anticonvulsant levetiracetam for the treatment of pain in polyneuropathy: a randomized, placebo-controlled, cross-over trial. Eur J Pain 2011; 15(6): 608-14.
[http://dx.doi.org/10.1016/j.ejpain.2010.11.007] [PMID: 21183370]

[227] Wright C, Downing J, Mungall D, *et al.* Clinical pharmacology and pharmacokinetics of levetiracetam. Front Neurol 2013; 4: 192.
[http://dx.doi.org/10.3389/fneur.2013.00192] [PMID: 24363651]

[228] Schulze-Bonhage A. Perampanel for epilepsy with partial-onset seizures: a pharmacokinetic and pharmacodynamic evaluation. Expert Opin Drug Metab Toxicol 2015; 11(8): 1329-37.
[http://dx.doi.org/10.1517/17425255.2015.1061504] [PMID: 26111428]

[229] De Caro C, Cristiano C, Avagliano C, *et al.* Analgesic and Anti-Inflammatory Effects of Perampanel in Acute and Chronic Pain Models in Mice: Interaction With the Cannabinergic System. Front Pharmacol 2021; 11: 620221.
[http://dx.doi.org/10.3389/fphar.2020.620221] [PMID: 33597883]

[230] https://clinicaltrials.gov/ct2/show/NCT005927742021.

[231] https://clinicaltrials.gov/ct2/show/NCT005052842021.

[232] https://clinicaltrials.gov/ct2/show/NCT005670862021.

[233] Gomez-Mancilla B, Brand R, Jürgens TP, *et al.* Randomized, multicenter trial to assess the efficacy, safety and tolerability of a single dose of a novel AMPA receptor antagonist BGG492 for the treatment of acute migraine attacks. Cephalalgia 2014; 34(2): 103-13.
[http://dx.doi.org/10.1177/0333102413499648] [PMID: 23963355]

[234] LaPenna P, Tormoehlen LM. The Pharmacology and Toxicology of Third-Generation Anticonvulsant Drugs. J Med Toxicol 2017; 13(4): 329-42.
[http://dx.doi.org/10.1007/s13181-017-0626-4] [PMID: 28815428]

[235] Gou X, Yu X, Bai D, *et al.* Pharmacology and Mechanism of Action of HSK16149, a Selective Ligand of $\alpha 2\delta$ Subunit of Voltage-Gated Calcium Channel with Analgesic Activity in Animal Models of Chronic Pain. J Pharmacol Exp Ther 2021; 376(3): 330-7.
[http://dx.doi.org/10.1124/jpet.120.000315] [PMID: 33293377]

[236] Abram M, Zagaja M, Mogilski S, *et al.* Multifunctional Hybrid Compounds Derived from 2-(2,--Dioxopyrrolidin-1-yl)-3-methoxypropanamides with Anticonvulsant and Antinociceptive Properties. J Med Chem 2017; 60(20): 8565-79.
[http://dx.doi.org/10.1021/acs.jmedchem.7b01114] [PMID: 28934547]

[237] Góra M, Czopek A, Rapacz A, *et al.* Synthesis, Anticonvulsant and Antinociceptive Activity of New Hybrid Compounds: Derivatives of 3-(3-Methylthiophen-2-yl)-pyrrolidine-2,5-dione. Int J Mol Sci 2020; 21(16): 5750.
[http://dx.doi.org/10.3390/ijms21165750] [PMID: 32796594]

[238] Kamiński K, Socała K, Zagaja M, *et al.* N-Benzyl-(2,5-dioxopyrrolidin-1-yl)propanamide (AS-1) with Hybrid Structure as a Candidate for a Broad-Spectrum Antiepileptic Drug. Neurotherapeutics 2020; 17(1): 309-28.
[http://dx.doi.org/10.1007/s13311-019-00773-w] [PMID: 31486023]

[239] Socała K, Mogilski S, Pieróg M, *et al.* KA-11, a Novel Pyrrolidine-2,5-dione Derived Broad-Spectrum Anticonvulsant: Its Antiepileptogenic, Antinociceptive Properties and *in Vitro* Characterization. ACS Chem Neurosci 2019; 10(1): 636-48.
[http://dx.doi.org/10.1021/acschemneuro.8b00476] [PMID: 30247871]

[240] Rapacz A, Rybka S, Obniska J, *et al.* Analgesic and antiallodynic activity of novel anticonvulsant agents derived from 3-benzhydryl-pyrrolidine-2,5-dione in mouse models of nociceptive and neuropathic pain. Eur J Pharmacol 2020; 869: 172890.
[http://dx.doi.org/10.1016/j.ejphar.2019.172890] [PMID: 31874144]

[241] Kamiński K, Mogilski S, Abram M, *et al.* KA-104, a new multitargeted anticonvulsant with potent antinociceptive activity in preclinical models. Epilepsia 2020; 61(10): 2119-28.
[http://dx.doi.org/10.1111/epi.16669] [PMID: 32929733]

[242] Abram M, Rapacz A, Mogilski S, *et al.* Multitargeted Compounds Derived from (2,5-Dioxopyrrolidi--1-yl)(phenyl)-Acetamides as Candidates for Effective Anticonvulsant and Antinociceptive Agents. ACS Chem Neurosci 2020; 11(13): 1996-2008.
[http://dx.doi.org/10.1021/acschemneuro.0c00257] [PMID: 32479058]

CHAPTER 7

A Review of the Impact of Testosterone on Brain and Aging-related Decline in Brain Behavioural Function

Adejoke Y Onaolapo[1,*] and **Olakunle J Onaolapo**[2]

[1] *Behavioural Neuroscience/Neurobiology Unit, Department of Anatomy, Faculty of Basic Medical Sciences, Ladoke Akintola University of Technology, Ogbomoso, Oyo State, Nigeria*

[2] *Behavioural Neuroscience/Neuropharmacology Unit, Department of Pharmacology, Faculty of Basic Medical Sciences, Ladoke Akintola University of Technology, Ogbomoso, Oyo State, Nigeria*

Abstract: The hormone testosterone is known to affect a variety of functions in the body, a number of which are related to behaviour, and preservation of brain neuronal integrity. Along this line, certain experimental evidences suggest that testosterone supplementation may be beneficial in the preservation of cognitive functions and neuronal integrity in aging and may be a valuable addition to a growing arsenal of medications that can be used to combat aging-related cognitive decline. However, some other studies have suggested instances where testosterone supplementation may be deleterious for the brain neurons, while under certain conditions, the likely effects of testosterone supplement may not be clear. Some studies had even suggested that race may be a major determinant of testosterone's effects on the brain. In this review, salient aspects of testosterone's effects on the brain are discussed with emphasis on its behavioural and morphological effects. The impacts of aging on the behavioural and brain morphological effects of testosterone are also discussed, with emphasis on its nootropic effects. The limitations to the clinical application of testosterone in mitigating aging related cognitive decline are also considered.

Keywords: Aging, Androgen, Cognition, Hypothalamopituitary Axis, Neurobehaviour, Neuroprotection.

INTRODUCTION

Worldwide, an exponential growth in the aging population has been projected. Currently, available data suggests that by the year 2050, approximately one in

---

[*] **Corresponding author, Dr. O.J Onaolapo:** Behavioural Neuroscience/Neurobiology Unit, Department of Anatomy, Faculty of Basic Medical Sciences, Ladoke Akintola University of Technology, Ogbomoso, Oyo State, Nigeria; Tel: 2349026346768, Email: olakunleonaolapo@yahoo.co.uk, ojonaolapo@lautech.edu.ng

Atta-ur-Rahman & Zareen Amtul (Eds.)

every six people worldwide will be aged 65 or over, a significant increase from the one in eleven reported in 2019 [1, 2]. In 2018, people aged 65 and above were reported to (for the first time ever) outnumber children aged five years and below globally [3]. An increase in the aging population is associated with a resultant social, economic and health burden [4].

Aging is associated with a general decline in bodily function but, specifically, a deterioration of brain function which is often reflected as cognitive decline, has become widely recognized [4 - 7]. Age-related decline has been observed in cognitive function (with a reduction in a number of memory components, processing speed, and reasoning ability) and socioemotional domains [7 - 11]. There has been extensive research examining the anatomical changes and mechanisms involved in aging-related decline in brain function [11 - 18].

Observations of the existence of gender-related differences in brain aging and the impact of gender on brain aging increased awareness of the possible roles played by sex hormones in brain development, neuronal growth and senescence [19 - 21]. Thereafter, the need to examine the impact of the neuroendocrine system hormones such as cortisol, oxytocin, testosterone and estrogen on brain development, as well as aging-related changes in brain structure and function began to generate considerable interest [5, 22]. Overall, there is considerable evidence demonstrating the impact of sex hormone (mainly estrogens and less commonly testosterone) deficiency in accelerating age-related cognitive decline as well as the neuroprotective effect of hormone-replacement therapy in reversing age-related cognitive decline in humans [5, 21, 23 - 25]

The male hormone testosterone has been associated with the regulation of diverse body functions, amongst which are functions related to behaviour and preservation of brain neuronal integrity. There is increasing experimental evidences that suggest the beneficial effects of testosterone supplementation in preventing cognitive decline or in the preservation of cognitive functions and neuronal integrity in health and in aging [27 - 29]. There have also been suggestions that in elderly men, testosterone supplementation could become a valuable addition to a growing arsenal of medications that can be used to combat aging-related cognitive decline. However, there have been reports of the possible deleterious or unclear effects of testosterone supplementation [30]. In this review, salient aspects of testosterone's effects are discussed with emphasis on its behavioural and neuromorphological effects. The impacts of aging on the behavioural and brain morphological effects of testosterone are also examined, with emphasis on its likely clinical applications. The limitations to the clinical application of testosterone in mitigating aging-related cognitive decline are also considered.

Testosterone

The hormone testosterone (Fig. **1**), which is mainly produced in the testes, is a male sex hormone that is responsible for male sexuality, the production of spermatozoa and body changes such as increased muscle mass, growth of facial hair, and libido, that are associated with masculinity [30]. As puberty is attained in males, there is an increase in the activities of the hypothalamic-pituitary axis, with increased secretion of gonadotropin releasing hormone (GnRH) stimulating the release of luteinizing hormone (LH) and follicle stimulating hormone (FSH), which in turn leads to a significant increase in testicular testosterone production and the appearance of the characteristics mentioned above [31].

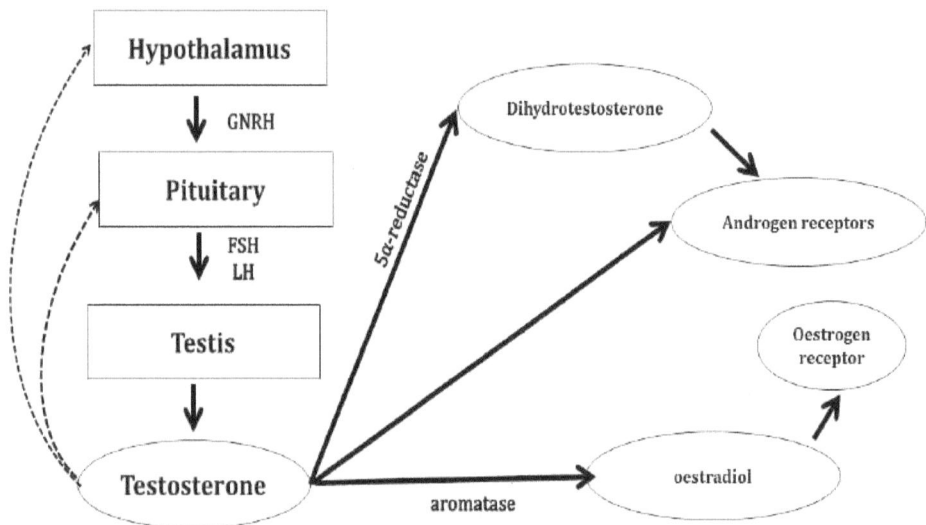

Fig. (1). Showing testosterone production and metabolism

While testosterone is not the only male sex hormone that is known in mammals, its principal role had been established before the mid-twentieth century [32]. Since time immemorial (before the discovery of testosterone or the characterization of its chemical structure), it had been observed that the absence of the testes was associated with loss of virility/fertility, and some characteristically easily recognizable differences or changes in physique [32]. Also, since its discovery and eventual synthesis in the first half of the 20th century, research has continued to reveal that both endogenous and exogenously administered testosterone can impact almost all body organs and body functions. While testosterone is a male sex hormone, it is the most important androgen in females. In women, approximately 50% of the circulating testosterone is produced primarily through the peripheral conversion of androstenedione, while the remaining half is secreted by the ovaries and adrenal gland [33].

Functions of Testosterone

Presently, the known functions of testosterone go beyond being regarded as just a male sex hormone, as it is an important contributor to the metabolic functions of multiple body organs and systems [31]. Testosterone's effects extend beyond the reproductive system to encompass the musculoskeletal system, integumentary system, hematological system, and central nervous system. It's effects on energy metabolism is beginning to be uncovered with suggestions (as well as some evidences) that it may play a role in the pathogenesis and management of disorders such as diabetes mellitus and dyslipidemia; via its ability to lower low density lipoprotein, cholesterol, blood sugar, glycosylated haemoglobin; and improve insulin resistance [31].

In the past few decades, the abuse of anabolic steroids that have effects on the musculoskeletal and cardiovascular systems have probably dissuaded researchers from focusing on acquiring an in-depth knowledge of the potential effects of testosterone on these systems. However, in recent times, there has been a renewed interest in this area, and it is known that in both men and women, testosterone decreases bone resorption while increasing bone mineral density through increased osteoclastic apoptosis [34, 35]. Testosterone is also aromatized with oestradiol (Fig. 1), which activates α and β oestrogen receptors in the bone, leading to a decrease in bone resorption, and an increase in bone mineral density [36]. It is also known that testosterone interacts with estradiol to achieve a balance in the relative proportions of lean muscle mass and fat in the body; with testosterone increasing the lean muscle mass, muscle strength and muscle size, while the estrogens cause an increase in body fat [35].

Testosterone influences the physiology of the skin and its appendages, especially the hair. It increases sebum production and can stimulate both hair growth and loss in variable regions of the body [37]. It is known that differences in the testosterone/estrogen ratio in males and females account for the sex-specific variation in skin thickness and texture, with the male epidermis being thicker, denser and more coarse [38]. Also, under the influence of testosterone and other androgens, males produce more sebum from their seborrheic glands [39].

Androgens such as testosterone have documented effects on the hemopoietic system, and before the era of recombinant hematopoietic growth factors, androgens were used for stimulation of erythropoiesis [40]. The major mechanisms that may underlie testosterone's effects on erythropoiesis include increased erythropoietin set point and increased iron utilization. However, the increased blood viscosity has made testosterone supplement to be associated with negative cardiovascular events such as thrombosis and risk of myocardial ischemia or infarction [40, 41].

Testosterone and the Mammalian Brain

The Role of Testosterone in Brain Development

The role of sex hormones on the brain (Table **1**) has been researched extensively with suggestions that sex hormones exert a strong influence on brain morphology and behavioral function [42 - 44]. These hormones cause transient and permanent neuroadaptations leading to structural and behavioral changes that begin in the prenatal period and persist throughout life [42 - 44]. Also, prenatal and postnatal exposure to surges in gonadal hormonal are crucial for sexual differentiation of the brain [43, 45].

Table 1. Effect of testosterone on the brain.

Actions	Ref
Crucial to the sexual differentiation of brain structure and functions	[46]
Perinatal exposure to testosterone causes masculinization and defeminization of sexual behavior and alters the morphology and function of the central nervous system	[47, 48]
Modifies the sexually dimorphic nuclei located in the amygdala, preoptic area of the hypothalamus or CA1 and CA3 regions of the hippocampus	[46]
Surges in testosterone levels during the prepubertal period modulates social memory	[49]
Testosterone impacts the development of aggressive behaviour	[50]
Prenatal exposure to testosterone is associated with fetal reprogramming that could manifest as adult pathologies especially in females such as the development of functional hyperandrogenism, hypergonadotropism and resultant defects in the neuroendocrine feedback mechanism	[53,54]
Testosterone has been linked to the differences in behaviours across the sexes	[55, 57]
Testosterone supplementation or replacement therapy has been associated with the improvement of spatial cognition in younger and older hypogonadal men [58]	[58]
Modulate mood and appetite behaviours, with low testosterone levels associated with depression	[59]
Testosterone supplementation of testosterone increases vigour and energy	[58]
Prenatal exposure to high levels of testosterone is associated with the impairment of neural development, derangement of mental functions, and the development of anxiety disorder especially in female offspring	[57, 64]
Testosterone supplementation in prepubertal gonadally-intact male mice was associated with a dose-dependent anxiogenic effect at higher doses and anxiolysis at the lowest dose of testosterone	[29]

Testosterone's ability to significantly impact brain development has been reported [46]. Studies have shown that the availability of testosterone or its metabolites during critical phases of development is crucial to the sexual differentiation of brain structure and functions. Exposure to testosterone during the perinatal period causes masculinization and defeminization of sexual behavior and alters the morphology and function of the central nervous system [47, 48]. There have been reports that testosterone particularly targets the amygdala, hypothalamus and hippocampus, impacting areas of memory consolidation. Testosterone modifies the sexually dimorphic nuclei located in the amygdala, preoptic area of the hypothalamus or CA1 and CA3 regions of the hippocampus through the activation of androgen receptors (following metabolism to dihydrotestosterone by 5-α reductase) or via the activation of estrogen receptors (upon conversion to estradiol by aromatase [46]. There have also been reports that surges in testosterone levels during the prepubertal period modulates social memory through the activation of androgen receptors which induces synaptic plasticity in the hippocampal CA1 region resulting in both short-term and long-term potentiation and long-term depression [49]. Testosterone also impacts the development of aggressive behaviours. The development of aggressive behavior is modulated by an interplay between the brain regions such as the subcortical structures in the amygdala and the hypothalamus that mediate emotive behaviours and the prefrontal cognitive centres that control and modulate the expression of our emotions. In the developmental phase, testosterone acting on androgen receptors causes genomic changes that play a role in the expression of aggressive behavior [50].

There have been suggestions that excessive exposure to androgens during periods of early development could cause epigenetic and epigenomic changes that have far reaching consequences in adulthood and across generations [51, 52]. There have been reports that the prenatal exposure to testosterone is associated with fetal reprogramming that could manifest as adult pathologies especially in females. Reports from studies in sheep revealed that prenatal exposure to testosterone was associated with intrauterine growth restriction and low-birth weight female offspring that went on to develop functional hyperandrogenism, hypergonadotropism and resultant defects in the neuroendocrine feedback mechanism [53, 54].

ACTIONS REF

Crucial to the sexual differentiation of brain structure and functions [46].

Perinatal exposure to testosterone causes masculinization and defeminization of sexual behavior and alters the morphology and function of the central nervous system [47, 48].

Modifies the sexually dimorphic nuclei located in the amygdala, preoptic area of the hypothalamus or CA1 and CA3 regions of the hippocampus [46].

Surges in testosterone levels during the prepubertal period modulates social memory [49].

Testosterone impacts the development of aggressive behaviour [50].

Prenatal exposure to testosterone is associated with fetal reprogramming that could manifest as adult pathologies especially in females such as the development of functional hyperandrogenism, hypergonadotropism and resultant defects in the neuroendocrine feedback mechanism]53,54].

Testosterone has been linked to the differences in behaviours across the sexes [55, 57].

Testosterone supplementation or replacement therapy has been associated with the improvement of spatial cognition in younger and older hypogonadal men [58] [58].

Modulate mood and appetite behaviours, with low testosterone levels associated with depression [59].

Testosterone supplementation of testosterone increases vigour and energy [58].

Prenatal exposure to high levels of testosterone is associated with the impairment of neural development, derangement of mental functions, and the development of anxiety disorder especially in female offspring [57, 64].

Testosterone supplementation in prepubertal gonadally-intact male mice was associated with a dose-dependent anxiogenic effect at higher doses and anxiolysis at the lowest dose of testosterone [29].

Effect of Testosterone on Neurobehaviour

In the last few decades or more, there has been increasing advocacy globally with respect to combating gender inequality. While the world continues to strive to achieve gender equality, a better understanding of the science has revealed that men and women differ significantly in several aspects [26, 55, 56]. The ability of testosterone to modulate brain function has been linked to the differences in behaviours across the sexes [55, 57]. This is supported by evidence of gender differences in spatial recognition, memory, anxiety behaviors, risk taking behaviors and age-related decline in brain function [26]. Testosterone has been reported to exert an important role in the modulation of physiological processes in the mammalian brain. By binding to androgen receptors, it modulates the expression of some genes; also, by acting through neurotransmitter receptors, testosterone and/or its metabolites have the potential to mediate non-genomic effects in the brain.

In adult humans, testosterone has the ability to activate cortical networks, the ventral processing stream, during spatial cognition tasks, with testosterone supplementation or replacement therapy being associated with the improvement of spatial cognition in younger and older hypogonadal men [58]. The ability of testosterone to modulate mood and appetite behaviours as well as reports associating reduced testosterone levels with depression [59] has led researchers to examine the possible role(s) of testosterone on mood and affective disorders [60 - 63]. While the link between testosterone and mood disorders has not been observed or reported consistently [62, 63]. The relationship between depression and testosterone is believed to depend on the genotype of the androgen receptor so that in some people with low testosterone levels, testosterone supplementation can increase positive mood and reverse depressed mood [58]. This opinion is also supported by the report of a metanalysis of randomized controlled trials of testosterone therapy which revealed that exogenous testosterone (especially at high doses) had the ability to reduce symptoms of depression in select males [61].

Testosterone has been reported to play a significant role in the arousal of aggression behaviors. There is now ample evidence of the presence of higher levels of testosterone in individuals with aggressive behaviors [50]. There are also reports showing increased levels of testosterone during the aggressive phases of some sports [50]. Aggressive behaviors have also been associated with athletes who supplement testosterone to excessively high levels without a prior diagnosis of hypogonadism, and in hypogonadal men, the supplementation of testosterone has been shown to increase vigour and energy which are the positive aspects of aggression [58].

The effects of testosterone on anxiety behavior have also been studied extensively, especially in anxiety models. There are reports that prenatal exposure to increased levels of testosterone is associated with the impairment of neural development, derangement of mental functions, and the development of anxiety disorder especially in female offspring [57, 64]. Also, exposure to high levels of circulating testosterone during prenatal development has been associated with the development of anxiety in adulthood [65]. In another study, using prepubertal gonadally intact male mice, a dose-dependent anxiogenic effect of testosterone was observed at higher doses; however, anxiolysis was observed at the lowest dose of testosterone examined [29].

Testosterone in the Female Brain

In the last few years, there have been reports that suggest that apart from gender-specific differences in males and females there are also noticeable differences in structure, volume, and functioning of the male and female brain. Differences which are believed to be influenced by sex hormones (estrogen and testosterone), which can impact cellular, signalling molecules and epigenetics in target regions of the brain. Studies have shown that in the brain, neurosteriods such as estradiol (which is synthesized from testosterone) is secreted by discrete regions, allowing it to influence brain function including memory and cognition. Also, more importantly, recent understanding that testosterone influences brain development, neurophysiology and behaviour especially in males is raising questions about the importance of testosterone in the female brain. There is some evidence now that the female brain synthesizes testosterone in quantities that are significant enough to regulate cognition, and behavior. Results from radioimmunoassay studies carried out on postmortem female brains revealed the presence of high testosterone concentrations in the hypothalamus, preoptic area and substantia nigra regions of the brain [66]. Also, brain imaging studies had also revealed that testosterone administered as part of sex reassignment therapy in adults resulted in structural and functional brain changes particularly in brain regions associated with speech and verbal fluency [67]. Another study carried out in the song bird reported that while testosterone did not fluctuate in the female songbird in response to song from other birds, there is a significant increase in the levels of testosterone in the caudomedial nidopallium of the birds following blockade of the local aromatization (conversion to oestrogens) suggesting that in the female brain testosterone levels could fluctuate in response to socially-relevant environmental stimuli [68].

Ageing and the Brain

Aging is a progressive change in physical and physiological attributes that starts in early adulthood. A gradual decline in bodily function is observed to coincide with the attainment of middle age. While there is no universal definition of the period old age begins However, the age of 60-65 years is widely accepted as the beginning of old age [1, 2]. Aging does not occur homogenously in all organs with evidence showing that within the same individual organs age at different rates. The ageing process is influenced by lifestyle choices, genetics, environmental exposures and disease [69 - 71]. According to the World Population Prospects 2019, the 2019 revision, by the year 2050 approximately 1 in 6 people world-wide would be aged 65 years and older compared to 1 1 in 11 in 2019 [2].

In the brain, ageing has been associated with a reduction in brain size with a decrease in brain volume and weight observed to occur with increasing age. There have been reports suggesting that a 5% decrease in brain volume occurs per decade after 40 years of age [72] with more marked changes in brain volume observed 70 years of age [73]. Changes due to brain ageing are also not homogenous with, reports showing that the [prefrontal cortex was the region most affected whilst the occipital cortex was least impacted by aging. Volumes of the striatum, hippocampus, temporal lobe, cerebellar hemisphere and cerebellar vermis were also reduced [74].

There have also been suggestions that men and women differ in the brain regions affected by aging. In women, the hippocampus and parietal lobes are the regions most affected by aging, however, in men, the frontal and temporal lobes are most affected [75, 76]. Ageing related changes in brain structure and function have been attributed to increased oxidative stress or a decrease in antioxidant status of the brain, an increase in programmed and non-programmed cell death, and also immune and endocrine system dysfunction [4]. In some individuals there have been reports of acceleration of the aging process. Accelerated aging has been described a rapid progression of the normal aging process resulting in the biological characteristics appearing older than expected when compared to the individual's chronological age. It has been associated with a number of disease processes [77]. Both normal aging or accelerated aging usually due to pathology have been associated with decline in various aspects of memory processes including executive function, verbal/visual memory and spatial ability [78, 79].

Age-associated memory impairment (AAMI) which is the degradation of memory with aging is a perturbing consequence of ageing. This natural process which has been reported to affect approximately 40% of people aged 60 years and older is usually not linked to any underlying medical conditions, although 1% of persons who have AAMI go on to develop dementia [80, 81]. In some instances, there is an acceleration of the memory decline associated with normal aging or a progression to dementia in situations of pathologic aging or neurodegenerative disorders including Alzheimer's disease and Parkinson's' disease. Dementia is characterized by the progressive loss of most of cognitive abilities, functional independence, and social relationships; representing a substantial physical, social and economic burden on the individuals, caregivers and society in general [82].

Testosterone and Aging–related Behavioural and Cognitive Decline

In men, aging has been associated with a decrease in serum testosterone levels. The prevalence of hypotestosteronemia has been reported to be about 20% in persons younger than 50 years of age, and 50% in those older [59]. Hypotestosteronemia has been linked to a decline in the function of the hypothalamo-pituitary-gonadal axis [57, 83, 84]. Also, there have been suggestions that this age-related hypogonadism contributes to a decrease in muscle mass, the development of osteoporosis, decreased sexual activity, depression and the development of age-related cognitive decline and anxiety [5, 57, 78, 85 - 87]. The result of a study carried out in 285 men aged 45 years and older with normo- or hypotestosteronemia but no prior history of depression revealed that the 2-year incidence of diagnosed depression was 21.7% in the men with low testosterone levels compared to 7.1% in those with normal testosterone [59]. Another study that examined the impact of circulating testosterone levels and age-associated hypotestosteronemia on spatial and cognitive abilities in men aged between 50 and 91 years showed that in those with age-related decline in testosterone levels, there was a concomitant decrease in memory and visuospatial performance with a faster rate of decline in visual memory [86]. A few studies have also associated androgen deprivation with a decrease in cell proliferation, loss of synaptic density and increased apoptosis in the hippocampus with a resultant decline in learning and memory function [88, 89]. Bussiere *et al.* [88] examined the differences in memory function between healthy men and those on androgen deprivation therapy and observed that in men on androgen deprivation therapy, there was a decline in retention ability compared to healthy men. Shin *et al.* [89] also examined the impact of chemical castration (using luteinizing hormone-releasing hormone agonists or anti-androgens) and surgical castration (bilateral orchiectomy) on structural changes in the hippocampus and spatial learning abilities. The authors observed that surgical castration, but not chemical castration was associated with a decline in spatial learning abilities, a decrease in cell proliferation and an increase in apoptosis in the hippocampus. This suggests that in addition to causing age-related cognitive decline, age associated hypotestosteronemia is also associated with the development of pathological cognitive aging [90]. There have been suggestions that low testosterone levels is a predisposing factor for the development of Alzheimer's disease (AD); and while the exact mechanisms are unclear, low testosterone levels were detected in men genetically predisposed to developing AD [91]. Questions have been raised regarding a baseline level for testosterone beyond which there is a decline in cognitive function in men with age-related hypogonadism. In a bid to answer this question, Muller *et al* [92] examined cognitive performances in a group of men between the ages of 40 and 80 years who lived independently and reported that while higher testosterone levels were associated with better cognitive performance in subjects in the oldest age category, a curvilinear relationship was observed between testosterone levels and certain aspects of cognitive function suggesting that for optimal cognitive functioning an optimal testosterone level was necessary particularly in the oldest age category [92]. The importance of testosterone in ensuring optimal cognitive functioning has also been affirmed by the result of studies that had demonstrated that testosterone supplementation (as occurs in testosterone replacement therapies) improves memory and spatial learning abilities particularly in aged men with age-related hypogonadism and memory impairment [87, 93, 94]. However, there have also been reports associating very high endogenous testosterone levels with poorer cognitive abilities in elderly men. Martin *et al.* [90] examined data from 1046 community-dwelling men (participants in the Florey Adelaide Male Ageing Study) aged between 35 and 80 years) observed that higher endogenous testosterone levels (free and total) were associated with both poorer executive functioning, and verbal memory but faster processing speed.

Exogenous Testosterone as a Possible Nootropic Agent

An increase in life-expectancy and a corresponding increase in the prevalence of age-related cognitive impairment, have increased efforts towards developing ways of preventing the development or reducing the incidence of age-related cognitive impairment and dementia [4]. Nootropics or cognitive enhancers are compounds or drugs that have the ability to improve cognitive functions (particularly memory, attention and intelligence) in health, and they are currently being evaluated for their effects in age-related cognitive decline [4, 95].

The effects of endogenous testosterone on brain structure, memory and mood had been reported severally, with studies demonstrating that a relationship exists between serum levels of testosterone and brain function, particularly in the aging population [87, 90, 92 - 94]. Also, there have been reports of the neuroprotective and neurotherapeutic effects of exogenous testosterone, although there are suggestions that these beneficial effects could be linked to the actions of its metabolites (dihydrotestosterone and estradiol) and not necessarily to testosterone itself [96 - 99]. More recently, there has been increasing advocacy for the use of exogenous testosterone as replacement therapy in men with age or disease-related hypogonadism, although the results of studies evaluating the possible nootropic effects of exogenous testosterone are at times conflicting [85, 100 - 102]. Differences of opinion have been attributed to variability in dosing/dosing schedules, route of administration, state of the gonads, and the memory tests used [29, 103]. In 2016, Onaolapo *et al.* [29] tested the hypothesis that age influences the effects of exogenous testosterone on neurobehavior in gonadally-intact male mice. He reported that compared to aged mice, a central inhibitory effect (open field) was observed in prepubertal mice following subchronic administration of testosterone. Regarding memory scores, while both prepubertal and aged mice showed a maze-dependent (radial arm maze vs. Y maze) nootropic response with testosterone supplementation; with prepubertal mice, memory scores were better than aged mice at the lower dose, while in aged mice, memory scores were better than observed in prepubertal mice at the higher dose. In humans, Jung and Shin [87] carried out a placebo-controlled study to evaluate the impact of testosterone replacement on cognition and depression in a hundred and six men with testosterone insufficiency and hypogonadism. They observed that after 8 months of intermittent testosterone supplementation, there was a significant increase in serum testosterone levels, a decrease in symptoms associated with aging and significant improvements in cognitive function and depression symptoms. This was attributed to the effect of testosterone on cognitive function and depression in men with testosterone insufficiency through its ability to enhance brain perfusion, by the action of androgen receptors in the brain [87]. The result of a review of the literature carried out by Mohamad *et al.* [104] also concluded that while low levels of endogenous testosterone were associated with poor cognitive function in healthy elderly men, supplementation with exogenous testosterone was associated with improvement in aspects of cognitive function in elderly men with testosterone insufficiency and hypogonadism [104]. The result of a recent systematic review by Buskbjerg *et al.* [105] examining the effects of testosterone supplementation on cognitive function in males with normal testosterone levels concluded that based on the reviewed studies, while there was some improvement in cognitive function following testosterone supplementation in this group of men, the effect was sufficient enough to warrant clinical relevance [105].

Other Benefits and Limitations of Testosterone Use

Currently, testosterone-replacement therapy (TRT) treatment is only indicated in the presence of low testosterone levels, resulting in symptoms and signs of hypogonadism. That being said, TRT continues to be an area of controversy, especially with regards to the major indication for testosterone supplementation in aging men and this is perpetuated as a result of a lack of large-scale, long-term studies targeted at assessing the benefits and risks of testosterone-replacement therapy in men. Reports from some studies indicate that TRT may provide pleiotropic benefits for men with hypogonadism, and such benefits extends beyond the central nervous system and core cognitive functioning to include enhancement of bone density/muscle mass, body composition and erythropoiesis [106]. Testosterone replacement can improve general outlook and body image in aging men, permitting a possible extension of the number of productive years for men in certain professions. It is also possible that a number of these non-central nervous system effects can converge to improve cognition or at least have permissive effects on the central cognitive processes.

While TRT may be appealing (if one considers the potential benefits), leading to an increase in its use in the United States over the past several years [106], there are warnings of risks that may be associated with it such as possible stimulation of prostate cancer, worsening symptoms of benign prostatic hypertrophy, liver toxicity, erythrocytosis/blood hyperviscocity and heart failure; however, factors that may modulate such risks include age and pre-existing medical conditions [107]. Perhaps, what is likely to yield the most desired result in terms of cognitive functioning is a medically supervised rather than an illicit application of TRT.

CONCLUSION

Testosterone's many-sided possible benefits in aging are supported by research, especially in areas relating to cognitive functioning. However, it is important that more studies are conducted in the areas of optimal dosing, safety and minimising side-effects.

ETHICS APPROVAL AND CONSENT TO PARTICIPATE

All procedures performed on the animals were in accordance with approved protocols of the Faculty of Basic Medical Sciences, Ladoke Akintola University of Technology, Ogbomoso, Oto State, Nigeria and within the provisions for animal care and were used as prescribed by the scientific procedures on living animals, European Council Directive (EU2010/63).

HUMAN AND ANIMAL RIGHTS

No humans were used in this study. All procedures performed on the animals were in accordance with approved protocols of the Ladoke Akintola University of Technology

CONSENT FOR PUBLICATION

Not applicable.

CONFLICT OF INTEREST

The author declares no conflict of interest, financial or otherwise.

ACKNOWLEDGEMENTS

Declared none.

REFERENCES

[1]   United Nations (2019a) Department of Economic and Social Affairs, Population Division (2019). World Population Ageing 2019: Highlights (ST/ESA/SERA/430) 2019.

[2]   United Nations (2019b). World Population Prospects 2019: Highlights (ST/ESA/SER.A/423). 2019. Available at https://population.un.org/wpp/Publications/Files/WPP2019_Highlights.pdf

[3]   United Nations (2020) Department of Economic and Social Affairs, Population Division World Population Ageing 2020. (ST/ESA/SER.A/444)

[4]   Onaolapo AY, Obelawo AY, Onaolapo OJ. Brain Ageing, Cognition and Diet: A Review of the Emerging Roles of Food-Based Nootropics in Mitigating Age-related Memory Decline. Curr Aging Sci 2019; 12(1): 2-14.
[http://dx.doi.org/10.2174/1874609812666190311160754] [PMID: 30864515]

[5]   Moffat SD. Effects of testosterone on cognitive and brain aging in elderly men. Ann N Y Acad Sci 2005; 1055: 80-92.
[http://dx.doi.org/10.1196/annals.1323.014] [PMID: 16387720]

[6]   Onaolapo OJ. Onaolapo AY Melatonin and Major Neurocognitive Disorders:Beyond the Management of Sleep and Circadian Rhythm Dysfunction Sleep Hypn 2019; 21(1): 73-96.
[http://dx.doi.org/10.5350/Sleep.Hypn.2019.21.0175]

[7]   Ebner NC, Kamin H, Diaz V, Cohen RA, MacDonald K. Hormones as "difference makers" in cognitive and socioemotional aging processes. Front Psychol 2015; 22(5): 1595.
[http://dx.doi.org/10.3389/fpsyg.2014.01595]

[8]   Ebner NC, Fischer H. Emotion and aging: evidence from brain and behavior. Front Psychol 2014; 5: 996.
[http://dx.doi.org/10.3389/fpsyg.2014.00996] [PMID: 25250002]

[9]   Salthouse TA. Selective review of cognitive aging. J Int Neuropsychol Soc 2010; 16(5): 754-60.
[http://dx.doi.org/10.1017/S1355617710000706] [PMID: 20673381]

[10]   Scheibe S, Carstensen LL. Emotional aging: recent findings and future trends. J Gerontol B Psychol Sci Soc Sci 2010; 65B(2): 135-44.
[http://dx.doi.org/10.1093/geronb/gbp132] [PMID: 20054013]

[11]   Park DC, Reuter-Lorenz P. The adaptive brain: aging and neurocognitive scaffolding. Annu Rev Psychol 2009; 60: 173-96.
[http://dx.doi.org/10.1146/annurev.psych.59.103006.093656] [PMID: 19035823]

[12]   Arenaza-Urquijo EM, Landeau B, La Joie R, *et al.* Relationships between years of education and gray matter volume, metabolism and functional connectivity in healthy elders. Neuroimage 2013; 83: 450-7.
[http://dx.doi.org/10.1016/j.neuroimage.2013.06.053] [PMID: 23796547]

[13]   Alexander GE, Ryan L, Bowers D, *et al.* Characterizing cognitive aging in humans with links to animal models. Front Aging Neurosci 2012; 4: 21.
[http://dx.doi.org/10.3389/fnagi.2012.00021] [PMID: 22988439]

[14]   Albert MS, Jones K, Savage CR, *et al.* Predictors of cognitive change in older persons: MacArthur studies of successful aging. Psychol Aging 1995; 10(4): 578-89.
[http://dx.doi.org/10.1037/0882-7974.10.4.578] [PMID: 8749585]

[15]   Bishop NA, Lu T, Yankner BA. Neural mechanisms of ageing and cognitive decline. Nature 2010; 464(7288): 529-35.
[http://dx.doi.org/10.1038/nature08983] [PMID: 20336135]

[16]   Henley JM, Wilkinson KA. AMPA receptor trafficking and the mechanisms underlying synaptic plasticity and cognitive aging. Dialogues Clin Neurosci 2013; 15(1): 11-27.
[http://dx.doi.org/10.31887/DCNS.2013.15.1/jhenley] [PMID: 23576886]

[17]   Tucker-Drob EM, Reynolds CA, Finkel D, Pedersen NL. Shared and unique genetic and environmental influences on aging-related changes in multiple cognitive abilities. Dev Psychol 2014; 50(1): 152-66.
[http://dx.doi.org/10.1037/a0032468] [PMID: 23586942]

[18]   Subramaniapillai S, Rajagopal S, Elshiekh A, Pasvanis S, Ankudowich E, Rajah MN. Sex Differences in the Neural Correlates of Spatial Context Memory Decline in Healthy Aging. J Cogn Neurosci 2019; 31(12): 1895-916.
[http://dx.doi.org/10.1162/jocn_a_01455] [PMID: 31393233]

[19]   Li R, Singh M. Sex differences in cognitive impairment and Alzheimer's disease. Front Neuroendocrinol 2014; 35(3): 385-403.
[http://dx.doi.org/10.1016/j.yfrne.2014.01.002] [PMID: 24434111]

[20]   Ding F, Yao J, Zhao L, Mao Z, Chen S, Brinton RD. Ovariectomy induces a shift in fuel availability and metabolism in the hippocampus of the female transgenic model of familial Alzheimer's. PLoS One 2013; 8(3): e59825.
[http://dx.doi.org/10.1371/journal.pone.0059825] [PMID: 23555795]

[21]   Duong P, Tenkorang MAA, Trieu J, McCuiston C, Rybalchenko N, Cunningham RL. Neuroprotective and neurotoxic outcomes of androgens and estrogens in an oxidative stress environment. Biol Sex Differ 2020; 11(1): 12.
[http://dx.doi.org/10.1186/s13293-020-0283-1] [PMID: 32223745]

[22]   Lu H, Ma K, Jin L, Zhu H, Cao R. 17β-estradiol rescues damages following traumatic brain injury from molecule to behavior in mice. J Cell Physiol 2018; 233(2): 1712-22.
[http://dx.doi.org/10.1002/jcp.26083] [PMID: 28681915]

[23]   Galea LA. Gonadal hormone modulation of neurogenesis in the dentate gyrus of adult male and female rodents. Brain Res Brain Res Rev 2008; 57(2): 332-41.
[http://dx.doi.org/10.1016/j.brainresrev.2007.05.008] [PMID: 17669502]

[24]   Saunders-Pullman R, Gordon-Elliott J, Parides M, Fahn S, Saunders HR, Bressman S. The effect of estrogen replacement on early Parkinson's disease. Neurology 1999; 52(7): 1417-21.
[http://dx.doi.org/10.1212/WNL.52.7.1417] [PMID: 10227628]

[25]   Zárate S, Stevnsner T, Gredilla R. Role of Estrogen and Other Sex Hormones in Brain Aging. Neuroprotection and DNA Repair. Front Aging Neurosci 2017; 9: 430.
[http://dx.doi.org/10.3389/fnagi.2017.00430] [PMID: 29311911]

[26]   Celec P, Ostatníková D, Hodosy J. On the effects of testosterone on brain behavioral functions. Front Neurosci 2015; 9: 12.
[http://dx.doi.org/10.3389/fnins.2015.00012] [PMID: 25741229]

[27]   Hua JT, Hildreth KL, Pelak VS. Effects of Testosterone Therapy on Cognitive Function in Aging: A Systematic Review. Cogn Behav Neurol 2016; 29(3): 122-38.
[http://dx.doi.org/10.1097/WNN.0000000000000104] [PMID: 27662450]

[28]   Beauchet O. Testosterone and cognitive function: current clinical evidence of a relationship. Eur J Endocrinol 2006; 155(6): 773-81.
[http://dx.doi.org/10.1530/eje.1.02306] [PMID: 17132744]

[29]   Onaolapo OJ, Onaolapo AY, Omololu TA, Oludimu AT, Segun-Busari T, Omoleke T. Exogenous Testosterone, Aging, and Changes in Behavioral Response of Gonadally Intact Male Mice. J Exp Neurosci 2016; 10: 59-70.
[http://dx.doi.org/10.4137/JEN.S39042] [PMID: 27158222]

[30]   Rahnema CD, Crosnoe LE, Kim ED.) Designer steroids-over-the-counter supplements and their androgenic component: Review of an increasing problem. Andrology 2015; 3: 150-5.

[31]   Bain J. The many faces of testosterone. Clin Interv Aging 2007; 2(4): 567-76.
[PMID: 18225457]

[32]   Nieschlag E, Nieschlag S. The history of testosterone and the testes: from antiquity to modern times. In: Hohl A, Ed. Testosterone: From Basic to Clinical Aspects. Berlin: Springer Verlag 2017; pp. 1-19.
[http://dx.doi.org/10.1007/978-3-319-46086-4_1]

[33]   Sowers MF, Beebe JL, McConnell D, Randolph J, Jannausch M. Testosterone concentrations in women aged 25-50 years: associations with lifestyle, body composition, and ovarian status. Am J Epidemiol 2001; 153(3): 256-64.
[http://dx.doi.org/10.1093/aje/153.3.256] [PMID: 11157413]

[34]   Roux S, Orcel P. Bone loss. Factors that regulate osteoclast differentiation: an update. Arthritis Res 2000; 2(6): 451-6.
[http://dx.doi.org/10.1186/ar127] [PMID: 11094458]

[35]   Tyagi V, Scordo M, Yoon RS, Liporace FA, Greene LW. Revisiting the role of testosterone: Are we missing something? Rev Urol 2017; 19(1): 16-24.
[PMID: 28522926]

[36]   Kalb S, Mahan MA, Elhadi AM, *et al.* Pharmacophysiology of bone and spinal fusion. Spine J 2013; 13(10): 1359-69.
[http://dx.doi.org/10.1016/j.spinee.2013.06.005] [PMID: 23972627]

[37]   Kopera D. Impact of Testosterone on Hair and Skin Endocrinol Metab Synd 2015; 4: 3.
[http://dx.doi.org/10.4172/2161-1017.1000187]

[38]   Markova MS, Zeskand J, McEntee B, Rothstein J, Jimenez SA, Siracusa LD. A role for the androgen receptor in collagen content of the skin. J Invest Dermatol 2004; 123(6): 1052-6.
[http://dx.doi.org/10.1111/j.0022-202X.2004.23494.x] [PMID: 15610513]

[39]   Baumann L. Acne.Cosmetic Dermatology. Mcgraw-Hill, New York: Principles and Practice, Baumann L, Weisberg E 2002.

[40]   Delev D, Rangelov A, Ubenova D, Kostadinov I, Zlatanova H, Kostadinova I. Mechanism of Action of Androgens on Erythropoiesis – A Review. International Journal of Pharmaceutical and Clinical Research 2016; 8(11): 1489-92.

[41]   Ruige JB, Ouwens DM, Kaufman JM. Beneficial and adverse effects of testosterone on the cardiovascular system in men. J Clin Endocrinol Metab 2013; 98(11): 4300-10.
[http://dx.doi.org/10.1210/jc.2013-1970] [PMID: 24064693]

[42]   Lenz B, Müller CP, Stoessel C, *et al.* Sex hormone activity in alcohol addiction: integrating organizational and activational effects. Prog Neurobiol 2012; 96(1): 136-63.
[http://dx.doi.org/10.1016/j.pneurobio.2011.11.001] [PMID: 22115850]

[43]   Huber SE, Lenz B, Kornhuber J, Müller CP. Prenatal androgen-receptor activity has organizational morphological effects in mice. PLoS One 2017; 12(11): e0188752.
[http://dx.doi.org/10.1371/journal.pone.0188752] [PMID: 29176856]

[44]   Liang X, Cheng S, Ye J, *et al.* Evaluating the genetic effects of sex hormone traits on the development of mental traits: a polygenic score analysis and gene-environment-wide interaction study in UK Biobank cohort. Mol Brain 2021; 14(1)
[http://dx.doi.org/10.1186/s13041-020-00718-x] [PMID: 33407712]

[45]   Talarovicová A, Krsková L, Blazeková J. Testosterone enhancement during pregnancy influences the 2D:4D ratio and open field motor activity of rat siblings in adulthood. Horm Behav 2009; 55(1): 235-9.
[http://dx.doi.org/10.1016/j.yhbeh.2008.10.010] [PMID: 19022257]

[46]   Filová B, Ostatníková D, Celec P, Hodosy J. The effect of testosterone on the formation of brain structures. Cells Tissues Organs 2013; 197(3): 169-77.
[http://dx.doi.org/10.1159/000345567] [PMID: 23306974]

[47]   Rhees RW, Kirk BA, Sephton S, Lephart ED. Effects of prenatal testosterone on sexual behavior,

reproductive morphology and LH secretion in the female rat. Dev Neurosci 1997; 19(5): 430-7.
[http://dx.doi.org/10.1159/000111240] [PMID: 9323463]

[48]   Negri-Cesi P, Colciago A, Celotti F, Motta M. Sexual differentiation of the brain: role of testosterone and its active metabolites. J Endocrinol Invest 2004; 27(6) (Suppl.): 120-7.
[PMID: 15481811]

[49]   Hebbard PC, King RR, Malsbury CW, Harley CW. Two organizational effects of pubertal testosterone in male rats: transient social memory and a shift away from long-term potentiation following a tetanus in hippocampal CA1. Exp Neurol 2003; 182(2): 470-5.
[http://dx.doi.org/10.1016/S0014-4886(03)00119-5] [PMID: 12895458]

[50]   Batrinos ML. Testosterone and aggressive behavior in man. Int J Endocrinol Metab 2012; 10(3): 563-8.
[http://dx.doi.org/10.5812/ijem.3661] [PMID: 23843821]

[51]   Xu N, Chua AK, Jiang H, Liu NA, Goodarzi MO. Early embryonic androgen exposure induces transgenerational epigenetic and metabolic changes. Mol Endocrinol 2014; 28(8): 1329-36.
[http://dx.doi.org/10.1210/me.2014-1042] [PMID: 24992182]

[52]   Salinas I, Sinha N, Sen A. Androgen-induced epigenetic modulations in the ovary. J Endocrinol 2021; 249(3): R53-64.
[http://dx.doi.org/10.1530/JOE-20-0578] [PMID: 33764313]

[53]   Padmanabhan V, Sarma HN, Savabieasfahani M, Steckler TL, Veiga-Lopez A. Developmental reprogramming of reproductive and metabolic dysfunction in sheep: native steroids *vs*. environmental steroid receptor modulators. Int J Androl 2010; 33(2): 394-404.
[http://dx.doi.org/10.1111/j.1365-2605.2009.01024.x] [PMID: 20070410]

[54]   Veiga-Lopez A, Steckler TL, Abbott DH, *et al*. Developmental programming: impact of excess prenatal testosterone on intrauterine fetal endocrine milieu and growth in sheep. Biol Reprod 2011; 84(1): 87-96.
[http://dx.doi.org/10.1095/biolreprod.110.086686] [PMID: 20739662]

[55]   Höfer P, Lanzenberger R, Kasper S. Testosterone in the brain: neuroimaging findings and the potential role for neuropsychopharmacology. Eur Neuropsychopharmacol 2013; 23(2): 79-88.
[http://dx.doi.org/10.1016/j.euroneuro.2012.04.013] [PMID: 22578782]

[56]   Cahill L. Equal ≠ the same: sex differences in the human brain. Cerebrum 2014; 2014: 5.
[PMID: 25009695]

[57]   Domonkos E, Hodosy J, Ostatníková D, Celec P. On the Role of Testosterone in Anxiety-Like Behavior Across Life in Experimental Rodents. Front Endocrinol (Lausanne) 2018; 9: 441.
[http://dx.doi.org/10.3389/fendo.2018.00441] [PMID: 30127767]

[58]   Zitzmann M. Testosterone and the brain. Aging Male 2006; 9(4): 195-9.
[http://dx.doi.org/10.1080/13685530601040679] [PMID: 17178554]

[59]   Shores MM, Sloan KL, Matsumoto AM, Moceri VM, Felker B, Kivlahan DR. Increased incidence of diagnosed depressive illness in hypogonadal older men. Arch Gen Psychiatry 2004; 61(2): 162-7.
[http://dx.doi.org/10.1001/archpsyc.61.2.162] [PMID: 14757592]

[60]   Amiaz R, Seidman SN. Testosterone and depression in men. Curr Opin Endocrinol Diabetes Obes 2008; 15(3): 278-83.
[http://dx.doi.org/10.1097/MED.0b013e3282fc27eb] [PMID: 18438177]

[61]   Walther A, Breidenstein J, Miller R. Association of Testosterone Treatment With Alleviation of Depressive Symptoms in Men: A Systematic Review and Meta-analysis. JAMA Psychiatry 2019; 76(1): 31-40.
[http://dx.doi.org/10.1001/jamapsychiatry.2018.2734] [PMID: 30427999]

[62]   Wu FC, Tajar A, Beynon JM, *et al*. Identification of late-onset hypogonadism in middle-aged and elderly men. N Engl J Med 2010; 363(2): 123-35.

[http://dx.doi.org/10.1056/NEJMoa0911101] [PMID: 20554979]

[63]   Kische H, Gross S, Wallaschofski H, *et al.* Associations of androgens with depressive symptoms and cognitive status in the general population. PLoS One 2017; 12(5): e0177272.
[http://dx.doi.org/10.1371/journal.pone.0177272] [PMID: 28498873]

[64]   Hu M, Richard JE, Maliqueo M, *et al.* Maternal testosterone exposure increases anxiety-like behavior and impacts the limbic system in the offspring. Proc Natl Acad Sci USA 2015; 112(46): 14348-53.
[http://dx.doi.org/10.1073/pnas.1507514112] [PMID: 26578781]

[65]   Xu XJ, Zhang HF, Shou XJ, *et al.* Prenatal hyperandrogenic environment induced autistic-like behavior in rat offspring. Physiol Behav 2015; 138: 13-20.
[http://dx.doi.org/10.1016/j.physbeh.2014.09.014] [PMID: 25455866]

[66]   Bixo M, Bäckström T, Winblad B, Andersson A. Estradiol and testosterone in specific regions of the human female brain in different endocrine states. J Steroid Biochem Mol Biol 1995; 55(3-4): 297-303.
[http://dx.doi.org/10.1016/0960-0760(95)00179-4] [PMID: 8541226]

[67]   European College of Neuropsychopharmacology (ECNP). "Testosterone changes brain structures in female-to-male transsexuals ScienceDaily 2015.www.sciencedaily.com/releases/2015/08/15083-1001124.htm

[68]   de Bournonville C, McGrath A, Remage-Healey L. Testosterone synthesis in the female songbird brain. Horm Behav 2020; 121: 104716.
[http://dx.doi.org/10.1016/j.yhbeh.2020.104716] [PMID: 32061616]

[69]   Blokzijl F, de Ligt J, Jager M, *et al.* Tissue-specific mutation accumulation in human adult stem cells during life. Nature 2016; 538(7624): 260-4.
[http://dx.doi.org/10.1038/nature19768] [PMID: 27698416]

[70]   vB Hjelmborg J, Iachine I, Skytthe A, *et al.* Genetic influence on human lifespan and longevity. Hum Genet 2006; 119(3): 312-21.
[http://dx.doi.org/10.1007/s00439-006-0144-y] [PMID: 16463022]

[71]   Mendenhall AR, Martin GM, Kaeberlein M, Anderson RM. Cell-to-cell variation in gene expression and the aging process. Geroscience 2021; 43(1): 181-96.
[http://dx.doi.org/10.1007/s11357-021-00339-9] [PMID: 33595768]

[72]   Svennerholm L, Boström K, Jungbjer B. Changes in weight and compositions of major membrane components of human brain during the span of adult human life of Swedes. Acta Neuropathol 1997; 94(4): 345-52. [doi].
[http://dx.doi.org/10.1007/s004010050717] [PMID: 9341935]

[73]   Scahill RI, Frost C, Jenkins R, Whitwell JL, Rossor MN, Fox NC. A longitudinal study of brain volume changes in normal aging using serial registered magnetic resonance imaging. Arch Neurol 2003; 60(7): 989-94.
[http://dx.doi.org/10.1001/archneur.60.7.989] [PMID: 12873856]

[74]   Peters R. Ageing and the brain. Postgrad Med J 2006; 82(964): 84-8.
[http://dx.doi.org/10.1136/pgmj.2005.036665] [PMID: 16461469]

[75]   Compton J, van Amelsvoort T, Murphy D. HRT and its effect on normal ageing of the brain and dementia. Br J Clin Pharmacol 2001; 52(6): 647-53.
[http://dx.doi.org/10.1046/j.0306-5251.2001.01492.x] [PMID: 11736875]

[76]   Murphy D, DeCarli C, McIntosh A. al Sex differences in human brain morphometry and metabolism: an in vivo quantitative magnetic resonance imaging and positron emission tomography study on the effect of ageing. Arch Gen Psychiatry 199653585-.

[77]   Christman S, Bermudez C, Hao L, *et al.* Accelerated brain aging predicts impaired cognitive performance and greater disability in geriatric but not midlife adult depression. Transl Psychiatry 2020; 10(1): 317.
[http://dx.doi.org/10.1038/s41398-020-01004-z] [PMID: 32948749]

[78]   Resnick SM, Matsumoto AM, Stephens-Shields AJ, *et al.* Testosterone Treatment and Cognitive Function in Older Men With Low Testosterone and Age-Associated Memory Impairment. JAMA 2017; 317(7): 717-27.
[http://dx.doi.org/10.1001/jama.2016.21044] [PMID: 28241356]

[79]   Schaie KW, Willis SL. Age difference patterns of psychometric intelligence in adulthood: generalizability within and across ability domains. Psychol Aging 1993; 8(1): 44-55.
[http://dx.doi.org/10.1037/0882-7974.8.1.44] [PMID: 8461114]

[80]   Small GW. What we need to know about age related memory loss. BMJ 2002; 324(7352): 1502-5.
[http://dx.doi.org/10.1136/bmj.324.7352.1502] [PMID: 12077041]

[81]   Pokorski RJ. Differentiating age-related memory loss from early dementia. J Insur Med 2002; 34(2): 100-13.
[PMID: 15305786]

[82]   Peracino A, Pecorelli S. The Epidemiology of Cognitive Impairment in the Aging Population: Implications for Hearing Loss. Audiol Neurotol 2016; 21 (Suppl. 1): 3-9.
[http://dx.doi.org/10.1159/000448346] [PMID: 27806351]

[83]   Harman SM, Metter EJ, Tobin JD, Pearson J, Blackman MR. Longitudinal effects of aging on serum total and free testosterone levels in healthy men. J Clin Endocrinol Metab 2001; 86(2): 724-31.
[http://dx.doi.org/10.1210/jcem.86.2.7219] [PMID: 11158037]

[84]   Feldman HA, Longcope C, Derby CA, *et al.* Age trends in the level of serum testosterone and other hormones in middle-aged men: longitudinal results from the Massachusetts male aging study. J Clin Endocrinol Metab 2002; 87(2): 589-98.
[http://dx.doi.org/10.1210/jcem.87.2.8201] [PMID: 11836290]

[85]   Cherrier MM, Asthana S, Plymate S, *et al.* Testosterone supplementation improves spatial and verbal memory in healthy older men. Neurology 2001; 57(1): 80-8.
[http://dx.doi.org/10.1212/WNL.57.1.80] [PMID: 11445632]

[86]   Moffat SD, Zonderman AB, Metter EJ, Blackman MR, Harman SM, Resnick SM. Longitudinal assessment of serum free testosterone concentration predicts memory performance and cognitive status in elderly men. J Clin Endocrinol Metab 2002; 87(11): 5001-7.
[http://dx.doi.org/10.1210/jc.2002-020419] [PMID: 12414864]

[87]   Jung HJ, Shin HS. Effect of Testosterone Replacement Therapy on Cognitive Performance and Depression in Men with Testosterone Deficiency Syndrome. World J Mens Health 2016; 34(3): 194-9.
[http://dx.doi.org/10.5534/wjmh.2016.34.3.194] [PMID: 28053949]

[88]   Bussiere JR, Beer TM, Neiss MB, Janowsky JS. Androgen deprivation impairs memory in older men. Behav Neurosci 2005; 119(6): 1429-37.
[http://dx.doi.org/10.1037/0735-7044.119.6.1429] [PMID: 16420147]

[89]   Shin MS, Chung KJ, Ko IG, *et al.* Effects of surgical and chemical castration on spatial learning ability in relation to cell proliferation and apoptosis in hippocampus. Int Urol Nephrol 2016; 48(4): 517-27.
[http://dx.doi.org/10.1007/s11255-015-1200-0] [PMID: 26781653]

[90]   Martin DM, Wittert G, Burns NR, Haren MT, Sugarman R. Testosterone and cognitive function in ageing men: data from the Florey Adelaide Male Ageing Study (FAMAS). Maturitas 2007; 57(2): 182-94.
[http://dx.doi.org/10.1016/j.maturitas.2006.12.007] [PMID: 17287097]

[91]   Hogervorst E, Bandelow S, Moffat SD. Increasing testosterone levels and effects on cognitive functions in elderly men and women: a review. Curr Drug Targets CNS Neurol Disord 2005; 4(5): 531-40.
[http://dx.doi.org/10.2174/156800705774322049] [PMID: 16266286]

[92]   Muller M, Aleman A, Grobbee DE, de Haan EH, van der Schouw YT. Endogenous sex hormone

levels and cognitive function in aging men: is there an optimal level? Neurology 2005; 64(5): 866-71.
[http://dx.doi.org/10.1212/01.WNL.0000153072.54068.E3] [PMID: 15753424]

[93]   Cherrier MM, Craft S, Matsumoto AH. Cognitive changes associated with supplementation of testosterone or dihydrotestosterone in mildly hypogonadal men: a preliminary report. J Androl 2003; 24(4): 568-76.
[http://dx.doi.org/10.1002/j.1939-4640.2003.tb02708.x] [PMID: 12826696]

[94]   Kenny AM, Bellantonio S, Gruman CA, Acosta RD, Prestwood KM. Effects of transdermal testosterone on cognitive function and health perception in older men with low bioavailable testosterone levels. J Gerontol A Biol Sci Med Sci 2002; 57(5): M321-5.
[http://dx.doi.org/10.1093/gerona/57.5.M321] [PMID: 11983727]

[95]   Frati P, Kyriakou C, Del Rio A, et al. Smart drugs and synthetic androgens for cognitive and physical enhancement: revolving doors of cosmetic neurology. Curr Neuropharmacol 2015; 13(1): 5-11.
[http://dx.doi.org/10.2174/1570159X13666141210221750] [PMID: 26074739]

[96]   Cai Y, Chew C, Muñoz F, Sengelaub DR. Neuroprotective effects of testosterone metabolites and dependency on receptor action on the morphology of somatic motoneurons following the death of neighboring motoneurons. Dev Neurobiol 2017; 77(6): 691-707.
[http://dx.doi.org/10.1002/dneu.22445] [PMID: 27569375]

[97]   Little CM, Coons KD, Sengelaub DR. Neuroprotective effects of testosterone on the morphology and function of somatic motoneurons following the death of neighboring motoneurons. J Comp Neurol 2009; 512(3): 359-72.
[http://dx.doi.org/10.1002/cne.21885] [PMID: 19003970]

[98]   Wilson RE, Coons KD, Sengelaub DR. Neuroprotective effects of testosterone on dendritic morphology following partial motoneuron depletion: efficacy in female rats. Neurosci Lett 2009; 465(2): 123-7.
[http://dx.doi.org/10.1016/j.neulet.2009.09.007] [PMID: 19735695]

[99]   Otzel DM, Lee J, Ye F, Borst SE, Yarrow JF. Activity-Based Physical Rehabilitation with Adjuvant Testosterone to Promote Neuromuscular Recovery after Spinal Cord Injury. Int J Mol Sci 2018; 19(6): 1701.
[http://dx.doi.org/10.3390/ijms19061701] [PMID: 29880749]

[100] Cherrier MM, Matsumoto AM, Amory JK, et al. Characterization of verbal and spatial memory changes from moderate to supraphysiological increases in serum testosterone in healthy older men. Psychoneuroendocrinology 2007; 32(1): 72-9.
[http://dx.doi.org/10.1016/j.psyneuen.2006.10.008] [PMID: 17145137]

[101] Vaughan C, Goldstein FC, Tenover JL. Exogenous testosterone alone or with finasteride does not improve measurements of cognition in healthy older men with low serum testosterone. J Androl 2007; 28(6): 875-82.
[http://dx.doi.org/10.2164/jandrol.107.002931] [PMID: 17609296]

[102] Snyder PJ, Bhasin S, Cunningham GR, et al. Effects of Testosterone Treatment in Older Men. N Engl J Med 2016; 374(7): 611-24.
[http://dx.doi.org/10.1056/NEJMoa1506119] [PMID: 26886521]

[103] Spritzer MD, Daviau ED, Coneeny MK, Engelman SM, Prince WT, Rodriguez-Wisdom KN. Effects of testosterone on spatial learning and memory in adult male rats. Horm Behav 2011; 59(4): 484-96.
[http://dx.doi.org/10.1016/j.yhbeh.2011.01.009] [PMID: 21295035]

[104] Mohamad NV, Ima-Nirwana S, Chin KY. A Review on the Effects of Testosterone Supplementation in Hypogonadal Men with Cognitive Impairment. Curr Drug Targets 2018; 19(8)
[http://dx.doi.org/10.2174/1389450118666170913162739] [PMID: 28914204]

[105] Buskbjerg CR, Gravholt CH, Dalby HR, Amidi A, Zachariae R. Testosterone Supplementation and Cognitive Functioning in Men-A Systematic Review and Meta-Analysis. J Endocr Soc 2019; 3(8): 1465-84.

[http://dx.doi.org/10.1210/js.2019-00119] [PMID: 31384712]

[106] Bassil N, Alkaade S, Morley JE. The benefits and risks of testosterone replacement therapy: a review. Ther Clin Risk Manag 2009; 5(3): 427-48.
[http://dx.doi.org/10.2147/TCRM.S3025] [PMID: 19707253]

[107] Bhasin S, Cunningham GR, Hayes FJ, *et al.* Testosterone therapy in adult men with androgen deficiency syndromes: an endocrine society clinical practice guideline. J Clin Endocrinol Metab 2006; 91(6)
[http://dx.doi.org/10.1210/jc.2005-2847] [PMID: 16720669]

SUBJECT INDEX

Atta-ur-Rahman & Zareen Amtul (Eds.)
All rights reserved-© 2022 Bentham Science Publishers

Processes 40, 41, 43, 72, 74, 75, 79, 80, 81, 82, 137, 138, 146, 147, 236, 237, 238, 247, 292, 322
  acetylation 82
  ageing 322
  dynamic 81
  herbal drug development 40
  neurodegenerative 147
  neuroinflammatory 247
  protein aggregation 146
  reactive astrogliosis 236
Production, neuromodulator 64
Prolactin-inhibiting factor (PIF) 64
Proteasome activity inhibition 36
Proteinaceous inclusions 81
Proteins 4, 5, 6, 10, 15, 18, 21, 27, 30, 34, 35, 73, 76, 77, 79, 80, 81, 82, 98, 99, 132, 134, 135, 136, 137, 141, 142, 153, 156, 287
  aggregates 79, 142
  amyloid precursor 4, 10, 27
  apoptotic 15
  binding 287
  expression 30, 34
  homeostasis 134
  huntingtin 35
  heat shock 82
  neurotrophic factor 132
  non-functional regulatory intracellular 80
  norepinephrine transporter 18
  nuclear-encoded 99
  oxidative phosphorylation-related 98
  synthesis 137

R

Reactions 73, 242
  inflammatory 73
  neurochemical 242
Reactive 5, 6, 18, 19, 21, 22, 26, 71, 72, 73, 90, 91, 92, 98, 233, 236
  astrogliosis 236
  nitrogen species (RNS) 98
  oxygen species (ROS) 5, 6, 18, 19, 21, 22, 26, 71, 72, 73, 90, 91, 92, 98, 233
Receptor tyrosine kinases 138, 139
Reduction 18, 21, 24, 30, 33, 38, 39, 151, 152, 244, 246, 248, 251, 288, 289
  migraine frequency 289
  of cell apoptosis 30

striatal MDA level 38
Release 20, 26, 63, 73, 80, 88, 228, 233, 235, 237, 240, 242, 282, 283, 293
  augmenting dopamine 88
  neurotransmitter 80, 293
  pro-inflammatory cytokine 237
Response 189, 227, 229, 231, 234, 235, 238, 239, 241, 251, 282, 283, 322
  glutamatergic 239
  neuroimmune 235
  switching immune 241
RET-downstream signaling 161
Reverse dopaminergic neurodegeneration 84
ROS 5, 12, 30, 73, 78, 79, 98, 99, 237
  excessive 73
  generation 12, 98
  -nitric oxide pathway 30
  production 5, 78, 79, 98, 99, 237

S

Schizophrenia 38
Sclerosis, multiple 230, 293
Scutellaria baicalensis 90, 91
SDH activity reduction 40
Secretase enzymes 5
Selective serotonin reuptake inhibitors (SSRIs) 93, 285, 286
Senescence-associated secretory phenotype (SASP) 78
Sepsis 241, 244
Severe cutaneous adverse reaction (SCAR) 279
Single ligature nerve constriction (SLNC) 240
Species 6, 16, 21, 22, 26, 83, 98, 233
  reactive nitrogen 98
  reactive oxygen 6, 16, 21, 22, 26, 83, 98, 233
Sphingolipids 136
Stem cell transplantation 62
Stevens-Johnson syndrome 289
Stroke 8, 94
Succinate dehydrogenase 37, 97
Superior frontal gyrus (SFG) 200
Symptoms 2, 3, 14, 83, 94, 143, 144, 145, 146, 188, 190, 278, 286, 290, 326, 327
  gastrointestinal 94, 278
  neuropsychiatric 144
  urinary 94
Synaptic transmission 236, 239, 276

www.ingramcontent.com/pod-product-compliance
Lightning Source LLC
Chambersburg PA
CBHW050804220326
41598CB00006B/116